Applied Electricity and Electronics

Clair A. Bayne

Electronics Technology Instructor

Venango County Area Vocational-Technical School

Oil City, Pennsylvania

Publisher

The Goodheart-Willcox Company, Inc.

Tinley Park, Illinois

Library of Congress Catalog Card Number 99-042167

International Standard Book Number 1-56637-707-2

1 2 3 4 5 6 7 8 9 10 00 03 02 01 00 99

Library of Congress Cataloging-in-Publication Data

Bayne, Clair.
    Applied electricity and electronics / Clair A. Bayne.
      p. cm.
    Includes index.
    ISBN 1-56637-707-2
      1. Electronics. 2. Electric engineering. I. Title.

TK7816 .B383 1999
621.381--dc21

                                        99-042167

Cover photo: ©Lester Lefkowitz, The Stock Market

INTRODUCTION

Rapid advancements in the field of electricity and electronics require students to learn a vast amount of information, update skills, and pursue higher education. To be competitive, a solid foundation in the basics is essential. Today's electrical/electronic technician or engineer must have a thorough understanding of electrical principles, use of a multimeter and oscilloscope, soldering techniques, assembly, and repair. Students and instructors in high schools, technical schools, and colleges will find **Applied Electricity and Electronics** provides the necessary preparation in an easy-to-follow format.

Electrical topics build from the simple to the complex. The textbook is divided into six sections:

- Fundamentals of Direct Current
- Electronic Assembly
- Fundamentals of Alternating Current
- Inductance, Capacitance, and RCL Circuits
- Ac Power and Motors
- Electronic Principles

Each section presents an activity for the classroom or lab. These activities are designed to peak your interest in some aspect of electricity and electronics and move you toward developing an investigative mind.

Each chapter begins with a graphic overview of the main ideas to be covered. These chapter "maps" help you visualize how the major concepts tie together. Learning objectives are stated up front and supported by summaries, review questions, and problems at the end of the chapter. Technical terms are clearly defined and emphasized throughout the text. Appendices, a glossary of important terms and definitions, and a comprehensive index enhance the usefulness of the textbook.

Introductory chapters prepare students for increasingly complex subjects. For example, *Math for Electricity* lays the mathematical groundwork for learning the principles of electricity. Because they have mastered the fundamentals, students are not confused later by mathematical equations.

Understanding graphs is a basic skill to master in electricity and electronics. Most students receive instruction on graphs only in algebra courses, and these graphs are typically associated with algebraic equations. The *Graphs* chapter provides all the information beginning students need to be able to interpret graphs for electrical and electronic purposes.

Unlike other textbooks that provide little or no "how-to" information, **Applied Electricity and Electronics** is a unique blend of theory and application. For example, *Using Electrical Meters* does not dwell on how a meter works or on calculating shunts and multipliers. Rather, the chapter concentrates on how to use a meter.

The Oscilloscope chapter covers the instrument in more depth than any other textbook. It provides a concise description of how the oscilloscope works and takes the student through detailed instructions on knob settings, normal operation, and waveform measurement. The subsequent chapter provides a complete description of waveforms and measurements.

The development, manufacture, production, and service of electronic products for government, industrial, and consumer use provide many opportunities for anyone interested in a career in electricity and electronics. The field is rapidly changing, so being able to apply established rules is more important than ever. By incorporating traditional electrical subjects with the practicality of a training manual, **Applied Electricity and Electronics** reaches beyond the common textbook and into tomorrow's world of technology.

About the Author

Clair A. Bayne holds a Bachelor of Science degree in industrial education from the University of Pittsburgh, Pittsburgh, Pennsylvania. He has been an electronics instructor for 30 years and devoted much time to curriculum development in basic electronics, digital electronics, microprocessors, soldering, and related math. He has taught classes in electronic communication, radio/television repair, and electronic technology, including the operation and maintenance of a school television studio. Recently he has worked to update and expand the electronics curriculum to serve the retraining needs of displaced workers.

Before entering the teaching profession, Mr. Bayne graduated from DeVry Technical Institute and Grantham School of Electronics. He was employed as chief engineer at a radio station and holds a general radiotelephone FCC license. He also was graduated from a U.S. Department of Defense school with certification in high reliability electrical soldering and is a certified examiner for the electronics technology competency test.

Acknowledgements

Special thanks to the following companies for their cooperation and contributions:

American Hakko Products, Inc.
Battery Warehouse
Delco-Remy Div., General Motors Corp.
Eveready
Liberty Electronics
Littlefuse
Matric Limited
McGraw-Edison Co.
Radio Shack
The Lincoln Electric Co.

Many people were involved in the development of this book. I would like to thank my wife Barbara for many hours of proofreading, my coworkers for their reviews and recommendations, and my students for their suggestions as to content.

Clair A. Bayne

CONTENTS

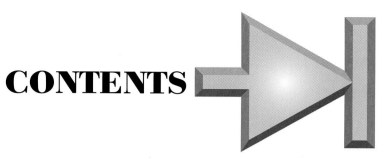

Section II:
Electronic Assembly

Section III:
Fundamentals of Alternating Current

Section IV:
Inductance, Capacitance, and RCL Circuits

Section V:
Ac Power and Motors

Section VI:
Electronic Principles

ELECTRICAL SAFETY AND FIRST AID

Electrical Safety Precautions

Because of the possibility of personal injury, fire, and damage to equipment, the following precautions should be taken before attempting any work on electrical or electronic devices.

✚ Do not use equipment unless you have been trained in its operation and safe handling.

✚ Do not work alone. Have someone with you who can go for help in an emergency.

✚ Wear insulated shoes.

✚ Do not wear loose clothing. It can get caught on objects and pull you into live electrical contact or cause other injury.

✚ Do not wear rings, earrings, wristwatches, bracelets, chains, tags, and similar metal items.

✚ Never stand in water or on a water-soaked floor when working with electricity.

✚ When working with high-voltage, use only one hand. Do not provide a complete circuit through your body.

✚ Never override safety interlock devices.

✚ Make sure all equipment is properly grounded.

✚ When working on equipment with a disconnect panel, do not just turn it off. Remove the fuses and put them in your pocket, or lock the disconnect in the OFF position.

✚ Use a shorting stick to remove charges on capacitors.

✚ Disconnect power from the circuit or equipment before working, including when you remove and replace fuses. Never assume the circuit power is off.

✚ Never use a wire or any other conductor to replace a fuse. Use an exact replacement fuse.

✚ Whenever you do testing, turn off the power if possible.

✚ Never use a metal table when working with electricity.

✚ Do not depend on a fuse or circuit breaker to protect you from electrocution. Always use caution.

✚ Use extreme caution when working with batteries that contain acid.

✚ In case of fire, turn off the power source and get help immediately. Fumes from burning materials can quickly cause death.

✚ Know where a fire extinguisher (dry chemical or CO_2) is located and how to operate it.

✚ Locate the fire alarm and the nearest exit in case you need to use them.

✚ Think before you act. Your actions could endanger the lives of others.

First Aid for Shock

The following procedures are recommended as first aid to a shock victim. More detailed information is provided in Chapter 1.

+ Turn off the power switch. Remove the victim from electrical contact using a dry stick, such as a broom handle, or another nonconductive object.

+ Stop any bleeding by applying direct pressure over the wound.

+ If the victim is not breathing, loosen clothing about the neck and abdomen, and give artificial resuscitation.

+ Cover the victim to conserve body heat. If the victim is on a cold surface, such as concrete or tile, "log roll" him or her onto another surface for thermal protection.

+ Treat for traumatic shock by raising the feet slightly.

+ Keep the victim from moving about.

+ Do not give stimulants.

+ Get medical help immediately.

First Aid for Burns

The following procedures are recommended as first aid to a burn victim. More information is provided in Chapter 1.

+ Immediately submerge the burned area in cold water.

+ Loosely wrap the burned area with sterile gauze to guard against infection.

+ See a doctor if the burn produces blisters or blackening of the skin.

Section I

Fundamentals of Direct Current

Section I Activity

Building an Electrical Circuit

Many different electrical circuits are used when working with electricity. Skill in the construction of circuits, both simple and complex, is vital to understand troubleshooting and testing of circuits. Although the circuit you will build is simple, it contains the basic principles of electrical circuitry. These principles must be fully understood for your study of electricity to be meaningful.

Objective

In this activity, you will build a simple electrical circuit.

Materials and Equipment

2–Batteries, Ray-O-Vac #945
2–Bulbs
2–Sockets to match bulb base
1–Switch (knife switch preferred)
1–Voltmeter
Miscellaneous wire

Procedures

1. Connect the circuit as shown in **Figure A.**
2. Open and close the switch and observe the results.
3. Ask the instructor for assistance measuring the voltage across each bulb.
   L_1 = _____ V
   L_2 = _____ V
4. Measure the voltage of the source.
   Source voltage = _____
5. How does the sum of the voltages across the lamps compare to the source voltage?

   _____

   _____

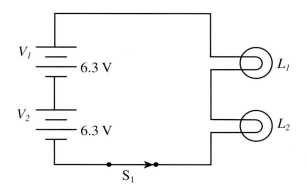

Figure A.

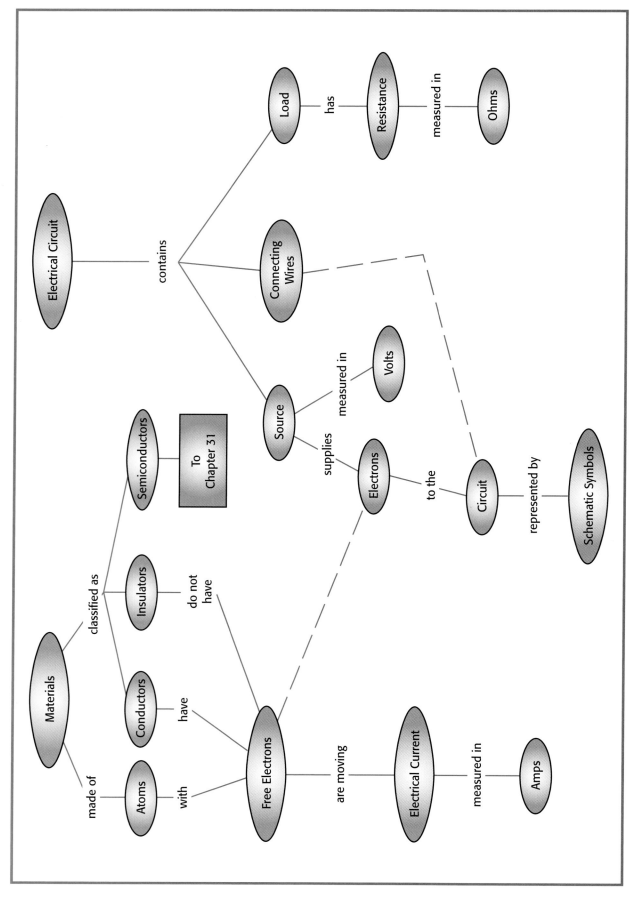

FUNDAMENTALS OF ELECTRICITY 1

Objectives

After studying this chapter, you will be able to:

- ○ Identify the parts of an atom.
- ○ State the differences between conductors, insulators, and semiconductors.
- ○ Identify the parts of an electrical circuit.
- ○ List the sources of voltage.
- ○ Identify schematic symbols for voltage source, switch, and lamp.
- ○ Identify types of electrical drawings.
- ○ Describe general safety precautions and first aid treatment for electrical shock and burns.

Introduction

This chapter develops the foundation upon which your electrical knowledge will be built. A good foundation will help ensure your success in the electrical field. The chapter investigates atoms and their association with electrical current, conductors, and insulators. It also introduces circuit principles, electrical safety, and first aid.

Elements, Atoms, and Compounds

Matter is anything that occupies space and has mass. All matter is made of elements. An *element* is a substance that cannot be reduced to a simpler material by chemical means. Elements are the building blocks of nature. Examples include silver, copper, carbon, hydrogen, zinc, and sulfur. The smallest particle of an element that retains the characteristic of that element is an *atom*.

The chemical combination of two or more elements makes a *compound*. Glass, plastic, wood, salt, ink, and detergent are examples of compounds. The smallest particle of a compound that retains the property of the compound is called a molecule. A *molecule* is the chemical combination of two or more atoms.

Atoms and Electricity

A simple atom contains a *nucleus* in the center and negatively charged *electrons* revolving around the nucleus. See **Figure 1-1.** The nucleus is made up of *protons,* which are positively charged particles, and *neutrons,* which have a neutral charge. The number of protons in an atom is called its *atomic number.* When an atom has the same number of electrons and protons, it is called a *balanced atom.*

Matter: Anything that occupies space and has mass.

Element: A material that cannot be broken down into a simpler form without destroying its identity.

Atom: The smallest particle of an element that retains the properties of the element.

Compound: A substance made of two or more chemical elements.

Molecule: A chemically bonded group of two or more atoms.

Nucleus: The center part of an atom containing protons and neutrons.

Electron: A small negative particle that revolves around the nucleus of an atom.

Proton: The positively charged particle of an atom.

Neutron: One of three elementary particles. The neutron has a neutral charge.

Atomic number: The number of protons in the nucleus of an atom.

Balanced atom: An atom with the same number of electrons as protons.

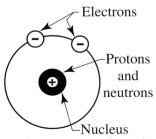

Figure 1-1. A balanced atom has the same number of protons and electrons.

Electrons travel in different shells or levels, **Figure 1-2.** Although the number of shells or levels is not of special interest in the study of electricity, the outer shell is important. It is called the *valence shell* and is the main factor in determining how well a material conducts electricity. If the valence shell is not full, electrons can be pushed into or out of it. Electrons that move in or out of an atom are called *free electrons.*

Valence shell: The outermost electron shell or level of an atom.

Free electron: An electron that is free to move from one atom to another.

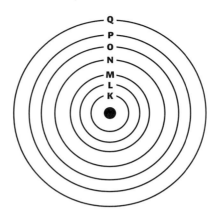

Figure 1-2. Each atom has a certain number of electron shells which have been assigned letters for identification.

Figure 1-3 illustrates free electrons in the valence shell of two atoms. An electron is pushed into atom A. Atom A becomes unbalanced, causing an electron to move out of A and into B. The unbalance in atom B causes an electron to leave atom B for another atom.

Conductors, Insulators, and Semiconductors

Materials can be classified as conductors, insulators, or semiconductors. Materials that contain free electrons are called *conductors.* The free electrons move from one atom to another, **Figure 1-4.** The movement of these electrons is called *electrical current.*

Insulators are materials that do not have free electrons. The valence shell is full and the electrons are held into the atom structure. With no free electrons available, no electric current can flow within the material.

Semiconductors are materials that are neither good conductors nor good insulators but are very important to the solid-state industry. Solid-state refers to such devices as diodes, transistors, and integrated circuits. Silicon and germanium are the two most

Conductor: Any material that contains free electrons and allows electrical current to flow.

Electrical current: The flow of electrons through a conductor.

Insulator: A material with few or no free electrons.

Semiconductor: A solid or liquid conductor with a resistivity between that of metals and insulators.

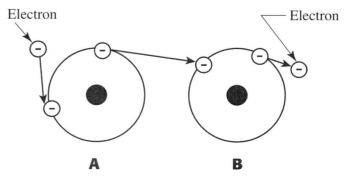

Figure 1-3. In the valence shell, free electrons are able to move from one atom to another.

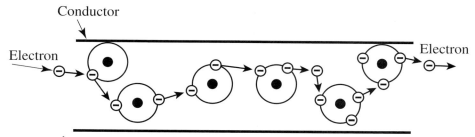

Conductor

Electron

Electron

Figure 1-4. The movement of electrons in a conductor results in electrical current.

common elements used; however, they are not good conductors in pure element form. Impurities are placed into the atomic structure of the semiconductor material to develop the conduction property.

Electrical Circuit Principles

The simplest circuit consists of a voltage source and a load. In **Figure 1-5,** the voltage source is a battery. It is connected to a lightbulb, which is the load. Any complete circuit must have a source, a load, and connecting wires. A switch that controls the current is also in the circuit. If the switch is closed (on), current flows and the bulb is on. If the switch is open (off), current does not flow and the bulb is off.

The voltage source supplies electrons to the circuit. In this case, the battery has a negative terminal with an excess of electrons. The other terminal is positive and has less than the normal amount of electrons. The battery is said to have a ***difference in potential*** (difference in charge) between the two terminals, which causes the electrons to move in the conductors. This difference in potential is referred to as ***electromotive force*** (emf), or the force that moves the electrons. Think of it as electron-moving force. This force is called ***voltage*** and is measured in ***volts*** (V).

As free electrons move from one atom to another, they come to atoms that are not willing to accept electrons readily. In other words, the atoms oppose or resist the movement of the electrons. This opposition to the movement of electrons is called ***resistance.*** It is measured in ***ohms,*** represented by the Greek letter omega (Ω). Every material has resistance to electron movement. The load also has a certain resistance to electron movement.

Difference in potential: The voltage difference between two points.

Electromotive force: The force that causes electrons to move through a conductor.

Voltage: Electrical pressure that causes electrons to move.

Volt: The unit of measurement for voltage, emf, or difference in potential.

Resistance: The property of conductors or materials that opposes the flow of current.

Ohm: The unit of measurement for resistance, inductive reactance, capacitive reactance, and impedance.

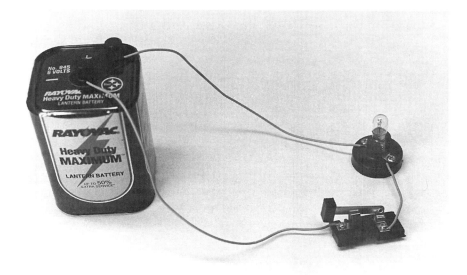

Figure 1-5. A simple light circuit controlled by a switch.

The wire conductors of the circuit provide a path for the electrons (current) to move from the negative terminal through the load to the positive terminal. The greater the number of electrons moving in the circuit, the greater the electrical current. Current is measured in **amperes** (A). When current of 1 A is maintained for 1 second, 6,240,000,000,000,000,000 (read 6.24×10^{18}) electrons are moving past a given point. This number of electrons is called a **coulomb.** One coulomb per second in a circuit is 1 ampere.

Ampere: The unit of measurement for current.

Coulomb: The unit of measurement for a quantity of electrons.

Voltage Sources

Where does electricity come from? One common source of electricity is magnetism. Although you may think of electricity as coming from water (hydroelectric power), coal, diesel fuel, or a nuclear plant, these sources use magnetism to generate electricity. For instance, nuclear energy, coal, or diesel fuel are used to make steam. The steam is used to turn a turbine that is connected to the shaft of a generator, **Figure 1-6.** The generator uses magnetism to make electricity.

Figure 1-6. These huge generators are powered by water from the Clarion River, Piney Dam in Clarion, Pennsylvania.

Cell: A chemical type of voltage source.

Battery: A dc voltage source consisting of two or more cells.

Other sources of electricity are chemical energy (cells), solar energy (from sunlight), thermal (heat) energy, and mechanical energy. The simple **cell** is a chemical type of voltage source. It produces electrical energy from chemical energy. Earlier you saw a battery used as a power source in a circuit. A **battery** is two or more cells.

Many solar cells are connected to form an arrangement that collects solar energy and converts it to electrical energy, **Figure 1-7.** A disadvantage of the solar source of electricity is the high cost of the solar cell.

Electricity can be generated by thermal power. A thermocouple is a device that operates by heat and is widely used for temperature sensing and control, **Figure 1-8.** Two different metals, such as copper and iron, are twisted together. When heat is applied to the twisted wires, a small voltage develops between the two connections. The output voltage of a thermocouple is often only a few thousandths of a volt.

Mechanical pressure is also a voltage source. When materials such as quartz or certain ceramics are subjected to mechanical stress or pressure, a voltage will develop across the crystal structure of the material. See **Figure 1-9.** When the material is sitting

Figure 1-7. Solar cell arrangements such as these harness electricity from the sun.

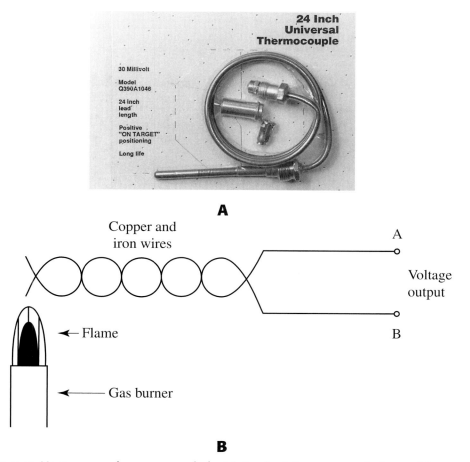

Figure 1-8. Heat energy is a source of electricity. A—A thermocouple is used in gas-operated appliances, such as furnaces, ovens, and hot water tanks. B—When heat is applied to dissimilar metals, a voltage is produced between connections A and B.

Piezoelectric effect: Characteristic of certain natural and synthetic crystals to produce a voltage when subjected to mechanical stress, such as compression, expansion, or twisting.

in a neutral condition, output is zero volts. When pressure is applied, a voltage appears at the output terminals. If a changing voltage is placed across the material, the opposite will occur; that is, the material will vibrate in step with the voltage change. This is called the *piezoelectric effect.* Piezoelectricity is used in phonograph pickup (the cartridge that contains the needle that rides in the record grooves), crystal microphones, and modern speakers.

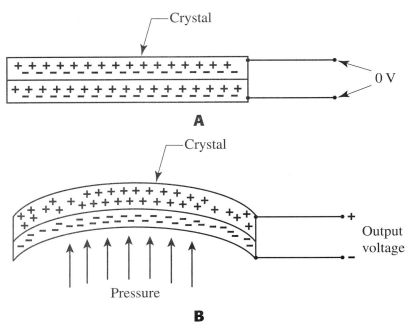

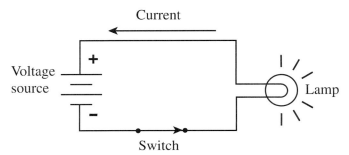

Figure 1-9. Mechanical energy is a source of electricity. A—A crystal with no pressure produces zero voltage. B—Pressure on the material produces a voltage across the crystal.

Schematic Symbols

Schematic: An electrical/electronic diagram that uses symbols to show how various components are electrically connected.

The more parts used in a circuit, the more difficult the circuit is to draw. *Schematic* symbols are used to represent the parts in drawings. **Figure 1-10** uses schematic symbols to represent the simple circuit shown in Figure 1-5. More schematic symbols are given in Appendix A.

Electrical current flows from the negative terminal of the battery through the lamp to the positive terminal. This negative-to-positive direction is referred to as the electron theory of electrical current. Some textbooks, engineers, and others still use the conventional theory that states current flows from positive to negative. This textbook, like most books published today, uses the electron theory proved by Alexander Graham Bell in 1883.

Types of Drawings

Two types of drawings have been introduced so far—pictorial and schematic. Other types of drawings used in electricity are the block diagram, **Figure 1-11,** and the ladder diagram, **Figure 1-12.** All these types of drawings will be used in your electrical training. The ability to read each type with speed and understanding is essential to electrical work.

Figure 1-10. A schematic diagram of a simple circuit showing symbols for voltage source, switch, and lamp. The current flows from negative to positive.

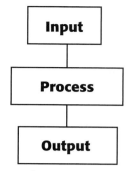

Figure 1-11. Block diagrams are made up of simple blocks.

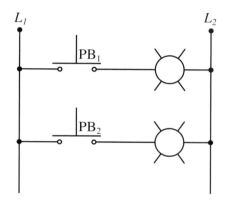

Figure 1-12. A ladder diagram is often used in alternating current power circuits.

Electrical Safety and First Aid

Because of the possibility of personal injury, fire, and damage to equipment, the following precautions should be taken before attempting any work on electrical or electronic devices:

- Disconnect power from the circuit or equipment before working, including when you remove and replace fuses. Never assume the circuit power is off.
- Never override safety interlock devices.
- Make sure all equipment is properly grounded.
- When working on equipment with a disconnect panel, do not just turn it off. Remove the fuses and put them in your pocket, or lock the disconnect in the OFF position.
- Use extreme caution when working with batteries that contain acid.
- In case of fire, turn off the power source and get help immediately. Fumes from burning materials can quickly cause death.
- Think before you act. Your actions could endanger the lives of others.
- When working with high voltage, use only one hand. Do not provide a complete circuit through your body.
- Do not work alone. Have someone with you who can go for help in an emergency.
- Do not wear loose clothing. It can get caught on objects and pull you into live electrical contact or cause other injury.
- Do not wear rings, earrings, wristwatches, bracelets, chains, tags, and similar metal items.
- Wear only nonconductive shoes (ones with rubberized soles).
- Use a shorting stick to remove charges on capacitors. A shorting stick is a long wooden stick with a grounding wire attached. It contains a resistor to limit the current going to ground.
- Do not use equipment unless you have been trained in its operation and safe handling.

First Aid for Electrical Shock

When working with electrical equipment, always guard against shock. The instant you look away from what you're doing, or fail to think before you act, a shock can occur and change your life forever. You may lose the use of your arms or legs, endure pain from burns for months, or even lose your life. A 440 V alternating current line can melt the soles right off your shoes. Think what that could do to your body!

The following procedures are recommended as first aid to a shock victim:

1. Turn off the power switch. Remove the victim from electrical contact. Do not become a victim yourself! Use a dry stick, such as a broom handle, or another nonconductive object, such as a yardstick, plastic handle, or several thicknesses of dry cloth, to remove the victim from contact with the electric wire.
2. If the victim is bleeding, stop the bleeding by applying direct pressure over the wound. Use a clean cloth or part of the victim's clothing to apply the pressure. If the cloth becomes saturated with blood, apply more layers. Do not remove the previous layers.
3. If the victim is not breathing, loosen clothing about the neck and abdomen, and give artificial resuscitation.
4. Cover the victim with a blanket, coat, or any covering that will conserve body heat. If the victim is on a cold surface, such as concrete or tile, "log roll" him or her onto another surface for thermal protection. An old wooden door would work well as an alternate surface.
5. Treat for traumatic shock by raising the feet slightly.
6. Keep the victim from moving about.
7. Do not give stimulants.
8. Get medical help immediately.

First Aid for Burns

As with electrical shock, everyone has experienced a burn at some time. Most electrical burns are caused by hot components or soldering. If you are burned, immediately submerge the burned area in cold water. The real danger with burns is infection, especially burns that produce blisters or blackening of the skin. The burned area should be loosely wrapped with sterile gauze to prevent infection. See a doctor.

 ## Summary

- Matter is anything that occupies space and has mass. All matter is made up of elements. The smallest particle of an element is an atom.
- Atoms are made up of protons in the nucleus and electrons revolving around the nucleus. Electrons have a negative charge and protons have a positive charge.
- A full valence shell will not give up electrons or allow additional electrons into the atom.
- A material that has a full valence shell is an insulator.
- Conductors contain free electrons. Electrical current is the movement of free electrons in a conductor.
- Semiconductors are neither good conductors nor good insulators. They are used in the solid-state industry.
- Any electrical circuit must contain at least one voltage source, a load, and connecting wires.

- A coulomb is a measurement of a quantity of electrons. One coulomb per second is 1 ampere of current.
- The unit of measurement for voltage is volts (V); current is measured in amperes (A); and resistance is measured in ohms (W).
- Drawings used in the electrical industry may be either pictorial, schematic, block, or ladder diagram.
- Always follow the safety rules when working with electricity. The consequences can be fatal.

Important Terms

Do you know the meanings of these terms used in the chapter?

ampere	insulator
atom	matter
atomic number	molecule
balanced atom	neutron
battery	nucleus
cell	ohm
compound	piezoelectric effect
conductor	proton
coulomb	resistance
difference in potential	schematic
electrical current	semiconductor
electromotive force	valence shell
electron	volt
element	voltage
free electrons	

Questions and Problems

Please do not write in this text. Write your answers on a separate sheet of paper.

1. _____ occupies space and has mass.
2. An electron has a(n) _____ charge.
3. The smallest particle of an element that retains the characteristic of that element is a(n) _____.
4. In electricity, the most important electron shell of an atom is the _____ shell.
5. Electrons that move from one atom to another are called _____.
6. What are the three classifications of electrical materials?
7. A complete circuit must have what three parts?
8. Current flows in a circuit when a switch is in the _____ (closed, open) position.
9. The opposition to current is called _____.
10. A unit of measure for a quantity of electrons is called a(n) _____.
11. List five common sources of electricity.
12. The electron theory states current flows from _____ to _____.
13. Draw the schematic symbols for battery, switch, and lamp (load).
14. List four types of drawings used in electricity.
15. Why should you use only one hand when working with high voltage?
16. How do you remove a shock victim from electrical contact?
17. What is the first aid for burns?

Chapter 2 Graphic Overview

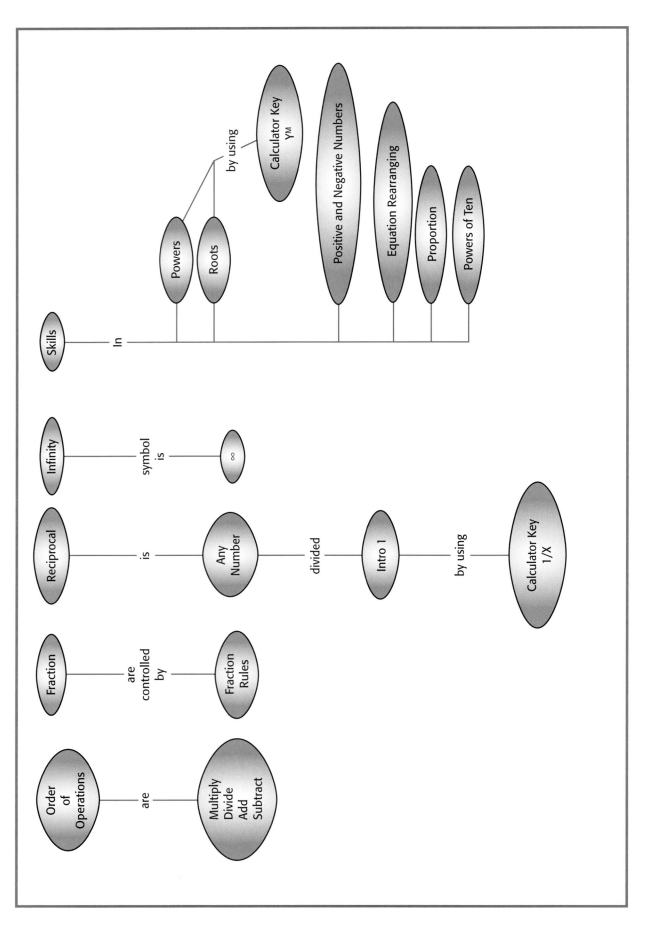

MATH FOR ELECTRICITY

Objectives

After studying this chapter, you will be able to:

- ○ Solve mathematical problems in the order of operation.
- ○ Change a fraction to a decimal.
- ○ Multiply and divide fractions.
- ○ Find the reciprocal of a number.
- ○ Raise a number to a power.
- ○ Find the root of a number.
- ○ Add and subtract positive and negative numbers.
- ○ Solve for unknown values using the basic rules for equations.
- ○ Solve proportion problems.
- ○ Identify direct and inverse relationships.
- ○ Use a scientific calculator to calculate powers of ten.

Introduction

Mathematics is often used to illustrate actions taking place in circuits. This chapter will provide the foundation in the math you need to pursue your electrical studies. The chapter can also serve as a reference for later use. It is assumed you can add, subtract, multiply, and divide whole numbers including decimals.

Use of Parentheses

If two people are given the problem 10 – 4 + 2, one person would give an answer of 10 – 4 = 6 and 6 + 2 = 8, and the other would give an answer of 4 + 2 = 6 and 10 – 6 = 4. To make sure the math is done in the proper order, parentheses are used to group operations. If the problem is written 10 – (4 + 2), you know the sum of 4 + 2 is to be subtracted from 10 for an answer of 4. Parentheses are often used to group numbers or operations that must be completed before any other math is done.

State the answer for each of the following problems:

20 – (4 + 8) = _____
15 + (15 – 7) = _____
6 + 7 – (3 + 2) = _____
(5 + 3) – 4 = _____
(35 – 15) = _____

Besides parentheses, brackets [ ] and braces { } are used for grouping. Brackets are often used with parentheses. The work inside the brackets must be completed before any operations outside the brackets are done.

▼ *Example 2-1:*

    20 – [5 + 6 – (16 – 8)] = 17
    16 – 8 = 8   (First do the operations inside parentheses.)
    5 + 6 = 11
    11 – 8 = 3   (The 8 is from the 16 – 8 above.)
    20 – 3 = 17   (The result inside the brackets is subtracted from 20.)   ▲

Parentheses are also used in multiplication. No multiplication sign is needed; it is an understood operation.

▼ *Example 2-2:*

3 (7 − 2) = _____

21 − 6 = 15   (Multiply everything inside the parentheses by 3.)   ▲

Order of Operations

If you were given the problem $7 + 4 \times 2$, you might ask, "Where are the parentheses?" Mathematicians have agreed upon the order of operations; that is the sequence in which the math is done. Multiplication is done first, division next, addition next, and subtraction last. Because of the order of operations, no parentheses are necessary.

▼ *Example 2-3:*

$7 + 4 \times 2$ = _____

$7 + 8 = 15$

and

$20 \div 5 + 8 + 5 \times 3$ = _____

$4 + 8 + 15 = 27$ ▲

Fractions

Skill with fractions is essential to working with electricity. For example, to find the current of a circuit, divide the voltage by the resistance.

▼ *Example 2-4:*

Assume a circuit has a voltage of 3 V and a resistance of 12 Ω.

$\dfrac{3\text{ V}}{12\text{ Ω}}$ = 0.25 A (3 is divided by 12, or the fraction $\dfrac{3}{12}$ has been changed to a decimal.)

▲

Fraction Operation Rules

Numerator: The top number in a fraction.

Denominator: The bottom number in a fraction.

Proper fraction: A fraction in which the numerator is smaller than the denominator.

Improper fraction: A fraction in which the numerator is equal to or greater than the denominator.

A ***numerator*** is the top number in a fraction. The ***denominator*** is the bottom number in a fraction. In a ***proper fraction,*** the numerator is smaller than the denominator. In an ***improper fraction,*** the numerator is equal to or greater than the denominator. Several rules can be applied when working with fractions:

- The value of a fraction is not changed when both the numerator and denominator are multiplied or divided by the same number.
- To reduce a fraction to its lowest terms, divide the numerator and denominator by the same number. When the numerator and denominator cannot be further divided by the same number, the fraction is expressed in its lowest (simplest) terms.
- To reduce an improper fraction to its lowest terms, divide the numerator by the denominator, and reduce the resulting fraction.
- To add or subtract fractions, change the fractions so they have a common denominator. Add the numerators. Using the common denominator, reduce the fraction to lowest terms.
- To multiply fractions, multiply the numerators and denominators; then reduce the fraction to lowest terms.
- To divide fractions, invert the second fraction; then, multiply the fractions and reduce to lowest terms.

Changing a Fraction to a Decimal

To change a fraction to a decimal, divide the denominator into the numerator. If you have difficulty remembering which number is divided into which, try the method shown in **Figure 2-1:**

a. Imagine the top number of the fraction is sitting on a table.
b. You read from left to right, so pick up the left side of the table and let the top number slide off.
c. Turn the tabletop over.
d. The fraction is now set up for division.
e. Place the decimal point after the 1. Add zeros as necessary to get the answer.

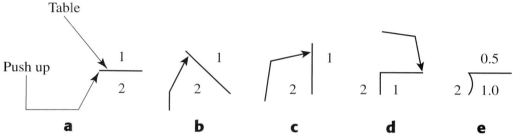

Figure 2-1. Setting up a division problem to change a fraction to a decimal.

Multiplying and Dividing Fractions

When multiplying two or more fractions, the numerators of each fraction are multiplied together; then the denominators are multiplied together. The fraction is reduced to its lowest terms.

▼ *Example 2-5:*

$$\frac{2}{3} \times \frac{6}{7} = \frac{12}{21} = \frac{4}{7}$$

▲

When dividing two fractions, invert the second fraction and multiply.

▼ *Example 2-6:*

$$\frac{4}{10} \div \frac{3}{5} = \frac{4}{10} \times \frac{5}{3} = \frac{20}{30} \text{ or } \frac{2}{3}$$

▲

Reciprocals

Many calculations involving electrical circuits use a math form called a reciprocal. Reciprocals are very simple. The *reciprocal* of a number is that number divided into 1.

▼ *Example 2-7:*

The reciprocal of 4 is $\frac{1}{4}$ or 0.25.

The reciprocal of R is $\frac{1}{R}$.

The reciprocal of 10.7 is $\frac{1}{10.7}$.

▲

Reciprocal: The reciprocal of a number is found by dividing one by that number ($\frac{1}{X}$).

Three-deep Fractions

The reciprocal of the fraction $\dfrac{3}{12}$ is $\dfrac{1}{\dfrac{3}{12}}$ and may be referred to as a "three-deep" fraction. A three-deep fraction involving the reciprocal may be simplified by throwing away the numerator 1 and inverting the remaining fraction.

▼ **Example 2-8:**

$$\frac{1}{\dfrac{3}{12}} = \frac{12}{3} = 4$$

▲

Calculator Reciprocal Key

Most of the calculations introduced in the remainder of the chapter require the use of a scientific calculator, **Figure 2-2.** These calculators have special keys for simplifying mathematical processes. To find the reciprocal of a number, enter the number on the display and press the **1/x** (reciprocal) key.

▼ **Example 2-9:**

Find the reciprocal of 20.

   Place the number 20 on the calculator display.

   Press the **1/x** key.

   The calculator display now shows the reciprocal of 20, which is 0.05.   ▲

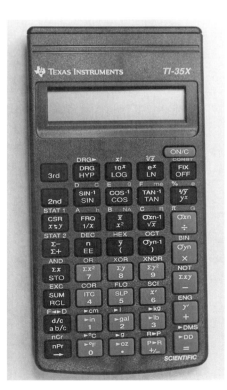

Figure 2-2. Scientific calculators come with different types of keyboards and function keys.

Infinity

Infinity is the highest number that can exist hypothetically. The letter symbol for infinity looks like the number 8 lying on its side (∞). **Figure 2-3** illustrates the number system with number 1 as the reference point. The upward direction goes to infinity and the downward direction to zero. The number 1 as reference will be discussed with powers of ten later in this chapter.

Infinity: A hypothetical measurement so large or small no number can be assigned to it.

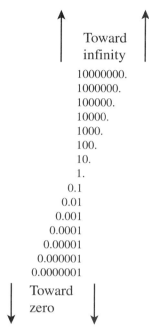

Toward infinity

10000000.
1000000.
100000.
10000.
1000.
100.
10.
1.
0.1
0.01
0.001
0.0001
0.00001
0.000001
0.0000001

Toward zero

Figure 2-3. The decimal numbers reach to infinity and zero.

Raising a Number to a Power

Raising a number to a power is repeated multiplication. If you wish to raise the number 6 to the third power, you would write it as 6^3. The 6 is called the *base* and the 3 is called the exponent. The *exponent* tells you how many times the base is to be multiplied by itself. So, 6^3 would be $6 \times 6 \times 6 = 216$.

Scientific calculators allow you to raise a number to a power very easily using the $\mathbf{y^x}$ key. The process may vary depending upon the type of calculator you are using.

Base: A number raised to a power.

Exponent: The number of times a number is to be multiplied by itself.

▼ *Example 2-10:*

Place the number 6 on the calculator display.
Press the $\mathbf{y^x}$ key. (This tells the calculator you are raising the number to a power.)
Press the number 3.
Press the equal key.
The answer 216 is shown on the display.   ▲

Try the process on the following numbers:
5^2
10^4
16^8
25^5

The calculator has another key, x^2, which is used in the powers operation when numbers are raised to the second power or squared. Try squaring the following numbers using the x^2 key:

4^2

7^2

16^2

100^2

Finding the Root of a Number

Finding the root of a number is the opposite process of raising a number to a power. The calculator key for the square root of a number is $\sqrt{}$.

▼ *Example 2-11:*

Find the square root of 625.

    Place the number 625 on the calculator display.

    Press the square root key.

    The answer 25 is shown on the display.   ▲

To find the root of a number other than the square root, use the y^x key.

▼ *Example 2-12:*

Find the fourth root of 2401.

    Place the number 2401 on the calculator display.

    Depending on your calculator, press the **inv** or **2nd** key.

    Press the y^x key.

    Press the number 4 and then the equal key.

    The answer 7 is shown on the display.   ▲

Positive and Negative Numbers

In electricity, plus (+) and minus (–) signs are used to indicate electrical polarities. In common arithmetic, it is not necessary to think of numbers as having different polarities. However, in more advanced math every number has a polarity.

Use of Plus and Minus Signs

A plus sign in front of a number indicates it is positive, such as +4, +235, or $+\frac{1}{2}$.

If a number does not have a sign in front of it, the number is considered positive. A negative number must always have the negative sign in front of it, for example, –7, –338, or $-\frac{1}{8}$. Numbers that are considered negative or positive are called *signed numbers*.

Signed number: A number preceded by a positive or negative sign.

Adding Numbers with Like Signs

To add numbers with positive signs, simply add the numbers together.

▼ *Example 2-13:*

    $(+5) + (+3) = +8$

or

    $5 + 3 = 8$ ▲

To add numbers with negative signs, add the numbers together. Note parentheses must be used.

▼ *Example 2-14:*

    $(-5) + (-8) = -13$ ▲

Adding Numbers with Unlike Signs

To add numbers of different signs, ignore the signs and subtract the smaller number from the larger number. Then, carry the original sign of the larger number to the answer.

▼ *Example 2-15:*

$$4 + (-9) = -5$$
$$7 + (-4) = 3$$
$$-12 + 5 = -7$$
$$18 + (-12) = 6$$
▲

Subtracting Signed Numbers

Subtracting numbers with positive signs is the same process you learned in grade school; for example, $12 - 7 = 5$. When negative numbers are used, it is a different situation. To subtract signed numbers, change the sign of the number being subtracted and continue as in addition of numbers with negative signs.

▼ *Example 2-16:*

$$-12 - 7 = -12 + (-7) = -19$$
$$-6 - (-5) = -6 + 5 = -1$$
$$14 - (-3) = 14 + 3 = 17$$
$$-36 - 22 = -36 + (-22) = -58$$
$$2 - 8 = 2 + (-8) = -6$$
▲

Try the equations in the previous example on your calculator. Use the **+/−** key to change the sign of the numbers. The first one would be 12 +/− (−7). Notice how the number 12 changed sign.

Try these for more practice:

$8 - (-6)$
$-16 - (-6)$
$-24 - (-7)$
$-9 - (-22)$
$12 - (-45)$
$10 - (-11)$

Solving for Unknowns

There are two kinds of numbers—absolute and literal. An ***absolute number*** has a definite value. For example, 0, 6, and 89 are all absolute numbers. A ***literal number*** is expressed by a letter of the alphabet or some other symbol. Numbers such as R, Y, and π are literal numbers. A literal number may represent a known or unknown value. Working with literal numbers will be a major part of your electronic training since most circuit action is explained with mathematical equations.

Equations

Two things with exactly the same value are said to be equal. The quantity on the left side of an equal sign is the same as the quantity on the right side. In any ***equation***, the problem revolves around the equal sign. The left side must always be equal to the right side, no matter what literal or absolute values are involved.

One of the basic skills in algebra is to be able to rearrange an equation to find the unknown. To rearrange an equation, the *opposite* operation is used, **Figure 2-4.** The process works for any equation regardless of its complexity.

Absolute number: A number having a definite value.

Literal number: A number with no specified value, symbolized by a letter of the alphabet or some other special character.

Equation: Two expressions separated by an equal sign.

Operation	Opposite
Add	Subtract
Subtract	Add
Multiply	Divide
Divide	Multiply
Square	Square root
Square root	Square
Log	Antilog
Antilog	Log

Figure 2-4. Each math operation has an opposite.

▼ *Example 2-17:*

Given $C = Y + T$, find Y.

$\qquad C - T = Y + T - T$ (To get Y by itself, subtract T from both sides of the equal sign.)

$\qquad C - T = Y$

or

$\qquad Y = C - T$

▼ *Example 2-18:*

Given $S = C - R$, find C.

$\qquad S + R = C - R + R$ (Add R to both sides.)

$\qquad S + R = C$

or

$\qquad C = S + R$

▼ *Example 2-19:*

Given $X = \dfrac{E}{B}$, find E.

$\qquad X \times B = \dfrac{E}{B}\ (B)$ (Multiply both sides by B.)

$\qquad X \times B = E$

or

$\qquad E = X \times B$

▼ *Example 2-20:*

Given $X = \dfrac{E}{B}$, find B.

$\qquad X \times B = E$ (Multiply both sides by B.)

$\qquad X \times \dfrac{B}{X} = \dfrac{E}{X}$ (To get B by itself, divide both sides by X. *Note: X* divided into X equals 1, no matter what number is used.)

$\qquad B = \dfrac{E}{X}$

▼ *Example 2-21:*

Given $G = \dfrac{R^2}{N}$, find R.

$G \times N = R^2$ (Multiply both sides by N.)

$\sqrt{G \times N} = R$ (Since R is squared, take the square root of both sides.)

or

$R = \sqrt{G \times N}$ ▲

▼ *Example 2-22:*

Given $X = 2\pi fL$, find L.

$\dfrac{X}{2\pi f} = \dfrac{2\pi fL}{2\pi f}$ (Divide both sides by $2\pi f$.)

$\dfrac{X}{2\pi f} = L$ ▲

Proportion

A ***proportion*** is an equation of ratios or fractions. The proportion $1/2 = 3/6$ can be read two ways: "One half equals three sixths," or "one is to two as three is to six." Proportions are used throughout electronics. Solving a proportion problem when one of the numbers is unknown is a simple process involving cross multiplication.

Cross Multiplication

Cross multiplication means to multiply the numerator of one fraction by the denominator of the other fraction to get a quotient on both sides of the equal sign.

Proportion: A relationship of four quantities where the product of the first and fourth quantities equals the product of the second and third quantities, or the first ratio is equal to the second ratio.

▼ *Example 2-23:*

$\dfrac{3}{4} = \dfrac{6}{8}$ (Cross multiply.)

$3 \times 8 = 4 \times 6$

$24 = 24$ ▲

▼ *Example 2-24:*

Solve for Z.

$\dfrac{R}{T} = \dfrac{Y}{Z}$ (Cross multiply.)

$RZ = TY$ (Divide both sides by R.)

$Z = \dfrac{TY}{R}$ ▲

▼ *Example 2-25:*

Solve for X.

$\dfrac{6}{X} = \dfrac{X}{6}$ (Cross multiply.)

$X^2 = 36$ (Take the square root of both sides.)

$X = 6$ ▲

Direct and Inverse Relationships

Directly proportional: The value of two numbers increases or decreases at the same rate.

Inversely proportional: The value of one number increases as the value of another number decreases or vice versa.

In **Figure 2-5,** R is *directly proportional* to Y and *inversely proportional* to Z. What does this mean? Two numbers are said to be directly proportional when the value of both numbers goes up. In the case of R, if R increases, Y increases, too. On the other hand, if R increases, Z decreases. They are inversely proportional because when one goes up, the other goes down.

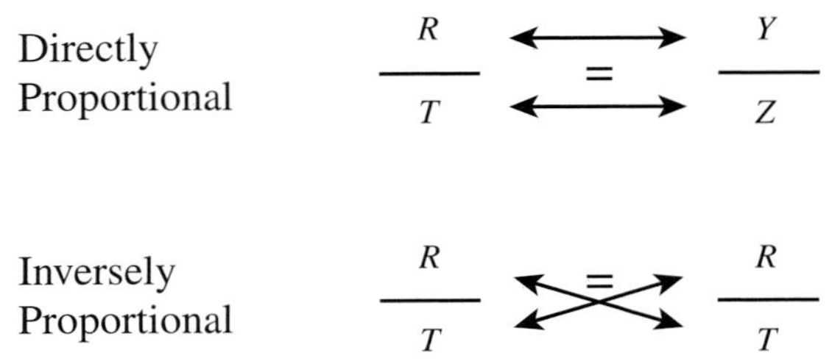

Directly
Proportional $\dfrac{R}{T} \xleftrightarrow{=} \dfrac{Y}{Z}$

Inversely
Proportional $\dfrac{R}{T} \;\begin{smallmatrix}=\end{smallmatrix}\; \dfrac{R}{T}$

Figure 2-5. Direct and inverse relationships.

Powers of Ten

What does 10^3 mean? It means raise the number 10 to the third power, or $10 \times 10 \times 10$. The exponent (3) tells you how many times 10 is to be multiplied by itself. The concept of powers of ten enables you to work with very large or very small numbers. It is important to master the use of powers of ten on your calculator.

What is the value of 10^0? Ten to the zero power is equal to 1. Except for zero, the value of any number raised to the zero power is equal to 1. This is called *unity.* Try raising several numbers to zero power on your calculator.

Unity: Any number raised to the zero power is equal to 1.

Powers of Ten as a Decimal Placer

Multiplying a number by 10 is the same as moving the decimal point one place to the right. Multiplying by 10 three times is the same as moving the decimal point three places to the right or multiplying by 1000. Therefore, multiplying a number by 10^3 can be done by moving the decimal point three places, the number of the exponent.

In **Figure 2-6,** notice how the decimal point is moved around the number 1. A positive exponent moves the decimal point to the right, and a negative exponent moves the decimal point to the left. Also, notice that 10^0 moves the decimal zero places.

▼ *Example 2-26:*

$3.44 \times 10^2 = 344$	(Two places to the right.)
$123 \times 10^4 = 1{,}230{,}000$	(Four places to the right.)
$1.23 \times 10^6 = 1{,}230{,}000$	(Six places to the right.)
$7 \times 10^{-3} = 0.007$	(Three places to the left.)
$1.8 \times 10^{-9} = 0.0000000018$	(Nine places to the left.) ▲

Powers of Ten and the Calculator Keys

How would you use your calculator to calculate powers of ten? The $\mathbf{y^x}$ key could be used; however, powers of ten are so useful that special keys have been placed on the calculator for just this purpose. Your calculator should have an **EE** or **Exp** key. This is the power of ten exponent key. Try the following operation:

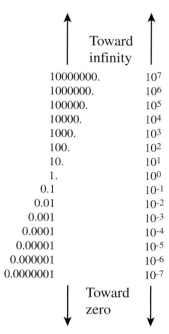

Toward
infinity

10000000.	10^7
1000000.	10^6
100000.	10^5
10000.	10^4
1000.	10^3
100.	10^2
10.	10^1
1.	10^0
0.1	10^{-1}
0.01	10^{-2}
0.001	10^{-3}
0.0001	10^{-4}
0.00001	10^{-5}
0.000001	10^{-6}
0.0000001	10^{-7}

Toward
zero

Figure 2-6. Decimal numbers and powers of ten used as a decimal placer.

Enter 1, 2, 3, **EE** (or **Exp**). The display may shift slightly to the left and two zeros should appear at the right side of the display. These zeros are the exponent position of the display.

Now enter 4. The display should be 123×10^4.

You have just entered the number 1,230,000 into the calculator. The exponents can range from 99 places to the right or left of the decimal point. This is one of the powerful uses of a calculator. The display is capable of showing only eight or ten numbers. By using the powers of ten key, you can work with very large or very small numbers.

Scientific and Engineering Notation

Two means for writing large or small numbers are using scientific notation and engineering notation. The scientific notation method is for general use in working with numbers. Engineering notation is very useful when working with numbers in electricity.

Scientific Notation

Scientific notation always has one digit to the left of the decimal point. The exponent is whatever value is required to indicate the correct decimal position.

Scientific notation: Writing numbers using powers of ten and maintaining one digit to the left of the decimal point.

▼ *Example 2-27:*

$234{,}570{,}000{,}000{,}000 = 2.3457 \times 10^{14}$

$0.000000100972 = 1.00972 \times 10^{-7}$

$4{,}683.8 = 4.683 \times 10^3$

$0.058\ 71 = 5.871 \times 10^{-2}$ ▲

Engineering Notation

Engineering notation can have more than one digit to the left of the decimal point, but the power of ten exponent is always a multiple of three.

Engineering notation: A notation using powers of ten in which the exponent is always a multiple of three.

▼ ***Example 2-28:***

$234,561,000,000,000 = 234.561 \times 10^{12}$

$4,783 = 4.783 \times 10^{3}$

$0.0000000822 = 82.2 \times 10^{-9}$ ▲

How Powers of Ten Are Used

Engineering notation makes it possible to express numbers with metric prefixes, **Figure 2-7.** Notice that each power of ten exponent is a multiple of three. You should know the letter symbol and metric prefix for each power of ten.

Letter Symbol	Prefix	Power of Ten
T	Tera	10^{12}
G	Giga	10^{9}
M	Mega	10^{6}
k	Kilo	10^{3}
–	Unity	10^{0}
m	milli	10^{-3}
μ	micro	10^{-6}
n	nano	10^{-9}
p	pico	10^{-12}

Figure 2-7. Memorize these letter symbols, prefixes, and powers of ten.

If you were working with a resistance of 1,500,000 ohms, you would not state it as "one million five hundred thousand ohms," nor would you write it as 1,500,000 Ω. Instead, it would be stated or written as 1.5 megohms. First, the number 1,500,000 is changed to engineering notation, 1.5×10^{6}. Then, the power of ten is changed to the metric prefix "mega," which is added to the word ohms (the unit of resistance). Thus, the resistance is expressed as 1.5 megohms. (The "a" is dropped from mega to avoid placing two awkward vowels together.)

To further illustrate, if you are working with a current of 0.015 (fifteen thousandths) amperes, you would state it as 15 milliamperes (milliamps, for short) or write 15 mA. Again, 0.015 is changed to engineering notation, 15×10^{-3}. The power of ten is changed to the metric prefix "milli" and added to the word amperes (the unit of current).

To multiply or divide two values using the power of ten, enter the first value on the calculator. Press the desired function; then, enter the second value, and finally the equal key. For instance, voltage equals current times resistance; therefore, using the values above, 1.5 megohms and 15 mA, find the voltage: $1.5 \times 10^{6} \times 15 \times 10^{-3} = 2.25 \times 10^{4}$.

Change 2.25×10^{4} to engineering notation, 22.5×10^{3}. Last, change the power of ten to the metric prefix kilo. The answer is 22.5 kilovolts (kV).

Try the above problem on your calculator. Enter 1.5 on the calculator display. Press **EE** (or **Exp**), enter 6, then the times function. Enter 15 on the calculator display. Press **EE** (or **Exp**), enter (–3), and the equal key. *Do not* make the mistake of pressing 15 × 10 and then the **EE** (**Exp**) key. The **EE** (or **Exp**) key tells the calculator the number that follows is a power of ten.

The measurement of time in waveforms is another example of the use of powers of ten. Rapid time is typically measured in seconds, but in electricity movement happens so quickly that time may be measured in fractions of seconds, such as 0.000000018 seconds. You would never write or state time in decimal form. Rather, you would write or state it as 18 nanoseconds (ns) or 18×10^{-9}. Try to put 0.000000018 on your calculator. Can you do it? The decimal number has too many digits for most calculator displays; however, the powers of ten make it possible to work this number on the calculator.

Try the following equations on your calculator:
1. 136 milliamps × 1.5 kilohms
2. 3 kilohms × 80 microamps
3. 40 picovolts ÷ 100 kilohms
4. 4 millivolts ÷ 10 milliamps
5. 8 nanoamps × 6.8 kilohms
6. 500 volts ÷ 2.5 megohms

Figure 2-8 lists some common expressions using the electrical units of measurement and metric prefixes. Using a calculator in the engineering mode makes it easy to use powers of ten, convert units of measurement with metric prefixes, and work with large and small numbers.

You may find yourself trying to use your old methods in working with numbers. It may seem like too much trouble to use powers of ten, but if you use them every time you are faced with a math problem, the process will become second nature. Once using powers of ten becomes a habit, working with electrical problems will be easier.

Answers to above equations:
1. 204 volts
2. 240 millivolts
3. 400×10^{-18} amps
4. 400 milliohms
5. 54.4 microvolts
6. 200 microamps

Electrical Unit	Abbreviation	Power of Ten	Metric Prefix
Volt	kV	10^3	kilovolt
	mV	10^{-3}	millivolt
	μV	10^{-6}	microvolt
Amp	mA	10^{-3}	milliamp
	μA	10^{-6}	microamp
	nA	10^{-9}	nanoamp
Ohm	kΩ	10^3	kilohm
	MΩ	10^6	megohm

Figure 2-8. Common electrical units and their metric prefixes.

 **Summary**

- Parentheses and brackets are often used to group numbers or math operations.
- The order of operations is: multiply, divide, add, subtract.
- The reciprocal of a number is that number divided into the number one. The reciprocal of a "three-deep" fraction can be simplified by throwing away the numerator 1 and inverting the remaining fraction.
- Infinity is the largest number that can exist.
- All numbers have a polarity. To add numbers of unlike signs, subtract the smaller number from the larger number and carry the sign of the larger number.
- The opposite operation is used to rearrange an equation to find an unknown value.
- Problems of proportion are solved by cross multiplication.
- Scientific notation always has one digit on the left side of the decimal point. In engineering notation, the power of ten is always a multiple of three.

 Important Terms

Do you know the meanings of these terms used in the chapter?

absolute number	inversely proportional
base	literal number
denominator	numerator
directly proportional	proper fraction
engineering notation	proportion
equation	reciprocal
exponent	scientific notation
improper fraction	signed number
infinity	unity

 Questions and Problems

Please do not write in this text. Write your answers on a separate sheet of paper.

1. Solve the following problems.
   a. $4 \times 3 + 1$
   b. $3 + 6 \div 2$
   c. $5 - 2 \times 2$
   d. $4 + 4 \div 4$
   e. $10 \div 5 + 1$
   f. $80 \div 12 + 6$

2. Change the following fractions to decimals.
   a. $\dfrac{1}{10}$
   b. $\dfrac{9}{16}$
   c. $\dfrac{5}{20}$
   d. $\dfrac{31}{100}$

3. Multiply the following fractions and reduce them to simplest terms.
   a. $\dfrac{4}{5} \times \dfrac{3}{4}$
   b. $\dfrac{6}{7} \times \dfrac{2}{5}$

   c. $\dfrac{3}{4} \times \dfrac{8}{9}$

   d. $\dfrac{12}{15} \times \dfrac{4}{21}$

4. Find the reciprocal of each of the following numbers.

   a. 10

   b. 5

   c. 25

   d. 4

   e. $\dfrac{1}{10}$

5. Solve the following powers.

   a. 5^4

   b. 12^3

   c. 15^6

   d. 12^2

   e. 2^8

   f. 7^3

   g. 33^2

   h. 6^4

   i. 3^5

   j. 8^0

6. Solve the following roots.

   a. $\sqrt{16}$

   b. $\sqrt{34}$

   c. $\sqrt{15{,}625}$

   d. $\sqrt[3]{144}$

   e. $\sqrt{89}$

   f. $\sqrt[4]{279{,}841}$

   g. $\sqrt[5]{32{,}768}$

   h. $\sqrt[8]{256}$

7. Add the following signed numbers.

   a. $5 + 6 + 3$

   b. $-23 + 5$

   c. $34 + (-20)$

   d. $5 + (-3) + (-4)$

8. Subtract the following signed numbers.

   a. $4 - (-6)$

   b. $-10 - (6)$

   c. $10 - (-6)$

   d. $-47 - (-12)$

   e. $-16 - (18)$

   f. $-102 - (-17)$

   g. $21 - (-12)$

   h. $4 - (-2) - 16$

   i. $-12 - (-3) - (-7)$

   j. $120 - (-16)$

   k. $5 - (-10) - 15$

   l. $-18 - (-28) - 6$

9. For each of the following problems, write the equation to find the unknown.

    a. $A + C = S$ $A = $ _____
    b. $R + V = B$ $R = $ _____
    c. $C + J = D$ $C = $ _____
    d. $A + N = Z$ $N = $ _____
    e. $F - L = A$ $F = $ _____
    f. $D - N = M$ $D = $ _____
    g. $X - T = Z$ $T = $ _____
    h. $P - L = Q$ $L = $ _____

10. For each of the following problems, write the equation to find the unknown.

    a. $V = I \times R$ $R = $ _____
    b. $P = I \times E$ $I = $ _____
    c. $E = P \times D$ $D = $ _____
    d. $W = R \times U$ $U = $ _____

    e. $I = \dfrac{P}{E}$ $E = $ _____

    f. $R = \dfrac{S}{Y}$ $S = $ _____

    g. $Q = \dfrac{V}{I}$ $I = $ _____

    h. $G = \dfrac{T}{A}$ $T = $ _____

11. Solve each of the following problems for the unknown.

    a. $A + 3 = 12$ $A = $ _____
    b. $G + 4 = 16$ $G = $ _____
    c. $R + 10 = 75$ $R = $ _____
    d. $D - 21 = 33$ $D = $ _____
    e. $K - 7 = 13$ $K = $ _____
    f. $V - 4 = 2$ $V = $ _____
    g. $F + 6 = 17$ $F = $ _____
    h. $F - 12 = 50$ $F = $ _____

12. Solve each of the following problems for the unknown.

    a. $16 = 4T$ $T = $ _____
    b. $5L = 55$ $L = $ _____
    c. $3Z = 18$ $Z = $ _____
    d. $90 = 2H$ $H = $ _____

    e. $54 = \dfrac{B}{3}$ $B = $ _____

    f. $500 = \dfrac{25}{C}$ $C = $ _____

13. Solve each of the following proportion problems for the unknown.

    a. $\dfrac{V}{L} = \dfrac{C}{E}$ $C = $ _____

    b. $\dfrac{W}{K} = \dfrac{Q}{L}$ $K = $ _____

    c. $\dfrac{P}{S} = \dfrac{R}{U}$ $U = $ _____

d. $\dfrac{CD}{X} = \dfrac{RT}{B}$ $B = $ _____

e. $\dfrac{JN}{L} = \dfrac{SA}{Q}$ $L = $ _____

14. Express the following numbers in scientific notation.
    a. 6800
    b. 12,000
    c. 0.0168
    d. 1,000,000
    e. 100,000,000
    f. 56,000
    g. 0.00956
    h. 12.45
    i. 0.00002905
    j. 0.013

15. Express the following numbers in engineering notation.
    a. 1,000,000
    b. 100,000,000
    c. 0.00001
    d. 0.01
    e. 120,000
    f. 0.00002905
    g. 0.013
    h. 0.590
    i. 23,900
    j. 0.00234

16. Multiply the following powers of ten.
    a. $(120 \times 10^3) \times (6 \times 10^5)$
    b. $(18 \times 10^{-9}) \times (7 \times 10^{-11})$
    c. $(224 \times 10^{-3}) \times (3 \times 10^{-5})$
    d. $(14 \times 10^{-14}) \times (3 \times 10^8)$

17. Divide the following powers of ten.
    a. $\dfrac{66 \times 10^{-5}}{3 \times 10^{-3}}$

    b. $\dfrac{350 \times 10^{-16}}{5 \times 10^5}$

    c. $\dfrac{224 \times 10^{11}}{2 \times 10^6}$

    d. $\dfrac{160,000 \times 10^5}{4 \times 10^{10}}$

18. Multiply the following numbers.
    a. $0.00000000001234 \times 0.000000000567$
    b. $4,500,000,000,000,000,000 \times 0.0000000000123$
    c. 12 nanoamps $\times$ 5000 megohms
    d. $680,000,000,000,000 \times 10^{-9} \times 2,000,000,000,000 \times 10^8$

Chapter 3 Graphic Overview

Black	0
Brown	1
Red	2
Orange	3
Yellow	4
Green	5
Blue	6
Violet	7
Gray	8
White	9

Resistors
- identified by → Resistor Color Code
- can be → Fixed — or — Variable
  - Variable are → Thermistor, Light-dependent, Potentiometer, Adjustable Tapped

Factors That Determine Resistance
- are → Type of Material
  - determines the → Resistivity
    - while → Insulators
      - have a → Breakdown Voltage
      - measured by → Dielectric Constant
- are → Temperature, Length, Area

Types of Conductors
- are → Cable
  - may be → Ribbon, Coaxial
- are → Superconductor
- are → Single Conductor
  - may be → Stranded, Solid
    - sized by → American Wire Gage — and — Circular Mils

CONDUCTORS, INSULATORS, AND RESISTORS

3

Objectives

After studying this chapter, you will be able to:

○ Explain how the size of a conductor relates to its resistance.
○ State the factors that determine the resistance of a conductor.
○ Explain the difference between an insulator and a conductor.
○ State the types of resistors.
○ Identify the resistance value of a resistor using a standard color code.

Introduction

Conductors have a low resistance to current. Special conductors are needed for the high-frequency alternating currents used in television and computer systems. Various types of insulators are available, such as plastic for general use or Teflon™ for special applications. Resistors are used to reduce current or provide a voltage drop.

Conductors

There are many types of conductors, **Figure 3-1.** A *cable* is made up of one or more insulated conductors in the same protective container. A *coaxial cable* is used for distribution of television signals, **Figure 3-2.** *Ribbon cables* are used in the connection of printed circuits, **Figure 3-3.** Conductors and cables often make up wiring harnesses used in the assembly of automobiles, aircraft, and computer systems, **Figure 3-4.**

Cable: A stranded conductor or group of conductors insulated from each other.

Coaxial cable: A concentric transmission line in which one conductor completely surrounds another conductor.

Ribbon cable: A flat, ribbon-shaped set of conductors.

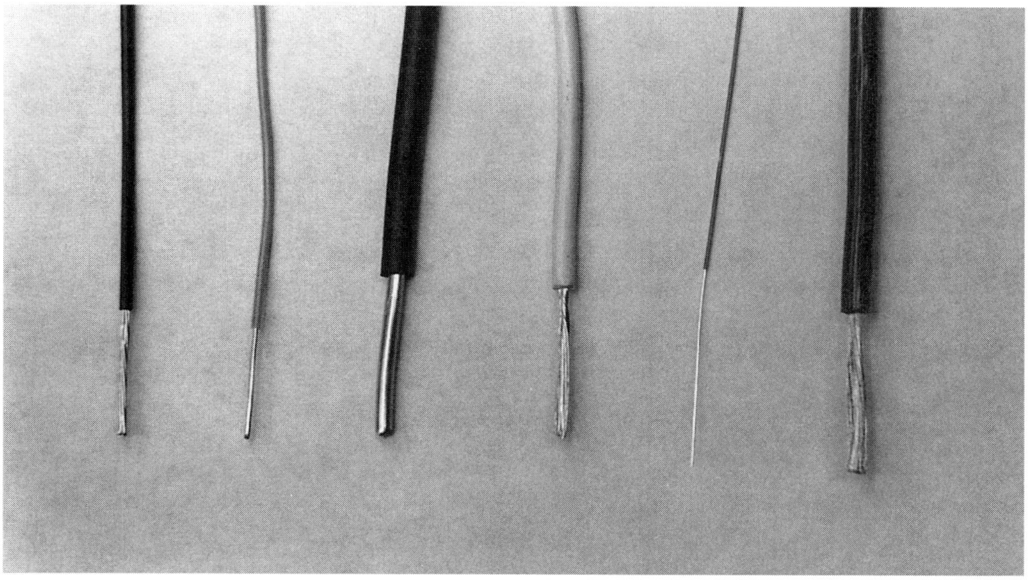

Figure 3-1. Single conductors come in various types and sizes with many choices of insulation.

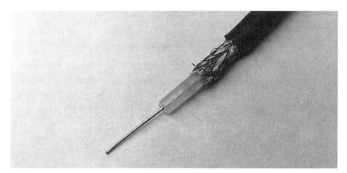

Figure 3-2. Coaxial cable can come with an aluminum layer over the braid as extra shielding for the center conductors.

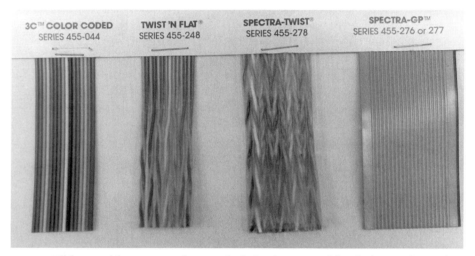

Figure 3-3. Ribbon cables are used extensively in the assembly of electronic equipment.

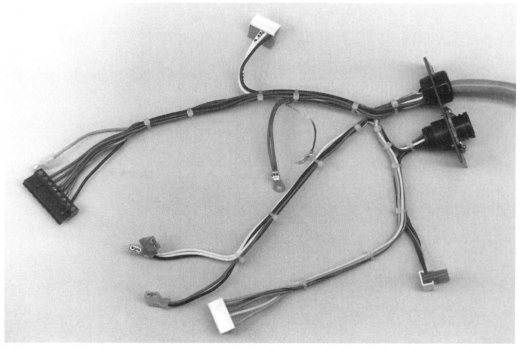

Figure 3-4. Wiring harnesses are used in all types of electrical and electronic equipment. (Liberty Electronics)

Wire Size

The amount of current a conductor can carry depends on the size or diameter of the conductor. The larger the conductor, the less resistance it has and the more current it can carry. The smaller the conductor, the more resistance it has and the less current it can carry. A wire gauge is an instrument used to measure the diameter of wire, **Figure 3-5.** The larger the wire number designation, the smaller the diameter. The size of the wire is inversely proportional to the diameter. Sizes 18, 20, and 22 are the most common in electronic equipment. No. 30 is often used for *wire wrap* assemblies. These assemblies have terminals in which the wire is wrapped tightly several times and no solder is required.

Wire wrap: Making an electrical connection by spinning (wrapping) the wire around a terminal and making the connection without soldering.

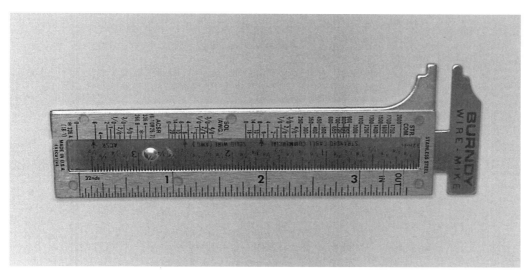

Figure 3-5. A wire gauge is used to measure the diameter of conductors.

Circular Mils

The English system of measurement is not used to measure wire size. Wire is sized by *circular mils* (cmils), a measurement of the cross-sectional area of the wire. One circular mil is equivalent to the area of a circle one mil (0.001″) in diameter, **Figure 3-6.** Do not confuse this with the cross-sectional area in square inches.

Circular mil: A universal term used to define the cross-sectional area of a conductor.

$$\text{Area} = \text{Diameter (in mils)}^2$$

$$1 \text{ mil} = 0.001''$$

Where:
A = Area
D = Diameter

0.001″

1 cmil

Figure 3-6. One circular mil.

▼ *Example 3-1:*

If a wire has a diameter of 0.05″, what is the cross-sectional area in circular mils?

Diameter in mils   = inches/0.001
                   = 0.05/0.001
                   = 50 mils

Therefore:

A = D^2
      = 50^2
      = 2,500 cmils   ▲

▼ *Example 3-2:*

If a wire has a diameter of 0.006″, what is the cross-sectional area in circular mils? See **Figure 3-7.**

Diameter in mils   = inches/0.001
                   = 0.006/0.001
                   = 6 mils

Therefore:

A = D^2
      = 6^2
      = 36 cmils   ▲

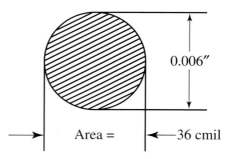

Figure 3-7. Calculating circular mils.

Figure 3-8 lists American Wire Gage (AWG) designations, diameters, circular mils, maximum conductor currents, and other information concerning conductors. For residential and industrial wiring regulations, refer to the National Electric Code.

Conductance

A conductor has a certain amount of resistance to the current. As you learned in Chapter 2, each mathematical function has an opposite. Likewise, each electrical unit of measurement has an opposite. The opposite of resistance is conductance.

Conductance: The ability of a circuit or conductor to carry an electrical current.

Conductance is the ability of a material to pass electrical current. The higher the conductance, the lower the resistance. Conversely, the lower the resistance, the greater the conductivity. The unit of measurement for conductance is mho (ohm spelled backwards) or the international standard unit siemens (S). It is represented by the letter G. Conductance is equal to the reciprocal of resistance (R). The equation is:

$$G = \frac{1}{R}$$

AWG	Bore Diameter	Cmils	Ohms per 1000'	Current Capacity
0000	0.4600	211600	0.04901	302
000	0.4096	167800	0.06182	240
00	0.3648	133100	0.07793	190
0	0.3249	105600	0.09825	151
1	0.2893	83690	0.1239	120
2	0.2576	66360	0.1563	95
3	0.2294	52620	0.1971	75
4	0.2043	41740	0.2485	60
5	0.1819	33090	0.3134	47
6	0.1620	26240	0.3952	37
7	0.1443	20820	0.4981	30
8	0.1285	16510	0.6281	24
9	0.1144	13090	0.7925	19
10	0.1019	10380	0.9988	15
11	0.0907	8230	1.26	12
12	0.0808	6530	1.59	9
13	0.0720	5180	2.00	7
14	0.0641	4110	2.52	6
15	0.0571	3260	3.18	5
16	0.0508	2580	4.02	4
17	0.0453	2050	5.05	3
18	0.0403	1620	6.39	2.9
19	0.0359	1290	8.05	1.8
20	0.0320	1020	10.1	1.5
21	0.0285	812	12.8	1.16
22	0.0253	640	16.2	0.914
23	0.0266	511	20.3	0.730
24	0.0201	404	25.7	0.577
25	0.0179	320	32.4	0.457
26	0.0159	253	41.0	0.361
27	0.0142	202	51.4	0.289
28	0.0126	159	65.3	0.227
29	0.0113	123	81.2	0.183
30	0.0100	100	104.0	0.143
31	0.0089	79.2	131	0.113
32	0.0080	64.0	162	0.091
33	0.0071	50.4	206	0.072
34	0.0063	39.4	261	0.057
35	0.0056	31.4	331	0.045
36	0.0050	25.0	415	0.036
37	0.0045	20.2	512	0.029
38	0.0040	16.0	648	0.023
39	0.0035	12.2	847	0.017
40	0.0031	9.61	1080	0.014

Figure 3-8. Specifications for wire gauge, diameter, and maximum current.

▼ *Example 3-3:*

If a conductor has a resistance of 25 Ω, what is its conductance?

$$G = \frac{1}{25}$$

$$= 0.04 \text{ S}$$ ▲

Conductance is used to find the total resistance of parallel resistances and is covered in Chapter 4.

Factors Determining Resistance of Conductors

The resistance of a conductor depends on four factors:
- Type of material
- Cross-sectional area
- Length
- Temperature

Type of Material

Materials that are poor conductors have fewer free electrons and higher resistance. Good conductors have more free electrons and lower resistance. **Figure 3-9** lists materials in their order of conductivity. Even though carbon is at the bottom of the list, it does not mean carbon is not a good conductor. It only means the materials above it are better conductors. In fact, the center electrode of a common dry cell used in a flashlight is made of carbon.

Conductors
Silver
Copper
Aluminum
Tungsten
Zinc
Nickel
Iron
Lead
Mercury
Carbon

Figure 3-9. Silver is the most conductive and carbon is the least conductive of these materials.

Resistivity: A measure of the resistance of a material to electric current.

Resistivity is the resistance (in ohms) to the flow of current exhibited by a certain length of material, **Figure 3-10.** The symbol for resistivity is the Greek letter rho (ρ).

Cross-sectional Area

The greater the cross-sectional area, the lower the resistance. Conversely, the smaller the cross-sectional area, the higher the resistance. More cross-sectional area simply provides more atoms (and more free electrons) for conducting current.

Length of the Conductor

Increasing the length of a conductor increases the total resistance and reduces the conductance. For example, if a conductor has a resistance of 1 Ω for every 12′, then 48′ of conductor would have a resistance of 4 Ω.

Material	Resistivity
Selenium	7.3
Silver	9.7
Copper	10.7
Gold	14.55
Aluminum	16.06
Tungsten	33.22
Zinc	37.4
Nickel	60
Iron	60.14
Tin	69.5

Figure 3-10. Resistivity of some common conductors.

Temperature of the Conductor

When the temperature of a conductor is increased, the atoms within the conductor convert the thermal energy to mechanical energy. The increase in temperature causes an increase in the random movement of the atoms, leading to a greater number of collisions between the moving electrons and neighboring atoms. The outcome is an increase in opposition to the current flow (resistance). Hence, the resistance is directly proportional to the temperature. The resistance of a conductor depends on all these factors and can be calculated by the following equation:

$$R = \rho \times \frac{l}{A}$$

Where:
R = Resistance of the conductor (in ohms)
ρ = Resistivity of the conducting material
l = Length of the conductor (in feet)
A = Area of the conductor (in circular mils)

▼ *Example 3-4:*

Calculate the resistance of 85′ of copper conductor with a conductor area of 4200 cmils.

$$R = \rho \times \frac{l}{A}$$

$$= 10.7 \times \frac{85}{4200}$$

$$= 0.2165 \ \Omega$$

▲

These calculations are for the standard temperature of 68°F (20°C). Resistance calculations involving temperature changes are covered in Chapter 6.

Solid and Stranded Conductors

Wire comes in two forms—solid and stranded. As the name implies, solid wire is a solid piece of material. *Stranded conductors* come in AWG sizes but are made up of several small, solid conductors twisted together. See **Figure 3-11.**

Stranded conductor: Consists of several small conductors twisted into a single conductor.

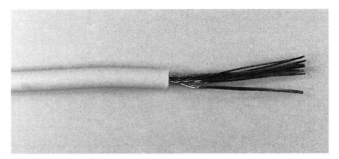

Figure 3-11. Stranded conductors are made of several solid conductors twisted together.

Solid conductors are used for general purpose wiring. Stranded conductors are used where some vibration or movement of the device may occur. They allow for movement without breakage.

Superconductivity

You now know that resistance increases as the temperature increases. What happens if the temperature decreases? In 1911, Dutch physicist Heike Onnes discovered that mercury (a liquid conductor) lost its resistance to electrical current when the temperature was decreased to –459.7°F (–273.1°C). The mercury became a ***superconductor,*** offering no resistance and allowing a supercurrent to flow in the conductor.

In 1986, two IBM scientists discovered that a conductor composed of barium, lanthanum, copper, and oxygen would superconduct at –406°F (–243°C). Today, scientists have increased the superconductivity to at least –300°F (–184°C) and continue to try to achieve room temperature. Such superconductivity would allow a 2″ thick conductor to be replaced by one the thickness of a human hair.

Insulators

Insulators are materials whose electrons are held tightly to the nucleus. The electrons are not free to move from one atom to another. Insulators are used to protect conductors and other devices from each other. They also protect people from electric shock. Some typical insulators include air, porcelain, Bakelite™, rubber, paper, Teflon™, glass, and mica.

If sufficient voltage is applied to an insulator, "breakdown" occurs and the insulator conducts current. This ***breakdown voltage*** must be great enough to dislodge and free the electrons. The amount of breakdown voltage depends on the type of material and thickness of the insulator.

Dielectric strength expresses how well a particular material serves as an insulator. Dielectric strength is the amount of voltage, applied across 1/1000 of an inch of the insulator material, that will cause the material to break down and conduct a large current. The value of breakdown voltage is given a rating called the ***dielectric constant.*** The rating is an average value that reflects how much better the material insulates than air.

Resistors

Conductors offer very little resistance to current flow. For devices such as transistors, integrated circuits, and other electronic circuits to operate correctly, ***resistors*** must be used to control the current or voltage in those circuits.

Superconductor: A conductor where resistance decreases as its temperature is reduced to near absolute zero.

Insulator: A material with few or no free electrons.

Breakdown voltage: The voltage at which a dielectric fails and a current flows through the dielectric.

Dielectric strength: The maximum voltage a dielectric can withstand without puncturing.

Dielectric constant: The value of breakdown voltage that reflects how much better the material insulates than air.

Resistor: Electrical component used to oppose the flow of electrical current.

Types of Resistors

There are two major types of resistors—fixed value and variable (adjustable) value. Fixed value types include:

- Carbon composition
- Carbon film
- Metal film
- Metal oxide
- Surface mount technology (SMT)
- Thick film
- Wirewound

Carbon composition. This was one of the first types of resistors to be developed. It is called carbon composition because it uses powdered carbon mixed with an insulation material. Resistance is controlled by adjusting the amount of carbon to insulation. See **Figure 3-12.**

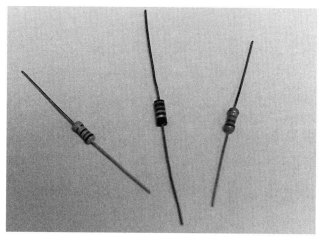

Figure 3-12. Carbon composition resistors are one of the most commonly used types. The size of the resistor increases with the wattage.

Carbon film. Carbon film resistors are constructed of a thin layer or film of resistive material deposited on a ceramic form. The film is cut in a spiral (helix) to adjust the resistance, **Figure 3-13.**

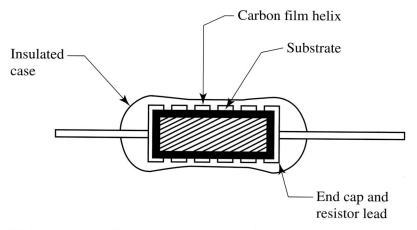

Figure 3-13. In the carbon film resistor, a substrate provides a surface to which the carbon film is glued. The resistance element is trimmed into a helix to adjust resistance.

Metal film. Metal film resistors are made in the same way as carbon film resistors. A thin layer of metal is sprayed on a ceramic form, and a spiral cut is made to adjust the resistance.

Metal oxide. This type of resistor is constructed by depositing an oxide of a metal, such as tin, on an insulated form. The ratio of metal to oxide determines the resistance, **Figure 3-14.**

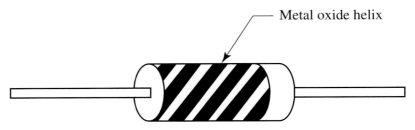

Figure 3-14. Construction of a metal oxide resistor.

Surface mount technology (SMT). SMT is a thick film type of construction. The size of the resistors can vary from the size of a dime to a fraction of that size, **Figure 3-15.**

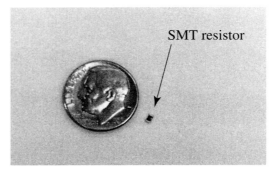

Figure 3-15. Surface mount resistors are tiny. (Dime shown for comparison.)

Thick film. Two types of thick film resistors are the single-in-line package (SIP) and the dual-in-line package (DIP), **Figure 3-16.** The leads of the SIP are laid out in a single line. The DIP is constructed by a screen printing process. A conductive layer is screened onto the form, and a resistive material (bismuth/ruthenate) is applied.

A   **B**

Figure 3-16. Thick film resistors. A—Single-in-line package (SIP). B—Dual-in-line package (DIP).

Wirewound. This type of resistor is constructed by winding a length of resistive wire on an insulated form, **Figure 3-17.** The length and diameter of the wire determines the resistance. Copper wire is not used.

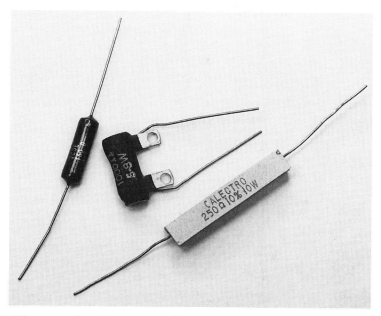

Figure 3-17. Wirewound resistors come in many sizes and shapes. They are used when more power dissipation is required.

Adjustable value resistors include the following types:
- Light-dependent or photoresistor
- Thermistor
- Adjustable tapped
- Potentiometer

Light-dependent resistor (LDR). A photoresistor is a thin piece of photoconductive material whose resistance decreases as light is increased, **Figure 3-18.** An increase in light causes the atoms within the photoconductive material to release their valence (free) electrons. The result is an increase in the current passing through the device. Resistance, in turn, is decreased. An ohmmeter can be used to verify the decrease in resistance.

Figure 3-18. Light-dependent resistors (LDR), also called photoresistors, come in various sizes and shapes.

Thermistor: A
resistive device in
which resistance
changes with
temperature.

Thermistor. **Thermistors** are used in ovens, freezers, and motors to sense temperature, **Figure 3-19.** These devices are made of semiconductor material. Their resistance can vary either directly or inversely with temperature.

Figure 3-19. Thermistors are often used where temperature-sensing is required. Resistance changes with temperature.

Adjustable tapped. In an adjustable tapped resistor, the total resistance is across the end connections. The moveable tap on the side of the resistor can be adjusted for the required resistance between the outside terminals and the tap. See **Figure 3-20.**

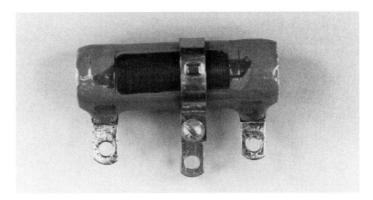

Figure 3-20. This wirewound resistor is an adjustable tapped type.

Potentiometer: A
variable resistor with
three terminals.

Potentiometer. A **potentiometer** ("pot" for short) is a variable resistor with three terminals, **Figure 3-21.** The outside terminals are connected to the ends of the resistive element, and the middle terminal is free to move across the resistive element. The resistance between the center terminal and the outside terminals varies as the shaft of the potentiometer is turned. Some potentiometer shafts are designed to turn only once, while others can be turned ten or 15 times before the whole resistive element is adjusted, **Figure 3-22.**

Resistor Color Coding

Many small resistors are identified by color bands. A color code system for resistors has been adopted by the Electronics Industries Association (EIA), **Figure 3-23.** A four-band resistor is standard. To read the value of a resistor, hold it so the color bands are on the left. Each band has a particular significance, **Figure 3-24:**
- Band 1 is the first digit.
- Band 2 is the second digit.
- Band 3 is the number of zeros or the multiplier.
- Band 4 is the tolerance.

Figure 3-21. Potentiometers are used in every area of electronics. They may look different but electrically are the same.

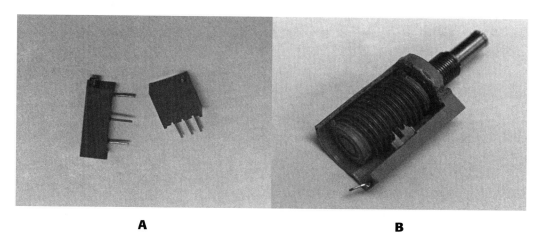

| A | B |

Figure 3-22. Multiturn potentiometers. A—These potentiometers require more than one turn of the adjustment shaft. B—Cutaway view of a multiturn potentiometer.

Number	Color	Multiplier	Tolerance
0	Black		
1	Brown		
2	Red		
3	Orange		
4	Yellow		
5	Green		
6	Blue		
7	Violet		
8	Gray		
9	White		
	Gold	0.1	5%
	Silver	0.01	10%

Figure 3-23. Resistor color code values.

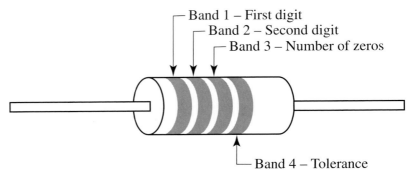

Figure 3-24. The meaning of resistor color code bands.

To determine the resistance value of the resistor in **Figure 3-25,** read the bands according to the values in Figure 3-23. The first band is orange, which corresponds to number 3 in the table. The second band is white; it corresponds to number 9. The third band is red; it corresponds to number 2 (meaning add two zeros). Therefore, the resistor has a value of 3900 Ω. See **Figure 3-26** for other examples.

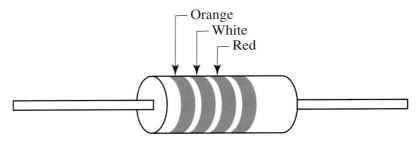

Figure 3-25. Color code example.

Figure 3-27 depicts a five-band transistor used by some manufacturers and the military for precision resistors. The fourth band represents tolerance, and the fifth band represents reliability.

Tolerance

What is tolerance as represented by the fourth band? Tolerance can best be demonstrated in a machine shop. The machinist is to turn a piece of metal on a lathe to a certain diameter. In the specifications for the piece, the machinist is allowed so many thousandths of an inch in diameter over or under the specified size. The allowable percentage of deviation over or under the specified diameter is known as the *tolerance.*

Tolerance: The allowable percentage of deviation over or under the specified value.

Tolerance works the same way in a resistor. The value of the resistor is allowed to be off by a certain percentage of the resistor color code value. For example, the resistor in **Figure 3-28** has a value of 200 Ω, and the tolerance is 10% (silver band). Ten percent of 200 equals 20. Therefore, the value of the resistor must be ±20 Ω, or 180 Ω to 220 Ω.

Reliability

The fifth band on a resistor indicates reliability. It gives the maximum failure rate percentage per 1000 hours of use. The color codes on the fifth band are as follows:

- Brown = 1%
- Red   = 0.1%
- Orange = 0.01%
- Yellow = 0.001%

This means, for example, that out of 5000 resistors with an orange fifth band, 50 of them will fail over a period of 1000 hours of use.

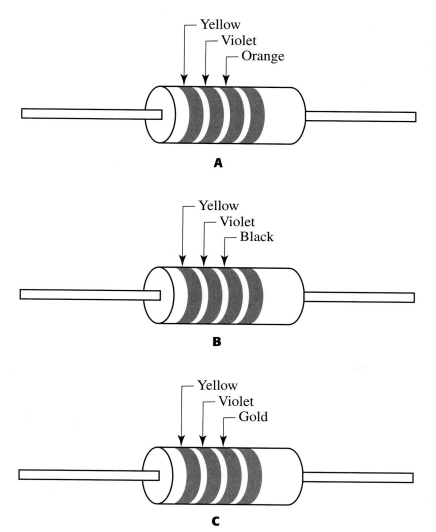

Figure 3-26. The first and second bands in all three resistors are yellow and violet. Thus, the first and second digits are 4 and 7. A—The third band is orange, which corresponds to the number 3. Add three zeros to 47 for a resistance value of 47,000 Ω. B—The third band is black, which corresponds to zero. Do not add any zeros for a resistance value of 47 Ω. C—The third band is gold, which has no corresponding number. Use the multiplier 0.1 to calculate a resistance value of 4.7 Ω.

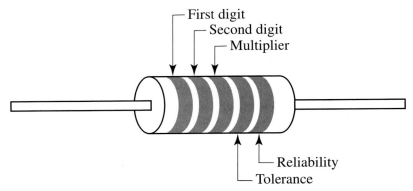

Figure 3-27. The fourth and fifth bands are important in precision resistor manufacturing.

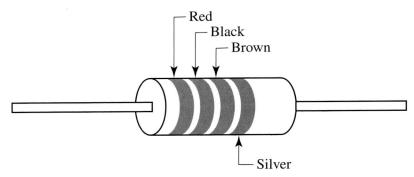

Figure 3-28. An example of resistance tolerance.

Summary

- Conductors may take the form of single wires and coaxial, ribbon, solid, and stranded cables, plus many other forms. The amount of current a conductor can carry depends on the size of conductor. Conductors are sized by the American Wire Gauge standard. The smaller the conductor the larger the number. Conductors are sized in circular mils.

- The resistance of conductors is determined by the type of material, cross-sectional area, length, and temperature of the conductor. Superconductivity presents a condition of zero resistance.

- Insulators are specified by their breakdown voltage. The dielectric strength is determined by the type of insulator material. The dielectric constant of materials is compared to air.

- Resistors are made to give either a fixed or variable value. Fixed value resistors include carbon film, metal film, wirewound, networks, and SMT. Variable value resistors include photoresistors, thermistors, variable tapped-type, and potentiometers. The first band of a resistor color code is the first digit in the value of the resistor. The second band is the second digit in the value of the resistor. The third band is the multiplier or number of zeros added to the two digits.

Important Terms

Do you know the meanings of these terms used in the chapter?

breakdown voltage
cable
circular mil
coaxial cable
conductance
dielectric constant
dielectric strength
insulator
potentiometer

resistivity
wire wrap
tolerance
thermistor
superconductor
stranded conductor
ribbon cable
resistor

Questions and Problems

Please do not write in this text. Write your answers on a separate sheet of paper.

1. _____ and _____ are used in the assembly of many types of equipment.
2. A set of conductors in the same protective container is called a _____.
3. If a resistor has a resistance of 1000 Ω, what is the conductance?
4. List the four factors that determine the resistance of conductors.
5. What is the resistance of 450′ of No. 24 copper wire?
6. What is the resistance of 379′ of copper conductor with a conductor area of 3206 cmils?
7. What is the resistance of 135′ of aluminum conductor with a diameter of 75 cmils?
8. Materials with no free electrons are called _____.
9. List the seven types of fixed resistors.
10. List the four types of variable resistors.
11. If a resistor value is 680 Ω with a tolerance of 5%, what is the range of its resistance?
12. Determine the resistance value of the following resistors:
    a. red, green, brown
    b. brown, gray, red
    c. orange, orange, orange
    d. white, brown, black
    e. yellow, violet, green
    f. red, green, orange
    g. blue, red, gold
    h. yellow, violet, yellow
    i. brown, black, blue
    j. brown, black, black
    k. green, blue, silver

Chapter 4 Graphic Overview

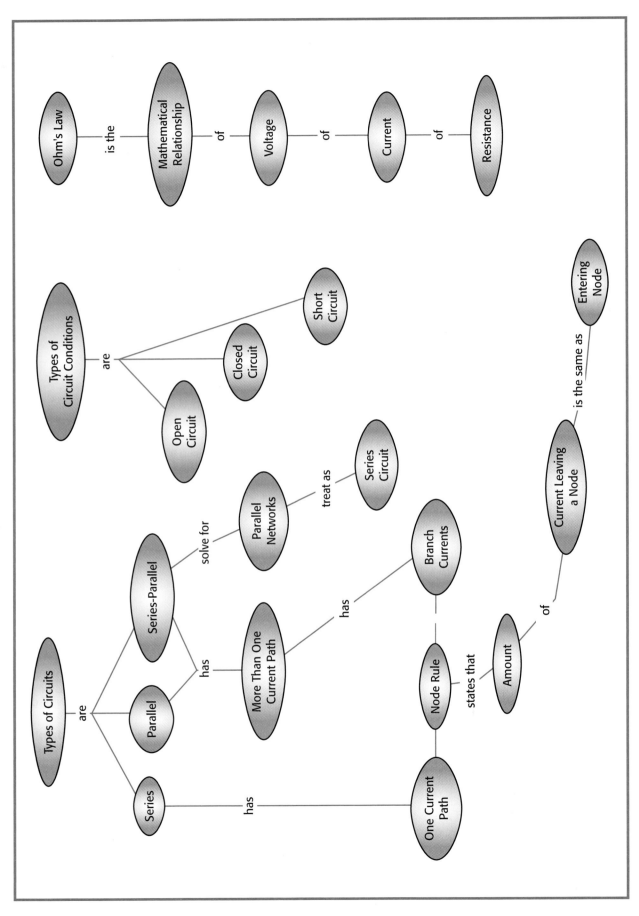

ELECTRICAL CIRCUITS

 **Objectives**

After studying this chapter, you will be able to:

○ Identify open, closed, and short circuit conditions.
○ Use Ohm's law to find the current, resistance, or voltage of a circuit.
○ Calculate the total resistance, total voltage, and voltage drop in a series circuit.
○ Calculate the total conductance and total resistance of parallel circuits.
○ Calculate the total resistance of series-parallel circuits.
○ Calculate the total voltage of sources connected in series and parallel.

Introduction

This chapter presents series, parallel, and series-parallel circuits and the rules for analyzing these circuits. Understanding the rules of circuit analysis is important to your study of electricity.

Types of Electrical Current

So far, the discussion of electrical current has shown the current flowing from negative to positive. Current that flows in one direction is called *direct current* (dc). *Alternating current* (ac) flows in one direction, then the other, or it is said to flow in both directions.

Current originates from a source. A *source* is any device, such as a battery, that supplies electrons or power to the circuit. A *load* is any device, such as a lamp, that uses the electrical power to change energy from one form to another. A *switch* is used to control the flow of current in the circuit. See **Figure 4-1.**

Circuits Conditions

When the switch in a circuit is in the OFF position, no current can flow, and the circuit is called an *open circuit.* When the switch is in the ON position, current can flow to illuminate the lamp. The circuit is called a *closed circuit.* See **Figure 4-2.**

Direct current: Current that flows in one direction.

Alternating current: A current of electrons that moves first in one direction, then the other.

Source: A device that supplies voltage to a circuit or piece of equipment.

Load: The device or circuit that consumes power.

Switch: A control device that either prevents or allows the flow of current in a circuit.

Open circuit: A circuit that does not have a complete path for current to flow.

Closed circuit: A circuit that has a complete current path.

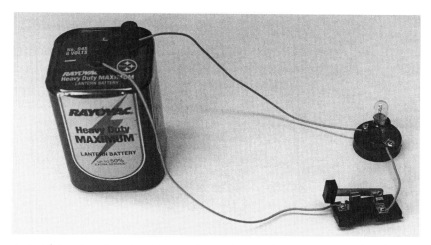

Figure 4-1. This simple circuit consists of a voltage source, load, switch, and connecting wires.

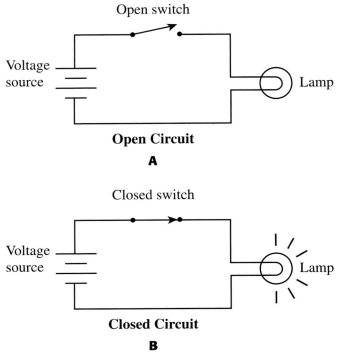

Figure 4-2. Switches control the flow of electrons in the circuit. A—In an open circuit, the switch is off and no current flows. B—In a closed circuit, the switch is on and current flows.

Continuity: An uninterrupted path for electrical current.

Short circuit: An unwanted path of current flow.

In a closed circuit, current has a continuous path of travel. The circuit is said to have *continuity.* If the circuit or device is open, the current path is broken and has no continuity.

One unwanted circuit condition is called a *short circuit.* Instead of going through the lamp, the current goes around it, **Figure 4-3.** This condition usually causes components to get hot and results in smoke, blown fuses, and total circuit failure. A short circuit also causes an increase in current.

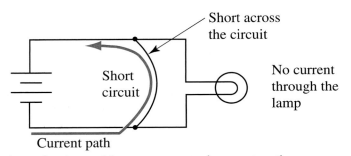

Figure 4-3. A short circuit provides an unwanted current path.

Ohm's Law

Ohm's law: The mathematical relationship of voltage, current, and resistance.

The relationship between volts, amperes, and ohms is basic to electricity and electronics, **Figure 4-4.** *Ohm's law* is the mathematical relationship between voltage, current, and resistance. The equation is typically stated as:

$$V = I \times R$$

(Voltage = Current $\times$ Resistance)

Quantity	Letter Symbol	Unit
Voltage	*E* or *V*	Volts
Current	*I*	Amps
Resistance	*R*	Ohms

Figure 4-4. Electrical units of measurement.

▼ *Example 4-1:*
Find the voltage of **Figure 4-5.**

$V = I \times R$
$= 5\ A \times 30\ \Omega$
$= 150\ V$

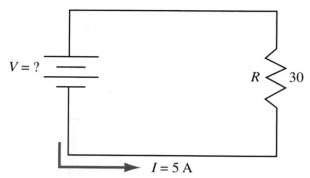

Figure 4-5. Calculating the circuit voltage.

The amount of current flowing in a circuit can be determined by rearranging the equation:

$$I = \frac{V}{R}$$

(Current = Voltage ÷ Resistance)

▼ *Example 4-2:*
Find the current (amperage) in **Figure 4-6.**

$I = \frac{V}{R}$

$I = \frac{6\ V}{3\ \Omega}$

$= 2\ A$

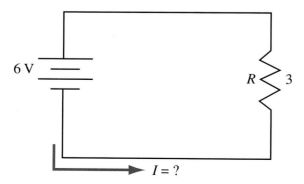

Figure 4-6. Calculating the amount of current.

The resistance of a circuit can be found by rearranging the equation:

$$R = \frac{V}{I}$$

(Resistance = Voltage ÷ Current)

▼ *Example 4-3:*

Find the resistance in **Figure 4-7.**

$$
\begin{aligned}
R &= \frac{V}{I} \\
&= \frac{50 \text{ V}}{5 \text{ A}} \\
&= 10 \ \Omega
\end{aligned}
$$

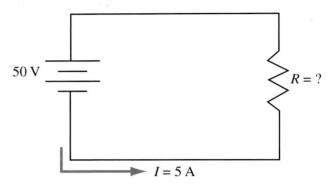

Figure 4-7. Calculating the circuit resistance.

When using Ohm's law, make sure you are using the correct voltage, current, or resistance values in the equation.

▼ *Example 4-4:*

What is the current through the indicated resistor in **Figure 4-8?**

$$
\begin{aligned}
I &= \frac{V}{R} \\
&= \frac{6 \text{ V}}{30 \ \Omega} \\
&= 0.2 \text{ A}
\end{aligned}
$$

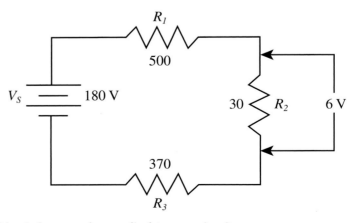

Figure 4-8. Ohm's law can be applied to any circuit.

If the source voltage of 180 V had been used to calculate the current through the indicated resistor, an *incorrect* current of 60 A would have been found.

$$I = \frac{V}{R}$$

$$= \frac{180 \text{ V}}{30 \text{ }\Omega}$$

$$= 60 \text{ A (wrong answer)}$$

The problem requires using the voltage *across* the resistor.

Series Circuits

A *series circuit* is a circuit with only one current path. When more than one resistor is found in a circuit, subscript numbers such as R_1, R_2, and R_3 are used to identify each resistor. See **Figure 4-9.**

Series circuit: A circuit with only one current path.

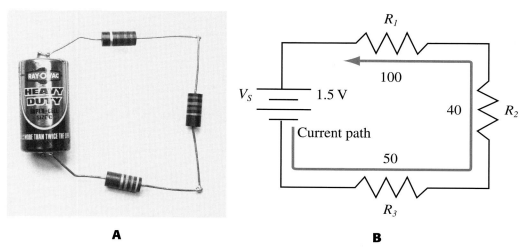

A   **B**

Figure 4-9. A series circuit. A—This circuit has three resistors. B—The resistors are identified as R_1, R_2, and R_3. The current follows only one path.

Certain rules must be followed when analyzing electrical circuits. If you violate the rules, any calculations will be wrong. Take the time to memorize, understand, and apply the following rules for series circuits:

- *Voltage.* The sum of the voltage drops around a series circuit is equal to the source voltage. The equation is:

$$V_{SOURCE}\ (V_S) = V_1 + V_2 + V_3$$

- *Current.* The current is the same in all parts of a series circuit. The equation is:

$$I_{TOTAL}\ (I_T) = I_1 = I_2 = I_3$$

- *Resistance.* The total resistance of a series circuit is equal to the sum of the individual resistances in the circuit. The equation is:

$$R_{TOTAL}\ (R_T) = R_1 + R_2 + R_3$$

First, let's look at the voltage rule. The voltage drop across a resistor is the result of the current flowing through an individual resistor.

▼ *Example 4-5:*
Calculate the voltage drop for each of the resistors in **Figure 4-10.**

$$V = I \times R$$

Therefore:

$$V_1 = 2 \times R_1$$
$$= 2 \times 6$$
$$= 12 \text{ V}$$

$$V_2 = 2 \times R_2$$
$$= 2 \times 4$$
$$= 8 \text{ V}$$

$$V_3 = 2 \times R_3$$
$$= 2 \times 5$$
$$= 10 \text{ V}$$

Add the voltage drops.

$$V_T = V_1 + V_2 + V_3$$
$$= 12 + 8 + 10$$
$$= 30 \text{ V}$$

The sum of the voltage drops equals the source voltage (30 V). Compare the above calculations with the circuit rules.   ▲

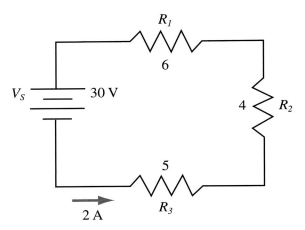

Figure 4-10. Series circuit example.

Next, let's look at an example of the current rule. The total current is found by dividing the total voltage by the total resistance.

▼ *Example 4-6:*
Find the total current in Figure 4-10.

$$I_T = \frac{V_T}{R_T}$$

$$= \frac{30}{15}$$

$$= 2 \text{ A}$$

▲

Finally, let's look at an example of the resistance rule. The total resistance of a series circuit is found by adding the resistors.

▼ *Example 4-7:*
Find the total resistance of the circuit in Figure 4-10.

$$R_T = R_1 + R_2 + R_3$$
$$= 6 + 4 + 5$$
$$= 15 \ \Omega$$

▲

Current Node Rule

A *node* is the junction where the components in a circuit are connected. The current leaving a node must be equal to the current entering a node. This rule applies to all circuits.

The nodes in **Figure 4-11** are identified as A, B, and C. The current coming from R_3 and entering node A is 6 A. The 6 A leave node A and go to R_2, then leave and enter node B. After leaving node B, the 6 A enter R_1. The amount of current is 6 A in and 6 A out.

Then, the 6 A flow out of R_1, enter and leave node C, and go to the positive terminal of the source. This flow illustrates the current node rule for series circuits: *The current is the same in all parts of the circuit.*

Nodes are used in another way. In **Figure 4-12,** 8 A enter the node; 5 A leave the node and flow on one path to R_2; 3 A leave the node and flow on another path to R_3. This is perhaps a better illustration of the node rule. Because there is more than one current path, the circuit is not a series circuit.

Node: A point where two or more components are connected.

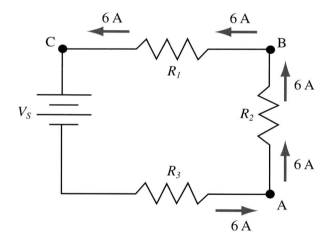

Figure 4-11. Nodes of a series circuit.

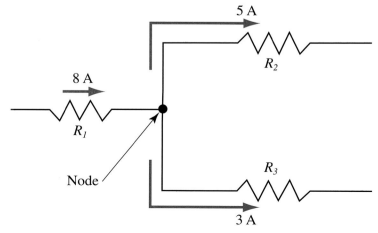

Figure 4-12. Nodes of a simple parallel circuit.

Figure 4-13 has three current paths, I_1, I_2, and I_3, flowing through the respective resistors. The total current of 15 A flows to node A and divides between the resistors as shown. The current entering node B is the sum of the three currents, and the current leaving node B is 15 A.

Another way to look at nodes is shown in **Figure 4-14.** The group of resistors R_1, R_2, R_3, and R_4 is seen as a node. The current entering and leaving the node is 12 A. Any point in a circuit can be looked upon as a node.

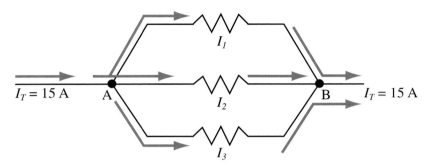

Figure 4-13. This parallel circuit has two nodes and three current paths.

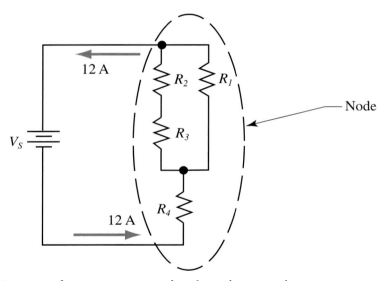

Figure 4-14. A group of components can be viewed as a node.

Parallel Circuits

Parallel circuit: A circuit that contains more than one path for current to flow.

A *parallel circuit* has more than one current path, **Figure 4-15.** The rules for parallel circuits are:
- *Voltage.* The voltage is the same for all parts of the circuit.
- *Current.* The total current is the sum of all the branch currents.
- *Resistance.* The total resistance is always less than the smallest resistance.

If you trace the source voltage in Figure 4-15, the positive terminal is connected to each of the resistors. Likewise, the negative terminal is connected to the other end of each resistor. If a meter were placed across each resistor, the same voltage would be found. The voltage is the same across a parallel circuit.

The total current (I_T) is the sum of the branch currents I_1, I_2, and I_3. See **Figure 4-16.**

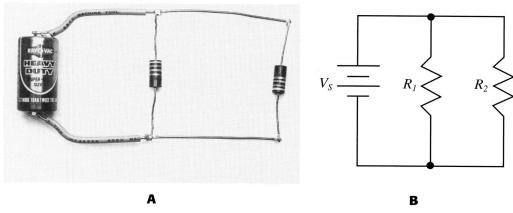

A   **B**

Figure 4-15. A parallel circuit. A—The circuit has more than one path. B—Schematic diagram.

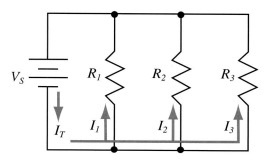

Figure 4-16. Various currents flow in a parallel circuit.

Two equal resistors. Resistors of the same value are easy to work with. If all the resistors in parallel have the same value, divide the value of one resistor by the number of resistors in parallel:

$$R_T = \frac{\text{Value of one resistor}}{\text{Number of resistors in parallel}}$$

▼ *Example 4-8:*

Find the total resistance of the circuit in **Figure 4-17.**

$$R_T = \frac{600}{3}$$

$$= 200 \ \Omega$$ ▲

Two unequal resistors. An equation often used in parallel resistances is referred to as the product over the sum formula. This equation can be used to find the resistance of no more than two resistors in parallel:

$$R_T = \frac{R_1 \times R_2}{R_1 + R_2}$$

▼ *Example 4-9:*

Find the total resistance of the circuit in **Figure 4-18.**

$$R_T = \frac{20 \times 80}{20 + 80}$$

$$= \frac{1600}{100}$$

$$= 16 \ \Omega$$ ▲

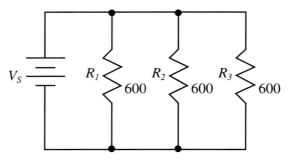

Figure 4-17. Resistance of equal resistors.

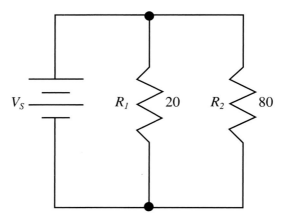

Figure 4-18. Resistance of unequal resistors.

Three or more unequal resistors. To find the total resistance of parallel circuits, it is often best to use the conductance of the circuit. *Conductance* (represented by G) is the ability of a material to pass electrical current. It is the opposite of resistance. The higher the conductance, the less the resistance. Conductance is equal to the reciprocal of resistance:

$$G = \frac{1}{R}$$

See **Figure 4-19.** The total conductance (G_T) is found by adding the reciprocal of each resistor:

$$G_T = G_1 + G_2$$

Thus:

$G_1 = \dfrac{1}{20}$

$\quad = 0.05$

$G_2 = \dfrac{1}{80}$

$\quad = 0.0125$

$G_T = 0.05 + 0.0125$

$\quad = 0.0625 \text{ S}$

After the total conductance is found, the total resistance of the parallel circuit can be calculated. The total resistance is equal to the reciprocal of the total conductance:

$$R_T = \frac{1}{G_T}$$

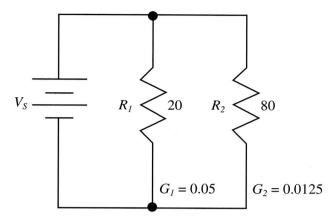

Figure 4-19. Conductance can be used to find the total resistance.

Thus, the total resistance of the parallel circuit is:

$$R_T = \frac{1}{0.0625}$$
$$= 16 \ \Omega$$

Notice the total resistance of the circuit is less than the smallest resistor R_1 of 20 Ω. This is the resistance rule for parallel circuits.

In most textbooks you will see an equation given for resistors in parallel as:

$$R_T = \frac{1}{\dfrac{1}{R_1} + \dfrac{1}{R_2} + \dfrac{1}{R_3}}$$

This is simply the reciprocal of the total conductance, which has just been given.

Series-Parallel Circuits

A *series-parallel circuit* is a combination of series and parallel circuits, **Figure 4-20.** To calculate the values of this circuit, find the total resistance of the parallel, then treat it as a series circuit.

The resistance of parallel resistors R_2 and R_3 is 240 Ω. The total resistance (R_{PAR}) is the sum of R_1 plus the 240 Ω of the parallel network. A *network* is any combination of two or more electrical components.

Series-parallel circuit: A circuit made up of both series and parallel circuit combinations.

Network: An electrical combination of two or more components.

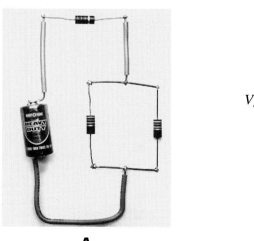

A

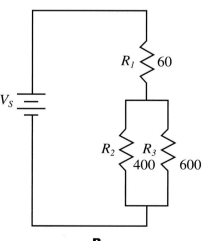

B

Figure 4-20. A series-parallel circuit. A—A series-parallel circuit is also called combination circuit. B—Schematic diagram.

Thus:

$$R_{PAR} = \frac{400 \times 600}{400 + 600}$$

$$= \frac{240,000}{1000}$$

$$= 240 \ \Omega$$

$$R_T = R_1 + R_{PAR}$$
$$= 60 + 240$$
$$= 300 \ \Omega$$

Voltage Sources

The connecting of voltage sources closely follows the voltage rules for series and parallel circuits. Voltage sources in series are added, **Figure 4-21.** Pay attention to the polarity of the voltage sources. Notice that V_4 is in opposite polarity; therefore, the value is a negative number.

The total voltage is:
$$V_T = V_1 + V_2 + V_3 + V_4$$
$$= 12 + 12 + 6 + (-8)$$
$$= 22 \ V$$

In **Figure 4-22,** the voltage sources are in parallel. The total voltage is the same as one of the sources. The total current capacity of the sources in parallel is the sum of the three currents:

$$I_T = I_1 + I_2 + I_3$$

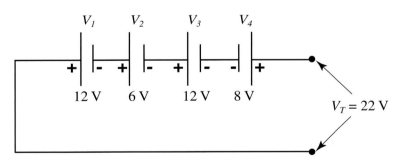

Figure 4-21. Voltage sources in series.

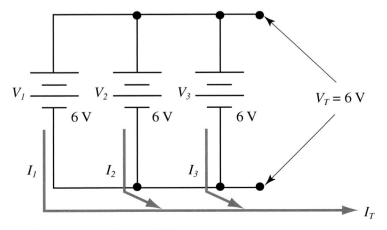

Figure 4-22. Voltage sources in parallel.

Summary

- The two types of electrical current are direct current and alternating current.
- A source is any device that supplies power to a electrical circuit. A load is any device that uses electrical power to change energy from one form to another. A switch is a device used to control the current in a circuit.
- The three circuit conditions are open circuit, closed circuit, and short circuit.
- Ohm's law is used to calculate the voltage, current, and resistance of a circuit.
- The three types of electrical circuits are series, parallel, and series-parallel. Each type of circuit has certain rules for calculating voltage, current, and resistance.
- A node is a junction where two or more components are connected. The current leaving a node must be equal to the current entering a node.
- A network is any combination of two or more electrical components.
- Voltage sources in series are added. In parallel, the total voltage is the same as one of the sources. The total current of the sources in parallel is the sum of the currents.

Important Terms

Do you know the meanings of these terms used in the chapter?

alternating current	Ohm's law
closed circuit	open circuit
conductance	parallel circuit
continuity	series circuit
direct current	series-parallel circuit
load	short circuit
network	source
node	switch

Questions and Problems

Please do not write in this text. Write your answers on a separate sheet of paper.

1. The two types of electrical current are _____ and _____.
2. A device used to control the current in a circuit is called a(n) _____.
3. A device that supplies electrons to a circuit is called a(n) _____.
4. Three types of circuit conditions are open circuit, closed circuit, and _____ circuit.
5. Solve for the unknown current(s) in **Figures 4-23A, 4-23B,** and **4-23C.**
6. Solve for the unknown values in **Figures 4-24A, 4-24B,** and **4-24C.**
7. State the current node rule.
8. Solve for the unknown currents in **Figures 4-25A, 4-25B, 4-25C, 4-25D,** and **4-25E.**
9. List the three rules for series circuits.
10. Solve for the unknown currents in **Figure 4-26.**
11. List the three rules for parallel circuits.
12. Conductance is equal to the _____ of resistance.
13. Solve for the unknowns in **Figures 4-27A, 4-27B,** and **4-27C.**
14. Solve for the output voltage in **Figures 4-28A, 4-28B, 4-28C,** and **4-28D.**

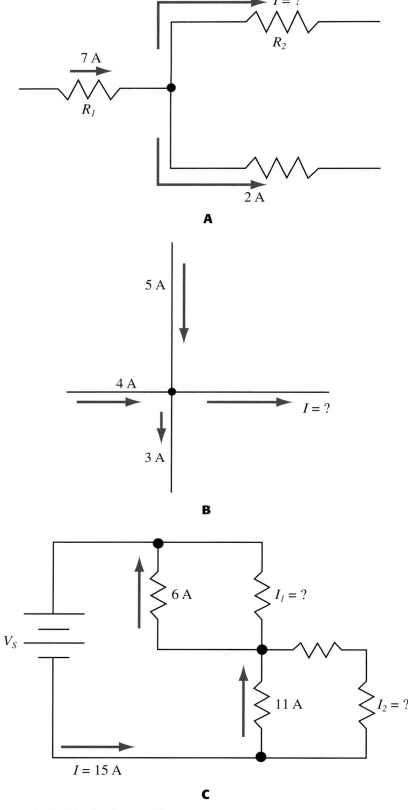

Figures 4-23A–C. Circuits for problem 5.

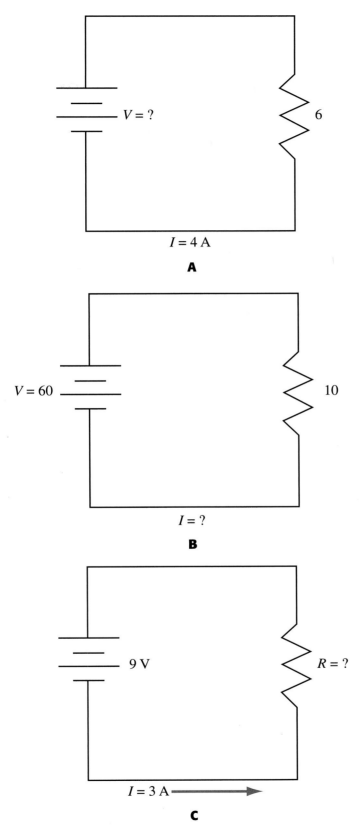

Figures 4-24A–C. Circuits for problem 6.

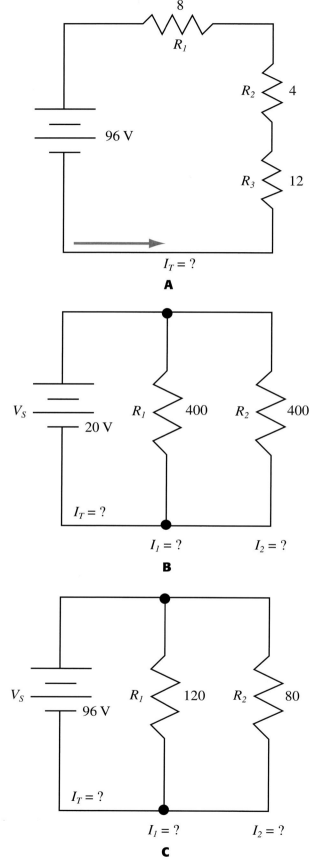

Figures 4-25A–C. Circuits for problem 8.

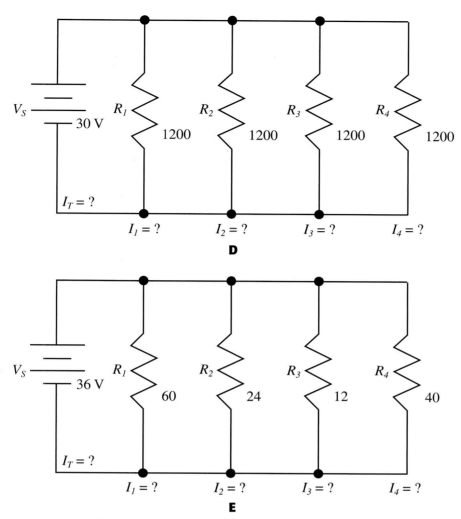

Figures 4-25D–E. Circuits for problem 8.

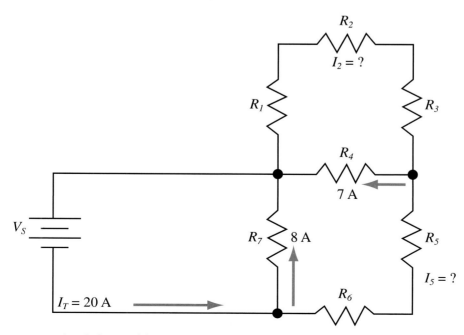

Figure 4-26. Circuit for problem 10.

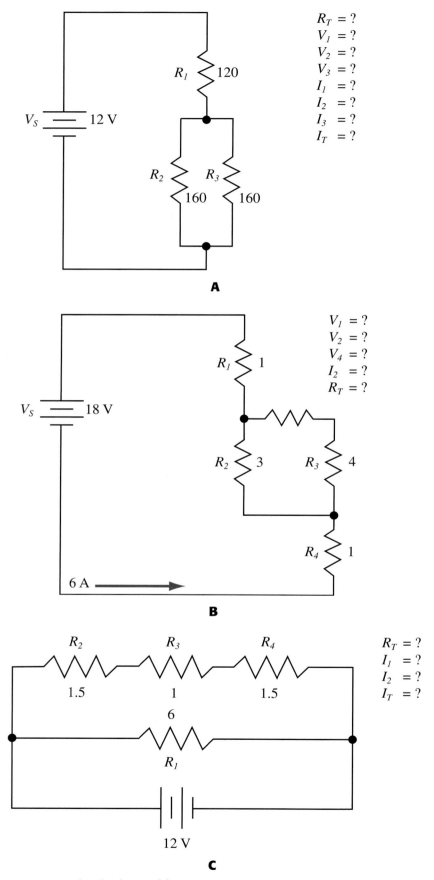

$R_T = ?$
$V_1 = ?$
$V_2 = ?$
$V_3 = ?$
$I_1 = ?$
$I_2 = ?$
$I_3 = ?$
$I_T = ?$

A

$V_1 = ?$
$V_2 = ?$
$V_4 = ?$
$I_2 = ?$
$R_T = ?$

B

$R_T = ?$
$I_1 = ?$
$I_2 = ?$
$I_T = ?$

C

Figures 4-27A–C. Circuits for problem 13.

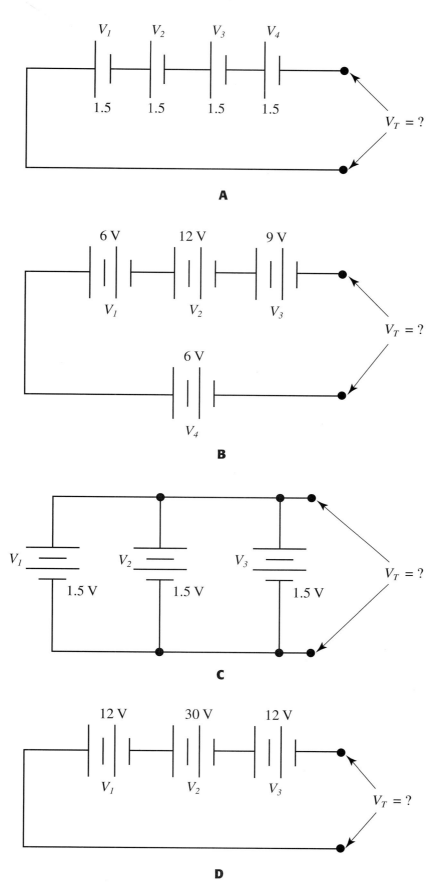

Figures 4-28A–D. Circuits for problem 14.

Chapter 5 Graphic Overview

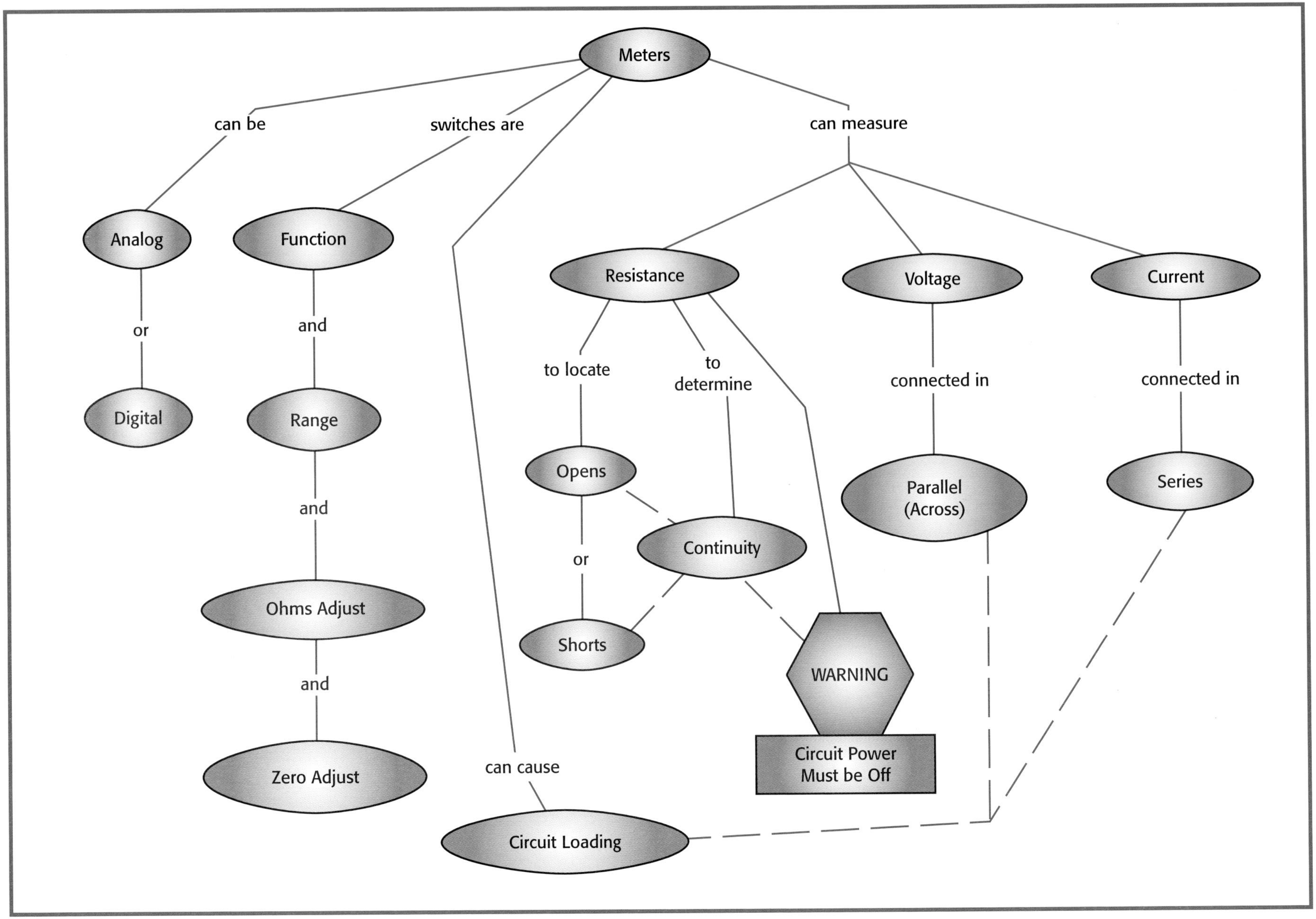

USING ELECTRICAL METERS

5

➤| Objectives

After studying this chapter, you will be able to:

○ Identify the types of meters and common switches.
○ Measure voltage, current, and resistance.
○ Explain the possible errors in making resistance tests.
○ Test components and circuits for continuity.
○ Use a meter to locate open and short circuits.
○ Describe the meter specifications of analog and digital meters.
○ State the safety precautions to observe when using a meter.

➤| Introduction

Many types of measuring devices are used by electrical technicians. Measuring instruments enable the technician to measure voltage, resistance, and current to determine if circuits are working properly.

Safety Precautions

When working with meters, the following precautions should be observed:

• Handle the test probes by their plastic handles, not their metal tips.
• Unless you know what voltage to expect, start measurements with the highest scale of a meter.
• Turn off the meter when it is not in use.
• Keep the meter away from any components that may have a strong magnetic field.
• Check the batteries in the meter at least every six months.
• Turn off the power of equipment that is to be tested before making resistance measurements.
• Never measure voltages above the maximum value specified for the voltmeter.
• To minimize the shock hazard from one hand to another, keep one hand in your pocket when measuring voltage.
• When using a high voltage probe, make sure the ground clip cannot become disconnected. If the ground clip disconnects, you may find yourself holding the full high-voltage in your hand.
• Never work alone.

Types of Meters

Meters are either analog or digital. An *analog meter* has a scale and a pointer and is called a volt-ohm-milliammeter, or VOM for short, **Figure 5-1.** A panel meter is a type of analog meter that takes a specific measurement, such as milliamperes, **Figure 5-2.**

In an analog meter, the pointer is attached to a moving mechanism. The magnetic field produced by the coil reacts with the permanent magnet field causing the mechanism to rotate. See **Figure 5-3.**

Analog meter: A meter that uses a magnetically operated pointer with a scale to indicate an electrical quantity.

Figure 5-1. Analog meters come in many types, sizes, and prices.

Figure 5-2. Single panel meters are mounted to continuously monitor and measure only one electrical value.

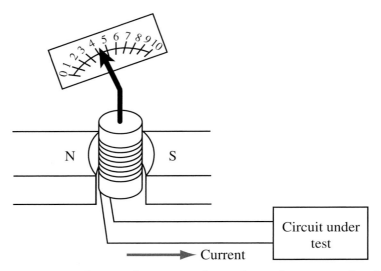

Figure 5-3. Operation of an analog meter depends on the current flowing through the meter.

Whether a voltmeter, ammeter or ohmmeter, all analog meters require current to create the magnetic field for operation of the meter movement. Current is taken from the circuit being tested. However, an ohmmeter contains a power source (usually a battery) that operates the meter.

The technician simply reads the numbers displayed on a ***digital meter,*** **Figure 5-4.** Although digital meters come in many types and sizes, their operation is the same.

The operation of a digital instrument can be illustrated in a block diagram, **Figure 5-5.** The circuit under test is connected to the input processing circuits, a set of very low-tolerance (0.1%) resistors with a selection switch for dc, ac, and ohms. The amplifier provides the necessary amplification of the voltages before they are applied to the analog-to-digital (A/D) converter. The amplifier also eliminates interference between the circuits by forming a buffer between the input processing circuits and the A/D converter. The digital signal from the A/D converter is connected to a

Digital meter: A meter that provides a digital readout of a measurement.

digital counter. The counter counts the digital pulses coming from the A/D converter, which are then connected to the digital display. The displayed numbers represent volts, amps, or ohms.

A *multimeter* is a type of digital meter, **Figure 5-6.** A multimeter can measure volts, amps, ohms, frequency, capacitance, transistor gain, and digital logic levels.

Multimeter: A meter that measures volts, ohms, and amps.

Figure 5-4. Digital meters like this one measure dc, ac, and ohms with great accuracy.

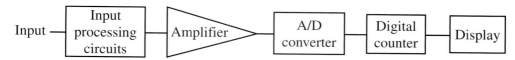

Figure 5-5. Block diagram of a digital instrument.

Figure 5-6. In addition to voltage, resistance, and current, this multimeter can measure transistor beta and capacitance, and has a diode test.

Test Leads and Probes

Test leads and probes are used to connect the meter to the circuit under test. Leads and probes come in a variety of sizes and types, **Figure 5-7.** Some have simple alligator clips on the end of a wire; others have a more complicated design for special purposes. Follow the manufacturer's instructions for use.

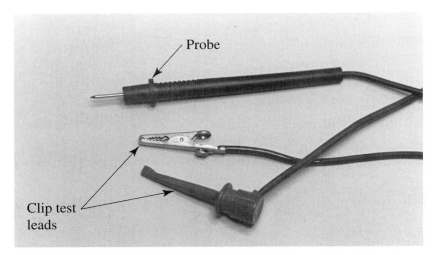

Figure 5-7. Various types of test leads and probes used with meters.

Meter Switches

Many types of meters with a variety of switches and knobs are in use. A discussion of all of them is not practical. However, four basic meter switches are common to most meters. **See Figure 5-8.**

Function switch:
The main switch on a meter that indicates what is to be measured.

The *function switch* indicates what is to be measured, for example, ac voltage, dc voltage, or resistance. Notice there are two dc voltage positions, one with a negative sign and one with a positive sign. Switching from one dc position to the other is the same as reversing the test leads. On this meter, the function switch also has an AC OFF position to turn off the meter.

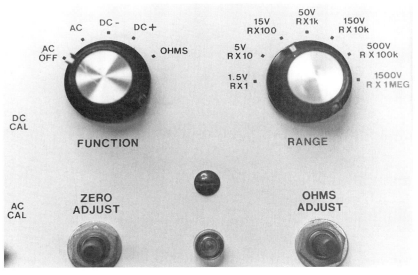

Figure 5-8. Some meters have several switches and knobs that are important for proper use of the meter.

The ***range switch*** determines the maximum value that can be measured (full-scale deflection of the pointer). For example, the range switch in Figure 5-8 is set in the 15 V position (upper numbers). This means the largest voltage that can be measured is 15 V. If the range switch were turned all the way to the right (clockwise), the maximum value that could be read is 1500 V.

The lower numbers on the range switch are used for resistance measurements. When the switch is set on the R × 100 position (as shown), the reading on the meter scale is multiplied by 100. If the switch were turned all the way to the right (R × 1 MEG), the reading would be multiplied by one million.

The ***ohms adjust*** switch is used when measuring resistance. It is adjusted to make the meter pointer rest on the infinity (INF or ∞) mark on the resistance scale.

The ***zero adjust*** switch is used to place the meter pointer on zero when using any of the meter scales. Most digital meters do not have a zero adjustment. However, the pointer in an analog meter must rest on zero. There are two zero adjustments on analog meters: the meter movement zero adjustment and the electrical zero adjustment. Both adjustments are made when the meter probes are not connected to a circuit.

 Caution: Do not confuse the electrical zero adjust with the meter movement zero adjustment. The meter movement zero adjustment screw is also shown in Figure 5-8. Improper adjustment of this screw can cause permanent damage to the meter movement. Once the meter movement is damaged, it is often too expensive to repair and the meter must be junked.

Figure 5-9 shows the back side of the meter movement zero adjustment screw. *This adjustment should not be made more than 1/4 turn.* Turn the screw adjustment a little and observe how the pointer moves. If it does not move, turn the adjustment back to its original position; then, turn it in the other direction. Only a small amount of adjustment is normally required.

The electrical zero adjustment will depend upon the make and model of the meter. Investigate the meter switches and locate the zero adjust. Adjust the control carefully until the pointer rests on zero of the meter scale. If possible, refer to the manual that came with the meter. Remember, this adjustment is made with the meter test leads *not* connected to any type of device or circuit. Some meters require the metal test lead probes be connected together during the zero adjustment.

Range switch: The switch on a meter or other instrument used to select the maximum range of measurement.

Ohms adjust: The control switch on an analog meter used to adjust the pointer to infinity or another designated mark.

Zero adjust: An adjustment to an electronic instrument so it is calibrated to a zero value.

Figure 5-9. The meter movement adjustment screw of an analog meter should not be adjusted more than 1/4 turn; otherwise, permanent damage to the meter movement may occur.

Reading Analog Meters

On an analog meter, the meter pointer usually does not stop exactly on a number. Therefore, an estimate of the value must be made. With some practice, the readings can be quite accurate. For example, in **Figure 5-10,** the pointer is on 1. In **Figure 5-11**, the pointer is halfway between 2 and 3; therefore, the reading is 2.5. More precisely, each line between 2 and 3 is equal to 0.2. The pointer is halfway between 0.4 and 0.6; thus, the reading is 2.5. In **Figure 5-12,** the pointer is at 4.63. In **Figure 5-13,** the pointer is at 26.7, and in **Figure 5-14,** it is at 35.

The previous voltage and current scales are linear. The distances between numbers on the scale are equal. The resistance scales of an analog meter are nonlinear, that is, the distance between numbers on the scale are not the same, **Figure 5-15.** Extra care must be taken when determining the pointer value of the resistance scale.

In **Figure 5-16,** the pointer at position A is indicating 1.5 Ω (halfway between 1 and 2). The pointer at position B is also halfway between two numbers, but is indicating 1200 Ω. In this case, the halfway mark is not equal to 0.5 because the distance between the numbers is different.

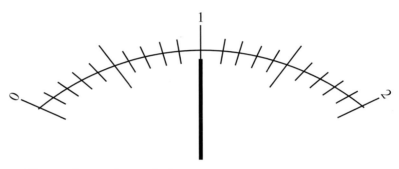

Figure 5-10. The analog scale reads 1.

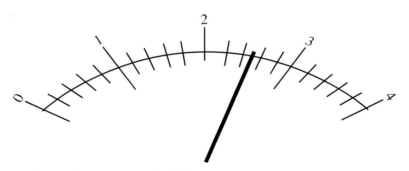

Figure 5-11. The analog scale reads 2.5.

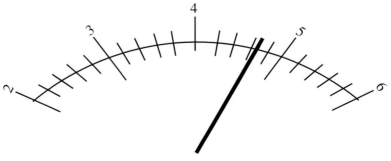

Figure 5-12. The analog scale reads 4.63.

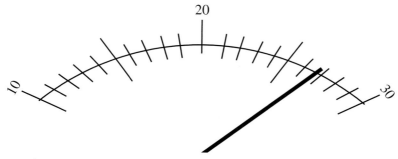

Figure 5-13. The analog scale reads 26.7.

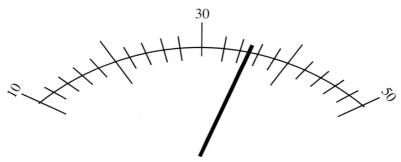

Figure 5-14. The analog scale reads 35.

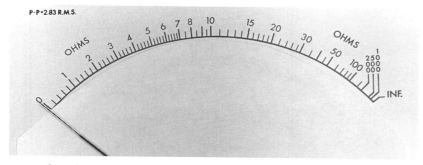

Figure 5-15. The ohms scale of an analog meter is nonlinear.

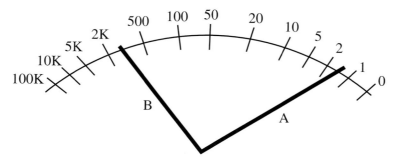

Figure 5-16. The halfway point between numbers is not necessarily half the value.

Another point of confusion can exist with the resistance scales. The zero can be on the right or left side of the scale, **Figure 5-17.** Also, the scales are usually combined, so you must sort out the one you will use. Furthermore, the function and range switches on analog meters may be combined into one switch, **Figure 5-18.**

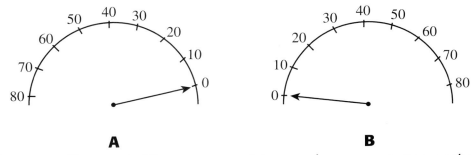

Figure 5-17. Placement of the zero on a resistance scale can vary. A—Zero on the right side. B—Zero on the left side.

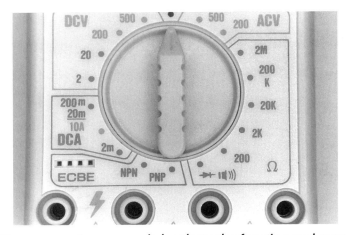

Figure 5-18. On many meters one switch selects the function and range.

Reading Digital Meters

Digital meters often combine range and function into one switch and have a few other switches for specific selections. Reading a digital meter may not be as simple as just reading the numbers. The meter in **Figure 5-19** displays 10.00 with M to the right of the numbers. This indicates the reading is in megohms. When the display shows only zeros, and the vertical bar at the left is flashing, infinity is being indicated.

Some digital meters have a switch to hold the last reading on the display. Some are auto-ranging, while others give you a choice of auto-ranging or not. Another feature of some digital meters is a bar graph display at the lower part of the meter display, **Figure 5-20**. The bar graph is useful in making adjustments to circuits while watching for a high or low voltage point.

Taking Meter Measurements

Taking measurements with any meter requires certain skills. First, you must understand the purpose of each switch adjustment. Second, you must be able to read the meter display (analog or digital). Finally, you must know how to take the basic measurements of voltage, current, and resistance.

Measuring Voltage

Three types of voltage measurements are taken: source voltage, voltage drop across some type of device, and the presence or absence of voltage.

Figure 5-19. A digital meter reading in megohms.

Figure 5-20. Some digital meters have bar graphs for peak or null voltage adjustments.

Figure 5-21 shows a simple series circuit consisting of a 12 V dc source and three resistors. To measure the voltage drop of the resistors, the meter is connected across (in parallel) with the resistors. To prevent meter damage when using an analog meter, always start with the highest range.

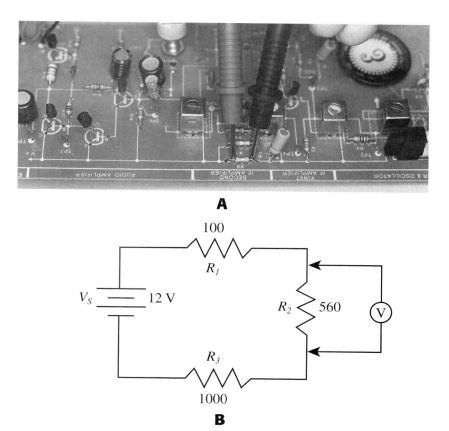

A

100
R_1
V_S = 12 V
R_2 560 (V)
R_3
1000

B

Figure 5-21. Measuring voltage. A—The voltage drop of R_2 is being measured. B—The voltage measurement of R_2 is shown on this schematic diagram.

Ground Connections

Two conductors are required to make a complete path in all circuits. In most practical circuits, one side of the circuit is grounded. In an automobile, for example, the negative side of the battery is usually connected to the frame of the vehicle. The frame is the ground connection for all the circuits in the vehicle. That is, the frame is one of the conductors for each of the electrical circuits; it is a common ground.

In 120 V residential wiring, the ground is actually earth, **Figure 5-22.** The white (neutral) side is connected to a common point and grounded by an 8' grounding rod, as specified by the National Electric Code.

Many wire connections can be eliminated by using common ground connections. The circuit grounds indicate each ground is connected to the same point. See **Figure 5-23.**

Most voltage measurements are taken with reference to the chassis ground or the common ground of a printed circuit board, **Figure 5-24.** Unless otherwise stated, the ground reference for measurements is assumed.

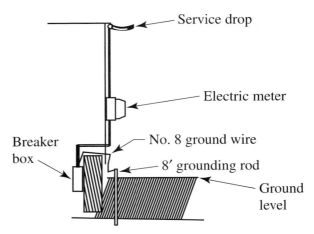

Figure 5-22. A residential power service is grounded with an 8' grounding rod. (National Electric Code)

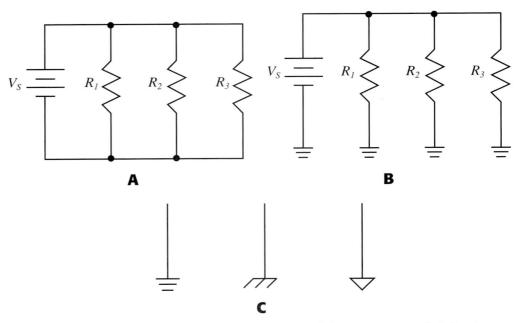

Figure 5-23. Common ground connections. A—Parallel circuit. B—Parallel circuit using common ground connections. C—Various ground symbols.

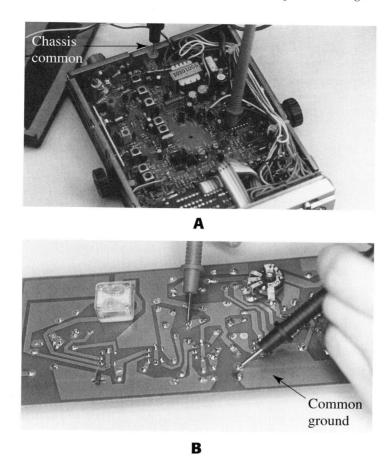

Figure 5-24. Ground references for voltage measurements. A—Chassis ground. B—Common ground of a printed circuit board.

Measuring Current

To measure current, turn the power off. The circuit must be opened and the meter placed in series with the circuit. Connect the negative ammeter terminal to the negative side of the circuit, and the positive terminal to the positive side of the circuit. The current must flow from the negative source terminal, through the ammeter, to R_2, R_1, and the positive source terminal. See **Figure 5-25.**

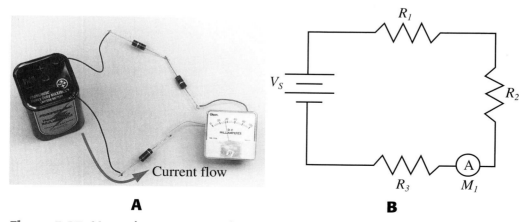

Figure 5-25. Measuring current. A—The meter must be in series with the circuit. B—Schematic diagram.

In **Figure 5-26,** the ammeter is *incorrectly* connected to the circuit. It is connected as a voltmeter instead of an ammeter. This improper connection causes the current to flow from the negative terminal, through the meter, to the positive terminal, and does not measure the circuit current. This method of connection allows the current to flow beyond the limits of the meter and results in meter damage.

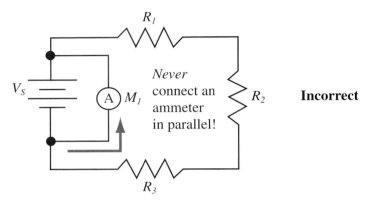

Figure 5-26. Incorrect connection. An ammeter is *never* connected across a component or source.

The schematic diagram in **Figure 5-27** illustrates the measurement of the total current of a parallel circuit. The schematic in **Figure 5-28** illustrates the measurement of the current of R_1. The circuit must be broken and the meter placed in series with the branch in which the current is to be measured.

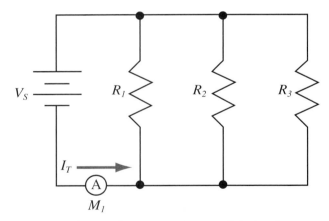

Figure 5-27. Measuring the total current of a parallel circuit.

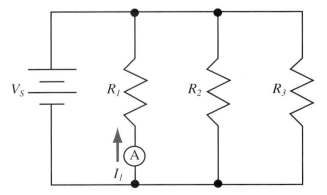

Figure 5-28. When measuring the current of an individual component, the component must be disconnected and the meter placed in series with the component.

Measuring Resistance

Resistance is measured to determine if a resistor is out of tolerance or has changed value. Resistance measurements are also taken to test continuity and locate shorts or opens. The proper method of taking the resistance measurement of a resistor is shown in **Figure 5-29.** When measuring resistance, the circuit power must be turned off.

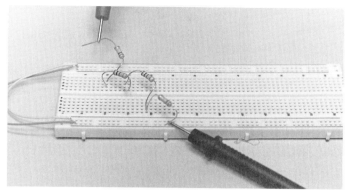

Figure 5-29. Measuring the resistance of a resistive current.

Zeroing the Meter

Zeroing the meter when taking resistance measurements is slightly different than zeroing as discussed previously. When taking resistance measurements, the meter pointer must rest on zero and correctly show infinity. To zero most meters, the test leads are connected together, and the meter is adjusted so the pointer rests on zero. See **Figure 5-30.**

The test leads are then separated and the meter pointer adjusted to infinity. This process may need to be repeated to ensure accuracy. The zeroing process is usually not necessary when using a digital meter.

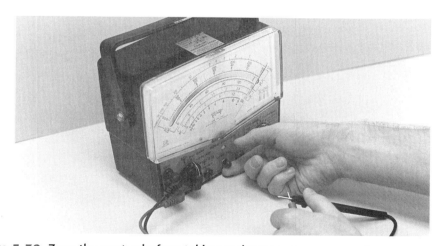

Figure 5-30. Zero the meter before taking resistance measurements.

Errors in Resistance Measurements

A common error when making resistance measurements is to hold the device and the meter's metal probes with both hands. This presents no problem with a low resistance. However, when the resistance value gets above 30,000 Ω, your body resistance factors into the reading. In **Figure 5-31,** the meter is measuring the resistance of the resistor and the person's body, which is in parallel with the resistor. The resistor should be held by an alligator clip, **Figure 5-32.**

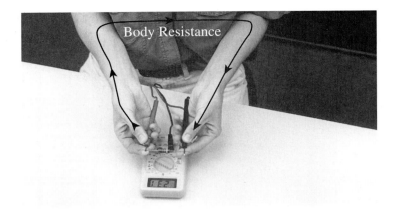

Figure 5-31. Human hands cause error in resistance measurements.

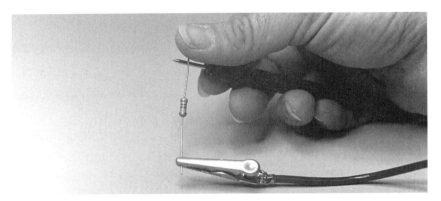

Figure 5-32. An alligator clip helps eliminate resistance errors caused by body resistance.

Another common error occurs when making a resistance test on a device that appears to have an incorrect reading. The meter is actually measuring back into another part of the circuitry that is in parallel with the device under test. To eliminate this type of error, one end of the device under test should be disconnected, **Figure 5-33.**

Resistance Bridge

Wheatstone bridge:
A type of bridge circuit used to measure unknown resistance.

A resistance bridge (called a ***Wheatstone bridge***) meter is used to make very accurate resistance measurements, **Figure 5-34.** It can accurately measure resistances as low as 0.001 Ω. The instrument dials are adjusted for a null or centered meter indication. The Wheatstone bridge is discussed further in Chapter 8.

Figure 5-33. Often a component must be disconnected to measure its resistance.

Figure 5-34. A resistance bridge can give extremely accurate measurements.

Continuity Testing

The purpose of a ***continuity test*** is to make sure a complete or continuous current path exists through a device, conductor, or circuit. An ohmmeter is used to check the continuity of a fuse, **Figure 5-35.** The meter will read 0 Ω if the fuse is good. If the fuse is open, the meter will read infinity. The same test is used for the other types of devices or circuits. The power must be turned off and the device or circuit disconnected before you make a continuity test.

Continuity test: A test performed to ensure a complete or continuous current path exists through a device, conductor, or circuit.

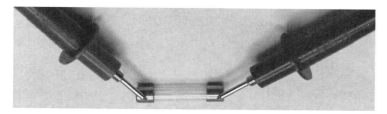

Figure 5-35. Technicians commonly test fuses for continuity.

A long piece of coaxial cable can be tested for continuity by connecting an alligator jumper to one end of the cable, **Figure 5-36.** This test can be used for any length of cable. Remember, the cable under test will have a certain amount of resistance, depending on its length and the size of the conductor. Another test of a coaxial cable is from the center conductor to the outer shield, **Figure 5-37.** In the assembly of the cable, it is possible that the outer shield conductors may come in contact with the inner conductor, causing a short circuit. With both ends of the coaxial cable open, connect an ohmmeter from the inner conductor to the outer conductor. The meter should indicate infinity.

Locating an Open Circuit

An open circuit can be found by making resistance or voltage measurements, **Figure 5-38.** The junctions of the resistors are labeled A, B, C, and D. If a voltage reading taken at point B indicates voltage present, and a reading taken at point C is zero, then an open exists between the two points. If a voltmeter reading at point A indicates voltage and shows zero at point B, then an open exists between points A and B. If the voltmeter reading is taken at point D, the reading would be zero, which is normal. The meter is reading the same point, from ground to ground.

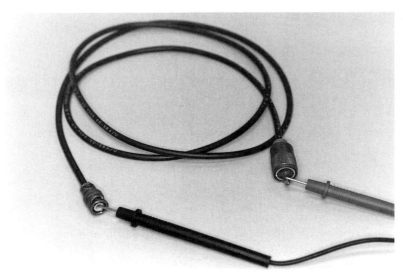

Figure 5-36. Testing a coaxial cable for continuity.

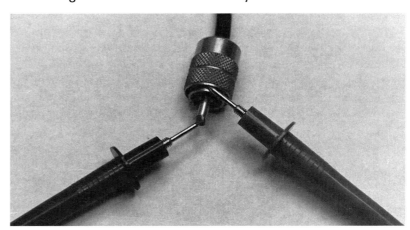

Figure 5-37. Testing for continuity of a coaxial cable from center to outer conductors verifies no short exists between the two conductors.

Locating a Short Circuit

Short circuit: An unwanted path of current flow.

A *short circuit* can be a direct (dead) short or a partial short. The direct short is a direct connection or unwanted path for current having a value of zero resistance. A partial short is one that has a certain amount of resistance in the unwanted circuit path.

A direct short in an electrical circuit often causes visible signs of trouble, such as blown fuses, smoke, odors, and burned components. These events are only symptoms of trouble and are not the problem. If a blown fuse or burned component is found, replacing it and turning on the power often results in the new components being destroyed as well. A partial short can be troublesome to locate because visible signs of the problem are often not present. Only resistance measurements and careful thought processes will aid in locating such shorts.

The instrument used to locate a short is the VOM. Taking resistance measurements to locate the cause of the short circuit requires an understanding of the circuits under test. Often, the only way a short can be found is through a process of elimination, as illustrated in **Figure 5-39.**

The parallel circuit shown consists of three current paths. If a short is indicated at point A, then one of the parallel circuits is disconnected, and the measurement taken again. If the same short is indicated at point A, the second parallel circuit is disconnected, and the circuit resistance is measured again. If the short is still indicated at point A, the third parallel branch, is disconnected.

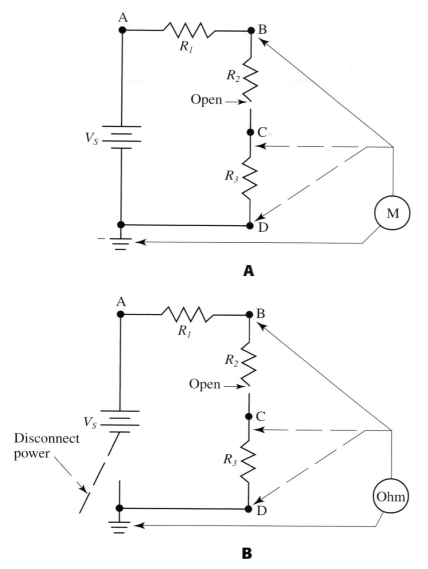

Figure 5-38. Locating an open circuit. A—A voltmeter can be used to locate an open circuit. B—Locating an open circuit using an ohmmeter requires the power to be disconnected.

Meter Specifications

Manufacturers of meters list specifications for accuracy, sensitivity, and input resistance. Each specification is described next.

Accuracy

The accuracy specification is plus or minus a percentage of the full scale reading. The full scale value depends on the range being used. For example, if the accuracy of a VOM is ±3%, and readings are taken on the 10 V range, the accuracy would be ±0.3V. On the 150 V range, the readings would be ±4.5 V.

Thus:

10 V × 0.03 (3%)

= 0.3 V

And:

150 V × 0.03 (3%)

= 4.5 V

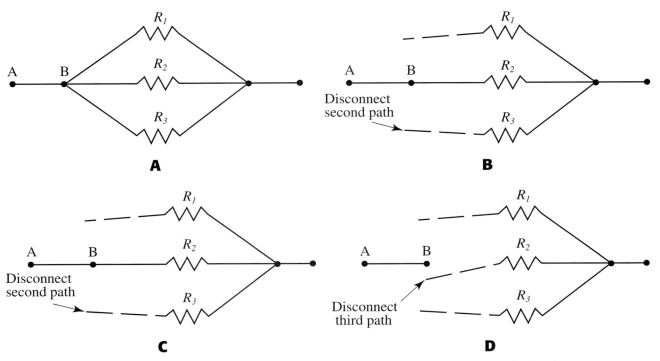

Figure 5-39. Locating a short in a circuit through a process of elimination. A—Example circuit. B—One circuit branch is disconnected to eliminate it as the possible cause of the short. C—The second branch is disconnected to further eliminate possible causes of the short. D—The third branch is disconnected to isolate the short to a specific part of the circuit.

Sensitivity

Sensitivity: A measure of how much current it takes to move the meter pointer to full scale position.

The *sensitivity* of a meter is a measure of how much current it takes to move the meter pointer to the full-scale position. Sensitivity is expressed as the number of *ohms per volt* (Ω/V) for a voltmeter and is equal to 1 V divided by the full-scale current. The larger the Ω/V rating, the smaller the current required for the meter to make voltage measurements.

Ohms per volt: The unit of measurement for sensitivity.

$$\text{Sensitivity} = \frac{1\ V}{\text{Full-scale current}}$$

The sensitivity of a voltmeter using a 20 μA meter movement would be:

$$\frac{1}{20\ \mu A}$$

$$= 50{,}000\ \Omega/V$$

Input Resistance

Input resistance: The resistance or impedance offered to a signal applied to an input circuit as "seen" by the signal source.

The *input resistance* determines how much current the meter will draw from a circuit. The voltmeter connected across the circuit becomes a current path, or load, on the circuit, **Figure 5-40.** A lower input resistance causes more current to go through the meter. Thus, the meter has a "loading effect" on the circuit.

The input resistance is determined by the meter sensitivity and range being used. The equation is:

$$\text{Input resistance} = \text{Sensitivity} \times \text{Range}$$

The input resistance of a meter with a sensitivity of 20,000 Ω/V on the 50 V range would be:

$$20{,}000 \times 50$$

$$= 1{,}000{,}000\ \Omega \text{ or 1 megohm}$$

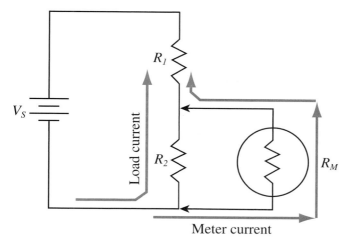

Figure 5-40. Illustration of input resistance. Any meter connected to a circuit becomes part of the circuit.

Circuit Loading

Whenever an ammeter, voltmeter, or other piece of equipment is connected to a circuit, the meter itself becomes part of the circuit. The internal components, resistances, reactances, and other characteristics of the meter movement are connected in series or parallel with the circuit under test. The meter is a load connected to the circuit. Under certain circumstances, this can affect the accuracy of the meter indications.

Ammeter Loading

In **Figure 5-41,** the circuit current is calculated at 6 mA:

$$I = \frac{V}{R}$$

$$= \frac{6}{1000}$$

$$= 6 \text{ mA}$$

If an ammeter is connected in the circuit, the resistance of the meter is placed in series with the circuit, **Figure 5-42.** Then, the total resistance is $R_1 + R_M$ or 1120 Ω. The measured current is less than expected because of the meter resistance:

$$I = \frac{6}{1120}$$

$$= 5.4 \text{ mA}$$

Ammeter loading is usually a factor in low-voltage, low-resistance circuits.

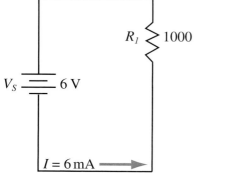

Figure 5-41. Circuit for ammeter loading.

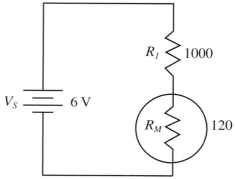

Figure 5-42. An ammeter connected in the circuit changes the total resistance and net total current.

Voltmeter Loading

Voltmeter loading is more common than ammeter loading and can occur in series or series-parallel circuits. Voltmeter loading usually occurs in high-resistance circuits where the meter input resistance is less than the circuit resistance.

Figure 5-43 indicates the values of voltages and current for the circuit without the meter connected to the circuit. A meter with a sensitivity of 1000 Ω/V set on the 5 V range has an input resistance of 5000 ohms:

$$\frac{1000\ \Omega}{V \times 5\ V}$$
$$= 5000\ \Omega$$

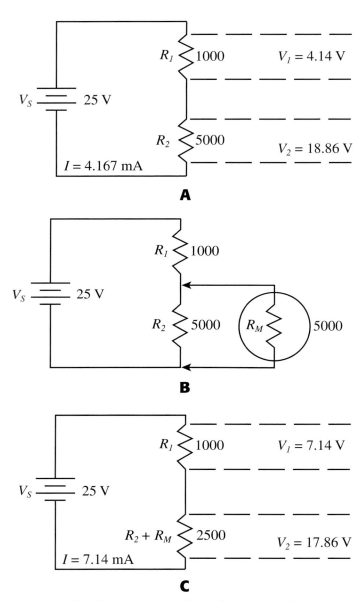

Figure 5-43. Voltmeter loading. A—Circuit for voltmeter loading. B—The input resistance of the voltmeter changes the total resistance of the circuit. C—The resulting voltages are different than expected because of voltmeter loading of the circuit.

The meter is connected across R_2, which makes a parallel circuit of R_2 and the meter circuit. The equivalent resistance of the parallel circuit is 2500 Ω. This changes the total resistance as well as the circuit voltages and current.

When using a meter, a technician must understand the circuits being measured and the effects of connecting instruments to the circuits. A voltmeter should have an input resistance 20 times greater than the circuit resistance being measured.

Summary

- Meters are classified as either analog or digital.
- Leads and probes are used to connect the meter to the circuit under test.
- The function switch determines what will be measured: dc volts, ac volts, resistance, or current.
- The range switch determines the maximum value the meter can measure on that setting. Many digital meters are auto-ranging.
- The ohms adjust control is used to adjust the meter into proper alignment for resistance measurements.
- The zero adjust control is used to adjust the meter to zero when nothing is connected to the meter test leads. The meter movement zero adjust is used to adjust the meter pointer to zero when nothing is connected to the meter and the meter power is off.
- When making measurements with analog meters, start with the highest range to prevent meter damage.
- Never connect an ammeter in parallel with a source. The ammeter must always be connected in series with the circuit under test.
- Testing for continuity is a form of measuring resistance.
- Always zero the meter before testing for resistance.
- Never hold both metal test probe tips with your hands when measuring.
- Meters are used to find open and short circuits.
- The input resistance of a meter determines how much of a load the meter will place on the circuit under test. Voltmeter loading is the most common.
- Follow proper safety precautions when working with a meter.

Important Terms

Do you know the meanings of these terms used in the chapter?

analog meter
continuity test
digital meter
function switch
input resistance
multimeter
ohms adjust

ohms per volt
range switch
sensitivity
short circuit
Wheatstone bridge
zero adjust

Questions and Problems

Please do not write in this text. Write your answers on a separate sheet of paper.

1. What are the two types of voltmeters?
2. Connections from the meter to the circuit under test are made using _____ or _____ .
3. List four common switches found on electrical measurement meters.
4. The ohms adjust is used to adjust the meter into proper alignment for _____ .
5. What are the two zero adjustments on an analog meter?
6. Which type of meter has auto-ranging?
7. Voltmeters are connected in _____ with the circuit and current meters are connected in _____ with the circuit under test.
8. When working with an analog meter, with what range should you begin working?
9. Describe a typical error made in making resistance measurements.
10. What test would you use to test a fuse or switch?
11. What are three important meter specifications?
12. What are two types of circuit loading?
13. How can you minimize shock hazard from one hand to the other when measuring voltage?
14. If the meter in **Figure 5-44** has a resistance of 18 Ω, how much current error will be made?
15. What are the voltage readings in **Figure 5-45?**
16. If a meter is in the 300 V range with an accuracy of 2.5%, what is the possible voltage range?
17. If the meter in problem 16 is in the 3 V range, what is the possible voltage range?
18. If a meter has a full-scale current of 50 μA, what is its sensitivity?
19. When used on the 25 V range, what is the input resistance of the meter in problem 18?
20. What are the resistance readings in **Figure 5-46?**

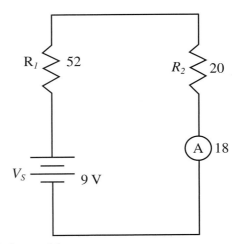

Figure 5-44. Circuit for problem 14.

A

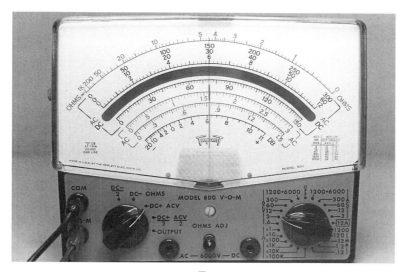

B

C

Figure 5-45A–C. Meters for problem 15.

D

E

F

Figure 5-45D–F. Meters for problem 15.

A

B

C

Figure 5-46A–C. Meters for problem 20.

Chapter 6 Graphic Overview

Units of Measurement
- classed in → Standard International Units (SI Units)
- by changing → Metric Prefix → to → Power of Ten → moving the → Decimal Point → moves in → Opposite Direction → as the → Exponent Chart Direction

Exponents → the difference between → (Decimal Point / Opposite Direction)

Temperature Coefficient
- can be:
  - Negative → means that → Quantity → will → Increase → with a → Decrease
  - Positive → means that → Quantity → will → Increase → with an → Increase
- states how → Quantity → will → Increase → or → Decrease → with → Temperature
- in

ELECTRICAL QUANTITIES ▶▶▶ 6

▶▌ Objectives

After studying this chapter, you will be able to:

○ Calculate the resistance of wire with changes in temperature.
○ Convert units of measurement to a lower or higher value.
○ State the standard international units of measurements.

▶▌ Introduction

This chapter discusses the effects of temperature on the conduction of materials. A method for quickly converting units of measurement to a lower or higher value is explained. Electrical properties and metric units of measurement are covered also.

Temperature Coefficient

Temperature coefficient is a factor that indicates whether electrical resistance will increase or decrease with temperature. It is symbolized by the Greek letter alpha (α). Some typical values of α are given in **Figure 6-1.**

Temperature coefficient: A factor that indicates whether electrical resistance will increase or decrease with temperature.

Material	Specific Resistance at 20° C	Temperature Coefficient ($\pm \ \Omega$ per °C)
Aluminum	17	0.0045
Carbon	*	−0.0003
Copper	10.4	0.0043
Gold	14	0.004
Iron	58	0.0065
Nichrome	676	0.0002
Nickel	52	0.0068
Silver	9.8	0.0041
Tungsten	33.8	0.0048

* The resistance of carbon depends on the quality of the carbon.

Figure 6-1. The temperature coefficient of some common materials.

Positive Coefficient

A *positive temperature coefficient* means a certain resistance will increase with an increase in temperature. All metals in their pure form, such as copper and lead, have a positive temperature coefficient. The temperature coefficient is specified as plus or minus ohms per degree centigrade.

Negative Coefficient

A *negative temperature coefficient* means a certain resistance will decrease with an increase in temperature. Notice carbon has a negative coefficient. As the temperature increases, the resistance of carbon decreases. The α is negative for most semiconductors, including germanium and silicon. All electrolyte solutions, such as sulfuric acid and water, have a negative α.

Lower resistance at higher temperatures can cause problems in semiconductor circuits, such as transistors and integrated circuits. As the temperature increases and resistance decreases, current flow through the device increases. This can result in overheating and destruction of the device.

Calculating Resistance Using Temperature Coefficient

The approximate resistance of a wire can be calculated by the equation:

$$R_T = R + R (\alpha \times \Delta T)$$

Where:

R_T = Total resistance

R = Specific resistance of the wire calculated by using:

$$R = \rho \times \frac{l}{A}$$

α = Temperature coefficient

ΔT = Change in temperature

▼ *Example 6-1:*

Find the resistance of 300′ of No. 20 aluminum wire at 131°F (55°C).

$$R = 17 \times \frac{300}{1022}$$

$$= 5 \ \Omega$$

$$R_T = R + R (\alpha \times \Delta T)$$

$$= 5 + 5 [0.0045 (35)]$$

$$= 5.788 \ \Omega$$

▲

Changing Units of Measurement

Many times a quantity is given in one unit and must be changed to another unit, such as seconds into microseconds or microamps into milliamps. Use the following procedure until you can easily convert units of measurement in your head:

Number of original units × Power of ten prefix = _____ × New prefix

---

▼ *Example 6-2:*

9 milliamps = _____ microamps

$9 \times 10^{-3} =$ _____ 10^{-6} (Change the metric prefix to a power of ten.)

The difference in the exponents is three. This tells you the decimal must move three places:

9 milliamps

= 9000 microamps

▲

Refer to the powers of ten chart in Figure 2-7. You will see the exponent value moves down from 10^{-3} to 10^{-6}. This tells you the decimal must be moved to the right to make the number larger. If the exponents go down on the chart, the number goes up, and the decimal point is moved to the right. If the exponents go up on the chart, the number goes down, and the decimal point is moved to the left. The answer will be larger or smaller by the difference in the exponents and will be inverse to the direction on the power of ten chart.

SI Units of Measurement

The original metric system was adopted in France in 1799. Various metric systems have been developed over the years. The newest system was adopted in 1960 with small changes since then. All metric units of measurement are written in Standard International units, or *SI units*. There are three classes of SI units, **Figures 6-2, 6-3,** and **6-4.**

SI units: The abbreviation for Standard International Units, common international units of measurement.

SI Base Units		
Property	**Name of Unit**	**Symbol**
length	meter	m
mass	kilogram	kg
electric current	ampere	A
thermodynamic temperature	kelvin	K
amount of substance	mole	mol
luminous intensity	candela	cd

Figure 6-2. SI base units.

Summary

- A quantity can increase or decrease depending on whether the temperature coefficient is positive or negative.
- The changing of a unit of measurement to an equivalent value unit is a common task for a technician.
- The three types of metric units are SI base units, SI derived units, and SI supplementary units.

Important Terms

Do you know the meanings of these terms used in the chapter?

negative temperature coefficient
positive temperature coefficient

SI units
temperature coefficient

SI Derived Units		
Property	**Name of Unit**	**Symbol**
area	square meter	m²
volume	cubic meter	m³
speed or velocity	meters per sec.	m/s
quantity of electric charge	coulomb	C
acceleration	meters per sec.	m/s
frequency	hertz	Hz
force	newton	N
energy or work	joule	J
power	watt	W
emf, potential difference	volt	V

Figure 6-3. SI derived units.

SI Supplementary Units		
Property	**Name of Unit**	**Symbol**
capacitance	farad	F
conductance	siemens	S
magnetic flux	weber	Wb
electric resistance	ohm	Ω
plane angle	radian	rad
solid angle	steradian	sr

Figure 6-4. SI supplementary units.

Questions and Problems

Please do not write in this text. Write your answers on a separate sheet of paper.

1. Aluminum has a _____ temperature coefficient.
2. Carbon has a _____ temperature coefficient.
3. A positive temperature coefficient means that the quantity will go _____ as the temperature goes down.
4. What is the resistance of No. 14 iron wire, 600′ long at 248°F (120°C)?
5. Convert the following units of measurement.
    a. 24,000 ohms = _____ kilohms
    b. 2400 volts = _____ kilovolts
    c. 0.015 amps = _____ milliamps
    d. 600 volts = _____ millivolts
    e. 155 microseconds = _____ seconds
    f. 0.00000167 amps = _____ microamps
    g. 0.420 microseconds = _____ nanoseconds
    h. 3,000,000 ohms = _____ megohms
    i. 2470 nanoamps = _____ microamps
6. Convert the following units of measurement. (Refer to Appendix C for a listing of standard abbreviations.)
    a. 4300 V = _____ mV
       = _____ kV
       = _____ MV
    b. 0.68 A = _____ mA
       = _____ μA
    c. 12,000 W = _____ kilohms
       = _____ megohms
    d. 0.001 μF = _____ F
       = _____ pF
    e. 220 mH = _____ H
       = _____ μH
    f. 430 μS = _____ S
       = _____ mS
    g. 23 kHz = _____ Hz
       = _____ MHz
    h. 0.0034 = _____ mA
    i. 0.00000076 = _____ ns
    j. 2200 = _____ kilohms
    k. 0.045 = _____ ms

Chapter 7 Graphic Overview

ELECTRICAL POWER

 **Objectives**

After studying this chapter, you will be able to:

○ Calculate the total power of series, parallel, and series-parallel circuits.
○ Calculate the power of individual resistances.
○ Calculate the efficiency of devices or circuits.

Introduction

Voltage by itself can do no work. A source has a voltage, but if no load is connected across it, no current flows, and no electrical work is accomplished. When a conductor and load are connected across a source, a current flows and work is done or electrical energy is used.

Energy and Power

Power refers to how fast energy is used or converted to another form of energy. The six basic forms of energy are mechanical, light, heat, magnetic, chemical, and electrical. For example, a lightbulb changes electrical energy into light and heat energy. An electric motor changes electrical energy into magnetic energy and changes the magnetic energy into mechanical energy.

The unit of power is the *watt* (W). Although devices are rated in different units, they can be expressed in watts. Mechanical devices, such as electric motors, are rated in *horsepower* (hp) and water heaters are rated in *British thermal units* (Btu) per hour.

$$1 \text{ hp} = 746 \text{ W}$$
$$1 \text{ Btu/h} = 0.293 \text{ W}$$

Resistors, diodes, transistors, integrated circuits, and other devices are all rated in watts. This rating refers to how fast electrical energy is converted into heat, which is useless work or energy lost in the form of heat. Electrical energy wasted as heat is known as *dissipated energy.*

Calculating Power

Electrical power (P) is calculated by multiplying the current by the voltage:

$$P = IV$$

In **Figure 7-1,** the circuit power is found by:
$$P = IV$$
$$= 3 \text{ A} (180 \text{ V})$$
$$= 540 \text{ W}$$

If the circuit consists of more than one resistance, the total resistance is found first. In **Figure 7-2,** the total resistance is 20 Ω; thus, the current is 2 A.
$$I = \frac{V}{R_T}$$
$$= \frac{40 \text{ V}}{20 \text{ Ω}}$$
$$= 2 \text{ A}$$

Power: How fast energy is used or converted to another form of energy.

Watt: The unit of measurement for power.

Horsepower: A measure of power used in motors. One horsepower is equal to 746 W.

British thermal units: The quantity of heat needed to raise one pound of water one degree Fahrenheit.

Dissipated energy: Electrical energy lost in the form of heat.

The total power is:

$$P_T = IV$$
$$= 2 \text{ A } (40 \text{ V})$$
$$= 80 \text{ W}$$

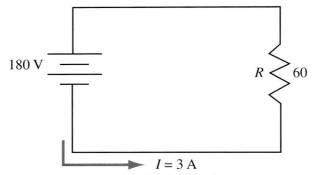

Figure 7-1. Calculating the power of a simple circuit.

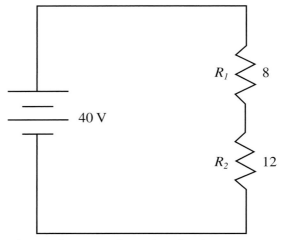

Figure 7-2. Calculating the total power of a series circuit.

Power of Series Circuits

The power dissipated for each individual resistor of a circuit can be calculated by multiplying the current flowing through the resistance by the voltage drop across the resistance.

The power of individual resistances can also be derived by another formula for power using resistance. As shown in **Figure 7-3,** since $V = IR$, then IR can be substituted for V in the power formula. For example, if the current is 2 A and resistance is 12 Ω, the power is found by:

$$P = I^2 R$$
$$= 2^2 \text{ A } (12 \text{ } \Omega)$$
$$= 4 \text{ } (12)$$
$$= 48 \text{ W}$$

$$P = IV \qquad\qquad P = I \, (IR) \text{ or } P = I^2 \, R$$
$$V = (IR)$$

Figure 7-3. Calculating circuit power using the formula $P = I^2R$.

▼ *Example 7-1:*

Find the total power and the power of each of the resistances in **Figure 7-4.**

First, find the total resistance:

$$R_T = R_1 + R_2 + R_3$$
$$= 10\ \Omega + 6\ \Omega + 8\ \Omega$$
$$= 24\ \Omega$$

Then, find the current:

$$I = \frac{V}{R}$$
$$= \frac{48\ V}{24\ \Omega}$$
$$= 2\ A$$

Next, calculate the individual powers:

$$P_1 = I^2 R_1$$
$$= 2^2\ A\ (10\ \Omega)$$
$$= 4\ (10)$$
$$= 40\ W$$

$$P_2 = I^2 R_2$$
$$= 2^2\ A\ (6\ \Omega)$$
$$= 4\ (6)$$
$$= 24\ W$$

$$P_3 = I^2 R_3$$
$$= 2^2\ A\ (8\ \Omega)$$
$$= 4\ (8)$$
$$= 32\ W$$

Finally, find the total power:

$$P_T = P_1 + P_2 + P_3$$
$$= 40\ W + 24\ W + 32\ W$$
$$= 96\ W$$

or

$$P_T = IV$$
$$= 2\ A\ (48\ V)$$
$$= 96\ W$$

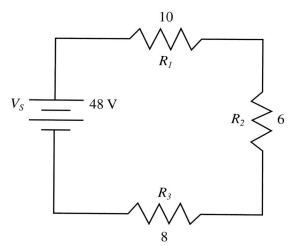

Figure 7-4. Circuit for Example 7-1.

Power of Parallel Circuits

The power dissipated for each individual resistor of a parallel circuit can also be calculated by multiplying the current flowing through the resistance by the voltage across the resistance. The voltage is the same across each of the resistances in the parallel circuit.

▼ *Example 7-2:*

Find the total power and the power of the individual resistances in **Figure 7-5.**
Find the individual and total currents:

$$I_1 = \frac{V}{R}$$
$$= \frac{36\ V}{30\ \Omega}$$
$$= 1.2\ A$$

$$I_2 = \frac{V}{R}$$
$$= \frac{36\ V}{20\ \Omega}$$
$$= 1.8\ A$$

$$I_T = I_1 + I_2$$
$$= 1.2\ A + 1.8\ A$$
$$= 3\ A$$

Find the individual powers:

$$P_1 = I^2 R_1$$
$$= 1.2^2\ A\ (30\ \Omega)$$
$$= 43.2\ W$$

$$P_2 = I^2 R_2$$
$$= 1.8^2\ A\ (20\ \Omega)$$
$$= 64.8\ W$$

Find the total power:

$$P_T = P_1 + P_2$$
$$= 43.2\ W + 64.8\ W$$
$$= 108\ W$$

or

$$P_T = IV$$
$$= 3\ A\ (36\ V)$$
$$= 108\ W$$

▲

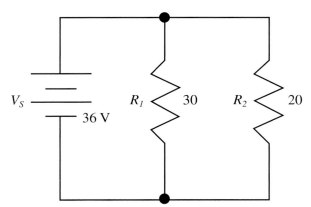

Figure 7-5. Circuit for Example 7-2.

▼ *Example 7-3:*

Find the total power and the power of individual resistances in **Figure 7-6.**

$$R_{PAR} = \frac{R_2 \times R_3}{R_2 + R_3}$$

$$= \frac{200\ \Omega \times 300\ \Omega}{200\ \Omega + 300\ \Omega}$$

$$= 120\ \Omega$$

$$R_T = R_1 + R_{PAR}$$
$$= 80\ \Omega + 120\ \Omega$$
$$= 200\ \Omega$$

$$I_T = \frac{V}{R}$$
$$= \frac{800\ V}{200\ \Omega}$$
$$= 4\ A$$

$$V_{R_1} = I_{R_1}$$
$$= 4\ A\ (80\ \Omega)$$
$$= 320\ V$$

$$V_{R_2} = V_S - V_{R_1}$$
$$= 800 - 320$$
$$= 480\ V$$

$$I_{R_2} = \frac{V_{R_2}}{R_2}$$
$$= \frac{480\ V}{200\ \Omega}$$
$$= 2.4\ A$$

$$I_{R_3} = \frac{V_{R_3}}{R_3}$$
$$= \frac{480\ V}{300\ \Omega}$$
$$= 1.6\ A$$

$$P_1 = I^2 R$$
$$= 4^2\ A\ (80\ \Omega)$$
$$= 1280\ W$$

$$P_2 = I^2 R$$
$$= 2.4^2\ A\ (200\ \Omega)$$
$$= 1152\ W$$

$$P_3 = I^2 R$$
$$= 1.6^2\ A\ (300\ \Omega)$$
$$= 768\ W$$

$$P_T = IV$$
$$= 4 \text{ A } (800 \text{ V})$$
$$= 3200 \text{ W}$$
or
$$P_T = P_1 + P_2 + P_3$$
$$= 1280 \text{ W} + 1152 \text{ W} + 768 \text{ W}$$
$$= 3200 \text{ W}$$

▲

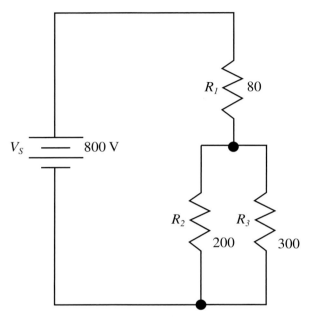

Figure 7-6. Circuit for Example 7-3.

Efficiency

If a device were perfect, the input power and output power would be the same, or the device would be 100% efficient. When a device converts energy from one form to another, a certain amount of power goes in (input power) and less power comes out (output power). This is usually the result of losses in the form of heat. A perfect device, which would have equal input and output power, has yet to be built. The percent of *efficiency* of a device is found by:

Efficiency: The ratio of the useful output to the total input of a device or circuit.

$$\text{Percent of efficiency } = \frac{P_{OUT}}{P_{IN} \text{ (100)}}$$

▼ *Example 7-4:*
What is the efficiency of a device if the input power is 60 W and the output power is 45 W?

$$\text{Percent of efficiency } = \frac{P_{OUT}}{P_{IN} \text{ (100)}}$$

$$= \frac{45 \text{ W}}{60 \text{ W } (100)}$$

$$= 0.75 \text{ W} \times 100$$

$$= 75\%$$

▲

Power Rating of Resistors

Resistors have a power rating specifying how much heat can be dissipated. The larger the resistor, the more heat dissipated. The largest resistor shown in **Figure 7-7** has a rating of 10 W, while the smallest one has a rating of 1/8 W.

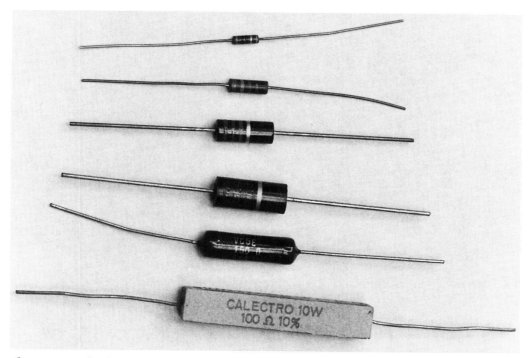

Figure 7-7. The larger resistor dissipates more power.

A resistor should be replaced with a resistor of the same power rating. If you are designing a circuit, a rule of thumb is to calculate the power dissipation and select a resistor with at least two times the calculated power.

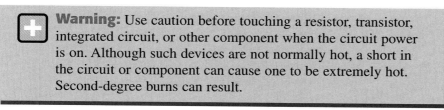

Warning: Use caution before touching a resistor, transistor, integrated circuit, or other component when the circuit power is on. Although such devices are not normally hot, a short in the circuit or component can cause one to be extremely hot. Second-degree burns can result.

Summary

- Power refers to how fast energy is used or converted to another form of energy.
- The six basic forms of energy are mechanical, light, heat, magnetic, chemical, and electrical.
- The watt is the unit of measurement for power.
- Electrical energy that is wasted or lost as heat is known as dissipated energy.
- To calculate power, multiply the current by the voltage.
- The total power of a circuit is the sum of all the individual powers of the components.
- The efficiency of a device is the ratio of the output power to the input power.
- A resistor should be replaced with one of the same power rating.

⬆️ Important Terms

Do you know the meanings of these terms used in the chapter?

British thermal units   horsepower
dissipated energy   power
efficiency   watt

⬆️ Questions and Problems

Please do not write in this text. Write your answers on a separate sheet of paper.

1. What are the six forms of energy?
2. What is the unit for power?
3. Write three formulas for calculating power.
4. A 35 W soldering iron has a resistance of 90 Ω. What is the current flowing in the iron circuit?
5. What will be the heat dissipation in watts of a 25 Ω resistor with 0.28 A flowing through it?
6. An 8 Ω speaker is driven by a 2 V signal and the input power is 0.7 W. What is its efficiency?
7. An incandescent bulb carries a current of 350 mA with a supply voltage of 120 V. How much power does it consume?
8. A radio receiver is connected across a 120 V power line and 0.65 A flows through it. How much power is being used?
9. How much current will a 1 W, 1500 Ω resistor safely pass?
10. How many ohms of resistance does an ordinary 50 W house lamp have when operating (assume 120 V)?
11. Solve for the following unknowns in **Figure 7-8.**
    a. $P_1 = $ _____
    b. $P_2 = $ _____
    c. $P_3 = $ _____
    d. $P_T = $ _____

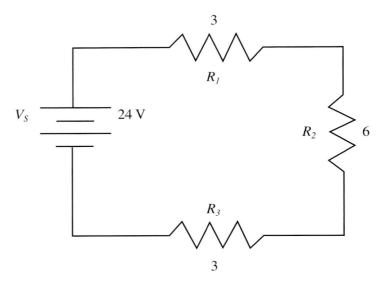

Figure 7-8. Circuit for problem 11.

12. Solve for the following unknowns in **Figure 7-9.**
    a. I_1 = _____
    b. I_2 = _____
    c. I_3 = _____
    d. P_1 = _____
    e. P_2 = _____
    f. P_3 = _____
    g. P_T = _____

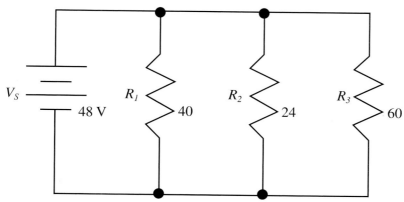

Figure 7-9. Circuit for problem 12.

13. Given:
    R_T = 18
    I_T = 2 A
    I_2 = 0.333 A
    I_4 = 1 A
    Solve for the following unknowns in **Figure 7-10.**
    a. P_1 = _____
    b. P_2 = _____
    c. P_3 = _____
    d. P_4 = _____
    e. P_5 = _____
    f. P_T = _____

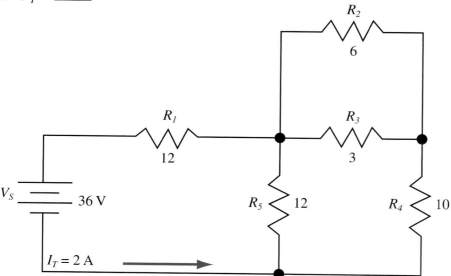

Figure 7-10. Circuit for problem 13.

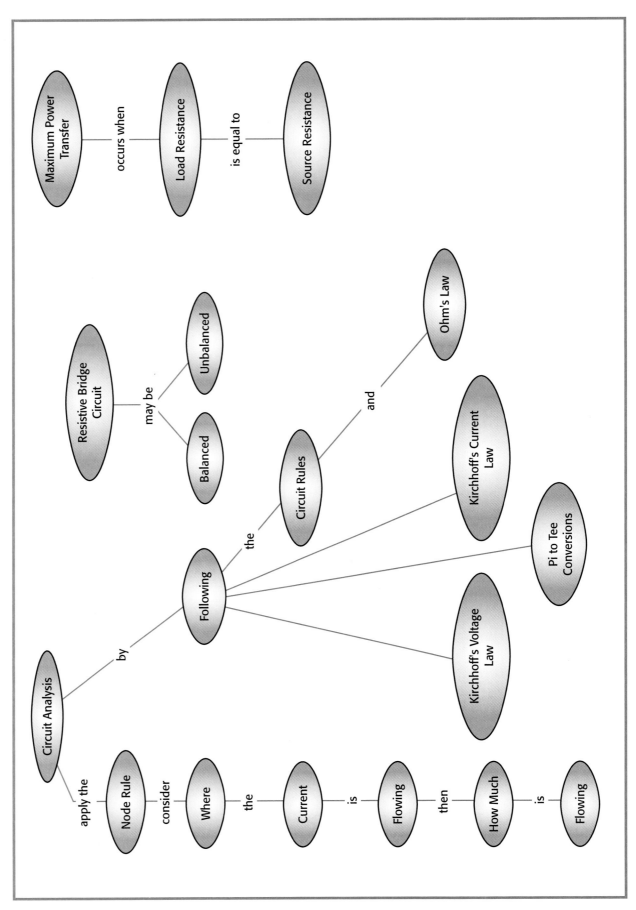

DC CIRCUIT ANALYSIS  **8**

Objectives

After studying this chapter, you will be able to:

- ○ Analyze series, parallel, and series-parallel circuits for unknown voltages, currents, and resistances.
- ○ Solve circuit voltages using the voltage divider equation.
- ○ Solve circuit currents using the current divider equation.
- ○ Analyze resistive bridge circuits.
- ○ Use pi and tee conversions to analyze resistive circuits.

Introduction

In many circuits, some components are connected in series and others are connected in parallel. This is true of resistors, capacitors, transistors, and integrated circuits. To understand the operation of complex electronic circuits, you must understand the operation of resistors in series, parallel, or series-parallel circuits. Analyzing dc circuits will be useful in troubleshooting other electronic problems.

Analysis of Series Circuits

A series circuit has only one current path. The series circuit rules are:

- *Voltage.* The sum of the voltage drops around a circuit is equal to the source voltage.

$$V_{SOURCE}\ (V_S) = V_1 + V_2 + V_3$$

- *Current.* The current is the same in all parts of the circuit.

$$I_{TOTAL}\ (I_T) = I_1 = I_2 = I_3$$

- *Resistance.* The total resistance of the circuit is equal to the sum of the individual resistances in the circuit.

$$R_{TOTAL}\ (R_T) = R_1 + R_2 + R_3$$

The analysis of a series circuit is a simple process when you consider the known and unknown variables in the circuit. By applying the series circuit rules and Ohm's law, you can determine the unknown variables.

▼ **Example 8-1:**

Find the total current, voltage drops, resistance, and power in **Figure 8-1.**

The current at R_2 is 3 A. The voltage of R_1 is found by applying Ohm's law:

$$V = IR$$
$$= 3\,A\ (5\,\Omega)$$
$$= 15\,V$$

The voltage of R_2 is found by applying the series voltage rule:

$$V_1 + V_2 + V_3 = V_S$$
$$15 + V_2 + 12 = 33$$
$$27 + V_2 = 33$$
$$V_2 = 6$$

The resistance of R_2 is found using Ohm's law:

$$R_2 = \frac{V_2}{I}$$

$$= \frac{6\ V}{3\ A}$$

$$= 2\ \Omega$$

The resistance of R_3 is found in the same manner:

$$R_3 = \frac{V_3}{I}$$

$$= \frac{12\ V}{3\ A}$$

$$= 4\ \Omega$$

The power of the circuit is found with the following equations:

$$P = IE$$

or

$$P = I^2R$$

$$P_1 = I^2R_1$$
$$= 3^2\ (5)$$
$$= 9\ (5)$$
$$= 45\ W$$

$$P_2 = I^2R_2$$
$$= 3^2\ (2)$$
$$= 9\ (2)$$
$$= 18\ W$$

$$P_3 = I^2R_3$$
$$= 3^2\ (4)$$
$$= 9\ (4)$$
$$= 36\ W$$

$$P_{TOTAL} = P_1 + P_2 + P_3$$
$$= 45 + 18 + 36$$
$$= 99\ W$$

or

$$P_{TOTAL} = IE$$
$$= 3\ (33)$$
$$= 99\ W$$

▲

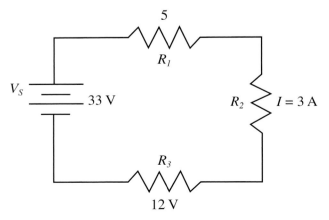

Figure 8-1. Series circuit example.

▼ *Example 8-2:*

Find the voltage and resistance of R_1 in **Figure 8-2.** In this circuit, the current is not given and must first be found.

The resistance and voltage drop of R_2 are given and can be used to find the current:

$$I_2 = \frac{V_{R_2}}{R_2}$$

$$= \frac{15\text{ V}}{3\text{K }\Omega}$$

$$= 5\text{ mA}$$

The voltage drop of R_1 can be found by subtracting the voltage drop of R_2 from the source voltage:

$$V_{R_1} = V_S - V_{R_2} \quad \text{(Applying the voltage rule.)}$$
$$= 25 - 15$$
$$= 10\text{ V}$$

The resistance of R_1 can be found using Ohm's law:

$$R_1 = \frac{V_1}{I}$$

$$= \frac{10}{5\text{ mA}}$$

$$= 2\text{K }\Omega$$

▲

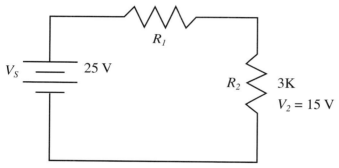

Figure 8-2. Series circuit example.

Voltage Levels of Series Circuit

Many beginning students look at a circuit and see voltage going into the parts of the circuit. This is the fastest way to cause problems in analyzing a circuit. Remember, voltage is applied *across* the circuit and current *flows* into the parts of the circuit. The voltage across resistors in a circuit is produced by the action of the current flowing through the resistance. Voltage does not go anywhere.

Figure 8-3 illustrates the voltage levels of a series circuit. The total resistance of the circuit is 25 Ω and the current is 3 A. As the current flows through R_4, a voltage drop of 15 V is produced across the resistor. A voltage drop of 30 V is produced across R_3, 12 V across R_2, and 18 V across R_1.

From the common reference point of ground, the voltage from ground to the top of R_3 is 45 V, the sum of the voltages of R_4 and R_3. From the top of resistor R_2 to ground, the voltage is 57 V, the sum of the voltages of R_4, R_3, and R_2. This 57 V added to the voltage drop across R_1 is equal to the source voltage of 75 V.

Kirchhoff's Voltage Law

The voltage rule for series circuits states the sum of the voltage drops around the circuit is equal to the source voltage. ***Kirchhoff's voltage law*** for series circuits states

Kirchhoff's voltage law: In a series circuit, the algebraic sum of all the voltages around the circuit, including the source, is equal to zero.

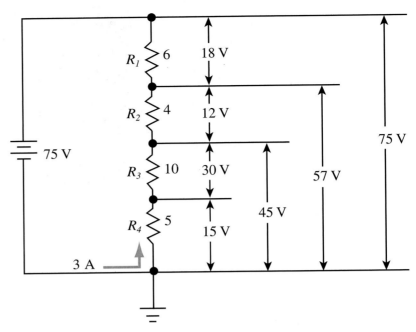

Figure 8-3. A series circuit has different voltage levels.

the same thing this way: *The algebraic sum of the voltages around a circuit is zero.* In Kirchhoff's voltage law, either the voltage source or the voltage drops are negative numbers.

▼ *Example 8-3:*

The source voltage is 30 V and the voltage drops are $V_1 = 15$, $V_2 = 8$, and $V_3 = 7$. Using Kirchhoff's law, the equation is:

$$V_S + (-V_1) + (-V_2) + (-V_3) = 0$$
$$30 + (-15) + (-8) + (-7) = 0$$

or

$$-V_S + V_1 + V_2 + V_3 = 0$$
$$-30 + 15 + 8 + 7 = 0$$

▲

Series Circuit as Voltage Divider

Any series circuit divides the voltage of the source proportionally between the series resistors. A special equation used for series circuits is called the *voltage divider equation:*

$$V = \frac{R}{R_T}(V_S)$$

Voltage divider equation: A special equation used for series circuits to find a voltage across any single resistance.

Where:

R = Resistance of voltage across which voltage drop is to be found
R_T = Total resistance of series circuit
V_S = Source voltage

By using this equation, a voltage across any single resistance can be found.

▼ *Example 8-4:*

Find the voltage across R_2 in **Figure 8-4.**

$$V = \frac{R}{R_T}(V_S)$$

$$= \frac{10\ \Omega}{15\ \Omega}(3\ V)$$

$$= 2\ V$$

The equation works no matter how many resistances are in the series circuit or where the resistor is located in the circuit.   ▲

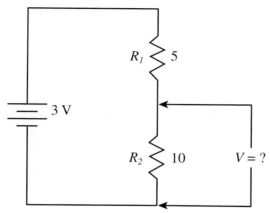

Figure 8-4. Series circuit as a voltage divider.

▼ *Example 8-5:*

Find the voltage of R_3 in **Figure 8-5.**

$$V = \frac{R}{R_T}(V_S)$$

$$= \frac{12\ \Omega}{70\ \Omega}(35\ V)$$

$$= 6\ V$$ ▲

Using the voltage divider equation has advantages. The calculation of voltages is easier and faster, especially when odd resistor values are used. In addition, the time involved in calculating the circuit current is eliminated.

Analysis of Parallel Circuits

A *parallel circuit* is a circuit that has more than one current path. When working with a parallel circuit, the first thing to consider is where the current is flowing. Then you can determine how much current is flowing in each branch.

The rules for parallel circuits are:

- *Voltage.* The voltage is the same for all parts of the circuit.
- *Current.* The total current is the sum of all the branch currents.
- *Resistance.* The total resistance is always less than the smallest resistance.

Parallel circuit: A circuit that contains more than one path for current to flow.

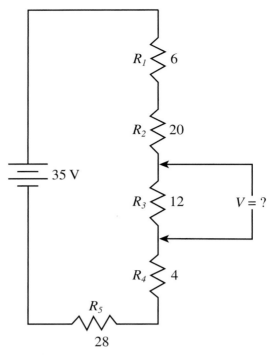

Figure 8-5. Applying the voltage divider formula to a circuit does not require the circuit current.

Node Rule

The node rule is important when working with parallel circuits. The rule states a current leaving a node must be equal to the current entering a node. This rule applies to all circuits.

Kirchhoff's Current Law

Kirchhoff's current law: In a parallel circuit, the algebraic sum of all the branch currents, including the source, is equal to zero.

Kirchhoff's current law for parallel circuits states the same thing as the node rule: At any junction of conductors in a circuit, the algebraic sum of the currents is zero. In Kirchhoff's current law, either the current leaving or the current entering is a negative number.

▼ *Example 8-6:*

The source current (I_S) is 7 A, and the two currents leaving the node are $I_1 = 2$ A and $I_2 = 5$ A. Written as Kirchhoff's law, the equation is:

$$I_S + (-I_1) + (-I_2) = 0$$
$$7 + (-2) + (-5) = 0$$

or

$$-I_S + I_1 + I_2 = 0$$
$$-7 + 2 + 5 = 0$$

▲

Parallel Resistances

Parallel resistances may be of three combinations:
- Resistors of the same value
- Two resistors of different values
- Three or more resistors of different values

Whatever the combination, the total resistance must be less than the smallest resistor in parallel.

▼ *Example 8-7:*

What is the total resistance and current flowing in each of the resistors in **Figure 8-6?**

$$R_T = \frac{\text{Value of one resistor}}{\text{Number of resistors in parallel}}$$

$$= \frac{100}{2}$$

$$= 50 \ \Omega$$

The current for I_1 is found using Ohm's law:

$$I = \frac{V}{R}$$

$$= \frac{100 \text{ V}}{100 \ \Omega}$$

$$= 1 \text{ A}$$

The current for R_2 is the same since the voltage and resistance values are the same. The process is used with any number of resistors of equal value in parallel. If the circuit had five resistors in parallel, the resistance would be:

$$R_T = \frac{\text{Value of one resistor}}{\text{Number of resistors in parallel}}$$

$$= \frac{100}{5}$$

$$= 20 \ \Omega$$

▲

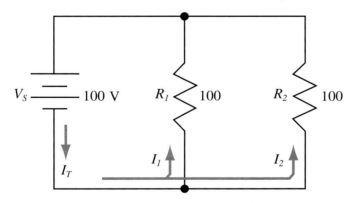

Figure 8-6. Parallel circuit with resistors of the same value.

▼ *Example 8-8:*

What is the total resistance and current flowing in each of the resistors in **Figure 8-7?**

Since the resistors are of different values, the product over the sum equation can be used:

$$R_T = \frac{R_1 \times R_2}{R_1 + R_2}$$

$$= \frac{20 \times 30}{20 + 30}$$

$$= \frac{600}{50}$$

$$= 12 \ \Omega$$

The current for each branch of the parallel circuit is found using Ohm's law:
For R_1:

$$I_1 = \frac{V}{R}$$

$$= \frac{24 \text{ V}}{20 \text{ }\Omega}$$

$$= 1.2 \text{ A}$$

For R_2:

$$I_2 = \frac{V}{R}$$

$$= \frac{24 \text{ V}}{30 \text{ }\Omega}$$

$$= 0.8 \text{ A}$$

For the total current, you can use Ohm's law with the total voltage and total resistance, or just add the branch currents:

$$I_T = \frac{V_T}{R_T}$$

$$= \frac{24}{12}$$

$$= 2 \text{ A}$$

or

$$I_T = I_1 + I_2$$

$$= 1.2 + 0.8$$

$$= 2 \text{ A}$$

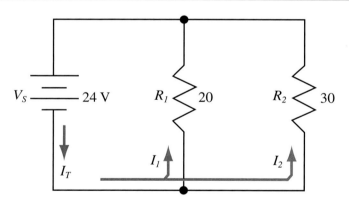

Figure 8-7. Parallel circuit with two resistors of different value.

Finding the total resistance of three or more resistors of different values requires the use of conductance (expressed in siemens or S). Conductance is the reciprocal of resistance. When using a calculator for conductance problems, it is best not to round off the numbers. Keep the decimal digits by storing the numbers in memory. Slight value errors will result if the numbers are rounded.

▼ *Example 8-9:*

Find the total resistance and currents of **Figure 8-8.**

$$R_T = \frac{1}{\dfrac{1}{R_1} + \dfrac{1}{R_2} + \dfrac{1}{R_3}}$$

$$= \frac{1}{\dfrac{1}{60} + \dfrac{1}{200} + \dfrac{1}{300}}$$

$$= 40 \text{ }\Omega$$

The individual branch currents are found using Ohm's law:

$$I_1 = \frac{V}{R_1}$$
$$= \frac{120\ V}{60\ \Omega}$$
$$= 2\ A$$

$$I_2 = \frac{V}{R_2}$$
$$= \frac{120\ V}{200\ \Omega}$$
$$= 0.6\ A$$

$$I_3 = \frac{V}{R_3}$$
$$= \frac{120\ V}{300\ \Omega}$$
$$= 0.4\ A$$

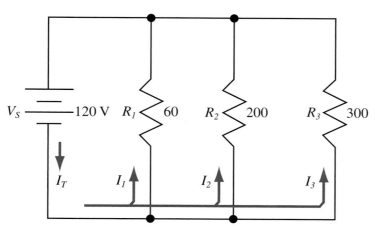

Figure 8-8. Parallel circuit with three or more resistors of different value.

Parallel Circuit as Current Divider

Just as a series circuit acts as a voltage divider, a parallel circuit acts as a current divider. The current is divided, inversely proportional, between the parallel resistances. Because of the inverse relationship of current and resistance (as the resistance increases, the current decreases), it is better to use conductance in the equation. Conductance can be used for any number of resistances as long as they are in parallel. The current divider equation is expressed as:

$$I = \frac{G}{G_T}(I_T)$$

Where:
G = Conductance in which the current is to be found
G_T = Total conductance of the parallel circuit
I_T = Total current

▼ *Example 8-10:*
Find the current of resistors R_1 and R_2 in **Figure 8-9A.** Changing resistances to conductances changes Figure 8-9A to **Figure 8-9B:**
Changing resistances to conductances.

$$I_1 = \frac{G_1}{G_T}(I_T)$$

$$= \frac{0.01\ \text{S}}{0.015\ \text{S}}(60\ \text{mA})$$

$$= 40\ \text{mA}$$

$$I_2 = \frac{G_2}{G_T}(I_T)$$

$$= \frac{0.005\ \text{S}}{0.015\ \text{S}}(60\ \text{mA})$$

$$= 20\ \text{mA}$$

The equation works, no matter how many resistances are in the parallel circuit. You could also arrive at 20 mA in R_2 by subtracting the 40 mA in R_1 from the total current of 60 mA.   ▲

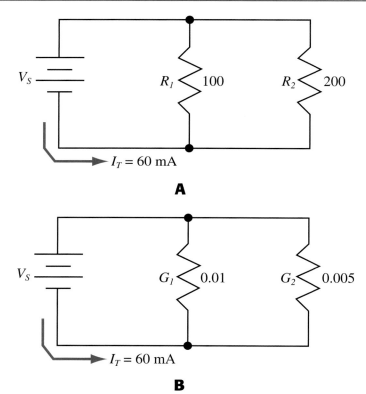

Figure 8-9. A parallel circuit operates as a current divider. A—Resistances in parallel. B—Changing the resistances to conductances.

▼ *Example 8-11:*
Find the current for G_4 in **Figure 8-10.**

$$I_4 = \frac{G_4}{G_T}(I_T)$$

$$= \frac{0.003\ \text{S}}{0.045\ \text{S}}(120\ \text{mA})$$

$$= 8\ \text{mA}$$ ▲

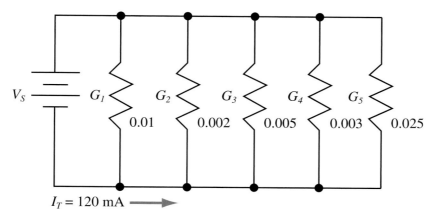

Figure 8-10. Applying the current divider formula to a circuit.

With the current divider equation, the current in the circuit branches can be calculated without knowing the circuit voltage.

Analysis of Series-Parallel Circuits

When solving a series-parallel circuit problem, calculate the parallel equivalent resistance. Then, treat the circuit as a series circuit.

▼ *Example 8-12:*

Calculate the currents and voltages in **Figure 8-11.**

Calculating the resistance of the parallel network changes **Figure 8-11A** to **Figure 8-11B:**

$$R_{PAR} = \frac{60}{2}$$
$$= 30 \ \Omega$$

Add the resistances for the total resistance:

$$R_T = R_1 + R_{PAR}$$
$$= 20 + 30$$
$$= 50 \ \Omega$$

Calculate the current:

$$I_T = \frac{V}{R_T}$$
$$= \frac{100 \text{ V}}{50 \ \Omega}$$
$$= 2 \text{ A}$$

Calculate the voltage drops of the circuit:

$$V_1 = IR_1$$
$$= 2 \times 20$$
$$= 40 \text{ V}$$

$$V_{PAR} = IR_{PAR}$$
$$= 2 \times 30$$
$$= 60 \text{ V}$$ ▲

Voltage Levels of Series-Parallel Circuit

The series-parallel circuit in **Figure 8-12A** is the same design as Figure 8-11A. The total resistance of the circuit is 90 Ω and the total current is 100 mA. The voltage drops are the voltage levels of the circuit.

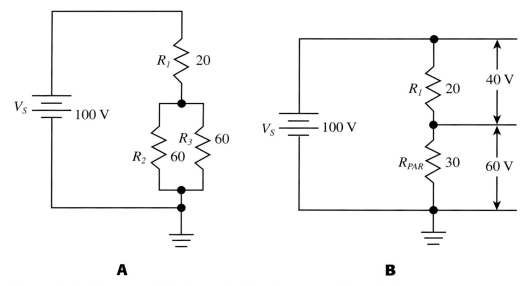

Figure 8-11. Series-parallel circuit. A—Calculating parallel equivalent resistance. B—Changing the circuit to a series circuit.

Figure 8-12B shows the voltage across the resistances. R_1 is 5 V ($V = IR$), while the voltage across the parallel network of R_2, R_3 is 4 V. The sum of these two voltages is equal to the source voltage. Although the voltage across both R_2 and R_3 is 4 V, you add only *one* of the 4 V to the voltage of R_1. The sum should be equal to the source voltage. The voltage across R_2 is the same voltage as across R_3.

Think of an automobile tire. If you look at the front and back of the tire, it is still the same tire. Likewise, in this case the same 4 V (not the voltage value) is being referenced.

Maximum Power Transfer

Maximum power transfer: The condition that exists when the resistance of the load equals the internal resistance of the source.

Maximum power transfer means getting the highest possible power from a source to its load. Maximum power occurs when the load resistance is equal to the source resistance.

Figure 8-12. Series parallel circuit. A—Voltage levels. B—Changing the parallel network to an equivalent circuit.

In **Figure 8-13,** the source has an internal resistance of 2 Ω. As the dashed lines indicate, this resistance is inside the source as one complete unit. The load resistance is 2 Ω; thus, the circuit has a total resistance of 4 Ω.

The circuit current is calculated by:

$$I = \frac{V}{R_T}$$
$$= \frac{10 \text{ V}}{4 \text{ Ω}}$$
$$= 2.5 \text{ A}$$

$$P_{INTERNAL} = I^2R$$
$$= 2.5^2 (2)$$
$$= 12.5 \text{ W}$$

The power for the load is the same since the values are the same. The total power would be the sum of the two powers, or 25 W.

Calculations for other values of load resistances are shown in **Figure 8-14.** Notice maximum power dissipation in the load occurs when the load is equal to the source resistance.

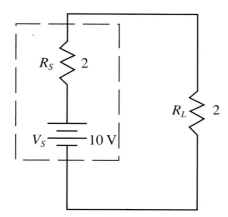

Figure 8-13. Maximum power transfer occurs when the load resistance equals the source resistance.

R_L	R_T	I_T	P_{SOURCE}	P_{LOAD}	P_T
0.5	2.5	4.0	32	0.8	32.8
1	3	3.33	22	11	33
2	4	2.5	12.5	12.5	25
3	5	2.0	8	12	20
4	6	1.67	5.6	11	16.6

Figure 8-14. P_L reaches a maximum of 12.5 W when R_L equals R_S.

Resistive Bridge Circuit

Figures 8-15 and **8-16** are two ways of illustrating a bridge circuit. The two drawings are the same electrically. Four current paths flow through a bridge circuit. The amount and direction of current through R_5 depends on the ratio of the other resistors. A common method of calculating voltages and currents in a bridge circuit is described next.

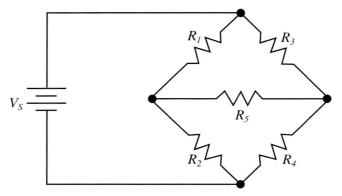

Figure 8-15. Basic bridge circuit.

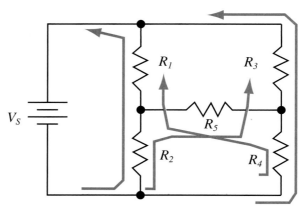

Figure 8-16. Various bridge circuit currents.

Balanced bridge:
A circuit whose components are adjusted so the output voltage is zero across the bridge.

Figure 8-17A is called a ***balanced bridge*** because the resistors are equal ratios and the voltages they produce cause no current through R_5. As shown in **Figure 8-17B,** resistors R_1 and R_2 form a voltage divider across the source, as do resistors R_3 and R_4. The voltage drop across resistors R_2 and R_3 can be calculated by the voltage divider equation:

$$V = \frac{R}{R_T} (V_S)$$

Where:
R = Resistance in which the voltage drop is found
R_T = Total series resistance
V_S = Source voltage
The voltage drop across R_2 is calculated by:

$$V_{R_2} = \frac{R_2}{R_T} (V_S)$$

$$= \frac{500}{500 + 1000} (15)$$

$$= 5 \text{ V}$$

The voltage drop across R_4 would be calculated the same way with the same result since the same values would be used.

The voltage across R_5 is the difference in potential between the voltage drops of R_2 and R_4:

$$V_{R_5} = V_{R_2} - V_{R_4}$$

$$= 5 - 5$$

$$= 0 \text{ V}$$

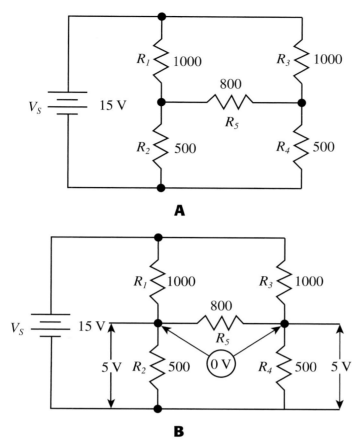

A

B

Figure 8-17. Balanced bridge. A—A balanced bridge has the same resistance ratios on each side of the bridge circuit. B— With no voltage difference across R_5, no current flows through R_5.

With no voltage difference across R_5, no current flows through R_5, and the bridge circuit is balanced.

Figure 8-18 shows an ***unbalanced bridge*** in which a voltage is across and a current through R_5. In the calculations, the voltage across R_2 is again 5 V. The voltage across R_4 is calculated as:

$$V_{R4} = \frac{R_4}{R_3 + R_4}(V_S)$$

$$= \frac{250}{250 + 1000}(15)$$

$$= 3 \text{ V}$$

The difference in potential across R_5 is:

$$V_{R5} = V_{R2} - V_{R4}$$
$$= 5 - 3$$
$$= 2 \text{ V}$$

With 2 V across R_5, a current through R_5 is:

$$I = \frac{V}{R}$$

$$= \frac{2 \text{ V}}{800 \text{ } \Omega}$$

$$= 2.5 \text{ mA}$$

The voltage of 5 V across R_2 is more positive than the 3 V across R_4; therefore, the current flows from the more negative point to the more positive point of R_5. The direction of the current through R_5 is from the R_4 side to the R_2 side (right to left).

Unbalanced bridge:
A voltage or current exists between the two branches of a bridge circuit.

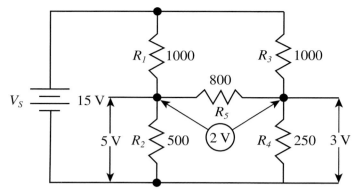

Figure 8-18. The resistance ratios of an unbalanced bridge are not the same, which causes a difference in potential across R_5.

Wheatstone bridge: A type of bridge circuit used to measure unknown resistance.

The **Wheatstone bridge** is a type of bridge circuit used to measure unknown resistance. See **Figure 8-19.** Resistors R_1 and R_2 form the usual voltage divider. R_S is an accurate set of resistor values which can be switched into the circuit. The unknown resistance is placed in the R_X position. As the value of R_S is changed, the meter indicates a balanced condition; that is, no current flows in either direction. The resistance set on R_S is read for the value of the unknown resistance.

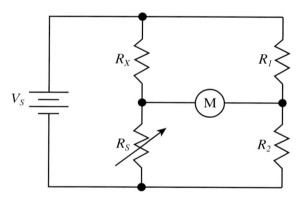

Figure 8-19. The Wheatstone bridge circuit is used to accurately measure unknown resistance.

Pi to Tee Circuit Conversions

A three terminal network can be depicted as a box, **Figure 8-20.** If the box consists of three resistors, the resistor network can be connected as either a *tee* (T) *network* or a *pi* (π) *network.* See **Figure 8-21.**

Tee network: A combination of components connected to form a "T" shape.

Pi network: A group of three components connected in the form of the Greek letter pi (π).

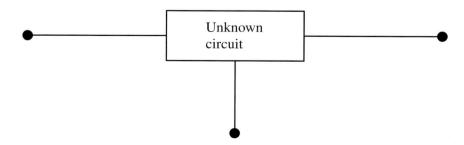

Figure 8-20. Many circuits are simply three-terminal networks.

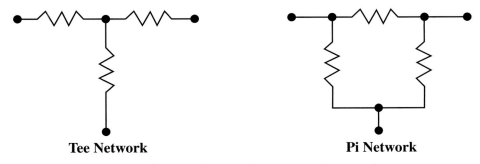

Tee Network   **Pi Network**

Figure 8-21. Three-terminal networks are often tee or pi networks.

These networks can be converted from one to another using the following equations:

For T to π:

$$R_A = \frac{R_1R_2 + R_2R_3 + R_1R_3}{R_2}$$

$$R_B = \frac{R_1R_2 + R_2R_3 + R_1R_3}{R_3}$$

$$R_C = \frac{R_1R_2 + R_2R_3 + R_1R_3}{R_1}$$

For π to T:

$$R_1 = \frac{R_A R_B}{R_A + R_B + R_C}$$

$$R_2 = \frac{R_B R_C}{R_A + R_B + R_C}$$

$$R_3 = \frac{R_A R_C}{R_A + R_B + R_C}$$

▼ *Example 8-13:*

Convert the pi network in **Figure 8-22** to a tee network.

$$R_1 = \frac{100 \times 400}{100 + 400 + 100}$$
$$= 66.7 \ \Omega$$

$$R_2 = \frac{400 \times 100}{100 + 400 + 100}$$
$$= 66.7 \ \Omega$$

$$R_3 = \frac{100 \times 100}{100 + 400 + 100}$$
$$= 16.7 \ \Omega$$

▲

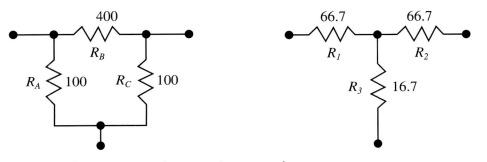

Figure 8-22. Pi-to-tee network conversion example.

Application of Tee and Pi Conversions

In **Figure 8-23A,** the combination of R_2, R_5, and R_4 makes a pi network. In **Figure 8-23B,** this network is converted to a tee network. The series resistances are added, changing the circuit to **Figure 8-23C.** With the R_1 and R_2 combination now connected in parallel, the equivalent resistance is calculated. As shown in **Figure 8-23D,** the equivalent resistance of 186 Ω is in series with R of 133.3 Ω. The total resistance of the circuit is 319.3 Ω. The current 50 mA is calculated. Using the current and the calculated resistances, the individual voltages and currents can be calculated.

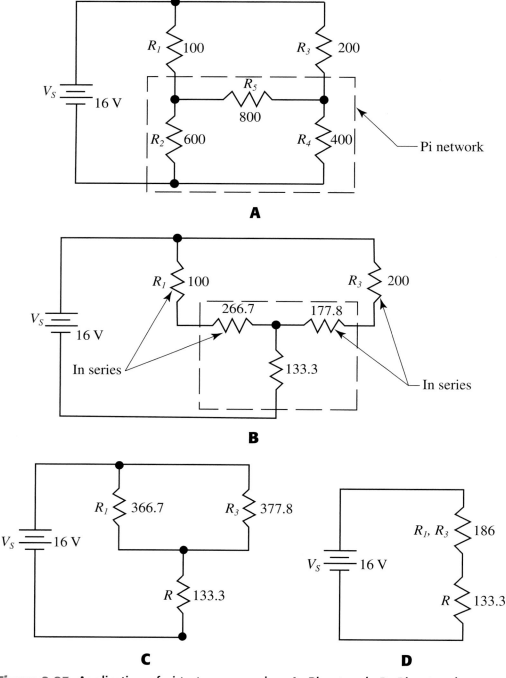

Figure 8-23. Application of pi-to-tee conversion. A—Pi network. B—Pi network converted to tee network. C—Circuit changes as series resistances are added. D—Circuit changes as the conversion is finalized.

 **Summary**

- When doing circuit analysis, always follow the circuit rules and consider what is known and unknown.
- The difference between a series circuit and a parallel circuit is the number of current paths.
- Voltage is applied *across* the circuit and current *flows* into the parts of the circuit.
- Kirchhoff's voltage law states the algebraic sum of all the voltages around a circuit is equal to zero.
- Kirchhoff's current law states at any junction of conductors the algebraic sum of all the currents in a conductor is equal to zero.
- Maximum power is transferred to a load when the source resistance is equal to the load resistance.
- A bridge circuit can be either balanced or unbalanced.
- The Wheatstone bridge is a type of bridge circuit used to measure unknown resistance.
- The solution of complex circuits can be simplified by using the tee to pi conversions.

 Important Terms

Do you know the meanings of these terms used in the chapter?

balanced bridge
Kirchhoff's current law
Kirchhoff's voltage law
maximum power transfer
parallel circuit

pi network
tee network
unbalance bridge
voltage divider equation
Wheatstone bridge

Questions and Problems

Please do not write in this text. Write your answers on a separate sheet of paper.

1. State the three rules for series circuits.
2. State Kirchhoff's voltage law.
3. State the three rules for parallel circuits.
4. State the node rule.
5. State Kirchhoff's current law.
6. Solve for the voltage drops, current, power of the individual resistors, and total power in **Figure 8-24.**
7. Solve for the voltage drops, current, power of the individual resistors, and total power in **Figure 8-25.**
8. Solve for the current through resistor R_5 in **Figure 8-26.**
9. Solve for the voltage drops in **Figure 8-27** using the voltage divider equation.
10. Solve for the currents in **Figure 8-28** using the current divider equation.
11. Is the bridge circuit in **Figure 8-29** balanced or unbalanced? Explain.
12. Using pi to tee conversions, solve Figure 8-29 for the total resistance and current.
13. What is the current through R_3 in **Figure 8-30?**
14. What is the current through R_5 in Figure 8-30?

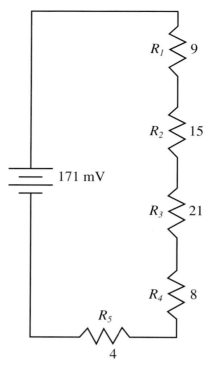

Figure 8-24. Circuit for problem 6.

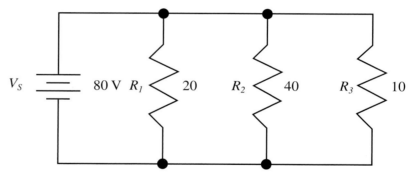

Figure 8-25. Circuit for problem 7.

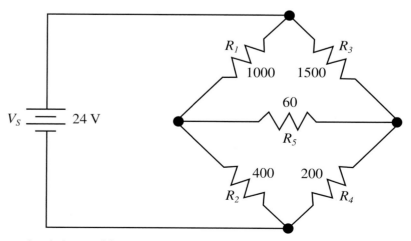

Figure 8-26. Circuit for problem 8.

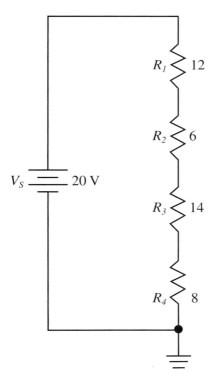

Figure 8-27. Circuit for problem 9.

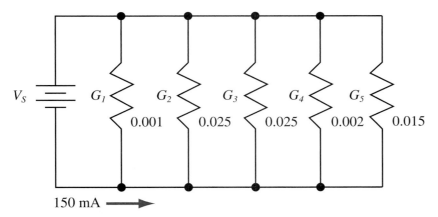

Figure 8-28. Circuit for problem 10.

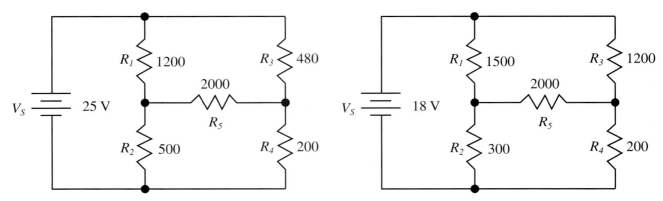

Figure 8-29. Circuit for problem 11.

Figure 8-30. Circuit for problems 13 and 14.

Chapter 9 Graphic Overview

Potentiometer

types are:
- Ganged
- Switch Attached
- 1:1
- 2:1
- 10:1
- 15:1

is a **Variable Resistor**

may be **Linear** or **Tapered**

Cells and Batteries

may be:
- **Secondary** — are **Rechargeable** — such as:
  - Lead-acid
  - Nickel-cadmium
  - Edison Cell
  - Alkaline Cell
- **Primary** — are **Nonrechargeable** — such as:
  - Carbon-zinc Cell
  - Mercury Cell
  - Lithium Cell
  - Alkaline Cell

produce **Direct Current** from a **Chemical Reaction**

Switch

may be:
- Slide
- Toggle
- DIP
- Push-button
- Lever
- Rotary
- Single-pole Single-throw (SPST)
- Double-pole Single-throw (DPST)
- Single-pole Double-throw (SPDT)
- Normally Open (N.O.)
- Normally Closed (N.C.)

CIRCUIT COMPONENTS

9

Objectives

After studying this chapter, you will be able to:
- ○ Identify various circuit components.
- ○ State typical problems associated with components.
- ○ Describe the testing procedures used with different components.

Introduction

Electrical circuits consist of such components as switches, fuses, miniature lamps, batteries, connectors, and potentiometers. This chapter will discuss typical problems associated with components and the testing of each component.

Switches

A switch is a device that controls an electrical circuit. There are six basic types of switches, **Figure 9-1.** Switches differ in appearance and operation. However, they have the same function in a circuit, to control the flow of current.

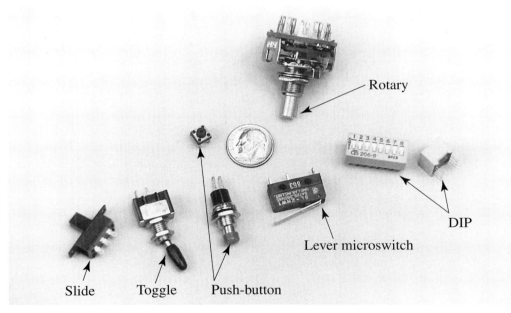

Figure 9-1. Various types of switches. (Dime shown for size comparison.)

Slide Switches

A *single-pole, single-throw* switch (SPST) controls only one current path (single pole), and the switch is either off or on (single throw). See **Figure 9-2.**

Single-pole: A switch or relay arrangement in which only one circuit can be controlled.

Single-throw: A switch or relay term used to indicate the number of different circuits a pole can control.

Figure 9-2. SPST slide switches are often found in electronic equipment.

Double-pole: In switch or relay contacts, indicates two circuits are controlled at the same time.

Double-throw: In switch or relay contacts, indicates switching will occur in two directions.

A *double-pole, single-throw* (DPST) switch controls two circuit paths at the same time, and the switch is either off or on. See **Figure 9-3.** When the switch is closed, both contacts move to turn the lamp and fan on. When the switch is open, current does not pass through either circuit path, and the lamp and fan are off.

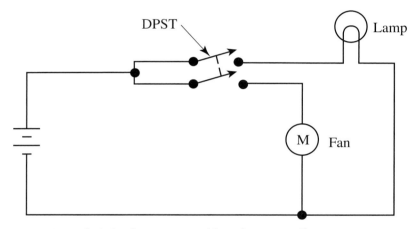

Figure 9-3. DPST switch in the open position does not allow current to pass through either circuit.

Three-pole: A switch or relay contact arrangement that can control three separate circuits at the same time.

Momentary switch: A switch that automatically returns to its original position after being activated.

Normally open: A type of switch or relay contact that closes when activated.

Normally closed: A type of switch or relay contact that opens when activated.

A *single-pole, double-throw* (SPDT) switch controls two circuits, but only one circuit at a time is on. See **Figure 9-4.** This type of switch has a center OFF position that turns off both circuits. **Figure 9-5** illustrates an application of the SPDT switch. Placing a SPDT switch in two locations can, for example, control a light from both sides of a room or from the top and bottom of a stairway.

A *three-pole, single-throw* (3PST) switch controls three circuits at the same time. See **Figure 9-6.**

A *double-pole, double-throw* (DPDT) switch has two ON positions, **Figure 9-7.** The switch is on in position A, controlling circuits 1 and 3; at the same time it is off for circuits 2 and 4. In position B, the switch is off for circuits 1 and 3, but it is on for circuits 2 and 4.

Toggle and Push-button Switches

Toggle and push-button switches come in standard sizes for all types of circuit switching, **Figure 9-8.** The toggle switch has a moveable switching lever. The push-button switch has a center button that operates the switching mechanism.

Another kind of switching device is the *momentary switch.* The contacts close or open when the button is pushed, then return to their original position when the button is released. In the original, at-rest position the switch contacts are said to be *normally open* (N.O.) or *normally closed* (N.C.), **Figure 9-9.**

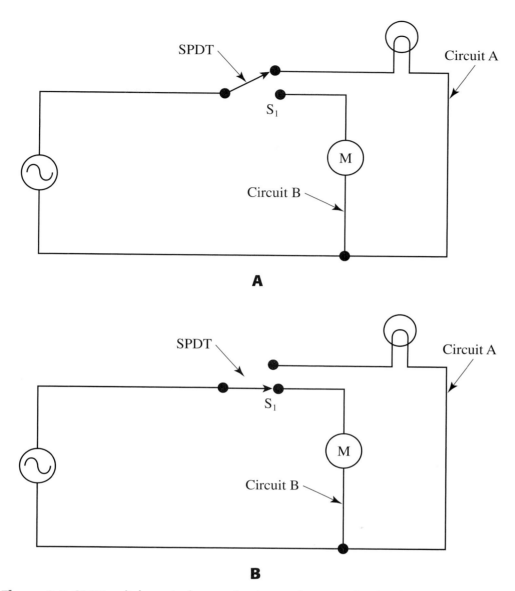

Figure 9-4. SPDT switch controls one circuit at a time. A—Circuit A is on and circuit B is off. B—Circuit A is off and circuit B is on.

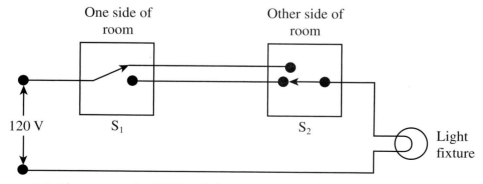

Figure 9-5. Placement of a SPDT switch on opposite sides of a room allows the light to be turned on and off from either location.

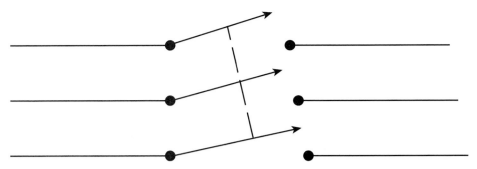

Figure 9-6. Schematic symbol for 3PST switch.

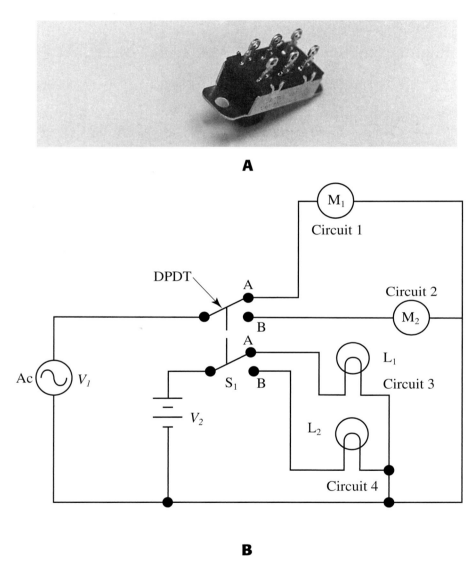

A

B

Figure 9-7. DPDT switch. A—Terminals of a DPDT slide switch. B—One pole is controlling dc lamp circuits, while the other is controlling ac motor circuits.

Lever Microswitch

The lever microswitch is a small SPST or SPDT switch, **Figure 9-10.** It is used in small current applications. For instance, the lever switch is used to detect the opening and closing of safety covers and to limit the travel of a mechanical device such as an automatic door.

Figure 9-8. Toggle and push-button switches.

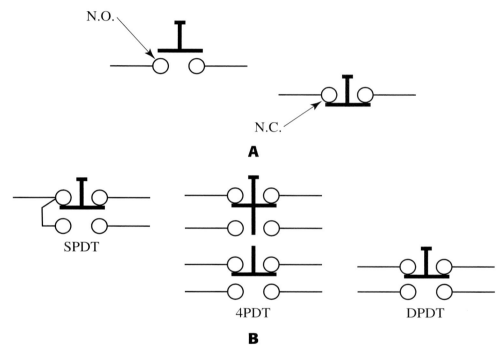

Figure 9-9. Momentary switch contacts can be normally open or normally closed. A—SPST contacts. B—Other switch contact arrangements.

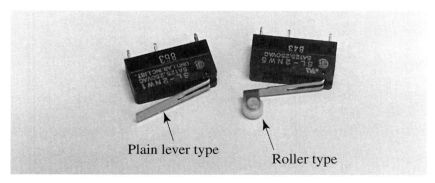

Figure 9-10. Lever microswitches. The roller type is used when the mechanism operating the switch, such as a door, has considerable movement.

Dual In-line Package Switch

The dual in-line package (DIP) switch is used on printed circuit boards for setting circuits for a specific operation or test setup. See **Figure 9-11.** A garage door opener transmitter can have a DIP switch to change the transmitting code for a particular receiver.

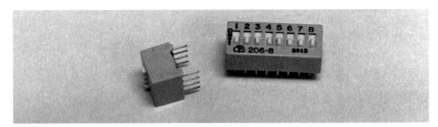

Figure 9-11. Dual in-line package (DIP) switches are used extensively in the computer industry.

Rotary Switch

Most rotary switches are the multiposition and *nonshorting* type, **Figure 9-12.** As the switch is moved to a new position, the common terminal breaks contact with the old terminal just before it closes the new contact. This is called ***break-before-make*** switch action, **Figure 9-13.** Some multiposition switches are the ***shorting*** type. The old terminal makes contact with the new terminal before the switch breaks contact with the old. This is called ***make-before-break*** switch action. Rotary switches can have any number of wafer segments controlling many different circuits at the same time. Troubleshooting these switches presents a challenge, even for an experienced technician.

Switch Ratings

Switch ratings are given for resistive loads and indicate the maximum current and voltage the switch can interrupt without damage to the switch contact surfaces. For example, if a switch is rated 4 A at 120 V ac, it can safely interrupt a maximum of 4 A when the source is 120 V ac. When an inductive load is operating, the current is more difficult to interrupt. The type of insulation used in the construction of the switch also determines the voltage rating of a switch. A switch should not be used in a circuit with a higher voltage than it can safely handle.

Common Switch Problems

Switch problems are usually mechanical. The switch actuator may not operate the switch, producing an open circuit. The switch contacts may be burned. The switch may have carried more current than the contact ratings allowed, or it simply may have failed due to age.

Nonshorting: In switch or relay contacts, contacts that break (open) completely before another set of contacts make (close).

Break-before-make: Refers to contacts that open before another set of contacts close.

Shorting: In switch or relay contacts, the new contacts make continuity before the old contacts break continuity.

Make-before-break: In switch or relay contacts, one set of contacts makes or closes its circuit before the other set breaks or opens.

Figure 9-12. A single wafer rotary switch.

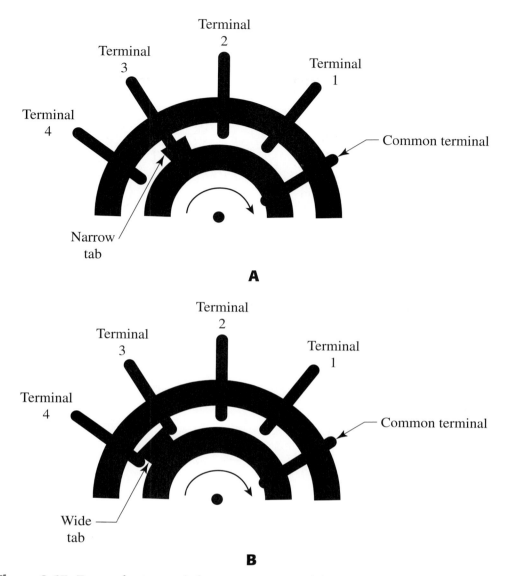

Figure 9-13. Types of rotary switch contacts. A—Break-before-make contacts (nonshorting). B—Make-before-break contacts (shorting).

Testing Switches

Switches can be tested using a voltmeter or an ohmmeter. With a voltmeter, when the switch is closed, no voltage is indicated on the meter. When the switch is open, the maximum voltage is indicated. See **Figure 9-14.**

The second method of testing uses an ohmmeter. With no power on the circuits, measure across the switch with the ohmmeter. When the switch is closed, the ohmmeter will indicate zero ohms; when the switch is open, the ohmmeter will indicate infinity. See **Figure 9-15.**

Circuit Safety Devices

Devices are used to protect equipment (and life) from excessive current flow. Fuses and circuit breakers are examples of overcurrent protection devices.

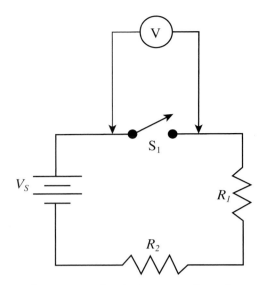

Figure 9-14. Using a voltmeter to check the operation of a switch. When the switch is open, the meter should indicate a voltage.

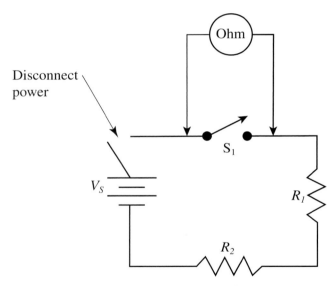

Figure 9-15. Using an ohmmeter to check a switch for continuity. When the switch is open, the ohmmeter will indicate infinity.

Fuse

A fuse is a safety device that protects the circuit from excessive current flow, **Figure 9-16.** When the current flowing in the circuit is greater than the current rating of the fuse, the conductor inside the fuse melts. This produces an open and stops the current flow in the circuit. A blown fuse indicates a circuit overload or overcurrent. An overload can be caused by a short circuit or by too many loads connected to the circuit.

Fuses are rated by the amount of current (amps) they can safely carry and the amount of voltage they can handle. As a general rule, a fuse is selected at 125% of the full load current of the protected circuit. Never replace a blown fuse with one of a larger current rating.

 Caution: An electrical fire can result from a current greater than the circuit conductors are able carry.

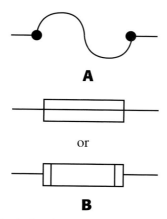

Figure 9-16. Schematic symbols for fuse. A—Common symbol. B—Industrial symbols.

Types of Fuses

Fuse types include glass body, cartridge, screw-in, or spade plug-in, **Figure 9-17.** The widely used glass body type can be fast-action, normal, or slow-blow. These are indicators of how quickly the fuse reacts to an overcurrent, **Figure 9-18.** The slow-blow type fuse is identified by the coiled or resistive element inside the glass. It may also have a ceramic body.

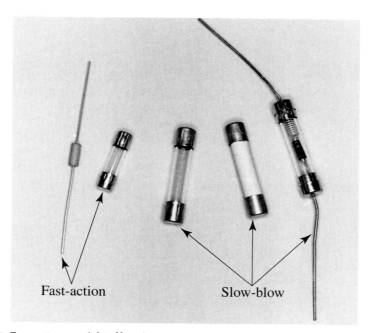

Figure 9-17. Fuses types. (Littelfuse)

Fuse Type	Blow Time
Fast-action	1 ms
Normal	5 ms
Slow-blow	30 ms

Figure 9-18. Approximate fail or blow time for fuses.

Both screw-in and cartridge fuses are used in residential wiring. Some cartridge fuses have a replaceable fuse element. The spade plug-in fuse is commonly used in the automotive industry.

Testing Fuses

To test a fuse, remove it from the circuit holder and test it for continuity. A good fuse will measure close to zero ohms, while an open fuse will measure infinity. Although a fuse may look good, only a meter test will verify its true quality.

> **Warning:** Fuses in an operating circuit can be hot enough to burn the skin even when operating at the fuse current rating.

Circuit Breaker

Circuit breaker: A safety device that automatically opens a circuit if it is overloaded.

A *circuit breaker* is another type of safety device, **Figure 9-19.** It has a mechanical switch contact controlled by either a thermal or a magnetic device. A circuit breaker does not have to be replaced when it opens. By pressing a button, the circuit breaker is reset and continuity restored. **Figure 9-20** shows a circuit breaker used in industrial and residential control panels. To reset this type of circuit breaker, the breaker must be pushed to the OFF position, then returned to the ON position.

Figure 9-19. Circuit breakers are available in many different types and current ratings.

Figure 9-20. A circuit breaker used in residential or industrial applications.

Miniature Lamps

Miniature lamps are used to indicate the power is on or an operation is taking place, **Figure 9-21.** Although light-emitting diodes have replaced lamps as indicators, miniature lamps are still in use.

Figure 9-21. Various types of miniature lamp bases.

Miniature lamps are identified by standard bulb numbers. Bulbs should not be replaced solely by matching the voltage and base. Current is also a major factor in replacement. Suppose a No. 47 bulb is to be replaced, and all you have is a No. 44. Because the No. 47 is rated at 150 mA and the No. 44 is rated at 250 mA, it would be unwise to use the No. 44. Although the No. 44 has the same voltage and base, it requires much more current to operate. The circuits may not be designed to supply the additional current. See Appendix F for a listing of miniature lamps and their characteristics.

Cells and Batteries

A *cell* is a single unit that produces a direct current voltage by converting chemical energy into electrical energy. Cells are either primary or secondary. A *primary cell* is nonrechargeable; a *secondary cell* can be recharged. Cells produce direct current from the chemical reaction of an electrolyte and two types of metal electrodes. The electrolyte may be a chemical solution or a paste.

Primary Cells

Primary cells are also called *dry cells.* Carbon-zinc, mercury, alkaline, and lithium cells are types of dry cells.

Carbon-zinc Cell

One common dry cell is made up of carbon and zinc electrodes. The electrolyte is an ammonium chloride mixture in paste form. This is the chemical composition of the common flashlight cell. Although called a flashlight battery, technically the source is a cell. A *battery* is made up of two or more cells connected in series, parallel, or combination.

The carbon-zinc cell is made up of a carbon center rod, which is the positive terminal of the cell, **Figure 9-22.** The outer container is a zinc can coated with a plastic cover. The electrolyte is a mixture of ammonium chloride, manganese dioxide, and granulated carbon.

During discharge of the cell, hydrogen bubbles collect around the carbon rod in an action called *polarization.* The manganese dioxide acts as a depolarizer, combining with the hydrogen bubbles to produce water and other by-products. Without the manganese dioxide, the hydrogen bubbles would collect around the carbon rod and shorten the life of the cell.

Mercury Cell

The mercury cell is a widely used dry cell available in different packages, **Figure 9-23.** The positive terminal consists of mercuric oxide. The negative terminal is the electrolyte, which is made up of zinc powder and potassium hydroxide.

Cell: A single unit that produces a direct current voltage by converting chemical energy into electrical energy.

Primary cell: A nonrechargeable electric cell that produces electric current through a chemical reaction.

Secondary cell: A rechargeable electric cell.

Dry cell: A voltage-generating cell in which the electrolyte is absorbed into a paper product or made into a paste.

Battery: A dc voltage source consisting of two or more cells.

Polarization: The increased resistance of an electrolytic cell as the potential of an electrode changes during electrolysis.

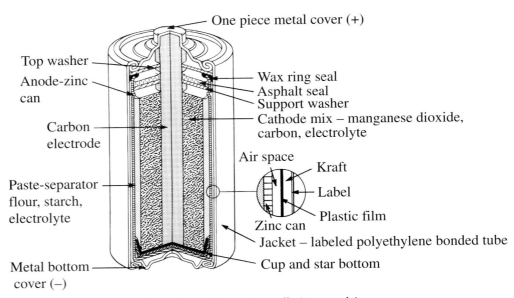

Top washer
Anode-zinc can
Carbon electrode
Paste-separator flour, starch, electrolyte
Metal bottom cover (−)

One piece metal cover (+)
Wax ring seal
Asphalt seal
Support washer
Cathode mix – manganese dioxide, carbon, electrolyte
Air space
Kraft
Label
Plastic film
Zinc can
Jacket – labeled polyethylene bonded tube
Cup and star bottom

Figure 9-22. Cutaway view of a carbon-zinc cell. (Eveready)

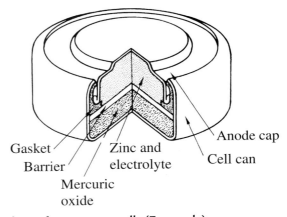

Gasket
Barrier
Mercuric oxide
Zinc and electrolyte
Anode cap
Cell can

Figure 9-23. Cutaway view of a mercury cell. (Eveready)

The mercury cell is small and has a good shelf life. It also has a reliable terminal voltage; that is, the cell voltage remains constant over the life of the cell. One disadvantage of the mercury cell is its high cost.

Alkaline Cell

The alkaline cell is constructed like the carbon-zinc cell, but it uses a potassium hydroxide mixture for the electrolyte. The alkaline cell has a higher current capacity and a longer shelf life than the carbon-zinc cell.

 Warning: Any leakage of the electrolyte from an alkaline cell is caustic and should be cleaned up immediately.

Lithium Cell

Lithium cells are used in watches, alarm systems, cameras, hearing aids, and other equipment. Although lithium cells are expensive, they are small, lightweight, and have a long life. See **Figure 9-24.**

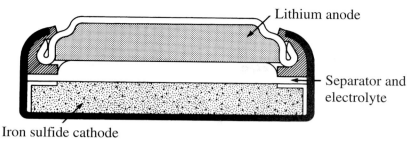

Lithium anode

Separator and electrolyte

Iron sulfide cathode

Figure 9-24. Cutaway view of a lithium cell. (Eveready)

Silver-oxide Cells

Silver-oxide cells are used in hearing aids and watches. Compared to other button-type cells, they have the highest electrical capacity and resistance to leakage. They have a stable discharge voltage of 1.5 V.

Secondary Cells

Many secondary cells are designed from the early voltaic cell. The voltaic cell consists of two different metals immersed in a diluted acid, **Figure 9-25.** It is called a *wet cell.* When the copper and zinc metal plates are placed into the acid electrolyte, a chemical reaction takes place between the metals and acid. Other chemicals can be used as the electrolyte of the wet cell. The actual chemical process is beyond the scope of this textbook.

Wet cell: A voltage source cell that contains a liquid electrolyte.

Nickel-cadmium Battery

The nickel-cadmium (ni-cad) battery is a secondary cell used in high "recurrent" applications where the battery is partly discharged and recharged many times. The ni-cad cell is made of nickel plates. The positive plate has a nickel salt solution, and the negative plate has a cadmium salt solution. The electrolyte is potassium hydroxide. Ni-cad cells have a low internal resistance and can deliver high current to a load.

Edison Cell

The Edison cell is a rechargeable cell used in heavy industrial applications. The negative plates are made of iron oxide and the positive plates of nickel oxide. The electrolyte is potassium hydroxide. Although more expensive than the lead-acid battery, the Edison cell is lighter, has a long life, and can be recharged many times.

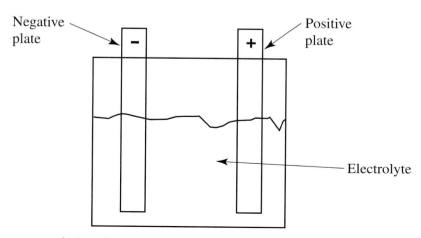

Negative plate

Positive plate

Electrolyte

Figure 9-25. A voltaic cell uses two different metals in a dilute acid.

Lead-acid Battery

The lead-acid battery consists of lead peroxide positive plates, lead negative plates, and a sulfuric acid electrolyte. See **Figure 9-26.**

The strength of the electrolyte is measured by its specific gravity. *Specific gravity* is the density ratio of a substance compared to water. It is measured using a hydrometer, **Figure 9-27.** The specific gravity of water is 1.00, while sulfuric acid has a specific gravity of 1.84. The electrolyte of the lead-acid battery is a diluted sulfuric acid and water solution.

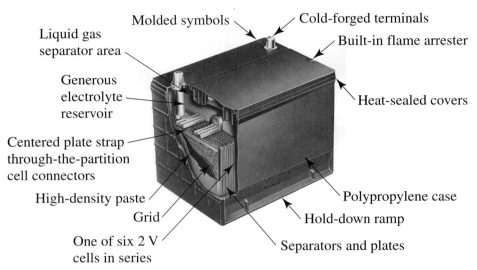

Molded symbols · Cold-forged terminals
Liquid gas separator area · Built-in flame arrester
Generous electrolyte reservoir
Heat-sealed covers
Centered plate strap through-the-partition cell connectors
High-density paste · Polypropylene case
Grid · Hold-down ramp
One of six 2 V cells in series · Separators and plates

Figure 9-26. An automotive 12 V lead-acid battery. (Delco-Remy Div., General Motors Corp.)

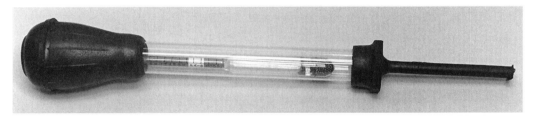

Figure 9-27. A hydrometer measures the specific gravity of the electrolyte in a lead-acid battery.

Lead-acid battery operation. When the lead-acid battery is at full charge, the positive plates are dark brown lead peroxide, and the negative plates are gray lead. The specific gravity of the electrolyte is between 1.27 and 1.30. The no-load cell voltage is about 2.2 V.

As the battery discharges, the negative plates deteriorate and form lead sulfate. The positive plates deteriorate and form lead sulfate and water. This dilutes the acid and causes the specific gravity to decrease.

When the battery has fully discharged, both the positive and negative plates are changed to lead sulfate. The specific gravity drops below 1.15, and the loaded cell voltage drops below 1.75 V. If the battery is not recharged for a long time, a condition called *sulfation* occurs. This happens when the lead sulfate forms a hard mass and the chemical reversal process becomes impossible.

Overcharging and overdischarging. Overcharging is the most destructive factor in the life of a battery. A battery needs to be charged to a point of returning all the sulfate to the electrolyte. Once this occurs, excess charging current causes the peroxide on the plates to be "stripped off" and collect at the bottom of the battery. Once removed, the peroxide is no longer active in the battery, and the charge capacity of the battery is reduced. Overcharging a sealed, maintenance-free battery can have the same results.

Overdischarging usually is caused by incomplete charging or insufficient battery (amp) capacity. A discharge greater than 50% shortens the cycle life of the battery. Incomplete charging causes overdischarging and sulfation.

Battery maintenance. Most automobile lead-acid batteries are maintenance-free. They contain lead, sulfuric acid, and other chemicals that prevent the loss of water and generation of hydrogen gas.

Maintenance and testing are required for batteries used in equipment such as fork-lifts, automated guided vehicles, railroad signals, emergency lighting, security and fire alarm systems, airline ground vehicles, and mine vehicles. **Figure 9-28** shows a large battery used in an electric crane or forklift.

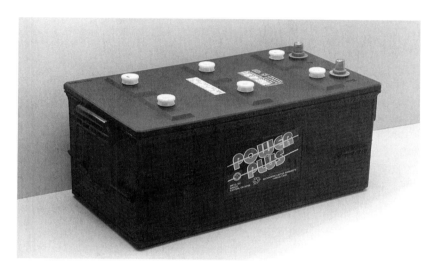

Figure 9-28. This large battery is 11″ wide, 20″ long, and 10″ high.

The lead plates in a lead-acid storage battery should always be covered with the liquid electrolyte. When exposed to air, the plates become oxidized and inactive to the electrolyte. Be sure to use distilled water to replace the liquid content. City tap water contains minerals and chemicals which will shorten the life of a lead-acid battery. Place a coating of gear grease or similar protectant on the battery terminals to prevent terminal corrosion.

 Caution: Do not charge a lead-acid battery too fast. Doing so will warp the plates and cause excess gassing and overheating.

When a battery is to be stored for a long time, attach a trickle charger on the battery to prevent sulfation. A trickle charger supplies a few milliamps of current to the battery and maintains the full charge.

Testing batteries. Figure 9-29 lists some of the specific gravity readings taken for a lead-acid battery at approximate levels of charge. When using a hydrometer to test a battery, be careful acid does not drip onto your hands, clothing, or other items.

Level of Charge	Specific Gravity
100%	1.26 – 1.28
75%	1.23
50%	1.20
25%	1.17
0%	1.15
Water	1.00

Figure 9-29. Specific gravity readings and approximate level of charge using a hydrometer.

A heavy load tester is also used to test a lead-acid battery. Such a tester provides a load of 100 A or more for a short period of time. If the voltage under load falls below 2 V, the meter pointer indicates a bad battery condition.

A primary cell can be tested simply by measuring the cell voltage. Sometimes conditions may indicate a cell is good when actually it is bad. Under a no-load condition the cell voltage is normal, but under a loaded condition the internal resistance of the cell may be high enough to cause the voltage to drop to almost zero. To check a cell, put it under a load and measure the cell voltage.

Battery disposal. Batteries that are no longer usable must be disposed of properly. Because of their poisonous effect on the natural environment, battery disposal is regulated by the federal government. Any chemical battery placed into landfills or just "dumped" will have an adverse effect on living organisms. Lead and other hazardous chemicals may find their way into drinking water supplies. Lead poisoning can cause brain damage and other illnesses in humans and animals.

Stores that sell new batteries usually take the old ones on trade. The old batteries are picked up by a battery company and taken to a central processing center where they are melted down for recycling. Because ni-cad batteries are made of different materials, they are separated from lead-acid batteries. A battery warehouse may even collect ordinary flashlight cells for processing.

Figure 9-30 lists some common types of cells and their voltages.

Solar Cells

The solar cell is a special cell made up primarily of silicon semiconductor materials, **Figure 9-31.** It typically has 0.3 V and a very low current rating.

Connectors

Connector: A coupling device that provides an electrical and mechanical connection or disconnection between one or more wires or between a cable and a piece of equipment.

A **connector** is a coupling device that provides an electrical and mechanical connection or disconnection between one or more wires or between a cable and a piece of equipment. Connectors are used in all areas of electricity and electronics.

Connectors can be classified into four groups: audio frequency (AF), radio frequency (RF), computer communications, and printed circuit board.

Type of Cell	Voltage
Carbon-zinc	1.5
Mercury	1.35
Alkaline	1.5
Lithium	3.0
Silver-oxide	1.5
Nickel-cadmium	1.2
Nickel-iron	1.4
Lead-acid	2.2

Figure 9-30. Common cells and their voltages.

Figure 9-31. A solar cell converts light energy to useful electrical energy.

Audio Frequency Connectors

Audio frequency (AF) connectors are used for audio amplifiers, compact disc and audio tape players, microphones, and other audio equipment. Phone plugs and jacks, phono (RCA) plugs, and XLR connectors are described here.

Phone Plugs and Jacks

One of the most commonly used audio connectors is the phone plug and jack combination, **Figure 9-32.** Phone plugs come in standard 1/4″ and 1/8″ diameter sizes. The connectors may be two-conductor (mono) or three-conductor (stereo), as indicated by the end of the plug, **Figure 9-33.** The large portion of the plug is the common ground and counts as one of the conductors. The schematic symbols for phone plug and jack are shown in **Figure 9-34.**

Figure 9-32. Common phone plugs are standard 1/4″ or 1/8″ diameter.

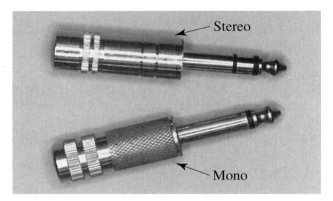

Figure 9-33. Phone connectors can be the two- or three-conductor type.

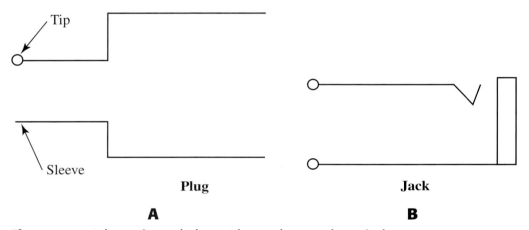

Figure 9-34. Schematic symbols. A—Phone plug. B—Phone jack.

Phone jacks can be either closed or open circuit, **Figure 9-35.** The closed circuit jack is used in speaker connections. When the phone plug is not in the jack, the speaker is connected to the output circuit of the jack. When the phone plug is inserted in the jack, it opens speaker circuit 1 and connects the plug circuit to the output circuit of the jack, speaker 2. See **Figure 9-36.**

A common problem with phone plugs is a broken conductor near the back end or inside the plug. The break is usually caused by pulling on the wire when disconnecting the plug from the jack.

Phono (RCA) Plug

Another commonly used audio connector is the phono (RCA) plug, **Figure 9-37.** It is used to connect speakers or compact disc or tape players to amplifiers, and for "line level" audio or video assemblies.

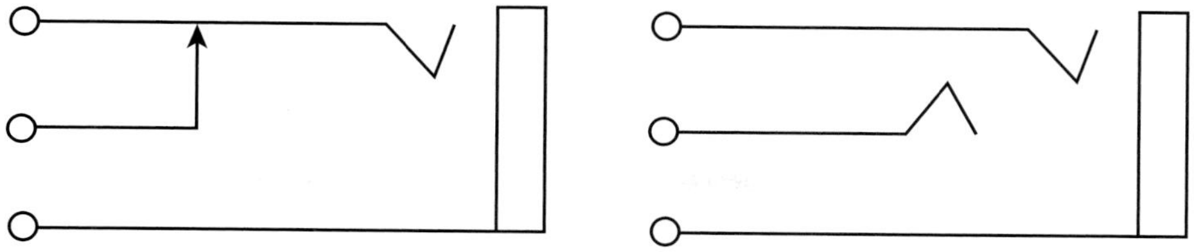

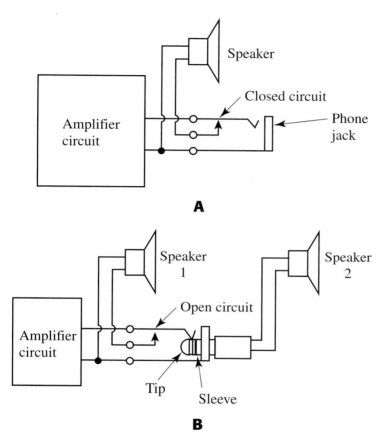

Closed Circuit   **Open Circuit**

Figure 9-35. Phone jacks can be closed or open circuit.

A

B

Figure 9-36. Closed circuit jack. A—The speaker operates when the plug is not inserted into the jack. B—The speaker is disconnected when the plug is inserted into the jack.

Figure 9-37. RCA phono plugs are used extensively in the consumer electronics industry.

XLR Connector

The XLR is used extensively in the audio industry for microphone and "line level" assembly connections, **Figure 9-38.** XLR connectors are used in balanced audio systems to help reduce noise and hum.

The XLR connector has a keyway that enables the connector to correctly line up the pins before the plug enters the socket. The socket has a locking mechanism that

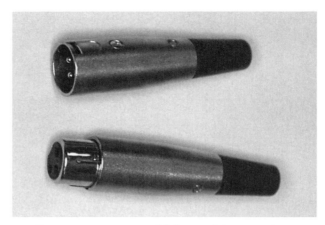

Figure 9-38. XLR audio connectors are widely used in sound systems.

keeps the connectors from separating accidentally. When disconnecting XLR connectors, hold the locking button in while pulling the connectors apart. *Do not pull on the wires.* This bad habit causes most of the problems with XLR connectors.

The XLR connector has three conductors numbered 1, 2, and 3, **Figure 9-39.** The cable connected to the XLR connector should be a three-wire shielded cable. The audio signal is taken from pins 2 and 3. Pin 1 is ground; pin 2 is the high (in-phase) signal conductor; pin 3 is the low (return or out-of-phase) signal conductor. The outer metal case of the XLR connector is the equipment ground and is connected to the shield of the cable. The XLR connector is also available in a five-pin arrangement.

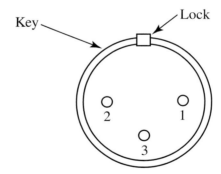

Figure 9-39. Terminals of XLR connector. Be careful not to reverse or exchange these connections.

 Caution: Do not reverse or exchange XLR connections. Doing so could result in a hum, an inoperable audio system, or even a shock.

Radio Frequency Connectors

RF connectors are used between video, television, FM, and UHF systems. The following are some common RF connectors. See **Figure 9-40.**

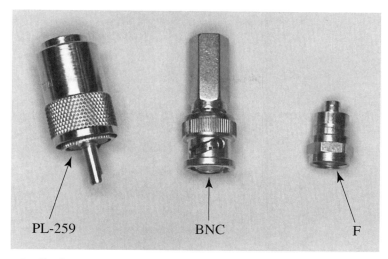

Figure 9-40. Radio frequency connectors.

PL-259 Connector

One of the most popular RF connectors is the PL-259 UHF connector. It is used in video systems, citizen band radios, televisions, and other high-frequency systems. The PL-259 comes in both soldered and solderless types.

BNC Connector

The BNC connector is another popular RF connector. It is used mostly in video, RF signal circuits, and test equipment. The small BNC connector is prone to broken wires from twisting and pulling on the wire cable instead of the connector body. Soldered, solderless, and crimped types are available.

F-connector

A popular connector in television cables, the F-connector is used extensively to interconnect television receivers, VCRs, and other equipment. The center conductor should be carefully inserted into its socket, never forced. Forcing can cause the center conductor to bend and eventually break off when you try to straighten it. The conductor may also short out the signal to the outer shield. F-connectors are usually a crimp-type assembly.

Computer Communications Connectors

The computer industry produces a set of connectors for the interconnection of equipment. The most often used connectors are D-connectors, DIN connectors, and various adaptors.

D-connector

The D-connector has a standard 9- or 25-pin configuration, **Figure 9-41.** Use care when inserting the D-connector into its socket as the pins are small and bend easily. If a D-connector will not go together, check for bent pins. Even the smallest misalignment of pins can prevent insertion. Unfortunately, when a D-connector pin is bent, any attempt to straighten it can cause the pin to break off. Computer communication using the standard RS-232 D-connector is shown in **Figure 9-42.**

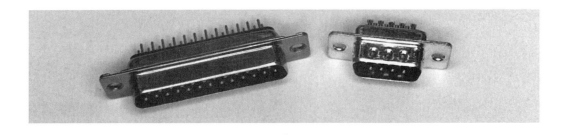

A

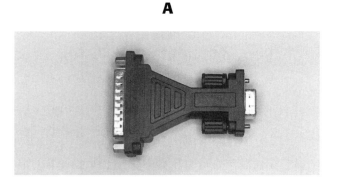

B

Figure 9-41. Common D-connectors. A—A 25-pin and 9-pin D-connector. B—A 9- to 25-pin adaptor.

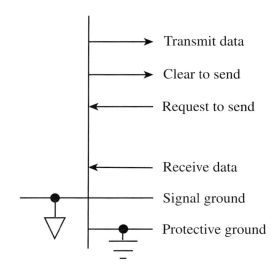

Figure 9-42. The RS-232 is used for communication between a computer and other equipment.

DIN Connector

DIN connectors are used to connect equipment such as a computer keyboard, mouse, page scanner, and graphic tablet, **Figure 9-43.** Extreme care should be used when plugging the DIN connector into its socket so the pins do not bend.

Printed Circuit Board Connectors

Often connectors for personal computers connect to the edge of the printed circuit board, **Figure 9-44.** Ribbon cables interconnect the printed circuit board with other boards or equipment, **Figure 9-45.** Connector headers may be protected or unprotected, **Figure 9-46.**

Figure 9-43. The pins of DIN connectors are susceptible to bending or breakage.

Figure 9-44. A printed circuit edge connector.

Schematic Symbols

Schematic symbols for connectors can vary depending upon the manufacturer. **Figure 9-47** shows some standard symbols for connectors.

Connector Problems

Connector problems, other than bent pins, often can be traced to broken conductors. A continuity test can locate a broken conductor. Intermittent operation can be caused by pin-socket connections that have become loose or corroded.

Figure 9-45. Printed circuit board header for a ribbon cable connector.

Figure 9-46. An unprotected connector header on a printed circuit board.

Potentiometers

A potentiometer is a variable resistor with three terminals, **Figure 9-48.** The complete resistive element is connected between the two outer terminals. The maximum resistance is measured across the two outer terminals, regardless of where the moveable arm is located. See **Figure 9-49.** The resistance from the center terminal to either of the outer terminals depends on where the moveable arm is positioned.

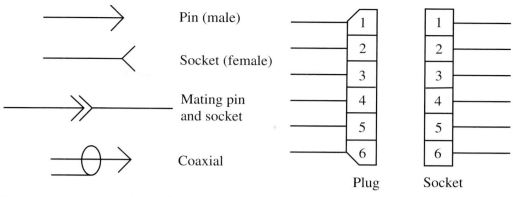

Pin (male)

Socket (female)

Mating pin
and socket

Coaxial

Plug   Socket

Figure 9-47. Schematic symbols for connectors.

Figure 9-48. Potentiometers come in various ohmic values and sizes to accommodate different circuits.

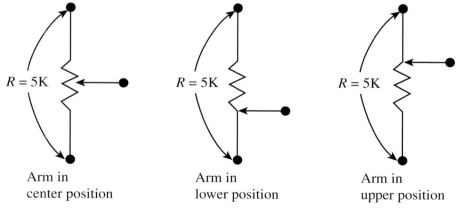

Arm in
center position

Arm in
lower position

Arm in
upper position

Figure 9-49. The total resistance of a potentiometer does not change no matter where the moveable arm is located on the resistive element.

Rheostat

A *rheostat* is a two-terminal variable resistor used to limit the current of a circuit. It has only two connections, one end terminal and the center moveable arm, **Figure 9-50.** When the moveable arm is at the maximum resistance position, a small current flows through the rheostat. When the moveable arm is at the minimum resistance position, a large current flows through the rheostat.

Rheostat: A variable resistor with two terminals, one fixed and one moveable.

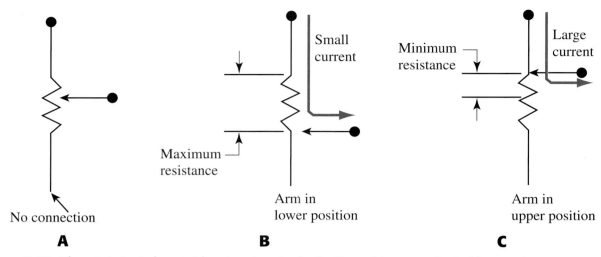

Figure 9-50. Rheostat. A—A rheostat has two terminals. B—Moveable arm adjusted for maximum resistance. C—Moveable arm adjusted for minimum resistance.

Types of Potentiometers

Potentiometers can have a shaft or screw adjustment, **Figure 9-51.** If the shaft of the potentiometer is turned once, the adjustable arm travels across the complete resistive element.

The screw adjustment must be turned 2, 10, 15 (2:1, 10:1, or 15:1) or more turns for the adjustable arm to travel across the complete resistive element. This type of potentiometer is used when careful adjustment of the resistance is necessary. Larger potentiometers may be ganged or have a switch attached, **Figure 9-52.**

Potentiometers are also classified as linear or tapered (nonlinear), **Figure 9-53.** In the ***linear*** potentiometer, the resistance of the moveable arm changes equally as the arm moves. If the shaft were moved 1/8 of a turn from one end of the resistive element, the resistance would change by a certain amount. If the shaft were moved to the center of the resistive element and moved 1/8 of a turn, the resistance would change by the same amount. The resistance changes proportionally with the shaft rotation; that is, the resistance is uniform throughout the length of the resistive element.

In the ***tapered*** potentiometer, the change in resistance is not the same through the length of the resistive element. The resistance varies more toward one end of the resistive element and less toward the other end as the shaft is turned. This type of potentiometer is sometimes called an ***audio taper*** because it is used in such equipment as audio volume control.

Linear: In a potentiometer, the resistance of the moveable arm changes equally as the arm moves.

Tapered: In a potentiometer, the resistance of the resistive element does not vary by the same amount throughout its range.

Audio taper: A control that has a nonlinear decrease in resistance and is used in audio applications.

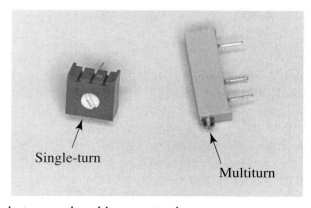

Figure 9-51. Single-turn and multiturn potentiometers.

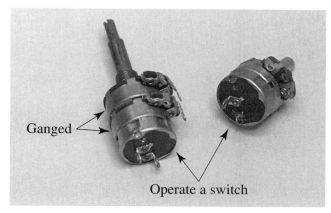

Figure 9-52. Some larger potentiometers are ganged or have more than one element adjusted by the same shaft. Some shafts operate a switch.

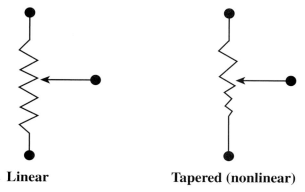

Linear   **Tapered (nonlinear)**

Figure 9-53. Potentiometers can be linear or nonlinear.

Potentiometers are used in circuits as voltage dividers and may consist of wire-wound, carbon, or cermet resistive elements. Potentiometers are rated by resistance value and power dissipation.

Potentiometer Problems

Problems with potentiometers are either mechanical or electrical. Mechanical problems may show up as broken leads connecting the resistive element to the terminals. Sometimes this problem can be corrected by careful soldering or replacement of a rivet. If a repair is not possible, the potentiometer must be replaced.

The moveable arm shaft of the potentiometer should turn freely. If the shaft becomes corroded, it will be difficult or impossible to turn. Placing a couple drops of turbine oil or heavy mineral oil on the shaft will help free it. Use care when applying oils to potentiometers and other components. Use only one or two drops, and allow the potentiometer to soak overnight. If too much oil is used, it can get on the resistive element and cause other problems. Most penetrating oils are petroleum-based and will react with some plastics. Work slowly and carefully to prevent damage or complete destruction of the component.

Dust and corrosion can build up where the moveable arm travels on the resistive element, causing intermittent operation. Control sprays are available to ·clean and lubricate the potentiometer. In some cases, the resistive element is worn or has a bad spot in the travel path. When this occurs, the spray will not cure the problem and the potentiometer must be replaced. Other possible electrical problems are an open terminal connection or a resistive element that has changed value.

Testing Potentiometers

Two methods can be used to test a potentiometer. With the first method, connect an ohmmeter to measure the resistance across the complete resistive element, **Figure 9-54.** Reconnect the ohmmeter and measure from the moveable arm to one of the outer terminals. Turn the shaft of the potentiometer. The ohmmeter should indicate the resistance is changing smoothly through the rotation of the shaft. Then, connect the ohmmeter to the other outer terminal and check for proper indication on the ohmmeter.

The second method is to connect a signal generator across the potentiometer, **Figure 9-55.** Using an ac meter (oscilloscope), measure the voltage at the moveable arm. The voltage should change continuously without jumps in value.

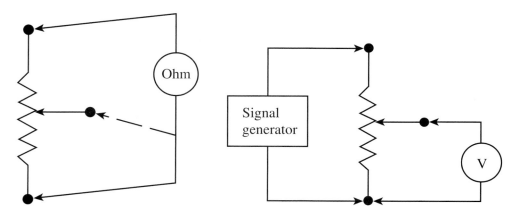

Figure 9-54. Testing a potentiometer using an ohmmeter.

Figure 9-55. Testing a potentiometer using an audio generator and meter.

Summary

- A switch controls the circuit operation. Switch types include slide, toggle, lever microswitch, push-button, DIP, and rotary.
- Switches are classified by the number of poles and throws.
- Switch contacts may normally be open or closed. Rotary switches may be shorting or nonshorting.
- Switches are rated by current and voltage.
- A fuse is a safety device. Fuses are tested using a continuity test.
- A miniature lamp should be replaced with one of the same part number.
- Cells are classified as either primary or secondary. The primary cell cannot be recharged. The secondary cell can be recharged many times.
- Primary (dry) cells include carbon-zinc, mercury, alkaline, and lithium types. Secondary (wet) cells include nickel-cadmium, Edison, and lead-acid types.
- Connectors are classified as audio frequency, radio frequency, computer communications, and printed circuit board.
- Audio frequency connectors include phone, XLR, and phono.
- Radio frequency connectors include PL-259, BNC, and F.
- The computer industry uses D- and DIN connectors.
- Printed circuit boards use edge connectors, protected-unprotected headers, and ribbon cable connectors.
- Connector problems often can be traced to open conductors or bent pins.

- A rheostat is a two-terminal variable resistor.
- Potentiometers are classified as single- or multiturn, linear or audio tapered, wirewound, carbon, or cermet.
- Potentiometers are rated in watts of dissipation and are tested with an ohmmeter.

Important Terms

Do you know the meanings of these terms used in the chapter?

audio taper	normally open
battery	polarization
break-before-make	primary cell
cell	rheostat
circuit breaker	secondary cell
connector	shorting
double-pole	single-pole
double-throw	single-throw
dry cell	specific gravity
linear	sulfation
make-before-break	tapered
momentary switch	three-pole
nonshorting	wet cell
normally closed	

Questions and Problems

Please do not write in this text. Write your answers on a separate sheet of paper.

1. What is meant by DPST, SPDT, 3PST, and DPDT?
2. List six basic types of switches.
3. What do the letters N.O. and N.C. stand for?
4. What is a common switch problem?
5. What type of switch has break-before-make contacts?
6. Describe how you would test a switch with an ohmmeter.
7. Draw the schematic symbol for the switches in question 1.
8. An open fuse will measure _____ on an ohmmeter.
9. List the different types of primary cells and secondary cells.
10. The specific gravity of a lead-acid storage battery measures 1.19. What does this tell about the state of charge on the battery?
11. Why should city tap water *not* be used in a lead-acid battery?
12. What can be done to prevent sulfation of a battery during storage?
13. List the different categories of connectors.
14. What are the most common problems found with connectors?
15. List the popular audio connectors and RF connectors.
16. What is the standard RS-232 connector used for?
17. What kind of test is commonly used in locating problems with connectors?
18. What is a linear potentiometer? A tapered potentiometer?
19. What are the two types of potentiometer problems?
20. What kind of oil should be used on a potentiometer shaft? Why should penetrating oil *not* be used?

Chapter 10 Graphic Overview

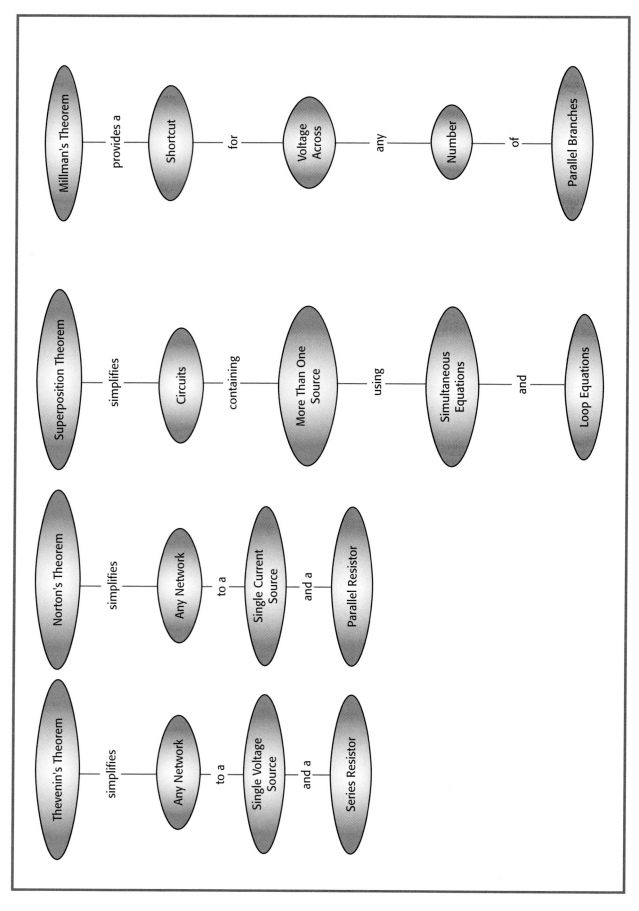

NETWORK THEOREMS 10

Objectives

After studying this chapter, you will be able to:
- ○ Apply Thevenin's and Norton's theorems to solve circuit problems.
- ○ Determine load voltage and current for various load values connected to a network.
- ○ Solve circuit problems for circuits with more than one voltage source.
- ○ Apply the superposition theorem to solve circuit problems.
- ○ Work with simultaneous equations.
- ○ Apply Millman's theorem to solve circuit problems.

Introduction

A *network* is a group of components, such as resistors, connected in any manner. Even though the rules for series and parallel circuits still apply, network theorems are a shortcut to solving the circuit. Theorems provide a process to convert the network to an equivalent circuit. The equivalent circuit can then be solved by following the series and parallel rules. Although the examples in this chapter use dc sources, the theorems apply to both dc and ac networks.

Network: An electrical combination of two or more components.

Thevenin's Theorem

M. L. Thevenin was a French engineer who developed a theorem that made it possible to reduce complicated networks to one simple resistance and voltage source. *Thevenin's theorem* states any network of voltage sources and resistors can be replaced by a single equivalent voltage source (V_{TH}) in series with a single equivalent resistance (R_{TH}).

In **Figure 10-1,** V_{TH} is the open-circuit voltage across the A and B terminals, and R_{TH} is the open-circuit resistance in series with the A and B terminals. V_{TH} and R_{TH} are the voltage and resistance a load "sees" when looking back into the circuit.

Thevenin's theorem: States any network can be reduced to a single resistance and a single voltage source.

Theveninizing Circuits

Figure 10-2A illustrates the application of Thevenin's theorem. To find the Thevenin equivalent of the circuit, disconnect the load resistance, **Figure 10-2B.** Removing the load leaves terminals A and B open. Now the task is to find the Thevenin voltage (V_{TH}) and resistance (R_{TH}).

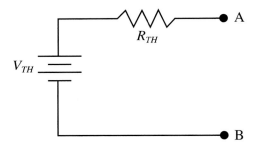

Figure 10-1. Thevenin equivalent circuit.

To find V_{TH}, use the voltage divider equation:

$$V_{TH} = \frac{R}{R_T} (V_S)$$

Where:

R = Resistance in which the voltage is to be found
R_T = Total resistance of the two resistors
V_S = Source voltage

Therefore:

$$V_{TH} = \frac{R}{R_T} V_S$$

$$= \frac{200}{600 + 200} (80)$$

$$= 20 \text{ V}$$

In calculating the Thevenin resistance (R_{TH}), the load remains disconnected and the source is shorted. This places R_1 and R_2 in parallel, **Figure 10-2C.**

To calculate the total resistance, use the parallel equation:

$$R_{TH} = \frac{R_1 \times R_2}{R_1 + R_2}$$

Therefore:

$$R_{TH} = \frac{R_1 \times R_2}{R_1 + R_2}$$

$$= \frac{600 \times 200}{600 + 200}$$

$$= 150 \ \Omega$$

The Theveninized circuit is shown in **Figure 10-2D.**

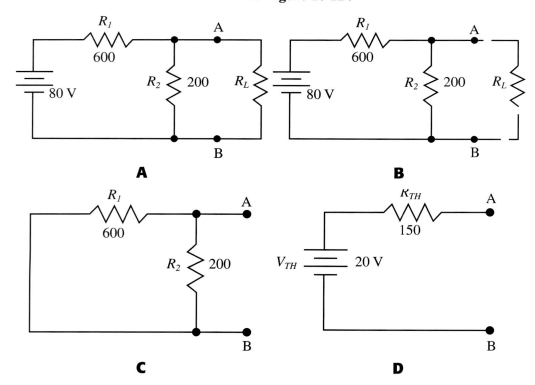

Figure 10-2. Theveninizing circuits. A—Circuit to be changed to a Thevenin equivalent. B—Remove the load. C—With the load terminals open, short the source that places the resistors in parallel. D—Thevenin equivalent circuit is the view from the load when looking back toward the source.

▼ *Example 10-1:*

Find the load current in **Figure 10-3A.** In **Figure 10-3B,** the load is disconnected. V_{TH} and R_{TH} are calculated.

$$V_{TH} = \frac{R}{R_T}\,(V_S)$$

$$= \frac{300}{200 + 300}\,(75\text{ V})$$

$$= 45\text{ V}$$

$$R_{TH} = \frac{R_1 \times R_2}{R_1 + R_2}$$

$$= \frac{200 \times 300}{200 + 300}$$

$$= 120\ \Omega$$

In **Figure 10-3C,** the load is connected to the Theveninized circuit. The load current can be calculated using Ohm's law.

$$R_T = R_{TH} + R_L$$

$$= 120 + 5$$

$$= 125\ \Omega$$

$$I = \frac{45\text{ V}}{125\ \Omega}$$

$$= 360\text{ mA}$$

▲

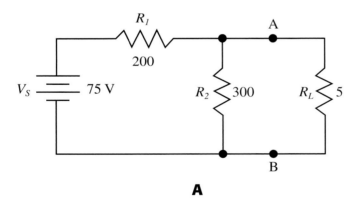

A

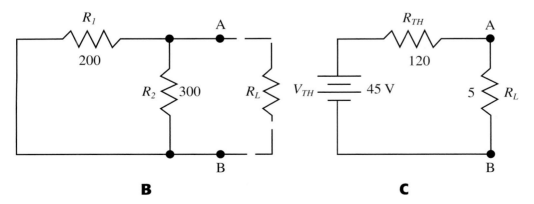

B   **C**

Figure 10-3. Example 10-1. A—Circuit. B—The load is disconnected. C—The load is reconnected to calculate the load current.

Most circuits are not as simple as the ones just shown. A circuit may consist of several resistors, **Figure 10-4.** However, the process is the same.

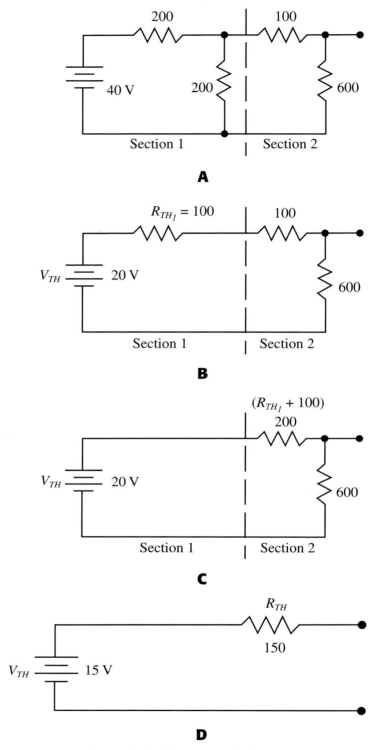

Figure 10-4. Complex circuit. A—Divide the circuit into sections. B—Make Thevenin calculations for the first section. C—Add the results of the first and second sections. D—Calculate Thevenin values for the Thevenin equivalent circuit.

▼ *Example 10-2:*

Divide the circuit into sections, **Figure 10-4A.**

Calculate V_{TH} and R_{TH} for Section 1 from the source, **Figure 10-4B.**

Add the result of Section 1 to Section 2 with V_{TH} as the source voltage, **Figure 10-4C.**

Calculate the V_{TH} and R_{TH} for the circuit again.

Draw the Thevenin circuit, **Figure 10-4D.** No resistor sections are left, so additional calculations are not necessary.   ▲

Source Resistance

In **Figure 10-5A,** the circuit has a source resistance of 10 Ω, which adds to the total resistance. Before any calculations begin, this resistance must be added to the first resistor in the network. The Thevenin voltage and resistance can be calculated by following the same process as before, **Figure 10-5B.**

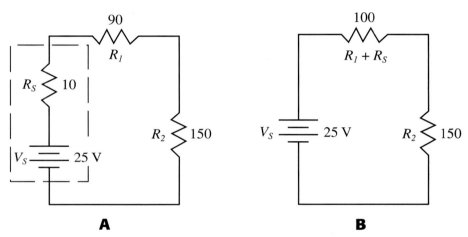

Figure 10-5. Source resistance. A—A source resistance must be added to the first resistance of the circuit. B—Calculate the Thevenin equivalent.

In **Figure 10-6,** R_3 does not enter into the V_{TH} calculations. For the Thevenin resistance, complete the R_1 and R_2 calculation; then add R_3 for the R_{TH}.

▼ *Example 10-3:*

Calculate the Thevenin voltage and resistance in **Figure 10-6A.**

$$V_{TH} = \frac{60}{40 + 60} \, (25 \text{ V})$$

$$= 15 \text{ V}$$

In **Figure 10-6B,** R_{TH} is:

$$R_{TH} = \frac{R_1 \times R_2}{R_1 + R_2 + R_3}$$

$$= \frac{40 \times 60}{40 + 60 + 100}$$

$$= 124 \, \Omega$$

The Theveninized circuit is shown in **Figure 10-6C.**   ▲

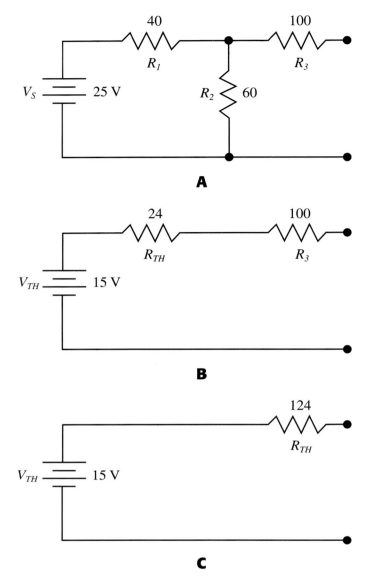

Figure 10-6. Example 10-3. A—Circuit. B—Calculate the Thevenin equivalent for the first section. C—Since R_3 is in series, it is simply added to the total resistance. It does not enter into the voltage calculations.

All the conditions discussed so far are shown in **Figure 10-7.** Divide the circuit into sections, **Figure 10-7A.** Add the source resistance to the first resistor in the network, **Figure 10-7B.** Calculate V_{TH} and R_{TH} for Section 1. This places R_{TH} in series with R_3 of Section 2, **Figure 10-7C.**

The sum of R_{TH} and R_3 makes a new R_{TH}, **Figure 10-7D.** The calculations start over again using Section 2. Notice R_5 is not part of Section 2.

Calculate V_{TH} and R_{TH} again, **Figure 10-7E.** Add R_5 to R_{TH} for the final results, **Figure 10-7F.**

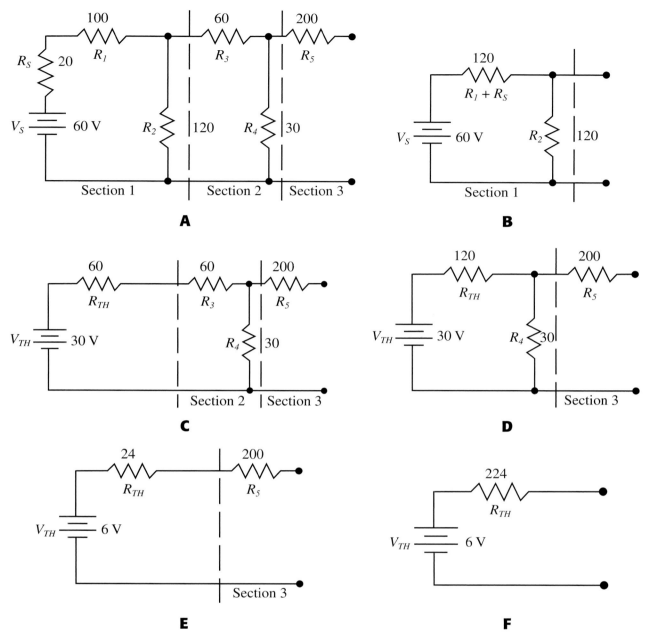

Figure 10-7. All circuit conditions. A—Divide the circuit into three sections. B—Add the source resistance to the first resistor and calculate V_{TH} and R_{TH} for the first section. C—Results of the first section. D—The first section is added to the second, and V_{TH} and R_{TH} are calculated again. E—Results of the second section. F—The last resistor is in series and added to the results of the second section.

Theveninizing a Bridge Circuit

In Chapter 8, you saw how the voltage divider equation was applied to a bridge circuit. Thevenin's theorem can be applied step by step to solve a bridge circuit as well.

▼ *Example 10-4:*

Figure 10-8 shows a typical bridge circuit. What is the current through R_5?

Divide the circuit into three sections, **Figure 10-8A.**

Calculate the R_{TH} of Section 1, **Figure 10-8B.**

$$R_{TH_1} = \frac{100 \times 100}{100 + 100}$$

$$= 50 \ \Omega$$

Calculate the V_{TH} of point A.

$$V_{TH_1} = \frac{100}{200} \ (40 \text{ V})$$

$$= 20 \text{ V}$$

Calculate the R_{TH} of Section 3, **Figure 10-8C.**

$$R_{TH_2} = \frac{60 \times 90}{60 + 90}$$

$$= 36 \ \Omega$$

Calculate the V_{TH} of point B.

$$V_{TH_1} = \frac{90}{150} \ (40 \text{ V})$$

$$= 24 \text{ V}$$

The total R_{TH} would be the sum of $R_{TH_1} + R_5 + R_{TH_2}$.

$$R_{TH} = 50 + 14 + 36$$

$$= 100 \ \Omega$$

The Thevenin equivalent circuit is shown in **Figure 10-8D.** The net voltage for V_{TH} is 4 V, which is the difference of potential between points A and B, or the difference between V_{TH_1} and V_{TH_2}. See **Figure 10-8E.**

The current through R_5 would be:

$$I_5 = \frac{V_{TH}}{R_{TH}}$$

$$I_5 = \frac{4 \text{ V}}{100 \ \Omega}$$

$$= 40 \text{ mA}$$

Since point B is more positive, the direction of the current through R_5 is from point A to point B.   ▲

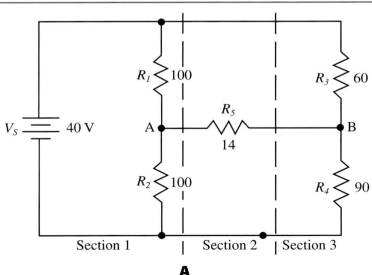

Figure 10-8. Bridge circuit for Example 10-4. A—Divide the circuit into sections. *(continued)*

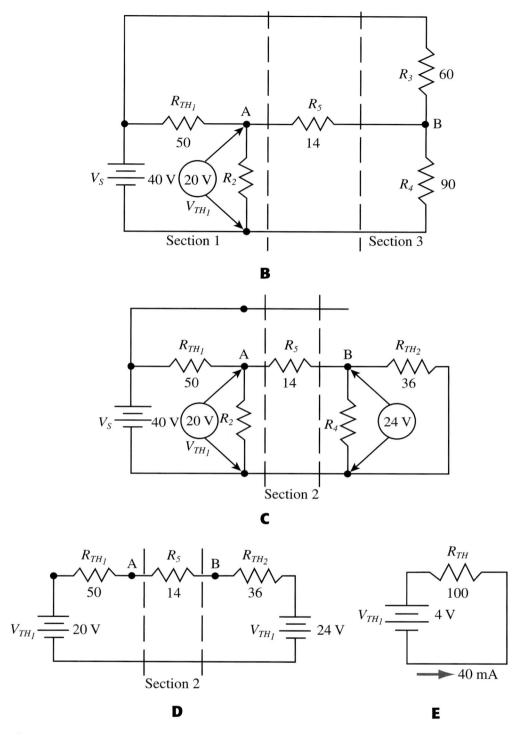

Figure 10-8. *(continued)* B–R_{TH} and V_{TH} are calculated for the first section. C–R_{TH} and V_{TH} are calculated for the third section. D–Since all resistances are in series, the total R_{TH} is the sum of the resistances. The net V_{TH} is the difference between the two voltages. E–The final Thevenin circuit allows R_5 to be calculated.

Changing Load Currents

The effect of changing load conditions can be calculated by reducing the circuit to a Thevenin equivalent circuit. Then, the output voltage can be calculated at different load currents.

▼ *Example 10-5:*

Calculate the V_{TH} and R_{TH} in **Figure 10-9A.**

$$V_{TH} = \frac{60}{100}(36)$$

$$= 21.6 \text{ V}$$

$$R_{TH} = \frac{40 \times 60}{100}$$

$$= 24 \ \Omega$$

The result is shown in **Figure 10-9B.**

Using the load current, calculate the voltage drop across R_{TH}.

Subtract the R_{TH} voltage drop from the source voltage. This is the voltage across the load.

Per the table in **Figure 10-9C,** the voltage across the load for 5 mA is:

$$V_{LOAD} = V_{TH} - I_{LOAD} \times R_{TH}$$

$$= 21.6 - 5 \text{ mA }(24)$$

$$= 21.48 \text{ V}$$

The remaining load currents in the table are calculated the same way. As the load current goes up, the voltage across the load becomes much lower.   ▲

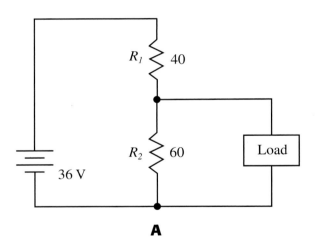

A

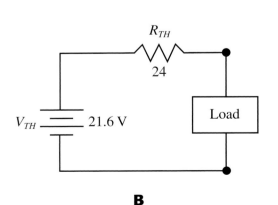

B

I_L	V_L
0 mA	21.6
5 mA	21.48
10 mA	21.36
20 mA	21.12
100 mA	19.2
400 mA	12.0
500 mA	9.6

C

Figure 10-9. Example 10-5. A—Circuit for calculating the output voltage at different load currents. B—V_{TH} and R_{TH} calculated. C—Output voltages at different load currents.

Norton's Theorem

E. L. Norton was an American scientist who developed another method of simplifying a complex circuit. **Figure 10-10** illustrates the differences between Thevenin and Norton circuits. Thevenin's theorem uses an equivalent voltage source (V_{TH}) and series resistance (R_{TH}). **Norton's theorem** simplifies the complex circuit to a current source (I_N) and parallel resistance (R_N).

Norton's theorem: Looking back from the terminals referenced, the current in any impedance connected to two terminals of a network is as though it were connected to a constant-current source.

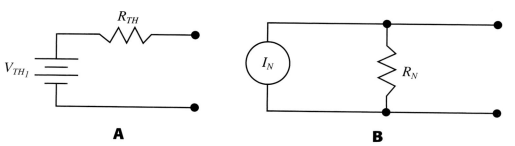

A **B**

Figure 10-10. Differences between Thevenin and Norton circuits. A—Thevenin's equivalent circuit. B—Norton's constant current source.

▼ *Example 10-6:*

Make an equivalent circuit in **Figure 10-11.**

In **Figure 10-11A**, place a short across terminals A and B, the output to the load. Calculate the Norton equivalent current using V_S and R_1. Resistors R_2 and R_L are not in the circuit because they are shorted by step 1.

$$I_N = \frac{V_S}{R_T}$$
$$= \frac{25 \text{ V}}{50 \text{ }\Omega}$$
$$= 500 \text{ mA}$$

Therefore, 500 mA is the total current available from the Norton current source. Remove the short across the load and remove the load; that is, open the load circuit and short circuit the source voltage. This places R_1 and R_2 in parallel, **Figure 10-11B.** The Norton resistance is calculated by:

$$R_N = \frac{R_1 \times R_2}{R_1 + R_2}$$
$$= \frac{50 \times 90}{50 + 90}$$
$$= \frac{4500}{140}$$
$$= 32 \text{ }\Omega$$

Figure 10-11C shows the result when R_L is returned to the circuit.

The currents for R_N and R_L can be calculated using the current divider equation.

$$I_1 = \frac{G}{G_T} (I_T)$$
$$= \frac{0.03125}{0.08125} (500 \text{ mA})$$
$$= 192.3 \text{ mA}$$

$$I_L = \frac{G}{G_T} (I_T)$$
$$= \frac{0.05}{0.08125} (500 \text{ mA})$$
$$= 307.7 \text{ mA}$$

▲

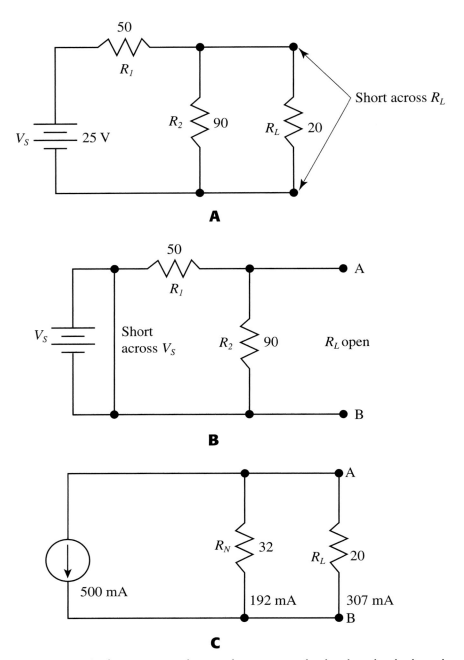

Figure 10-11. Norton's theorem. A–Place a short across the load and calculate the maximum current. B–Remove the load and calculate R_N the same as R_{TH}. C–Norton circuit result. Calculate the current for R_N and R_L using the current divider equation.

Superposition theorem: When a number of voltages in a linear network are simultaneously applied to the network, the current that flows is the sum of the component currents that would flow if the same voltages had acted individually.

Superposition Theorem

The ***superposition theorem*** simplifies the solution of circuits containing more than one source. The effect of one source at a time is calculated; then, the results of all the sources are superimposed by algebraic addition. To use one source at a time, all the other sources are shorted across and the effect of the active source calculated.

▼ *Example 10-7:*

Find the current through R_3 in **Figure 10-12A.**

Short out the voltage source V_2, **Figure 10-12B.** This places R_2 and R_3 in parallel. The total resistance is equal to:

$$R_T = \frac{R_1 \times R_2}{2}$$

$$= \frac{30 + 30}{2}$$

$$= 45 \ \Omega$$

The circuit current with V_1 is:

$$I_T = \frac{V_1}{R_T}$$

$$= \frac{12}{45}$$

$$= 0.2667 \ A$$

This current flows through R_1. Since the resistances of R_2 and R_3 are equal, the current will divide in half for each of them. Thus, the current for R_3 is 0.1333 A.

The voltage source V_1 is shorted, and the effect of V_2 is calculated by using the same process, **Figure 10-12C.**

$$R_T = \frac{R_2 + R_1}{2}$$

$$= \frac{30 + 30}{2}$$

$$= 45 \ \Omega$$

$$I_T = \frac{V_2}{R_T}$$

$$= \frac{54}{45}$$

$$= 1.2 \ A$$

The current through R_3 is again half the total current since resistances R_1 and R_3 are equal.

$$I_{R_3} = \frac{1.2}{2}$$

$$= 0.6 \ A$$

Both sources V_1 and V_2 produce current in the same direction through R_3; thus, the two currents have the same algebraic sign. The total current through R_3 is calculated by:

$$I_{R_3} = I_{V_1} + I_{V_2}$$
$$= 0.2 + 0.6$$
$$= 0.8 \ A$$

See **Figure 10-12D.** As shown in **Figure 10-12E,** if the direction of V_1 were reversed, the current through R_3 would be:

$$I_{R_3} = -I_{V_1} + I_{V_2}$$
$$= -0.2 + 0.6$$
$$= 0.4 \ A$$

This also indicates the current through R_3 flows from R_2 to R_3.   ▲

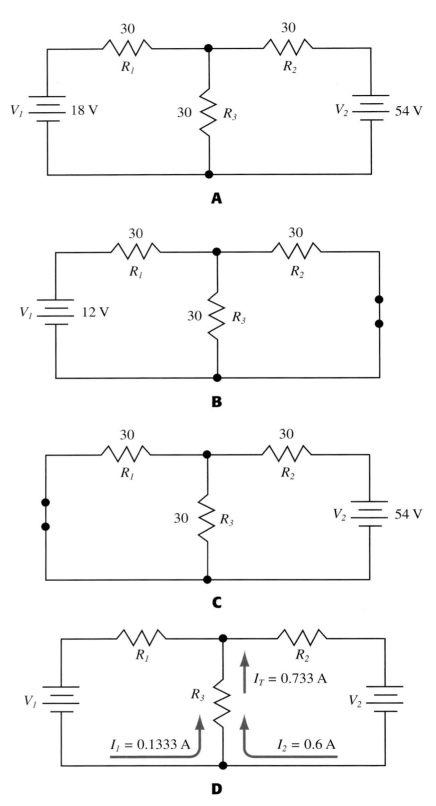

Figure 10-12. Example 10-6. A—Circuit. B—Short out V_2 and calculate the effect of V_1 on the circuit. C—Short out V_1 and calculate the effect of V_2 on the circuit. D—Add the resulting currents. *(continued)*

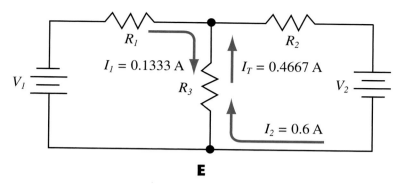

Figure 10-12. *(continued)* E—Current through R_3 if V_1 were reversed.

Simultaneous Equations

The equations so far have contained only one unknown value. When more than one unknown value is to be found, simultaneous equations are used. *Simultaneous equations* are two or more related equations. The objective is to eliminate one of the unknowns.

Simultaneous equation: Two or more equations that contain the same variables.

▼ *Example 10-8:*

Given the equations $6x + 4y = 32$ and $x - y = 7$, find the values of x and y.
Arrange the equations so similar unknowns are lined up vertically.

$$6x + 4y = 32$$
$$x - y = 7$$

Coefficients are the numbers that precede the unknowns. Determine if two similar unknowns have equal coefficients. In the first equation, 6 and 4 are the coefficients; in the second equation both coefficients are 1. Thus, the coefficients are not equal.

Coefficient: The number that precedes an unknown in an algebraic equation.

Match one of the unknowns with the opposite sign so it can be eliminated by algebraic addition. Since y has a negative sign, it will be easier to use. The coefficient in the first equation is 4; therefore, the coefficient in the second equation must be multiplied by 4 to cancel out the coefficients.

$$4 (x - y = 7)$$
$$4x - 4y = 28$$

Add the two equations and solve for x.

$$6x + 4y = 32$$
$$+ \quad 4x - 4y = 28$$
$$\overline{10x \quad\quad = 60}$$
$$x \quad\quad = 6$$

Substitute the value of x into either of the original equations and solve for y.

$$6x + 4y = 32$$
$$6 (6) + 4y = 32$$
$$36 + 4y = 32$$
$$4y = 32 - 36$$
$$4y = -6$$
$$y = -1.5$$

▲

▼ *Example 10-9:*

Find the values of r and w in the following equations.

$$5r = -w - 24$$
$$r + 8 = 3w$$

Rearrange to:

$$5r + w = -24$$
$$r - 3w = -8$$

Multiply the second equation by –5, and add the two equations. Solve for w.

$$
\begin{array}{rl}
5r + w & = -24 \\
+ \quad -5r + 15w & = 40 \\
\hline
16w \quad\quad\; & = 16 \\
w \quad\quad\; & = 1
\end{array}
$$

Substitute w into either original equation and solve for r.

$$
\begin{array}{rl}
5r + 1 & = -24 \\
5r & = -25 \\
r & = -5
\end{array}
$$

or

$$
\begin{array}{rl}
r - 3 & = -8 \\
r & = -5
\end{array}
$$

▲

▼ *Example 10-10:*

Find the values for s and c.

$$3c + 2s = 11$$
$$2c - 7s = -51$$

Multiplying the first equation by 2 and the second equation by –3 will eliminate c.

$$
\begin{array}{rl}
6c + 4s & = 22 \\
+ \quad -6c + 21s & = 153 \\
\hline
25s \quad\quad\; & = 175 \\
s \quad\quad\; & = 7
\end{array}
$$

Solve for c.

$$
\begin{array}{rl}
3c + 2s & = 11 \\
3c + 2\,(7) & = 11 \\
\\
3c + 14 & = 11 \\
3c & = 11 - 14 \\
3c & = -3 \\
c & = -1
\end{array}
$$

▲

Loop Equations

In **Figure 10-13**, each voltage source (V_1 and V_2) causes a loop of current (I_1 and I_2). For each of these current loops, a loop equation can be written.

The currents I_1 and I_2 flow through resistor R_3. This makes the resistor common to both sources; that is, the two currents are related to each other. Because of this relationship, simultaneous equations can be used to solve for the currents.

The loop I_1 equation is:

$$V_1 = I_1R_1 + I_1R_3 - I_2R_3$$

The loop for I_2 is in the opposite direction of current flow; therefore, I_2 is a negative number. If it were in the same direction as current flow, it would be a positive number.

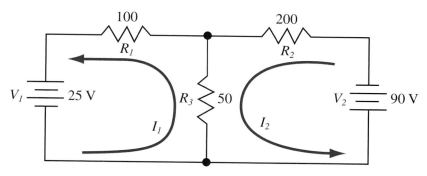

Figure 10-13. The voltage sources are related because both currents flow through R_3.

The loop I_2 equation is:

$$V_2 = I_2 R_2 + I_2 R_3 - I_1 R_3$$

Place the coefficients (resistor values) in the equations, and move the current variables after the coefficients for easier reading.

$V_1 = I_1 R_1 + I_1 R_3 - I_2 R_3$
$25 = 100 I_1 + 50 I_1 - 50 I_2$

and

$V_2 = I_2 R_2 + I_2 R_3 - I_1 R_3$
$90 = 200 I_2 + 50 I_2 - 50 I_1$

Combine like terms.
$25 = 150 I_1 - 50 I_2$
$90 = -50 I_1 + 250 I_2$

Multiply the second equation by 3 to cancel the I_1 values. Then, solve for the I_2 current.

$$\begin{aligned} 25 &= 150 I_1 - 50 I_2 \\ + \quad 270 &= -150 I_1 + 750 I_2 \\ \hline 295 &= 700 I_2 \\ I_2 &= \frac{295}{700} \\ &= 0.421 \end{aligned}$$

Place the known value for I_2 in either of the original equations, and solve for the I_1 current.

$90 = -50 I_1 + 250 I_2$
$90 = -50 I_1 + 250 \ (0.421)$
$90 = -50 I_1 + 105.357$
$-15.357 = -50 I_1$
$0.307 = I_1$

The net current through R_3 is:
$I_{R_3} = I_2 - I_1$
$\quad = 0.421 + 0.307$
$\quad = 0.114 \ \text{A}$

The voltage across R_3 can be calculated by:
$V = I_{R_3}$
$\quad = 0.114 \times 50$
$\quad = 5.7 \ \text{V}$

The equations can be divided by a number to get one of the unknowns to cancel. Multiplying or dividing by −1 can change the sign of the numbers without changing the coefficients.

Millman's Theorem

Millman's theorem provides a shortcut for finding the voltage across any number of parallel branches with different voltage sources. **Figure 10-14** is the same circuit as shown in Figure 10-13 but is drawn differently. The voltage from points A and B in relation to ground can be found using the equation:

$$V_{AB} = \frac{\dfrac{V_1}{R_1} + \dfrac{V_2}{R_2} + \dfrac{V_3}{R_3}}{\dfrac{1}{R_1} + \dfrac{1}{R_2} + \dfrac{1}{R_3}}$$

The equation is the sum of the parallel current sources divided by the total conductance, or the total current divided by the total conductance:

$$V_{AB} = \frac{\dfrac{25}{100} + \dfrac{90}{200} + \dfrac{0}{50}}{\dfrac{1}{100} + \dfrac{1}{200} + \dfrac{1}{50}}$$

$$= \frac{0.2}{0.035}$$

$$= 5.7 \text{ V}$$

This is the same voltage calculated before. Note the polarity of $\dfrac{90}{200}$ is negative because the current is in the opposite direction of the V_1 current.

This method can be used for any number of branches, but they must all be in parallel. No series resistance can exist between branches. If a branch has more than one voltage source, the sources can be combined algebraically for one voltage.

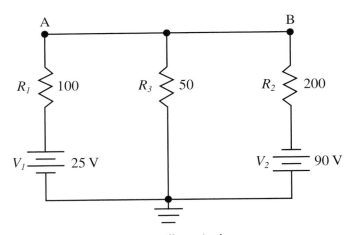

Figure 10-14. Figure 10-13 redrawn for Millman's theorem.

Summary

- Thevenin's theorem states any network with two open terminals can be reduced to one voltage source with a single resistance.
- Norton's theorem states any network with two terminals can be reduced to one current source in parallel with a single resistance.
- The superposition theorem states in a linear, bilateral network with more than one voltage source, the current and voltage in any part of the network can be found by adding the effects of each source algebraically. All the other sources are disabled by short-circuiting voltage sources and opening current sources.
- Millman's theorem states the common voltage across parallel branches with different voltage sources can be determined by the equation of the sum of the currents divided by the total parallel conductance. All branches must be in parallel.

Important Terms

Do you know the meanings of these terms used in the chapter?

coefficient

Millman's theorem

network

Norton's theorem

simultaneous equation

superposition theorem

Thevenin's theorem

Questions and Problems

Please do not write in this text. Write your answers on a separate sheet of paper.

1. Solve **Figure 10-15** for the current through R_3.
2. Write Thevenin's theorem.
3. Solve **Figures 10-16A–B** for V_{TH} and R_{TH}.
4. Solve **Figure 10-17** for V_{TH} and R_{TH}.
5. Write Norton's theorem.
6. Solve **Figure 10-18** for V_{TH} and R_{TH}.
7. Solve **Figure 10-19** for V_{TH}, R_{TH}, R_N, and I_N.
8. Solve **Figure 10-20** for V_{TH}, R_{TH}, R_N, and I_N.
9. Solve **Figure 10-21** for V_{TH} and R_{TH}.
10. Solve **Figure 10-22** for V_{TH} and R_{TH}.
11. Solve **Figure 10-23** for V_{XY} and I_{R_2} using Millman's theorem.
12. Using pi to tee conversions, solve **Figure 10-24** for the current through R_L.

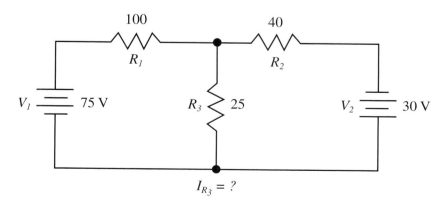

Figure 10-15. Circuit for problem 1.

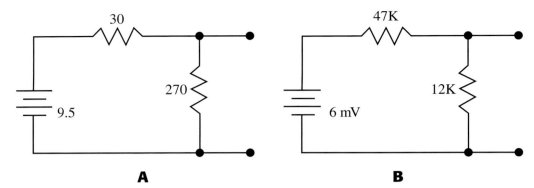

A   **B**

Figure 10-16A–B. Circuits for problem 3.

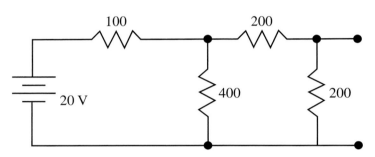

Figure 10-17. Circuit for problem 4.

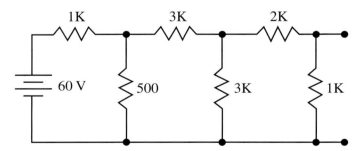

Figure 10-18. Circuit for problem 6.

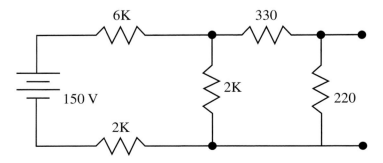

Figure 10-19. Circuit for problem 7.

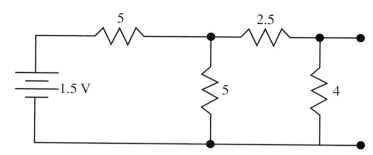

Figure 10-20. Circuit for problem 8.

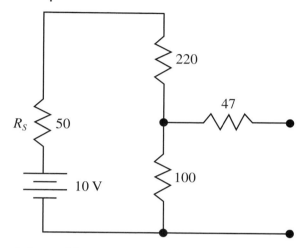

Figure 10-21. Circuit for problem 9.

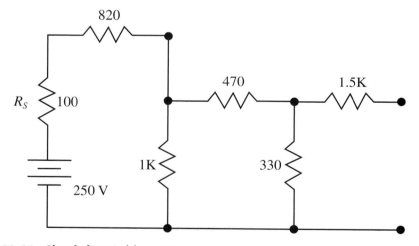

Figure 10-22. Circuit for problem 10.

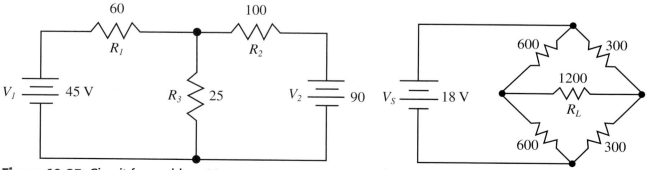

Figure 10-23. Circuit for problem 11.   **Figure 10-24.** Circuit for problem 12.

Section II

Electronic Assembly

Section II Activity

Tinning a Wire

Tinning a wire is a routine task in the soldering and assembly of electronic equipment. Often, stranded wire in an electrical assembly must be tinned. This activity provides the necessary information to properly tin a stranded wire using a soldering iron.

Objective

In this activity, you will tin the end of a stranded wire by hand.

Materials and Equipment

Wire stripper
Diagonal wire-cutting pliers
20-gage stranded wire
Soldering iron
Solder, 60/40 or 63/37
Soldering cleaning supplies
Small vise

Procedures

1. Cut a piece of stranded wire approximately 5″ long.
2. Carefully strip 1″ of insulation from one end. See **Figure A.**
   a. Do not strip the insulation in one motion; this will disturb the wire-lay of the strands. Cut the insulation and make a ¹⁄₁₆″ gap.
   b. Remove the insulation with your fingers, using a twisting motion as you pull it off the wire. Twist in the direction of the wire-lay. See **Figure B.**
3. Inspect the wire for:
   a. Damage to the wire or insulation.
   b. Disturbed wire-lay.
   c. Clean, even cut of the insulation and wire end.
4. Place the wire in a vise to hold it steady. Starting near the insulation, apply the soldering iron to the top of the conductor. Feed a wire solder to the soldering iron.
5. Slowly move the iron down and follow through off the end of the wire.
6. Clean the tinned wire end with rubbing alcohol.
7. Inspect for:
   a. Exposed copper.
   b. Contour soldering. You should be able to see the wire strands.
   c. Smooth, shiny surface.
   d. Wire or insulation damage.
   e. Excessive wicking of solder under the insulation.

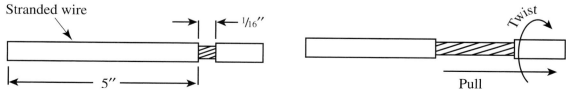

Figure A.   **Figure B.**

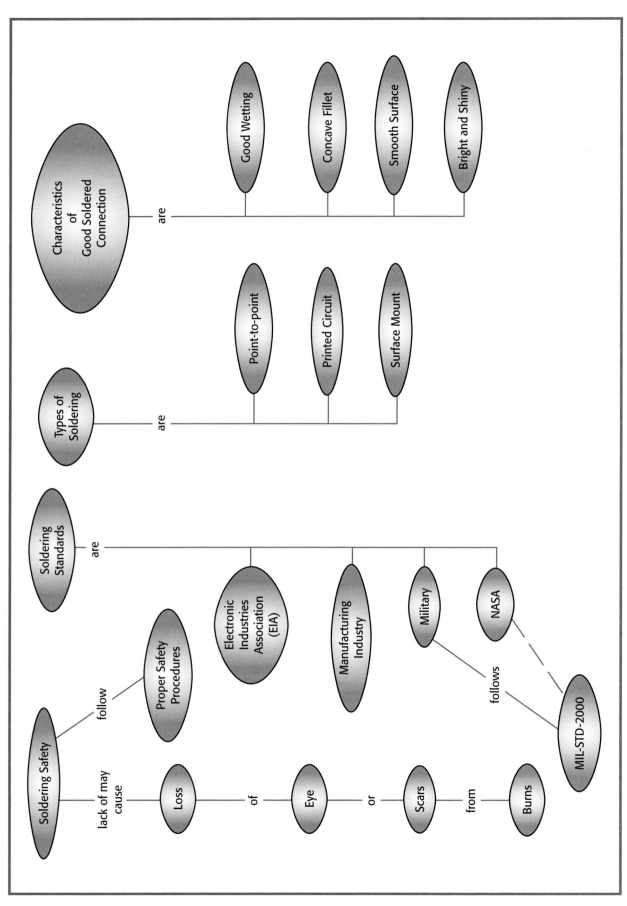

ELECTRICAL SOLDERING ▷ 11 ▷▷

⟩| Objectives

After studying this chapter, you will be able to:
- State the safety precautions to take when soldering.
- Describe the various forms of solder and soldering equipment available.
- Explain the three methods of wire stripping.
- Describe the procedures for tinning.
- Explain the soldering process.
- Compare or contrast point-to-point soldering, printed circuit board soldering, and surface mounted device soldering.
- Describe the procedures for common soldering and drag soldering.
- Distinguish a good soldered connection from a poor one.
- Explain the three methods of removing solder.

⟩| Introduction

Soldering is the process of joining two metals. A soldering iron is used to melt the solder and heat the metals to be joined. When cooled, the soldered connection produces a strong bond between the metals and forms a very low-resistance electrical connection.

Soldering: The process of joining two metals.

Soldering and desoldering are important skills for the assembly and trouble-shooting of electronic components. These skills require patience, practice, and experience. Making a satisfactory solder connection requires good hand-eye coordination and keen judgment.

Soldering Safety

Soldering poses hazards that can result in the loss of an eye or burns to the face or other parts of the body. Proper safety precautions must be followed to avoid these and other serious injuries.

General Safety

- Always wear safety glasses when soldering. Hot solder can splatter into the eyes.
- Keep the soldering iron in a place that does not require you to reach across or around it.
- When the soldering iron is not in use, keep it in its holder.
- When cutting wires, keep the open side of the cutter away from your body. Do not allow the wire pieces to fly into the air. They could cause injury to someone working nearby.
- Lead poisoning causes serious long-term health problems. Never place wire solder in your mouth.
- Wash your hands when you have finished soldering.

Solder Pot Safety

- Do not wear loose clothing and keep long hair pulled back or wear a cap. Loose clothing and long hair can end up in the molten solder or catch on the solder pot and cause spillage.

- Keep your eyes on your work.
- Keep only necessary work items in the solder pot area.
- Never store objects above a solder pot. Reaching across the solder pot could cause burns.
- Keep all liquids except flux away from the solder pot area.
- Place the solder pot on a level, sturdy, noncombustible surface.
- Use warning signs that say HOT, CAUTION, or DANGER when using the solder pot.
- Use good ventilation to exhaust fumes.

Soldering Standards

Soldering standards are written requirements that maintain quality by establishing the proper materials and procedures necessary to make soldered electrical/electronic connections. Standards address soldered connections for component leads, wires inserted in holes or attached to terminals, and surface mounted components. Standards also specify requirements for mounting, inspection, and evaluation.

Soldering standards are referenced by the Electronic Industries Association (EIA), the manufacturing sector, and the military. Much of the electronic industry uses EIA standards for soldering. Soldering standard publications are available from EIA.

Most manufacturing industries develop their own sets of soldering standards. Often an industry standard is a modified version of an EIA standard, tailored to meet individual industry requirements.

The U.S. Department of Defense has established soldering standards that all industries must follow in the manufacture of electronic equipment for government applications. These standards are identified by a military standard number such as MIL-STD-2000. All branches of the armed forces must comply with these standards. The National Aeronautics and Space Administration (NASA) follows MIL standards but also has its own set of standards.

Soldering Materials and Equipment

Sometimes wires must be connected to an electrical part. This is done by soldering. *Solder* is a metal mixture used to connect metal parts or components. Usually a tin/lead mixture is used.

Solder: A metal mixture used to connect metal parts or components.

Solders

Solder comes in bar, ribbon, paste, or wire forms. Solid bar solder is used for filling solder pots. Ribbon solder is wrapped around the joint to be heated. Paste solder is mixed into a flux and used extensively in surface mounted device (SMD) assembly.

The most commonly used solder is wire. Wire solder can be solid or have a flux-filled center. It comes in various diameters and different amounts of alloys. An *alloy* is a mixture of two or more metals, typically tin and lead.

Alloy: A mixture of two or more metals.

The melting point of wire solder is determined by the percentage of each metal in the alloy. See **Figure 11-1.** Temperature is given on the left side of the graph while tin/lead percentages are given across the top. The lower portion indicates the solder in the solid state, while the upper portion indicates the liquid state. Solder is in a liquid state when it is completely melted. The plastic state is between the solid and liquid states. In the plastic state, the metal is no longer solid but not yet liquid. A 50/50 tin/lead solder has a melting point of 414°F (212°C), while a 60/40 tin/lead solder has a melting point of 370°F (188°C). At a 63/37 mixture, the tin/lead solder is an *eutectic alloy,* meaning it has the lowest melting point possible.

Eutectic alloy: A mixture of metals with the lowest possible melting point.

Solder can be an alloy of other metals such as silver, bismuth, or antimony, **Figure 11-2.** Lead-free solder is supportive of environmental protection, **Figure 11-3.**

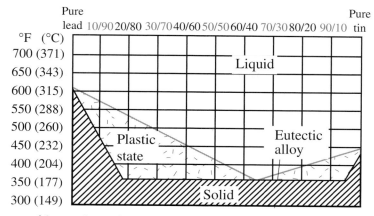

Figure 11-1. Melting points of various percentages of tin/lead solder alloy.

Percent of Metals Alloyed					Temperature
Sn	**Pb**	**Ag**	**Sb**	**Bi**	**°F (°C)**
96	0.1	3.5	—	—	430 (221)*
62	REM	2	0.3	0.25	354 (179)*
70	REM	—	0.3	0.25	379 (193)
94	0.2	—	5	—	464 (240)
0.25	REM	2.5	0.4	0.25	579 (304)*

* Eutectic alloy

Key:

Sn = tin   Sb = antimony

Pb = lead   Bi = bismuth

Ag = silver   REM = remainder of percentage

Figure 11-2. Melting points of various solder alloys.

Figure 11-3. Federal and state laws require lead-free solders for drinking water systems and other applications where lead would pose a health hazard.

Flux

Flux: A liquid or solid substance applied to metals before soldering to remove oxides and protect the hot metal from contamination by atmospheric oxygen.

Rosin flux: A common flux used for electrical soldering. It is inert at room temperature but becomes a mild acid when hot.

Acid flux: A special flux used in soldering sheet metal or plumbing.

Flux is a liquid or solid substance applied to metals before soldering to remove oxides and protect the hot metal from contamination by atmospheric oxygen. Soldering flux comes in liquid, powder, and paste forms. *Rosin flux* is inert at room temperature but becomes a mild acid when hot. *Acid flux* has acidic characteristics at all times and is used in plumbing and sheet metal work. It is never used for electrical soldering, **Figure 11-4.**

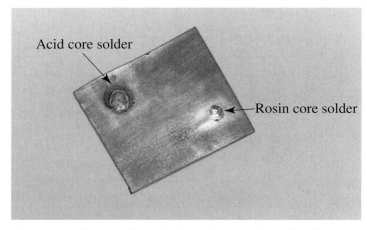

Figure 11-4. This piece of printed circuit board was soldered with rosin core solder in one area and acid core solder in another. Notice the corrosion where acid core solder was used.

Soldering Irons and Tips

Various types of soldering equipment are available. The pencil-type soldering iron is sold in hardware stores, **Figure 11-5.** This type of iron is uncontrolled, so tip temperature cannot be adjusted. The pencil soldering iron is inexpensive and good for general soldering. It is not used in manufacturing. Higher quality soldering irons can be obtained at electronic equipment dealers and companies that specialize in soldering equipment. Although the iron is uncontrolled, the tip temperature is more stable.

Figure 11-5. The tip temperature of this pencil-type soldering iron cannot be controlled.

A controlled soldering iron allows the operator to adjust tip temperature. This type of iron provides a temperature control, a necessary feature for making high-quality soldered connections.

Different soldering tips are available for soldering irons, **Figure 11-6.** The tip size must match the soldering job. A mid-sized cone or screwdriver-type tip is used to solder a small component. A much larger tip is required to solder a high-mass component such as a ground plane on a PC board.

The soldering tip stores heat. If a small tip is used to solder a large device, heat will quickly drain from the tip leaving no heat for soldering. A larger tip provides a bigger heat reservoir and sufficient heat to complete the job. See **Figure 11-7.** Between the time the two terminals are soldered, the heating element replaces the heat used to make the soldered connection.

Wire Stripping

Stripping: The process of removing the insulation from a conductor.

Stripping is the removal of insulation from a conductor or cable. Wire stripping can be done mechanically, thermally, or chemically.

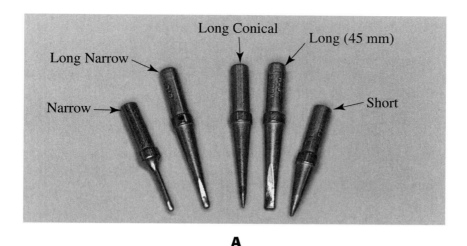

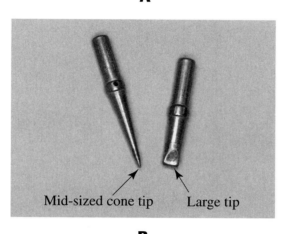

Figure 11-6. Soldering tips. A—Various screwdriver tips. The short screwdriver is often used for general soldering. B—A mid-sized cone tip is used for soldering small components. A large tip is used to solder components high in mass.

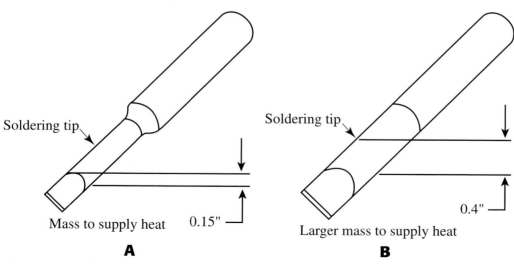

Figure 11-7. The soldering iron tip is a reservoir for heat. A—Heat is quickly drained from a tip 3/16″ or smaller. B—A larger tip supplies more heat because it has a mass of nearly 1/2″ in diameter.

Two types of mechanical strippers are shown in **Figure 11-8.** The hardware store type can do an acceptable job, but care must be taken not to nick or scrape the wire. Do not use a knife for stripping.

The stripper separates the insulation by approximately 1/16″. The insulation is removed by hand using a pulling and twisting motion so the wire lay is not disturbed, **Figure 11-9.**

Thermal strippers are typically tweezer-type. Each end of the tweezers has a loop, **Figure 11-10.** The loops are closed around the wire to melt the insulation. The temperature of the loops is adjusted by the operator so the insulation is melted but not burned. The insulation is also carefully removed by hand. For Teflon™ insulation, the temperature of the stripper should be increased gradually until the insulation melts.

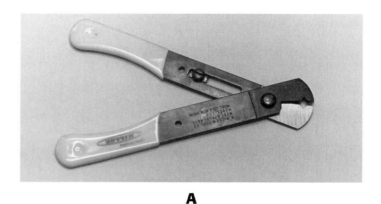

A

B

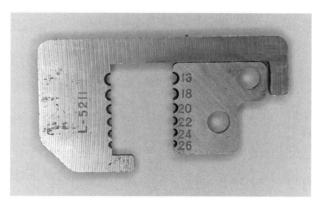

C

Figure 11-8. Strippers. A—A wire stripper purchased in a hardware store is acceptable for general use. B—This wire stripper has a set of fixed dies for different wire sizes. C—Replacement wire dies are used in military specification soldering.

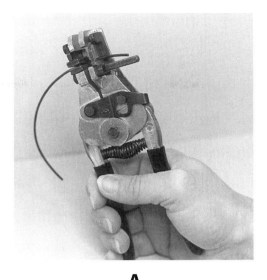

A

B

Figure 11-9. Using a wire stripper. A—The insulation is cut and separated. B—Removing the insulation with a pulling and twisting motion keeps the wire lay intact.

> **Warning:** Do not allow Teflon to burn. Burning Teflon produces toxic fumes.

Chemical stripping is used for removing the thin insulation on very small gage wire, such as the varnish coating on magnet wire. To chemically remove insulation, coat the wire with the chemical (such as Strip-X™) and wait two minutes. Use a paper towel to remove the chemical and insulation. If some insulation remains, repeat the process.

Using a solder pot to burn off the insulation is not recommended. Not only does this practice cause corrosion on the wire, it often causes contamination of the solder pot.

Stripping must be done carefully to prevent damage to wires or insulation. A nicked wire will create a weak spot in the conductor, causing the conductor to break if it is bent. Broken conductors can occur on machines that vibrate when operating.

A

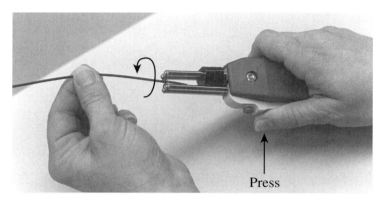

Press

B

Figure 11-10. Thermal wire stripper. A—This device uses heat to melt the insulation. B—The heated loops are clamped around the insulation to melt it.

Tinning

Tinning: Coating conductors with molten solder.

Birdcaging: The separation of a stranded conductor.

Tinning is the process of coating the metal surface of a wire with a thin layer of solder. Tinning is used mainly on stranded wires causing the strands to act as one solid wire. This prevents *birdcaging,* the separation of the wire strands, **Figure 11-11.** The solid end of the wire can be placed around a terminal or screw without the strands separating when the wire is formed or the screw is tightened.

When an untinned wire is under a tightened screw, the loose strands eventually relax and become reshaped. This causes a loose connection and high resistance to current flow. With sufficient current flowing, the loose connection produces heat at the screw terminal, potentially resulting in an electrical fire.

A fire hazard can also be caused by connecting untinned wires to a power source with wire nuts, **Figure 11-12.** No matter how much the wire nuts are tightened, the strands will eventually relax and make a high-resistance point in the circuit. The heat generated by the connection can be enough to start a fire. When inspecting wiring with wire nuts, be aware of changes in the shade of the colored nuts indicating a buildup of heat.

Wicking

Wicking: The process of solder being drawn underneath the wire insulation.

Wicking is the capillary action of the liquid solder being drawn upward on a stranded wire. When tinning wires, know the standard the tinning must meet. Some standards do not allow the solder to touch the insulation, while others state no solder may be wicked under the insulation. Others allow a maximum wicking of one wire diameter. For this text, the standard of no wicking is desired with a maximum of one wire diameter allowed. The wicking point can be determined by bending the wire, **Figure 11-13.**

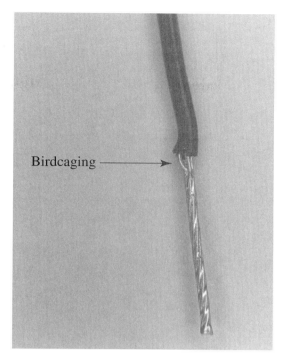

Figure 11-11. Birdcaging occurs when the wire is not tinned all the way to the insulation.

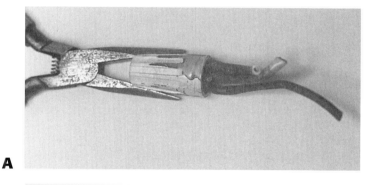

A

B

Figure 11-12. Untinned lamp wires. A— This piece of 18-gage stranded lamp cord was connected to a 120 V ac source with wire nuts. B—The heat generated by the connection was enough to burn the wire nut in half. Fortunately, the connections were in a metal junction box and did not cause a fire.

For most assemblies tinned wires present no problems, but stranded wire allows for bending and vibration. The bending stops at the point of tinning; therefore, the weakest point of the wire is where the tinning ends, **Figure 11-14.** Over time, as the wire bends at the edge of tinning, the strands break and current flow is reduced.

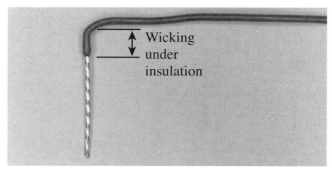

Figure 11-13. The point where the wire bends easily is the edge of the solder under the insulation. This shows excessive wicking well above the edge of the insulation.

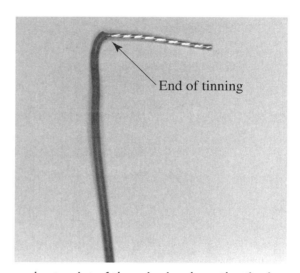

Figure 11-14. The weakest point of the wire is where the tinning ends.

Furthermore, as solder wicks under the insulation of a wire, it pushes flux ahead of it. The combination of chemicals in the flux and insulation eventually causes a breakdown of the insulation. The insulation near the connection crumbles, leaving bare wires to short out to other connections.

Solder Pot Tinning

A common method of tinning wires is with a solder pot, **Figure 11-15.** The solder pot temperature should be maintained at 500°F (260°C). A liquid rosin-based mildly activated (RMA) flux should be used for the tinning process.

The following procedure is used for solder pot tinning:

1. Properly strip the wire insulation. The amount of insulation to be stripped depends on the job.
2. Place the stripped end of the wire 1/8″ into the liquid flux.
3. **Dross** is the oxide residue on the surface of molten solder that occurs when the hot solder combines with oxygen in the air. Remove the dross from the surface of the solder in the solder pot, pushing the dross *away* from you.
4. Lower the stripped wire end into the molten solder. How far the end is lowered into the solder depends on the wicking standard. After a couple seconds, slowly remove the wire from the solder pot.

Dross: The film found on the top surface of molten solder in a solder pot caused by the surface of the solder combining with oxygen.

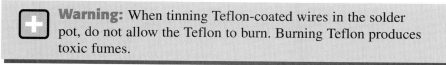

Warning: When tinning Teflon-coated wires in the solder pot, do not allow the Teflon to burn. Burning Teflon produces toxic fumes.

5. Clean the wire to remove any flux residue.
6. Inspect the tinned wire for defects.

Figure 11-15. Tinning using a solder pot. Keep your hands away from the 500°F (260°C) molten solder.

Tinning with a Soldering Iron

If a solder pot is not available, a wire can be tinned with a soldering iron. If only one wire is to be tinned, the wire is held by a vise or third hand, **Figure 11-16.**

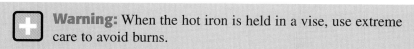

Warning: When the hot iron is held in a vise, use extreme care to avoid burns.

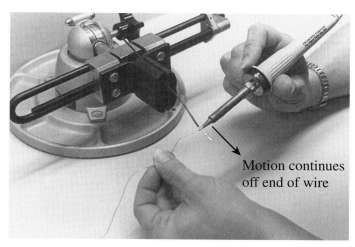

Motion continues off end of wire

Figure 11-16. If only one or two wires are to be tinned, a vise can be used to hold the wire while soldering is performed.

The following procedure is used for tinning with a soldering iron:

1. Melt a little solder on the soldering iron tip to ensure good heat conduction to the wire.
2. Place the soldering iron under the wire 1/4″ from the insulation.
3. Apply the solder at the top of the wire. As the solder melts, move the solder and iron tip downward toward the end of the wire. Add solder during the entire movement. Do not hurry. Allow time for the solder to go through to the inner strands of the wire.
4. As you get to the end of the wire, follow through with a downward movement off the end of the wire. If the proper amount of solder is used, any excess will follow the hot iron tip right off the end.
5. Clean and inspect the wire.

If several wires are to be tinned, the soldering iron can be placed at an angle in a vise. Each wire is drawn across the soldering tip and solder added as just described. See **Figure 11-17.** This speeds up the process when many wires are involved.

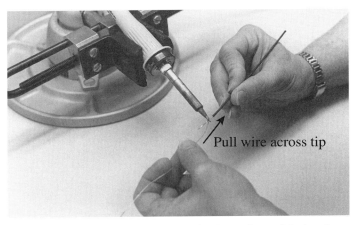

Figure 11-17. When tinning a large number of wires, the soldering iron can be held in a vise.

Tinning Components

Tinning increases the solderability of component leads. This is necessary when components have been stored for a long time and oxidation has occurred. Oxidation is the process of oxygen in the air combining with the surface of the metal and causing corrosion. Tinning of components can be done by hand using a solder pot or with an automated tinning machine.

The component is cleaned with alcohol (or another approved cleaner) and liquid flux is applied. Then, the component is carefully dipped into the solder pot, **Figure 11-18.** Resistors and other axial components can be tinned using the same procedure as for wires. Transistors and integrated circuits require some care to avoid bridging and excessive submerging of the component. If solder gets on the body of a component, the component usually becomes scrap.

Soldering Process

In the soldering process, the flux melts and cleans the surface to be soldered. It also protects the hot metal surfaces from atmospheric oxygen. The solder melts at a certain temperature depending on the type of alloy. In the liquid state, the solder becomes a solvent.

Figure 11-18. Components such as transistors and integrated circuits can be tinned using a solder pot.

Wetting is the attraction between the solder and a base metal (the metals to be soldered). The solder wets and flows out across the hot metal surfaces to be joined. The heavier liquid solder displaces the flux as it flows. A small contact angle indicates a good wetting action, **Figure 11-19.** A higher tin content also aids in the wetting action.

The solder is drawn into the joint by capillary action. The liquid solder dissolves some of the metals to be joined and produces a new alloy. The quality of this alloy determines the strength of the soldered joint. The solder then cools and solidifies.

Wetting: The attraction between the solder and a base metal.

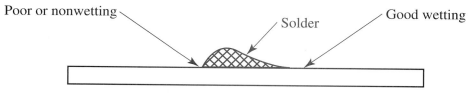

Poor or nonwetting   Solder   Good wetting

Figure 11-19. Good wetting is indicated by a small contact angle.

Hard soldering: Soldering in which metals are heated to over 800°F (427°C).

Soft soldering: Soldering done below 800°F (427°C).

Types of Soldering

Soldering can be classified as either hard or soft depending on the temperature of the metals. In *hard soldering,* the metals are heated over 800°F (427°C). Below 800°F (427°C) it is called *soft soldering.* Three types of soldering are done: point-to-point, printed circuit board, and surface mounted devices.

Point-to-point soldering involves soldering wires and components at the point of connection to the terminals. See **Figure 11-20.** The amount of heat and solder depends on the joint to be soldered.

Printed circuit board soldering uses various types of boards with copper traces (lines) used as the conductors. See **Figure 11-21.** The wires and components are set on the board with the conductor ends placed through holes and soldered on the copper side.

Surface mounted devices are placed on the surface of the printed circuit board and soldered, **Figure 11-22.** Most electronic manufacturing uses automated machines for soldering SMD-PC boards. Repairs and production changes, however, require hand soldering. Good hand-eye coordination is necessary for working with these small components.

A

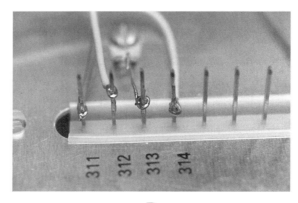

B

Figure 11-20. Point-to-point wiring. A—Components are soldered to terminals in various positions. B—The backs of plug connectors are sometimes used as soldering terminals.

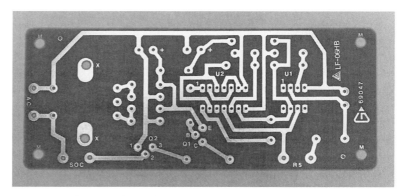

Figure 11-21. Printed circuit boards use copper traces for conductors. The width of the trace (tracks) determines the maximum current the trace can carry.

Soldering Techniques

Two techniques for soldering are described next. Common soldering is used for most electrical components. Drag soldering can be used with integrated circuits.

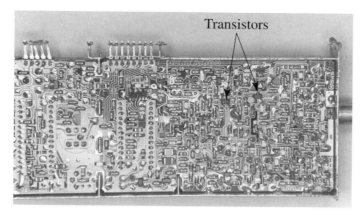

Figure 11-22. Surface mounted devices can be very small. They require a small soldering tip and a steady hand.

Common Soldering

The following procedure is used for common soldering:

1. Prepare the work for soldering by making sure the components are correctly mounted. Improper mounting is often the cause of a rejected soldering job; the soldered joint is satisfactory, but the physical mounting of the component lead or wires does not meet the standard.
2. Remove the soldering iron from the iron holder and clean the soldering tip by wiping it across a damp sponge.
3. Lightly place the soldering iron tip on the workpiece. (Pressing hard will not produce more heat). See **Figure 11-23.**
4. Place the solder at the iron tip. Melt some solder to make a heat bridge in a very small area in which heat is transferred from the iron to the workpiece. Once some of the solder has melted, a larger area is created for heat transfer. This makes it possible for the soldering iron to heat the work to be soldered in the shortest possible time. See **Figure 11-24.**
5. As soon as the heat bridge is formed, the solder is moved to the other side of the work and solder is applied to the wire ends. The amount of solder depends on the standard. As a rule, you should be able to see the contour of the wire or component lead in the solder joint, **Figure 11-25.** The solder joint should be concave and present good wetting. The concave junction formed between the two surfaces is called a ***fillet.*** Lack of a fillet is often the result of insufficient solder.
6. Quickly remove the solder and the soldering iron.
7. Clean and inspect the soldered work.

Fillet: The solder that solidifies between two or more items being soldered.

Drag Soldering

A soldering technique used in multipin devices, such as integrated circuits, is called ***drag soldering.*** The following procedure is used for drag soldering:

1. Position a length of solder across the pins of the device. Apply a liberal amount of liquid flux to the connections to be soldered.
2. Position the soldering iron as shown in **Figure 11-26.** Drag the tip of the iron across the connections to be soldered and the pins at the same time. You may have to fix a bridge, but in most cases the job will be completed.
3. Clean and inspect the soldered work.

Drag soldering: A soldering technique used in multipin devices.

Figure 11-23. The soldering iron is placed at approximately a 45° angle on the printed circuit board pad.

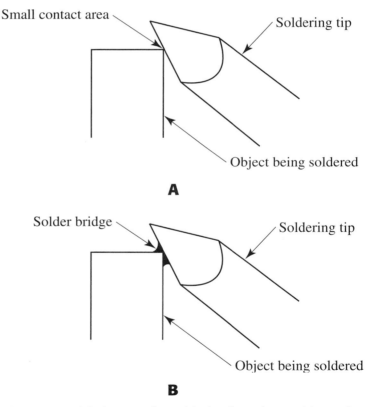

Figure 11-24. Common soldering. A—The soldering iron tip provides only a small surface for conducting heat to the joint being soldered. B—Make a heat bridge with a small amount of molten solder.

Quality of the Soldered Connection

A properly soldered joint exhibits good wetting, has a concave fillet, is smooth, and has a bright and shiny appearance. A poorly soldered connection is unreliable and may exhibit one or more of the following characteristics:

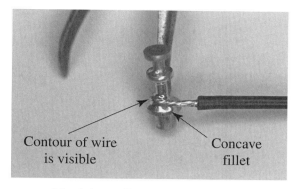

Figure 11-25. A proper solder joint will show the contour of the wire lead and a concave fillet.

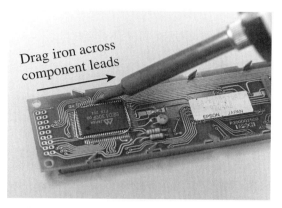

Figure 11-26. One continuous stroke is used in drag soldering.

- *Broken, nicked, or scraped conductor.* This reduces the cross-sectional area of the conductor.
- *Burned, scorched, or damaged wire insulation.* This is caused by excess heat or improper mechanical stripping.
- *Cold solder.* A **cold soldered joint** is a soldered connection made with insufficient heat. The solder appears dull and gray. It has poor or no wetting.
- *Excess solder.* If a component lead or wire contour is not visible, too much solder was used.
- *Excessive insulation clearance.* This can lead to shorting to other conductors or components.
- *Excessive wicking.* The wire does not bend at the proper point.
- *Fractures.* A fractured joint has a dull, chalky, or granular appearance and may have a crack between the soldered metals. It may also have stretch marks or resemble a cold soldered joint.
- *Heat damage to adjacent area.* This can cause unwanted conductive paths.
- *Insufficient solder.* If a PC board hole pad is not covered, the copper end is not sealed, or it contains voids or pinholes, insufficient solder was used.
- *Nonwetting.* This can cause a high-resistance connection.
- *Overheating.* When too much heat is applied to the joint, the solder has a dull or crystalline appearance and shows evidence of a grainy, lumpy, sandy, or pitted surface. The joint may appear fractured. Reworking a soldered connection without applying liquid flux can also cause these symptoms.
- *Solder bridges.* This can cause a short circuit.
- *Voids and pinholes.* These defects are caused by dirty soldering surfaces, insufficient solder, or improper heat.

Cold soldered joint: An undesirable soldered connection formed by insufficient heat.

Inspection and Grading

To ensure quality control, manufacturing industries may inspect all soldered connections. Other industries may require only a random sample inspection. Many industries and military applications require careful documentation of all processes and inspections.

A magnifying glass or microscope can be used for the inspection of soldered connections. Use a magnification of $1.5\times$ to $3\times$ for the inspection of soldered connections. A maximum of $10\times$ magnification can be used; however, if a defect cannot be seen with $10\times$ magnification, the defect does not exist.

Soldering should be done once without any rework. Manufacturing cannot make money by producing scrap or work that must be redone. Also, there is no profit when equipment is returned by the customer for repair under warranty agreements.

Removing Solder

Desoldering: The process of removing solder from a soldered connection.

Desoldering, or the removal of solder, is a subject that does not get much attention until it is needed. You may find more damage occurs during the desoldering operation than during soldering. Proper desoldering can mean the difference between restoring a piece of scrap to a working device or making scrap out of a good piece. The desoldering operation is done with solder wick braid or vacuum devices.

Solder Wick Braid

The term "wicking" was introduced earlier in the discussion of tinning conductors. However, wicking also applies to the removal of solder. In this case, *wicking* is the process of soaking up molten solder with a piece of stranded wire or solder wick braid. Solder wick braid consists of many small copper wires braided together and lightly coated with rosin flux. A spool of solder wick braid comes in different widths and lengths, **Figure 11-27.** A proper solder wick should be the same width as the connection to be desoldered. See **Figure 11-28.**

Wicking: The process of removing solder from a connection.

To properly use solder wick, place the wick on the soldered connection and the soldering iron on top of the wick. When the solder melts, pull the wick across the connection and watch the wick fill up with solder. See **Figure 11-29.** If another application is necessary, cut off the used end of the wick and repeat the process.

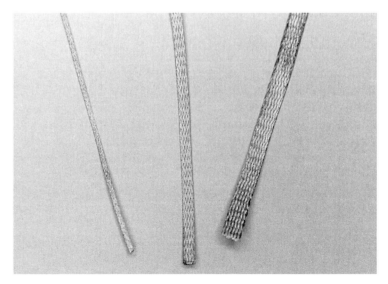

Figure 11-27. Solder wick comes in different widths on spools up to 100′ long.

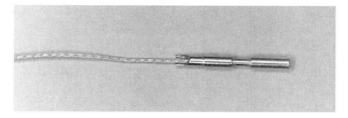

A

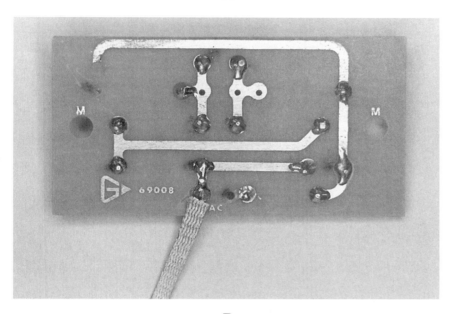

B

Figure 11-28. Solder wick. A—A small solder wick is used for a cup terminal. B—The larger width is 1/4″ in width.

Vacuum Devices

There are four basic types of vacuum devices used to remove solder: the common vacuum bulb, common iron bulb, common vacuum pump, and motor-driven vacuum pump.

A bulb-type vacuum device is used by hobbyists and service technicians, **Figure 11-30.** The bulb is collapsed and the soldered joint is heated using a soldering iron. When the solder melts, the bulb is released and the molten solder is drawn into the reservoir.

Figure 11-31 shows a common iron-bulb type of vacuum device, also used by hobbyists and service technicians. As with the previous device, this one combines the bulb and iron into one tool. The tip of the iron has a hole through which air draws the solder into a reservoir. The bulb is collapsed by thumb pressure.

 Caution: Do not collapse the bulb when the desoldering iron is over the workpiece. Molten solder from the reservoir could be blown onto the work.

The tip of the iron is placed over the soldered connection. When the solder of the connection melts, the bulb is released and the molten solder is drawn into the reservoir. After each use, the vacuum iron should be cleaned by depressing the bulb and forcing the molten solder out of the reservoir.

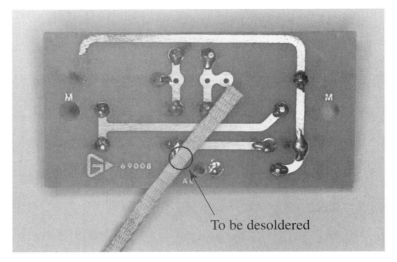

To be desoldered

A

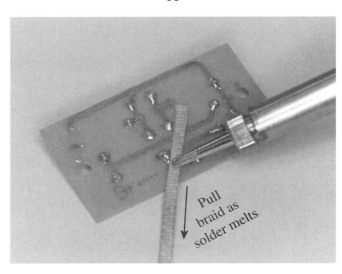

Pull
braid as
solder melts

B

Figure 11-29. Desoldering with solder wick braid. A—Place the solder wick over the connection. B—Pull the wick across the connection as the solder melts under the heat of the soldering iron.

Figure 11-30. Bulb-type vacuum device used for desoldering. The reservoir is used for collecting solder.

Figure 11-31. Desoldering iron with the bulb attached.

> ✚ **Warning:** Use care to prevent the hot solder from splattering and burning you or damaging nearby equipment.

Although a desoldering bulb is effective, it is not approved against electrostatic discharge. Static electricity can damage components. Electrostatic discharge is described in Chapter 12.

A common vacuum pump is also used for desoldering, **Figure 11-32.** The plunger is pushed into the top of the device, compressing a spring assembly inside. The device is placed next to the soldered connection at a 45° angle, and the connection is heated with a soldering iron. When the solder melts, the release button is pressed and the internal spring assembly creates a vacuum. This draws the molten solder into the vacuum device.

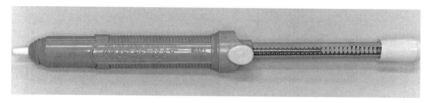

A

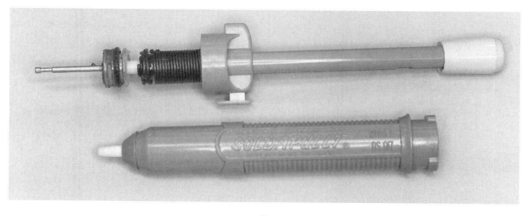

B

Figure 11-32. Vacuum desoldering devices. A—This desoldering tool has a plunger that compresses a spring assembly to make a vacuum. B—The desoldering tool can be disassembled for cleaning and replacement of internal parts.

The device can be disassembled for cleaning. A small amount of petroleum jelly on the assembly threads will make the process easier. The O-ring should also be lubricated. Replacements for the Teflon tip and O-ring are available. The plunger vacuum device comes in both ESD-approved and unapproved types. The approved type is usually black.

A soldering station vacuum pump is found in manufacturing plants or repair stations. **Figure 11-33** shows a surface mount rework station using a vacuum pickup to handle and place the components on a printed circuit board. A resistance heater with a temperature sensor is used for consistent soldering. The station can be operated in either a manual or automatic mode for repeated installation or removal of components. The nozzle vacuum, air temperature, hot air dwell time, and airflow can be controlled by the operator.

Figure 11-33. A fully self-contained surface mount rework station. Specially designed stainless steel nozzles direct the heated air onto the component leads, preventing the heating of nearby components. (American Hakko Products, Inc.)

Resoldering

Reflowed: To remelt a soldered connection.

A connection that has to be resoldered can sometimes be ***reflowed,*** that is, reheated and the present solder allowed to do the job. To properly reflow a connection, apply liquid flux and reflow the connection with the soldering iron. The flux prevents the connection from appearing overheated. If reflowing does not correct the soldering problem, it is best to remove the solder from the connection, clean the surfaces, apply liquid flux, and resolder.

Wave Soldering

Most soldering of printed circuit boards in manufacturing is done by wave soldering, **Figure 11-34.** The PC boards are mounted in a holder and placed on a moving conveyor track. The boards travel across a fluxing device. The flux removes light oxidation, prevents reoxidation, and reduces the surface tension of the solder, aiding in the wetting action.

The PC boards are preheated to 120°F (49°C). The preheater drives off the flux solvents and conditions the boards for soldering. Preheating also reduces warping caused by sudden temperature changes when coming in contact with the molten solder. Finally, the PC boards travel across a molten solder wave which solders all of the connections in one pass.

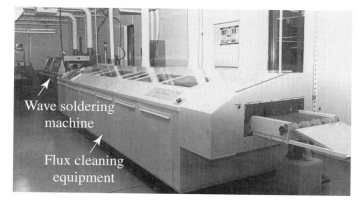

A

B

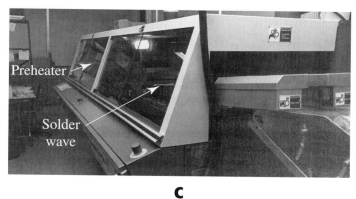

C

Figure 11-34. Wave soldering. A—All PC board connections can be soldered at the same time with a wave soldering machine. B—Printed circuit boards are being processed through a wave solder machine. C—The PC board travels across a fluxing device.

The success of wave soldered connections depends on several factors:
- Speed of the conveyor
- Angle of the conveyor rails
- Quality of the flux, usually controlled by a specific gravity test
- PC board immersion depth on the solder wave
- Preheat temperature

It takes experience and practice to operate a wave soldering machine. Any adjustment of the machine to correct defects should be made under the supervision of an experienced operator. Individual machines have different variables and "personalities." Corrections in the soldered joints must be made one at a time.

Soldering Equipment Maintenance

The maintenance of soldering equipment is often overlooked. Good maintenance consists of:

- Keeping the soldering equipment clean.
- Replacing the worn sponge used to clean the soldering tip. The sponge must be kept damp during soldering.
- Keeping the soldering tip tinned or replaced when necessary. To prevent oxidation, keep a blob of solder on the tip and store it this way.
- Cleaning a badly corroded soldering tip with sandpaper.

> **Warning:** Do not try to clean the tip while it is hot or you will get burned.

- Reshaping a copper tip with a file. Tin the tip using a solder pot. Do not file a nickel or iron plated tip. Use emery cloth to remove the corrosion.
- Applying petroleum jelly to the O-ring seals to prevent solder from sticking to them. Replacing the O-rings if needed.
- Emptying the solder chamber occasionally.

▶| Summary

- Follow safety procedures to prevent eye injuries and burns.
- Soldering standards are controlled by the Electronic Industries Association, the manufacturing industry, and the military.
- Solder is an alloy of two or more metals and comes in bar, ribbon, paste, and wire forms.
- Flux cleans and prevents oxidation of the hot metal to be soldered. Acid flux is never used in electrical work.
- Soldering equipment ranges from a simple uncontrolled pencil iron to a sophisticated, totally controlled manufacturing system.
- The size of the soldering tip must be adjusted for the type of soldering to be done.
- Wire stripping may be done by chemical, mechanical, or thermal methods. Use caution when working with Teflon insulation which emits toxic fumes when burned.
- Tinning makes a stranded wire into one solid unit thereby increasing the solderability of the wire or component.
- Wicking is the capillary action of the wire strands that causes the solder to be drawn under the insulation. Some soldering standards do not allow any wicking under the insulation.
- Tinning of wires can be done with a solder pot or soldering iron.
- Soldering types include point-to-point, printed circuit board, or surface mounted devices.
- The characteristics of an acceptable soldered connection include good wetting, a concave fillet, smoothness, and a bright and shiny appearance.
- Desoldering can be done with solder wick braid or a vacuum device.
- When resoldering a connection, an attempt can be made to reflow the connection using liquid flux. Often the best resoldered connection is made by removing the original solder and starting the process over again.
- Wave soldering is used in manufacturing to solder printed circuit boards.
- Proper maintenance of soldering equipment minimizes halts in production.

Important Terms

Do you know the meanings of these terms used in the chapter?

acid flux
alloy
birdcaging
cold soldered joint
desoldering
dross
eutectic alloy
fillet
flux
hard soldering

reflowed
rosin flux
soft soldering
solder
soldering
stripping
tinning
wetting
wicking

Questions and Problems

Please do not write in this text. Write your answers on a separate sheet of paper.

1. List six general safety precautions to follow when soldering.
2. List five safety precautions to observe when working with a solder pot.
3. What are the three standards used in soldering?
4. What metals are used in making solder alloys?
5. What is the purpose of soldering flux?
6. What type of flux should *not* be used in electrical soldering?
7. What are the three methods of wire stripping?
8. Why are stranded wires tinned?
9. Define *wicking*.
10. What are the two methods of tinning a wire?
11. Why are components tinned before soldering?
12. What is the difference between hard soldering and soft soldering?
13. List the three types of soldering.
14. List the characteristics of a properly soldered connection.
15. List five characteristics of a poorly soldered connection.
16. What are the two methods of removing solder?
17. When desoldering, why should a plain desoldering bulb *not* be used to remove the solder on circuit boards?
18. When reflowing a soldered connection, what would cause the finished soldered connection to have an overheated appearance?
19. What type of automated soldering process solders multiple connections in one pass?
20. List five considerations in the maintenance of soldering equipment.

Connectors — types are:
- Phone Plug
- Phono Plug
- XLR
- PL-259
- F-connector
- D-connector
- DIN
- PC Edge Board

Terminals — such as:
- Eyelet
- Hook
- Turret
- Cup
- Bifurcated
- Solderless

Board Assembly — can use Terminals

Board Assembly — may be:
- Stud-through Board
- Clinched
- Swagged
- Surface Mount

Board Assembly — may need:
- Stress Relief
- Electrostatic Protection

Electronic Assemblies — may be:
- Point-to-Point
- Breadboarding
- Perfboard
- Printed Circuit Board

Printed Circuit Board — may need:
- Track and Pad Replacement
- or Burn Repairs

Printed Circuit Board — may have Conformal Coating — protects from Dirt and Moisture

ASSEMBLY AND REPAIR TECHNIQUES ▶▶ **12** ◀

▶◀ Objectives

After studying this chapter, you will able to:

○ Discuss the causes, effects, and precautions against electrostatic discharge.
○ Describe the various types of electronic assemblies.
○ Describe the four methods of mounting component leads on a printed circuit board.
○ Describe the types of terminals used in the electronic industry.
○ Explain the procedures for making a splice and a coax breakout.
○ Explain how to properly assemble various connectors.
○ Explain how to make track and pad replacements and burn repairs on circuit boards.

▶◀ Introduction

This chapter covers the assembly and repair of electronic components, printed circuits, and connectors. The effect of electrostatic discharge on components, electronic assemblies, and printed circuits is also covered. Knowledge of basic tools and equipment is assumed.

Electrostatic Protection

Almost everyone has experienced a small dose of static electricity when walking across a carpet or sliding across an automobile seat and touching the door handle. Snap! The resulting jolt of static electricity is called *electrostatic discharge* (ESD). As you will learn, components and equipment must be protected from damage by ESD.

Electrostatic discharge: The movement of charges from one point to another.

Causes and Effects of Static Electricity

Static charge is caused by the transfer of electrons within a substance or between different materials. The amount of charge depends upon the type, size, and shape of the material, and the temperature and humidity of the air. The transfer of electrons from one material to another is accomplished by rubbing the two materials together. The generation of static electricity is called the *triboelectric effect.* Some materials tend to give up electrons easily, while others are better at accumulating electrons. Thus, some materials may be positively charged, while others are negatively charged, **Figure 12-1.** A material higher on the list is positively charged when rubbed with a material that is lower on the list. This is because materials higher on the list have more free electrons. Rubbing together two pieces of the same material, such as plastic, can also generate an electrostatic charge.

Triboelectric effect: Rubbing two materials together to generate static electricity.

An electrostatic field exists between a charged body and a body at a different electrostatic potential, **Figure 12-2.** Any conductive or insulative material in this field will be polarized by induction. Electrons near the negative part of the field are repelled, creating a positively charged area. Electrons nearest the positive part of the field are attracted, creating a negatively charged area. Negatively and positively charged areas are created while the net charge on the material remains at zero. Electrons will flow to or from the material, depending on the area grounded.

Charge	Material
Most positive (+)	Air
	Human hands
	Glass
	Mica
	Human hair
	Nylon wool
	Silk
	Paper
	Cotton
	Wood
	Sealing wax
	Acetate rayon
	Polyester
	Orlon
	Polyurethane
	Polyethylene
	PVC (vinyl)
	Silicon
Most negative (–)	Teflon™

Figure 12-1. Common materials and the triboelectric range. Materials higher on the list give up electrons, making them positively charged, while materials lower on the list become negatively charged.

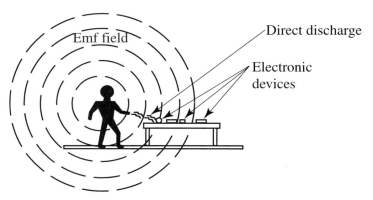

Figure 12-2. When an emf field surrounds a person, everything within the field is affected. A static current unnoticed by a person can be strong enough to damage components or equipment.

When the capacitance of a charged material, compared to another material, is reduced, the electrostatic field is affected. As the capacitance is decreased, the voltage is increased until the electrostatic charge discharges by an apparent arc. If two plastic bags are rubbed together and placed on a workbench they may have only a few hundred volts of charge potential. If a person were to pick them up off the bench, several thousand volts may be suddenly available because of the decrease in capacitance.

If a device, such as a transistor or integrated circuit, that is sensitive to ESD comes in contact with an electrostatic charge, the currents generated in the device can destroy or shorten its life. See **Figure 12-3.** For example, if a device has a resistance across its terminals of 5 megohms, and a 15,000 V ESD comes in contact, a current of 3 mA will flow through the device.

Thus:

$$I = \frac{V}{R}$$

$$= \frac{15 \times 10^3 \text{ V}}{5 \times 10^6 \text{ }\Omega}$$

$$= 3 \text{ mA}$$

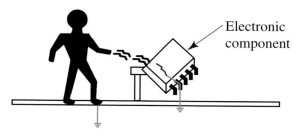

Figure 12-3. Electronic components can be damaged by electrostatic discharge. Even small film resistors can be damaged by ESD currents.

Although this is a small current, modern devices may have internal conductors that can only carry a maximum of 10 microamps or less. This results in a burnout of the device. Other components that may be sensitive to ESD are metal oxide semiconductor (MOS) transistors, piezoelectric crystals, film resistors and other film devices, surface mounted devices, and electronic printed circuit assemblies.

Prime Sources

The electrostatic voltage levels generated by some common materials can be extremely high, **Figure 12-4.** A voltage of 15,000 or more is not unusual for common plastics. **Figure 12-5** lists some typical voltages generated by various materials. Notice how low humidity affects the amount of voltage generated.

ESD Grounding

Two types of grounding are considered in electrostatic discharge. ***Hard ground*** is a connection that is direct or through a very low impedance. ***Soft ground*** is a connection through an impedance high enough to limit the current flow to a safe level (usually 5 mA or less). A soft ground is also used for dissipating ESD to ground. If a hard ground is used, the current will be high enough to cause damage to ESD-sensitive components and equipment. More importantly, injury to people could result if a hard ground is used.

Hard ground: A connection directly to ground.

Soft ground: A connection to ground through an impedance high enough to limit the current to a safe value.

Sources	Types
• Floors	• Vinyl tile • Sealed concrete • Waxed, finished wood
• Work surfaces	• Vinyl or plastic • Waxed, painted, or varnished wood
• Chairs	• Finished wood • Vinyl or upholstering • Fiberglass
• Assembly, test, and repair areas	• Spray cleaners, cleaning brushes • Soldering irons – ungrounded • Electrostatic copiers
• Packaging, shipping, and parts bins	• Plastic bags, wraps, and envelopes • Plastic bubble and foam packaging • Plastic trays and tote boxes
• Clothing and personal items	• Personal synthetic garments • Nonconductive shoes • Combs, hair brushes, and coats

Figure 12-4. Common sources of electrostatic voltage.

Voltage Generated By	Electrostatic Voltage	
	Relative Humidity	
	10%–20%	**65%–90%**
• Walking across a carpet	35,000	1,500
• Walking across a vinyl floor	12,000	250
• Plastic envelopes in repair areas	7,000	600
• Plastic bag picked up from the workbench	20,000	1,200
• Workbench chair • Clothing and personal items	18,000	1,500

Figure 12-5. Typical voltages generated by various materials.

Protected Areas

ESD can be controlled by protecting the area where electronic work is done. In a manufacturing facility, the protected area may be as small as a single workbench and its surrounding area or as large as an entire building. A repair service should set aside at least one work station that contains sufficient protection to ensure components and equipment will not have uncontrolled ESD exposure.

Figure 12-6 illustrates the essential components of an ESD workstation. The top of the workbench should have a static dissipative top or mat, a wrist strap for the bench worker, and a grounded mat or floor. An additional wrist strap should be available for anyone else that may come to the workbench. All these items are grounded to the common ground. Chairs should be made of an antistatic upholstery and may have a drag chain on a dissipative floor. A dissipative floor is coated with conductive paint or tiles. The lighting in the soldering area should be 100 foot-candles or more, diffused over the complete work surface. A combination of fluorescent and incandescent lighting is best.

Controlling temperature and humidity can help reduce ESD problems. An ionizer can be used to dissipate electrostatic charges. Nonstatic-generating or static-dissipative clothing, such as specially made workcoats, should be worn.

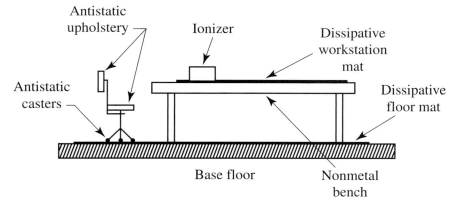

Figure 12-6. Typical ESD workstation. The workstation mat and floor mat should be connected to ground through a ground cable. These surfaces should limit the current to a maximum of 5 mA to protect personnel.

Why ESD Protection?

An ESD-sensitive component or PC board received for repair must be protected. Otherwise, a defective part may unknowingly be used. The repair may last for a time but eventually will break down. If the item being repaired is a television receiver or videocassette recorder, the customer will be inconvenienced. Suppose, however, the repair is a heart monitor, a guidance device for a passenger aircraft, or a guided missile. In such cases, the outcome will be disastrous. Chip resistors, surface mounted components, metal oxide semiconductor field-effect transistors (MOSFET) and the printed circuit boards on which they are mounted are a few examples of ESD-sensitive devices.

Protecting equipment and devices from ESD is not just the responsibility of the worker at the bench. Everyone who comes in contact with ESD-sensitive devices must be educated in proper protective procedures. Service repair technicians, engineers, shipping personnel, department supervisors, even the secretary who enters the manufacturing floor or service bench to deliver paperwork must all be aware of ESD and trained in preventative measures.

Types of Electronic Assemblies

Most electronic assembly is accomplished by printed circuit board construction. Other types of assembly are used as well in manufacturing, product development, repair work, or hobby projects.

Point-to-Point Wiring

In point-to-point wiring, the wires and component leads are connected to switch terminals, terminal strips, connectors, wiring to and from printed circuit boards, and other types of devices. See **Figure 12-7.**

Breadboarding

Breadboarding is an ideal assembly for performing lab experiments and exploring changes in the design of a circuit, **Figure 12-8.** It is named for the flat wooden boards that once were used to mount components. The circuit is neatly assembled and color coded. Careful breadboarding makes it easier to troubleshoot or find your way around the circuits. A haphazard placement of wires makes it impossible to change the integrated circuits without disassembling some of the wiring. Disassembly could introduce errors into the circuit.

Breadboarding: Using a flat plastic board to mount components.

Perfboard

Perfboard or Vectorboard™ is good for constructing one or two circuits. **Figure 12-9** shows a simple one-transistor project constructed on perfboard. Construction is accomplished with point-to-point wiring, using flea clips to mount the transistor. This type of construction makes it easy to modify the wiring and requires only basic tools.

Wire wrap is a type of perfboard construction in which the pins of integrated circuit sockets and other devices are wrapped with 30-gage conductor wire. See **Figure 12-10.** Perfboard construction makes for easy corrections or changes in the wiring.

A

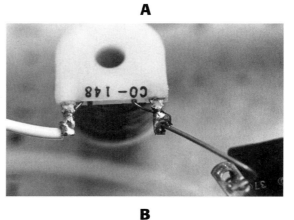

B

Figure 12-7. Point-to-point wiring. A—Point-to-point is used between printed circuit boards and other areas of equipment. B—Point-to-point wiring of the terminals of a coil to other parts of the circuitry.

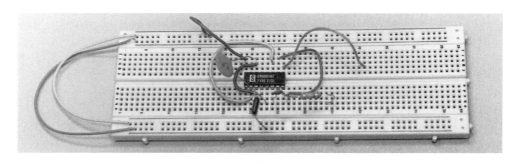

A

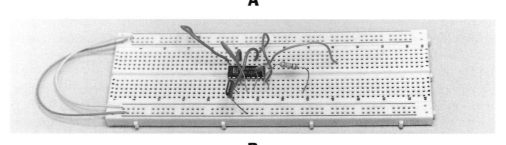

B

Figure 12-8. Breadboarding. A—Proper breadboarding saves time and eliminates errors. B—An unorganized assembly with wires over integrated circuits could cause errors to be introduced.

Figure 12-9. A perfboard assembly is useful for a one-time project. This type of construction makes connections easy and allows for modifications to the project.

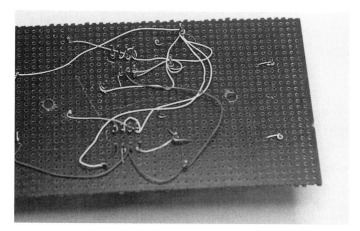

Figure 12-10. A wire-wrap assembly does not require soldering. Care must be taken in routing the many wires that could cross each other.

Printed Circuit Boards

A printed circuit (PC) board typically includes the board identification number, orientation marks, conductor runs (trace or tracts), pads, and substrate (PC board material). See **Figure 12-11.** Although PC boards can be made of various materials, the glass epoxy type is the industry standard. Boards made of phenol or other materials have problems with warping, which can cause circuit problems over time, **Figure 12-12.**

Conformal Coatings

Conformal coatings are placed on printed circuit boards to protect the circuitry from dirt and moisture. Various types of coatings are available and each requires careful consideration if it must be removed. For example, what would happen to a glass epoxy PC board if chemicals were used to remove the epoxy coating? Hot air should be used first to remove epoxy and polyurethane coatings. Acrylic coatings can be removed with a mild solvent like isopropanol. In any case, remove only the coating necessary to make the repair.

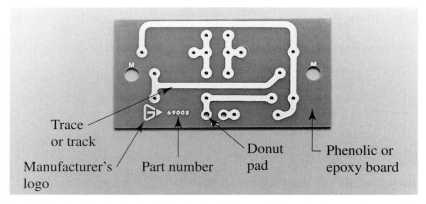

Figure 12-11. Parts of a typical printed circuit board. The design and manufacture of printed circuit boards is a major part of the electronics industry.

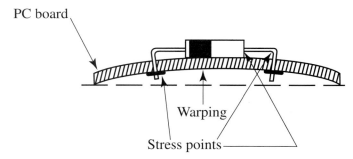

Figure 12-12. An inexpensive printed circuit board can easily become bent or warped, causing loose soldered connections.

Repair and Assembly

Correct assembly of a device may be the most important step in making a successful, lasting repair. The proper assembly of printed circuit boards, terminals, resistors, capacitors, and other parts depends on the standard to be met. When a standard is not available, certain guidelines can be followed.

Mounting Component Leads on PC Boards

The labels of components mounted on PC boards or other equipment should be readable, and the color bands of resistors should be mounted in the same direction. See **Figure 12-13.** There are four methods of mounting component leads on a printed circuit board: stud, clinched, swagged, and surface mounted.

Stud Mounting

Figure 12-14 shows the stud or straight-through mounting of an axial lead component. In this type of mounting, the component leads go straight through the PC board and are soldered. Depending on the standard, the excess lead length can be cut before or after soldering. The component should be down against the PC board for support. If the component has high heat dissipation, it may need to be mounted away from the PC board to prevent heat damage to the board and provide air circulation around the component.

Figure 12-15 illustrates a common error made when mounting axial lead components. The component lead on the right has been pulled through the mounting hole. This causes stress at the point where the lead comes out of the component. The lead can easily break loose from inside the component, resulting in an open or an intermittent connection which will be difficult to locate.

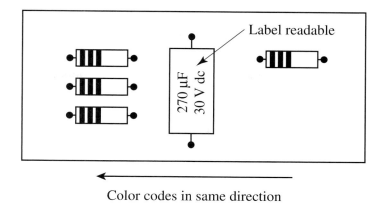

Color codes in same direction

Figure 12-13. Resistors should be mounted with colors in the same direction. The capacitor and other component values should be readable.

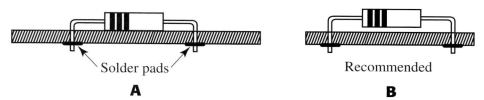

Solder pads

Recommended

A

B

Figure 12-14. Stud mounting. A—The leads of stud-mounted components go straight through the PC board. B—Components should be mounted against the PC board unless they have high heat dissipation.

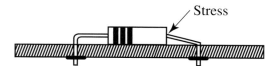

Stress

Figure 12-15. Improper mounting of a component places stress where the leads connect.

Stress relief. Stress relief is necessary because of the mechanical vibration, or G-forces, the components may endure. For example, a machine that vibrates a certain way all the time will cause the components to repeatedly bend back and forth the same way. Eventually, a component lead will break. Likewise, electronic components in fighter aircraft will have recurrent G-forces placed upon them. Because of G-forces, components should not have any part of the solder fillet in the bend radius, **Figure 12-16.** This makes the bend rigid, and the component lead will tend to break at the bend.

Various methods of stress relief are shown in **Figure 12-17.** Such methods eliminate the solder fillet in the bend radius and are especially useful for small-diameter components close to the PC board.

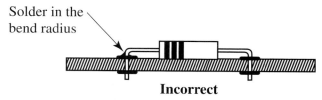

Solder in the bend radius

Incorrect

Figure 12-16. Solder in the bend radius will make the mounting rigid and cause breakage over time.

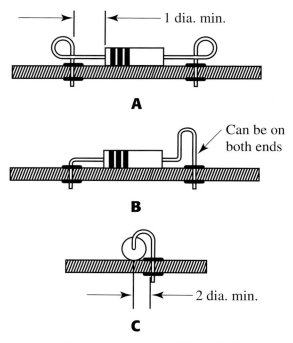

Figure 12-17. Stress relief. A—A loop on each end (double-loop) is a popular method for providing stress relief. B—The hump-type of stress relief is often used on very small components to keep solder out of the bend radius. C—A variation of the hump-type stress relief is to the side rather than in line with the component.

Another method of reducing stress is to secure the component to the PC board with a mounting bracket or adhesive. See **Figure 12-18.** Other component mountings are shown in **Figure 12-19.** The vertical mounting is not used in high G-force applications.

Clinched Mounting

Component leads can be clinched inward or outward, depending on the direction of the PC tracings, **Figure 12-20.** The lead should be clinched in the direction of the trace, **Figure 12-21.**

The component leads should be cut off before soldering. By placing a piece of plastic between the PC board and the lead before bending, the lead can be cut without damage to the board. The lead should be clinched as close to the PC board as possible, using a wooden (lemon) stick, **Figure 12-22.** Lemon wood is well-suited to clinching because it will not mar the PC board.

Swagged Mounting

Swagged mounting is a method of securing the component lead on the solder side of the lead hole, **Figure 12-23.** The swagging tool crushes the lead and cuts it off in one operation. The lead is soldered in the normal manner.

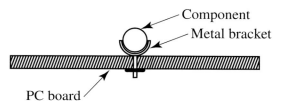

Figure 12-18. To relieve stress of larger components, brackets or adhesive are used where vibration may cause failure.

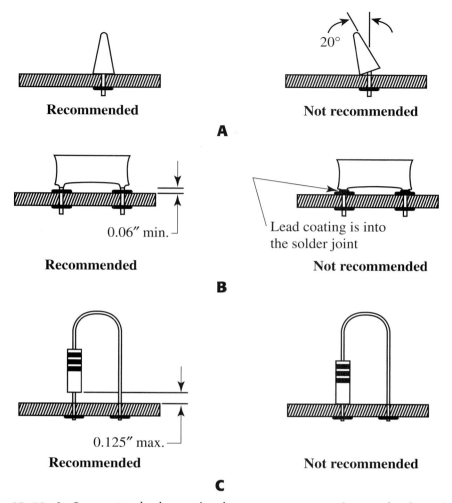

Figure 12-19. A—Some standards require the components to sit completely upright. B—The coating of a component should not be in the solder joint. C—Some assemblies allow vertical mounting of components.

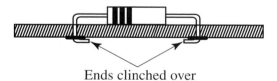

Ends clinched over

Figure 12-20. Some standards require the component leads be clinched for proper mounting. After clinching, the component is soldered.

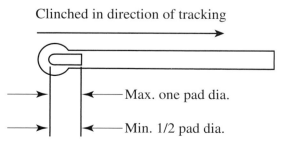

Figure 12-21. The clinch should be in the direction of the trace. The length of the cut component lead should not be greater than one diameter or less than half of the wire diameter.

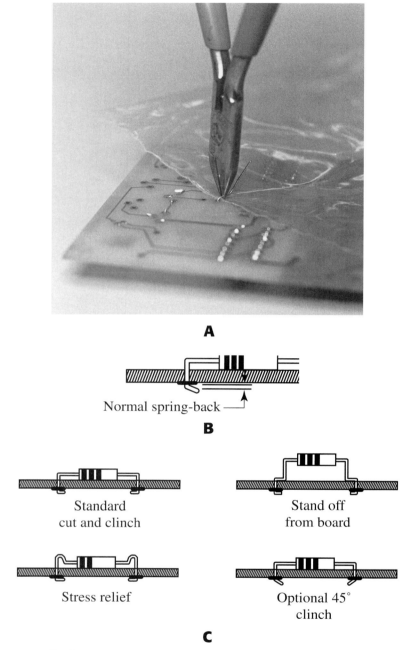

A

Normal spring-back

B

Standard
cut and clinch

Stand off
from board

Stress relief

Optional 45°
clinch

C

Figure 12-22. Clinch mounting. A—Some standards require the component leads be cut before soldering. B—Some spring-back of the lead is normal. C—Various clinch patterns.

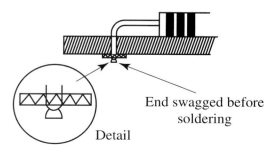

End swagged before
soldering

Detail

Figure 12-23. Swagged mounting. This type of mounting allows the component to fit loosely before soldering.

Surface Mounting

Surface mounted devices (SMD), resistors, transistors, and integrated circuits are often tiny and require soldering under magnification. Sometimes it is best to glue the component in position and use a very low-wattage, grounded soldering iron. Surface mounted devices may require modified soldering tips and special hand tools.

A common hand-soldering method places solder paste (solder beads in liquid flux) on the footprint areas, **Figure 12-24.** A hot air tool is used to heat the solder paste and complete the soldered connection. Glue secures the SMD so it does not move out of position.

MIL-STD 2000 specifies the surface mounted component can be skewed out of alignment by a maximum of 25% of the lead width, **Figure 12-25.** The ends of the SMD leads cannot extend beyond the PC land by more than 25% of the lead width or 0.5 mm, whichever is less. The component leads can be bent up or down in the same manner.

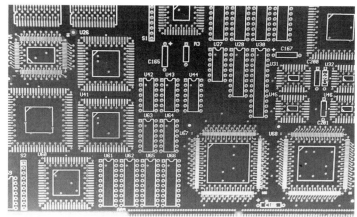

Figure 12-24. Surface mounted component with solder paste on the printed circuit footprint. The solder paste contains small balls of solder which melt as the PC board is put through an oven, completing the soldering process.

Terminals

Several types of terminals are used in the electronic industry. They include eyelet or pierced, hook, turret, cup, bifurcated, and specialized terminals.

Eyelet Terminal

The eyelet terminal is commonly found on switches, connectors, and other devices. The number of wires that can be connected to a single eyelet depends on the soldering standard.

The following procedure is used to install a wire on an eyelet terminal, **Figure 12-26**:
1. Place the wire through the opening, leaving an insulation clearance above the top or side of the terminal equal to one diameter of the wire insulation.
2. Using needle nose pliers, bend the end of the wire over the terminal.
3. Cut the wire end flush with the edge of the terminal. The wire should touch the sides of the terminal.
4. Place the soldering iron at the back of the terminal and make a heat bridge. Bring the solder out to the end of the wire and finish the soldered connection.

Various wire mounting positions and insulation clearances are shown in **Figure 12-27.**

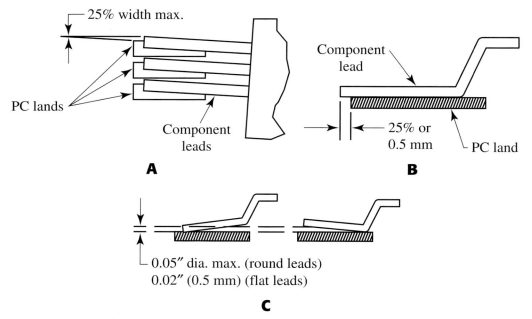

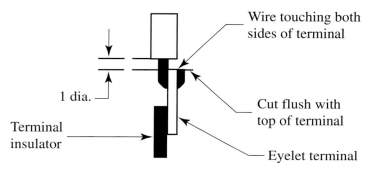

Figure 12-25. Allowable alignment skew. A—A small misalignment is allowed for surface mounting. B—Surface mount lead overhang. C—Surface mounted lead tipped.

Figure 12-26. Installing a wire on an eyelet terminal. The conductor should touch the sides of the terminal. A one-diameter insulation clearance prevents conductors from shorting out against each other.

Hook Terminal

Soldering a hook terminal is similar to soldering an eyelet terminal, **Figure 12-28.** No more than three wires should be attached to it.

Turret Terminal

The turret terminal is used for connecting wires and components to the PC board, **Figure 12-29.** The turret is swagged onto a PC board and soldered. Use care when swagging the terminal onto the PC board as the wire shaft can bend easily.

A turret terminal has several parts, **Figure 12-30.** The common wire wrap on a turret terminal is 180°; however, the wrap can go completely around the terminal. The end of the wire should not overlap the conductor. See **Figure 12-31.**

The following procedure is used to solder a turret terminal:

1. Form the wire around the terminal with the required wrap.
2. Place the soldering iron at the side of the terminal, **Figure 12-32.**
3. As soon as the solder begins to flow, carefully add solder to the connection by wiping across the cut end of the wire.

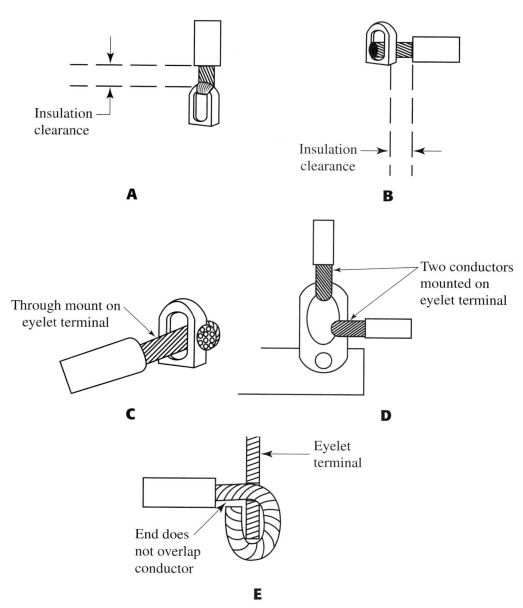

Figure 12-27. Mounting positions and insulation clearances. A—Insulation clearance should be no greater than one diameter of the wire insulation. B—Side mount of a wire on an eyelet terminal. C—Through mount of a wire on an eyelet terminal. D—Two conductors mounted on an eyelet terminal. E—When using a 360° wrap on an eyelet terminal, the wire end should not overlap the conductor.

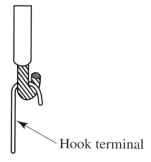

Figure 12-28. The usual 180° bend around the J-hook. Solder must not completely fill the J-hook and cover the contour of the conductor.

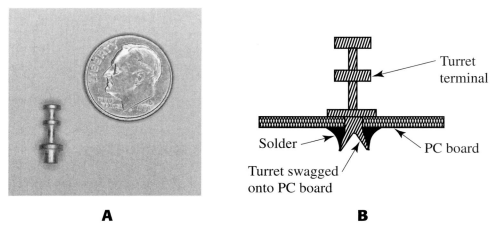

A   **B**

Figure 12-29. Turret terminal. A—A turret terminal can be quite small. (Dime shown for comparison.) B—A turret terminal is inserted into a hole on the PC board, swagged in place, and soldered.

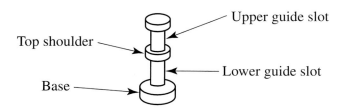

Figure 12-30. Parts of a turret terminal. Assembly instructions specify the guide slot on the terminal and the number of conductors to place on each.

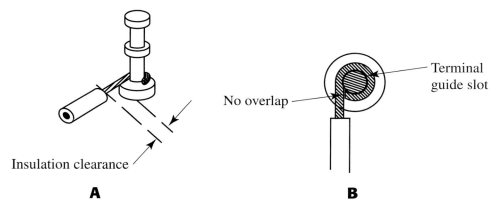

A   **B**

Figure 12-31. Wire wrap on turret terminal. A—Common wrap. B—The end of the wrap should not overlap the conductor.

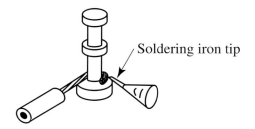

Figure 12-32. Placement of a soldering iron for a turret terminal.

Cup Terminal

Many cup terminals are gold plated. Some standards require the gold plating be removed before soldering a wire into the cup. In theory, the gold combines with the solder alloy to make a different alloy that results in an inferior soldered connection.

To remove the gold plating, fill the cup with molten solder. Using a small solder wick, remove the solder from the cup, **Figure 12-33.** Some standards require this procedure be done twice.

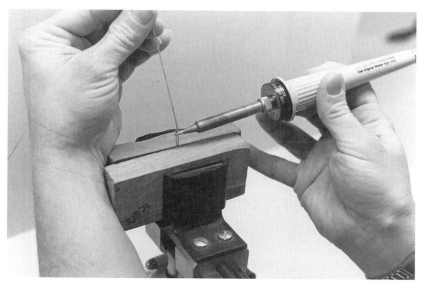

Figure 12-33. Remove the gold plating inside the gold cup terminal. If the gold stays in the soldered connections, a new alloy is created, which produces an inferior soldered joint.

The tinned wire should be the proper size to fit the inside cavity of the cup terminal, with the end of the wire touching the bottom of the cavity, **Figure 12-34.** The soldering iron is placed at the front of the cup, and solder is added to fill it. A fillet should be present at the backside of the top. Some standards specify no solder outside the cup reservoir, while others allow a thin film of solder with no visible contact angle at the edge of the solder.

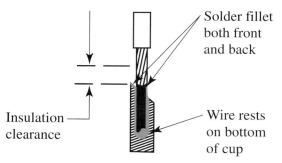

Solder fillet both front and back

Insulation clearance

Wire rests on bottom of cup

Figure 12-34. The wire should fit the cavity of the gold cup. If the wire is too large for the cup cavity, the solder joint may have insufficient solder and come loose.

Bifurcated Terminal

Figure 12-35 shows two types of top conductor connections to bifurcated terminals: a single conductor and a terminal with a larger opening. The conductor is doubled back on itself to take up the extra space in the terminal. Another type of connection to the bifurcated terminal is shown in **Figure 12-36.**

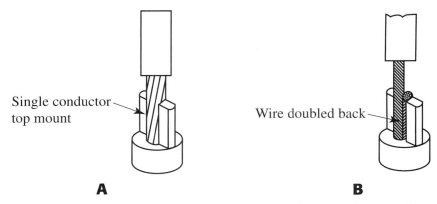

Single conductor
top mount

Wire doubled back

A   **B**

Figure 12-35. Bifurcated terminal. A—Top mount of a bifurcated terminal. B—Bifurcated terminals with larger openings may require the conductor to be bent back on itself.

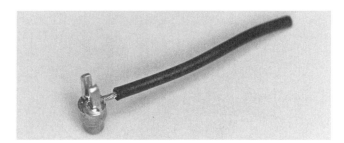

Figure 12-36. A popular method of connecting to a bifurcated terminal is the side connection.

Solderless Connectors

Sometimes connections are made using solderless connectors, **Figure 12-37.** These connectors are simply crimped onto the conductor.

The following procedure is used to make a solderless connection:
1. Cut a piece of heat shrink and place it on the wire.
2. Strip the wire. The stripped end should not extend beyond the connector crimping area.
3. Crimp the connector onto the wire using a crimping tool of the correct size, **Figure 12-38.**
4. Pull the piece of heat-shrink over the crimped end of the connector.
5. Using a heat gun and keeping the gun in motion, shrink the heat-shrink tubing, **Figure 12-39.** The heat-shrink keeps the conductor from breaking at the edge of the connector.

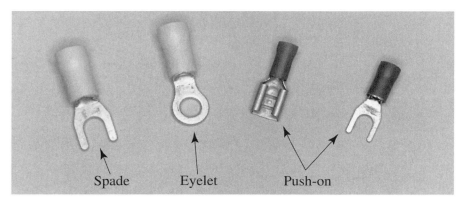

Spade   Eyelet   Push-on

Figure 12-37. Solderless connections. Popular spade, eyelet, and push-on terminals are crimped onto the wire without soldering.

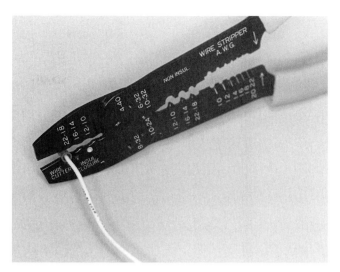

Figure 12-38. Crimping with a crimping tool. Using the incorrect die or excessive crimping pressure can cause the terminal to break or the terminal wire connection to fail after time.

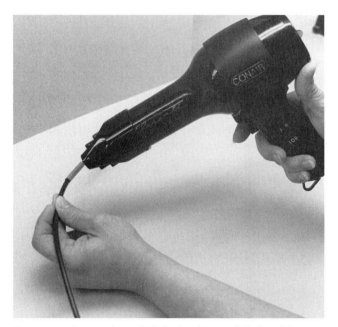

Figure 12-39. A heat gun is used to shrink the heat-shrink tubing. To properly use a heat gun, keep it moving continuously.

Splices

Splices on wire conductors are usually not permitted. In some cases, though, splicing may be necessary or required by the standard.

One method of making a splice is shown in **Figure 12-40**:

1. Place a piece of heat-shrink over the wire. Be sure the heat-shrink tubing is large enough for the finished splice.
2. Remove a portion of the insulation where the splice is to be made. Do not damage the conductor.
3. Strip the end of the wire to be spliced.
4. Tightly wrap the wire around the conductor.
5. Solder the splice connection.
6. Pull the heat-shrink tubing over the splice and use a heat gun to finish the splice.

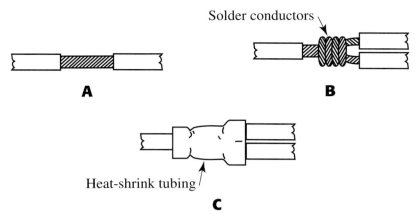

Figure 12-40. Making a splice. A—Carefully strip the insulation from the wires. Place heat-shrink tubing over the main conductor. B—Wrap the stripped ends together and solder. C—After the solder joint cools, push the heat-shrink tube over the splice and shrink it using a heat gun.

Making a Coax Breakout

When working with coax cable, often you must make a breakout. That means the center conductor must be brought out through the outside braid (shield).

The following procedure is used to make a coax breakout, **Figure 12-41**:

1. Carefully remove the required amount of outer insulation. Be careful not to cut or nick the wire braid under the insulation.
2. Push the wire braid back on itself, making a bulge close to the outer insulation.
3. Use a small, thin tool to make an opening in the bulged portion of the braid.
4. Work the tool under the inner conductor, and ease the conductor through the opening in the braid.
5. Check your work and inspect for damage.

Connector Assembly

Some common connectors and their assembly are discussed next. The skills for working with these connectors are transferable to other types of connectors. Presuming soldering is done properly, the main problem with connectors is sloppy physical assembly.

Phone Plug

The phone plug is one of the more common connectors you will be required to assemble. Phone plugs come in both soldered and screw-terminal types.

The following procedure is used to assemble phone plugs, **Figure 12-42**:

1. Carefully cut the wires to the proper length to avoid excess wire inside the connector.
2. Tin the ends of the wires. Stranded conductors are often used for audio. Tinning the ends will allow the wire strands to go through the terminal holes of the connector without missing the hole or fanning out from under the screw terminal. This is a common problem with improper assembly, which causes shorts and intermittent operation of the audio cable.
3. Before soldering, make sure the wires are on the proper terminals. Make a sketch of the wiring for later reference so you do not forget how you wired the plug.
4. Make sure the plug cover fits properly, and check the quality of soldering.
5. Crimp the strain-relief over the cable, and put the plug cover in place.

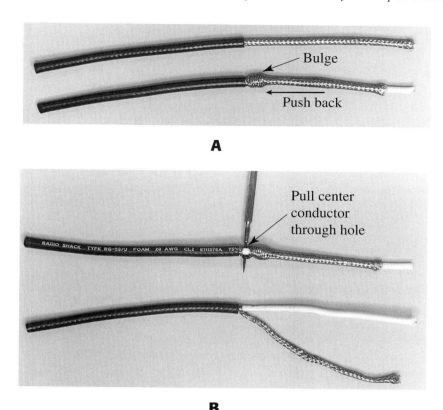

Figure 12-41. Making a coax breakout. A—Carefully remove the outer insulation and check for damage (top). Push the wire braid back on itself, making a bulge close to the outer insulation (bottom) B—Using a small thin tool, make an opening in the bulged portion of the braid (top). Work the tool under the inner conductor and ease the conductor through the opening in the braid (bottom).

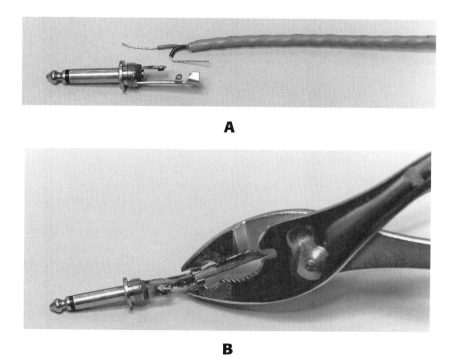

Figure 12-42. Phone plug assembly. A—Cut the wires to the correct length and tin the ends. B—Crimp the strain relief.

RCA Phono Plug

Though the RCA phono plug is a very simple connector, special care must be taken in its assembly. The following procedure is used to connect a phono plug, **Figure 12-43**:

1. Place a piece of heat-shrink tubing over the cable.
2. Cut the wires to proper length. Make sure the insulation of the center conductor will reach the center terminal.
3. Tin the ends of the conductors to keep the strands together.
4. Place the center wire into the plug, making sure the insulation reaches to the center terminal. Carefully form the other conductor around the outer terminal.
5. Hold the center wire away and solder the conductor onto the outer terminal. The center conductor must not touch the outer terminal; otherwise, the heat generated by soldering will melt the insulation and cause a short circuit.
6. When the outer terminal has cooled, pull the heat-shrink tubing up over it, and use a heat-gun to shrink the tubing.
7. Solder the center conductor at the end of the center terminal. If the outer terminal is soldered last, the process of connecting the outer terminal usually places stress on the center terminal connection.

 Caution: The soldered connection should be made in two or three seconds; otherwise, excess heat may travel up the center terminal and damage the plug insulation.

8. Cut the center conductor off, and carefully file the end of the center terminal round and smooth.
9. A continuity test will verify proper cable operation.

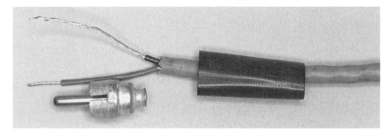

A

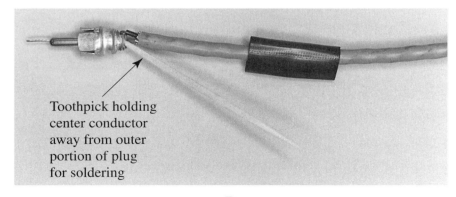

Toothpick holding center conductor away from outer portion of plug for soldering

B

Figure 12-43. RCA phono plug assembly. A—Cut the wires to the correct length. B—Hold the center conductor and solder.

XLR Connector

The XLR connector usually has an equipment grounding lug and three cup-type terminals; therefore, it requires a three-conductor cable with ground (shield). See **Figure 12-44.** Two procedures are important in the assembly of the XLR connector. First, the conductors must be connected to the same pin on both ends of the cable. Second, the conductor insulation must be very close to the cup terminal to eliminate any possibility of their shorting out one another.

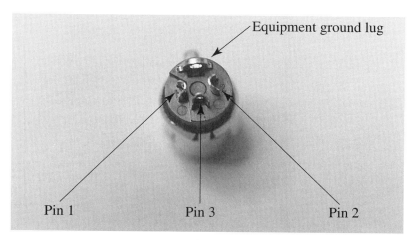

Figure 12-44. Terminals of an XLR connector. With a shielded three-conductor cable, terminals 1 and 2 are used for the signal conductor. Terminal 3 is used for the signal ground and equipment ground lug for the cable shielding.

The following procedure is used for XLR connectors:
1. Disassemble the connector, and place the outer cover portion on the cable.
2. Cut and strip the wire to align with the terminals, **Figure 12-45.** Proper physical alignment will make the soldering easier.
3. Tin the conductors.
4. Place the appropriate conductor into terminal 3 and solder.
5. Place the conductor into terminals 1 and 2, and solder. Make sure stress is not being placed on conductor 3.
6. Connect the shield to the equipment ground lug and solder.
7. Inspect all connections and the physical assembly.
8. Use continuity tests to verify proper connections from one end of the cable to the other.
9. Reposition the XLR connector cover over the assembly.

PL-259 and F-Connector

The solder-type PL-259 and F-connectors have been replaced with a crimp-type connector. As with any connection, correct assembly, placement of the cable into the terminal, and crimping are important factors.

Other Connectors

Other connectors, such as the D-connector, DIN, and printed circuit edge board connector contain either cup or eyelet terminals. Use the preceding methods to assemble these connectors.

With a D-connector, a small piece of heat-shrink can be placed on the eyelet terminals to prevent shorting. See **Figure 12-46.**

Figure 12-45. As with other connectors, cut, align, and tin the ends of the wires for an XLR connector.

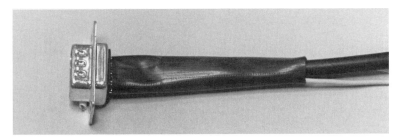

Figure 12-46. Heat-shrink tubing is placed over the connections to relieve stress in this D-connector.

Printed Circuit Repairs

Track and pad repairs are needed for several reasons. These include improper desoldering techniques, lack of proper desoldering equipment, technician's lack of experience and training, repeated repair of the same track or pad, or excess current flowing in the track due to a circuit overload.

Tracks and pads are glued to the surface of the printed circuit board. A lifted track is usually caused by excessive heat applied to the track area or repeated heating and cooling of the track or pad. The glue loses approximately 20% of its integrity each time it is heated. Often a pad comes loose when someone forces a component lead through the lead hole without first removing the old solder from the hole.

To make the best repair on a single-sided board, the pad should be fastened to the board using an eyelet or funnelette. See **Figure 12-47.** Simply regluing the pad to the board will not make a lasting repair. The next time the pad is heated, it will come loose again. A proper repair is made by using the correct size eyelet through the pad. The hole may need to be enlarged to the correct size. The eyelet is swagged flat onto the board surface. Then, the component lead is placed through the eyelet and soldered. The connection should be inspected to ensure the pad is soldered to the eyelet.

A funnelette should be used on a double-sided board, but an eyelet can also be used, **Figure 12-48.** A broken track can be repaired by splicing with a piece of track or flattened tinned copper wire, **Figure 12-49.**

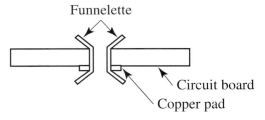

Figure 12-47. A funnelette used in PC board repair. The funnelette is swagged in place and soldered. Care must be taken not to swag too much and crack the PC board.

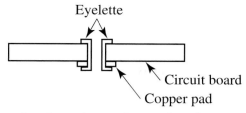

Figure 12-48. Some track and pad repairs can be made using an eyelet. After being swagged into place, the component is inserted and both the component and eyelet are soldered.

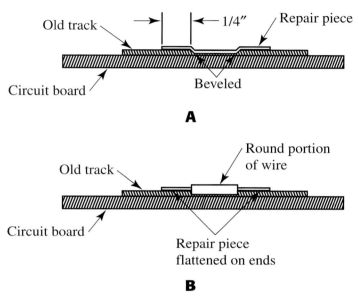

Figure 12-49. Repairing a broken track. A—When replacing a track with a piece of new track, the new piece should extend at least 1/4″ over the ends of the old track. B—A piece of flattened copper wire can be used to repair a bad track. Flattening the wire increases the surface area and provides better support.

The following procedure is used to repair a broken track:

1. Remove the loose portion of the old track by scoring it with a knife. Bend it back and forth until it breaks off. If the old track is broken but still glued to the board, removal is not necessary.
2. Clean the ends of the old track at least 3/8″ each way from the break.
3. Tin the clean portion of the old track.
4. Place the splice, piece of track, or flattened wire across the break. Hold it in place with a wooden stick.
5. Solder the splice.
6. Clean the spliced area and apply a coat of epoxy if needed.

Track and Pad Replacement

If damage to the track or pad is so extensive that repair is not possible, it should be replaced. Score the track with a knife. Bend the track back and forth until it breaks off. The soldering iron should be kept at a sufficient distance from the pad to prevent the heat from unsoldering the splice between the old and new track. The length of the overlap should be between 1/8″ and 1/4″. See **Figure 12-50.**

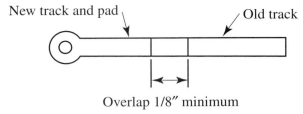

Overlap 1/8″ minimum

Figure 12-50. In some cases the only way to make a repair is to replace the pad and track. When soldering the new track to the old one, care must be taken that the glue remains intact.

The following procedure is used to solder the splice:

1. Clean and tin the old track where the splice is to occur.
2. Place an eyelet through the pad and component hole of the board. Check for proper alignment of the new track with the old track. Swag the eyelet in place. Before the eyelet is completely swagged, check the track alignment again.
3. Applying very light pressure, use a wooden stick to hold the new track in place just behind the splice. Apply liquid flux if available.
4. Place the soldering iron next to the wooden stick and toward the splice.
5. Apply solder. As the solder melts, move the iron down toward the splice and off the end of the overlap.
6. Clean the repair job twice.
7. Apply a coat of epoxy to hold the new track to the printed circuit board.

Burn Repairs

A burned circuit board is usually caused by a hot soldering iron or an overheated component. Most small burns are not critical; however, a burned area is carbon (a conductor) and can cause an undesired circuit to exist on the board.

The following procedure is used to repair a burned area:

1. Clean the charred material from the burned area. If the burn goes into the board, cut a beveled edge around the burned area. If the burn goes clear through the board, bevel both sides, **Figure 12-51.**
2. Make a mixture of 9 parts epoxy to 1 part fiberglass powder.
3. Place the mixture into the prepared area and allow it to cure. Smooth the repair until it is even with the board surface.
4. Replace any tracks or pads.

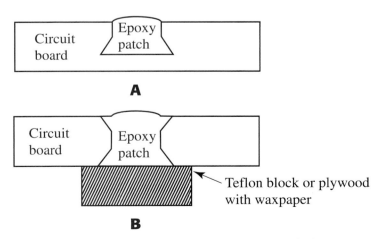

Figure 12-51. Burn repairs. A—The repair of a board burn can be made by using an epoxy patch. B—Repair of a burn when it goes completely through the PC board.

Summary

- Everyone who comes in contact with ESD-sensitive devices must be educated in procedures and policies for preventing ESD.
- Temperature and humidity affect the amount of ESD.
- Components and printed circuit boards can be affected by electrostatic fields.
- Plastics are a common source of ESD.
- Hard ground is a direct connection to ground. Soft ground is a connection through an impedance to limit the current.
- A protected area is an area in which the ESD is controlled.
- Electronic assemblies may be of the point-to-point wiring, breadboard, perfboard, or printed circuit board type.
- The most common type of printed circuit board material is glass epoxy.
- Conformal coating is a coating used to protect a printed wiring board from dirt and moisture.
- Track and pad problems often occur because of excessive or repeated heating.
- The carbon of printed circuit board burns can cause an unwanted circuit path.
- Different types of printed circuit component mounting include stud, clinched, swagged, or surface mounted.
- To prevent component lead breakage, stress relief is often required.
- Soldering terminal types include eyelet, hook, turret, cup, bifurcated, and solderless (crimp).
- In the assembly of connectors, conductors must be connected to the same pin on both ends of a cable, and the conductor insulation must be very close to the terminals to eliminate any possibility of their shorting out one another.
- Tinning the very ends of a wire will keep the strands from fanning outside the terminals and causing shorts.

Important Terms

Do you know the meanings of these terms used in the chapter?

breadboarding
electrostatic discharge
hard ground

soft ground
triboelectric effect

Questions and Problems

Please do not write in this text. Write your answers on a separate sheet of paper.

1. What causes static discharge?
2. Any conductor or insulator that enters an electrostatic field is polarized by _____.
3. List the different components that may be sensitive to ESD.
4. List the prime sources of ESD.
5. Define *hard ground* and *soft ground.*
6. What are the various types of electronic assembly?
7. What is the purpose of conformal coatings?
8. What causes tracks and pads to lift off the printed circuit board during soldering?
9. Why should the burned area of a printed circuit board be repaired?
10. List the four methods of mounting components on printed circuit boards.
11. Why should component stress relief be a concern?
12. List the six types of terminals.
13. What can be done to keep the strands of wire from fanning out from under a screw terminal?
14. What are common errors made during the assembly of electronic devices?

Section III

Fundamentals of Alternating Current

Section III Activity

Making a Circuit Graph

Graphs are used throughout electricity and electronics to show circuit action. Graphs show the strength of the voltage or current in relationship to time. For example, when you turn on a light, it does not come on instantly. A fraction of time (millisecond) lapses before the light reaches maximum brilliance.

Objective

In this activity, you will graph the current of a resistive dc circuit.

Materials and Equipment

1–Pen or pencil
1–Calculator (optional)
1–Ruler (optional)

Procedures

1. Calculate the total maximum current for **Figure A.**
2. Write the scale in amps on the Y-axis of the graph in **Figure B.**
3. Write the scale in milliseconds on the X-axis of the graph.
4. Assuming it takes 3.5 ms for the current to reach maximum, plot the current of Figure A on the graph.

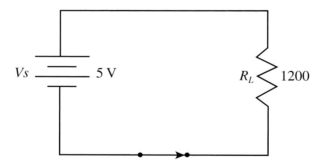

Figure A.

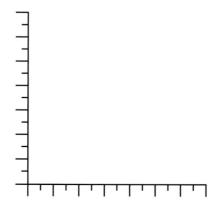

Figure B.

Chapter 13 Graphic Overview

A Vector —is an— Arrowed Line —length— Represents —the— Magnitude —the— Angle —indicates the— Direction

Graphs
- —scales can be— Linear, Semilogarithmic, Logarithmic, Polar Coordinate
- —described as— Linear —or— Nonlinear
- —use— Rectangular Notation —or— Polar Notation

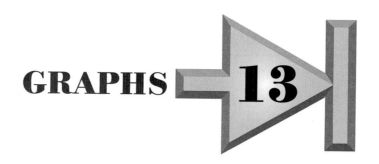

GRAPHS 13

Objectives

After studying this chapter, you will be able to:
○ Identify X and Y axes and their polarities.
○ Describe the differences between rectangular and polar notation.
○ Explain the differences between linear and nonlinear graphs.
○ Draw a line graph.
○ Identify the types of graph paper scales.

Introduction

Many circuits or components can be better understood with a visual graph. The importance of graphs was not stressed in your study of dc circuits. However, alternating current, RCL circuits, and the oscilloscope are topics in which graphs are widely used to illustrate circuit actions. Reading and constructing graphs are skills you will begin to develop in this chapter.

Graph Fundamentals

A graph and its reference points begin as a circle, and the circle is divided into four equal parts called **quadrants,** **Figure 13-1.** The horizontal line through the circle is called the **X-axis.** The vertical line is called the **Y-axis.**

The center of the circle where the axes cross is called the **origin,** **Figure 13-2.** The origin is the zero reference point for all other measurements. The circle is divided into 360 degrees, starting on the right side at zero and moving counterclockwise to 90°, 180°, 270°, and full circle to 360°.

Each of the axes and quadrants has a polarity, **Figure 13-3.** From the origin, the polarity of the X-axis is positive to the right and negative to the left. The Y-axis is positive toward the top and negative toward the bottom.

Quadrants: The divisions of a circle. Four quadrants of 90° make up a complete circle.

X-axis: The horizontal (left-to-right) axis of a graph.

Y-axis: The vertical (up-and-down) axis of a graph.

Origin: The intersection of the X and Y axes in a rectangular coordinate system.

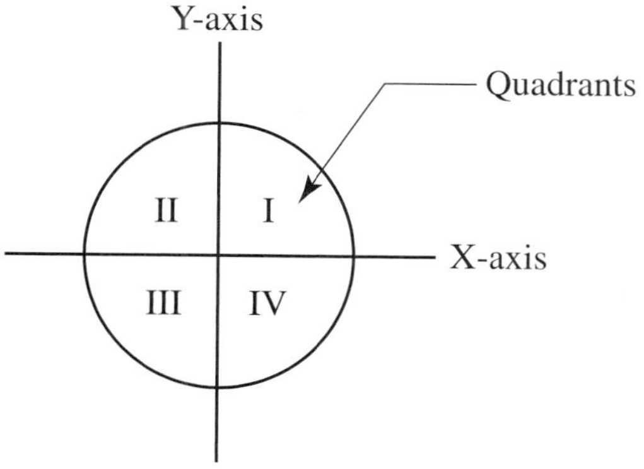

Figure 13-1. A circle of any size can be divided into four equal parts (quadrants) by the X and Y axes.

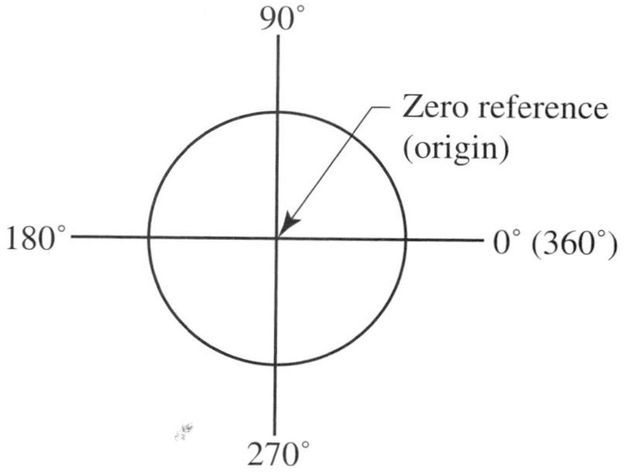

Figure 13-2. A circle can be divided into 360 degrees, 90 degrees for each quadrant. The origin is the zero reference point for all measurements.

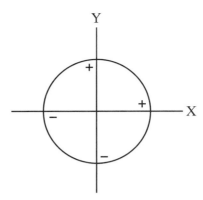

Figure 13-3. The X and Y axes have negative and positive values depending on the direction on the axes.

Locating a point on a graph is done by locating the values for X and Y. To locate the point where values are –8, 6, locate –8 on the X-axis and 6 on the Y-axis, **Figure 13-4.** Draw a dashed line from each of these values. The place where the two dashed lines cross is labeled point A. The values for point A are –8, 6. The values for point B are 5.5, 3. The values for point C are 3, –6. Notice the dashed lines and axes form rectangles at all three points. This method of locating points on a graph is known as *rectangular notation* or the *Cartesian coordinate system.*

Rectangular notation: A rectangular coordinate system using two axes specified by X, Y.

Cartesian coordinate system: A rectangular coordinate system using the X-axis for the horizontal direction and the Y-axis for the vertical direction.

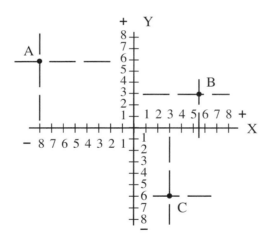

Figure 13-4. Cartesian coordinates are points located on a graph using rectangular notation.

▼ *Example 13-1:*
Using rectangular notation and given the values 4.5, –3, determine the location of the values on a graph. See **Figure 13-5.**
1. Draw a graph and number the scales of the X and Y axes.
2. On the X-axis, place a dot at the value 4.5.
3. On the Y-axis, place a dot at the value –3.
4. Draw a dashed line from the X-axis to any point beyond the Y-axis value.
5. Draw a dashed line from the Y-axis to any point beyond the X-axis dashed line.
6. Make a pencil point where the two dashed lines cross. This is the location of the X, Y values.   ▲

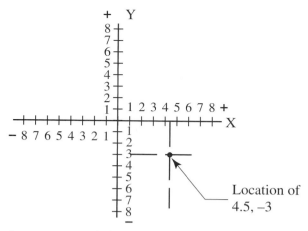

Figure 13-5. Locating values on a graph for Example 13-1.

Many graphs do not use all four quadrants and omit the circle. Often the first and fourth quadrants are used, **Figure 13-6.** Sometimes the graph consists only of the first quadrant.

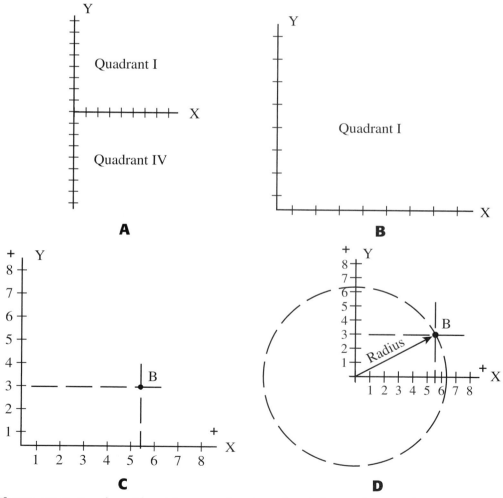

Figure 13-6. Graphs without four quadrants and a circle. A—This graph uses only the first and fourth quadrants. B—This graph uses only the first quadrant. C—Locating point B using quadrant I. Compare to quadrant I in Figure 13-4. D—Point B can be considered the radius of a circle.

Polar notation: In trigonometry, it is written as the radius and angle (r, ∠).

Theta: A letter of the Greek alphabet used to indicate a phase angle.

Another term used for point location is *polar notation.* This notation is written giving the radius and angle (r, ∠), **Figure 13-7.** An angle is identified by the Greek letter *theta* (θ). The polar notation for the figure shown is 6∠42°. An angle in the first quadrant is a positive theta, and an angle in the fourth quadrant is a negative theta. See **Figure 13-8.**

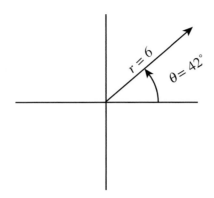

Figure 13-7. The radius and angle for a point on this graph are specified using polar notation (6∠42°).

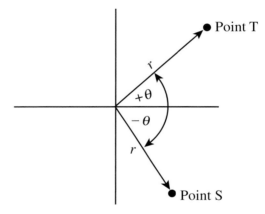

Figure 13-8. A positive angle (+θ) goes toward the positive side of the Y-axis. A negative angle (−θ) goes toward the negative side of the Y-axis.

▼ *Example 13-2:*

Given the notation 2.5∠25° (radius in inches), locate the point on a graph. See **Figure 13-9.**
   1. Draw lines for the X and Y axes. Make equal divisions along each axis.
   2. Using a protractor, measure and mark the angle of 25°.
   3. Draw a light line from the origin through the angle mark.
   4. Use a ruler to measure 2.5≤ from the origin.   ▲

Linear: Pertaining to a straight line. Something that changes equally.

Nonlinear: Not the same throughout a given distance, time, resistance, or temperature.

Graphs can also be *linear* or *nonlinear,* **Figure 13-10.** In electronics, the X-axis is used for time, and the Y-axis is used for voltage or current. In the linear graph, the line showing a rise in current-to-time is straight. This ratio of rise-to-time is the same throughout the graph. In the nonlinear graph, the rise-to-time ratio is not the same in all parts of the graph. Nonlinear graphs are useful for illustrating direct and alternating current, **Figure 13-11.**

The standard oscilloscope has a special calibrated screen called a **graticule,** **Figure 13-12.** The graticule is divided into one centimeter squares. The X-axis is divided into ten major divisions for time, and the Y-axis is divided into eight major divisions for volts. Details of the oscilloscope graticule are covered in Chapter 17.

Graticule: The graduated scale on the screen of an oscilloscope.

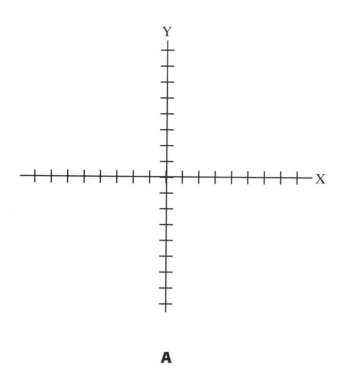

A

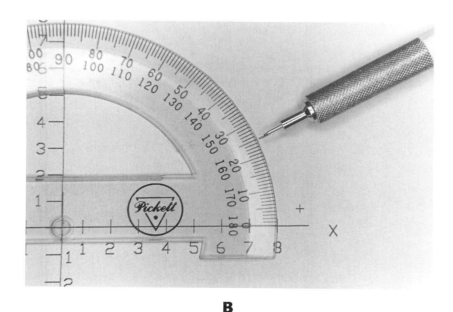

B

Figure 13-9. Locating points on a graph using polar notation. A—Axes for Example 13-2. B—The protractor is aligned on the graph with the origin of the graph at the 0° and 90° points on the axes. Mark the 25° point. (continued)

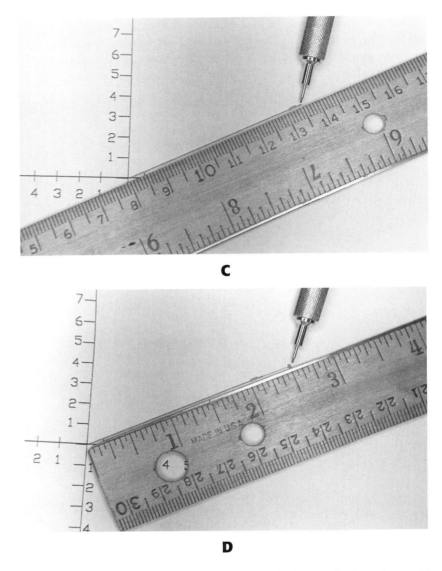

Figure 13-9. C—Draw a line from the origin through the marked angle. D—Use a ruler to measure the radius.

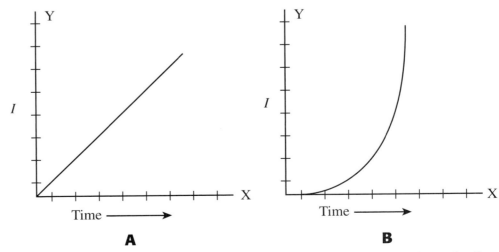

Figure 13-10. Linear and nonlinear graphs are often simple line graphs in the first quadrant. A—Linear graph. B—Nonlinear graph.

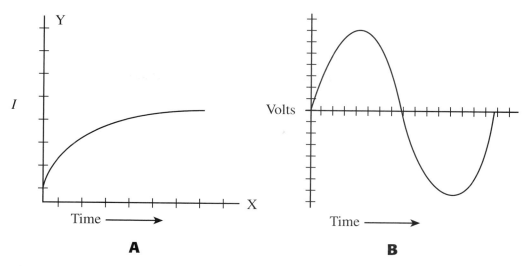

Figure 13-11. Nonlinear graphs. A—Graph illustrating direct current. B—Graph illustrating alternating current.

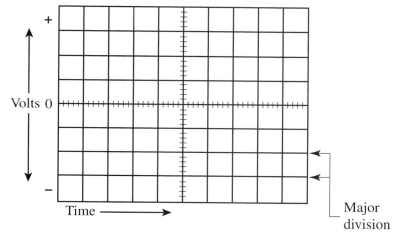

Figure 13-12. The standard graticule of an oscilloscope has 1 cm² divisions, ten horizontal and eight vertical.

Vectors

A *vector* is an arrowed line drawn from the origin to a point of radius, **Figure 13-13.** The length of the line represents the magnitude or strength of the vector. The angle, with reference to the X-axis, indicates the direction of the vector. Vector θ_1 has a magnitude of 15 and direction of 35°. Vector θ_2 has a magnitude of 8 and direction of 120°. A combination of two or more vectors on a graph is called a *vector diagram.*

Vector diagrams have a great number of uses beyond the scope of this text. The "who is leading who" vector diagram is one basic application. The angles of each quantity can be estimated, but no magnitudes are given, **Figure 13-14.**

Figure 13-15 is an example of the use of a vector diagram. A person is to swim across a stream to the other bank. The distance between the two banks is 500′. The force of the water causes the swimmer to go downstream 900′. The resulting vector shows the swimmer went 1030′ downstream to the other bank.

Stated another way, the force of the swimmer is one vector (*S*), while the force of the moving water (*W*) is another vector. The combined forces produce the resulting vector (*R*). The calculation is covered in later chapters.

Vector: An arrowed line indicating a certain magnitude and direction.

Vector diagram: A diagram using vectors to indicate the vector relationship of the quantities.

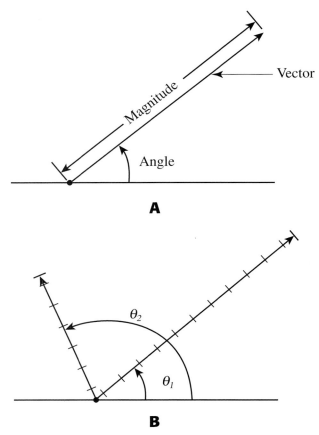

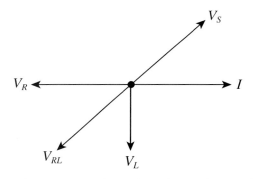

Figure 13-13. Vectors. A—A vector has magnitude (length of the line) and direction (from zero reference). B—Vector diagram.

Figure 13-14. A vector diagram can show the relationship of different quantities.

▼ *Example 13-3:*

Given the values R = 3, X = 4 (in inches), find the value of the Z vector in **Figure 13-16.**

   1. Using a ruler, draw the *R* value of 3″ at 0°.
   2. Using a ruler, draw the *X* value of 4″ at 90°.
   3. Draw dashed lines perpendicular to the *R* and *X* values.
   4. Draw a line from the origin to the point where the dashed lines cross.
   5. Using a ruler, measure the value of the *Z* vector (*Z* = 5″).
   6. Using a protractor, measure the angle of the *Z* vector (θ = 53°).
   7. The value of the Z vector in polar notation is 5 ∠53°.   ▲

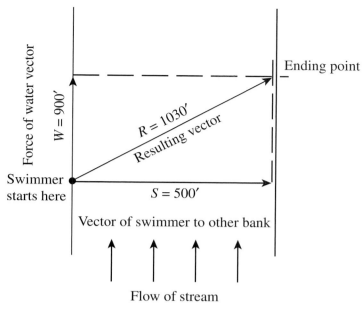

Figure 13-15. A vector diagram shows the forces involved and the results.

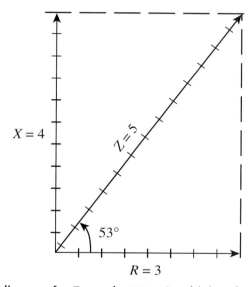

Figure 13-16. Vector diagram for Example 13-3. Combining the values for *R* and *X* results in a value of *Z*.

Selecting Scales for Graphs

How you interpret a graph will depend on the type of scale used as well as the numerical division of the scale. The same information presented on two types of graph scales appears very different. Both graphs in **Figure 13-17** are based on the same information; however, the numerical scales are different. One is scaled in units of ten, and the other is scaled in units of one. The graphs are visually different even though they contain the same information. Another factor that tends to distort the appearance of graphs is lack of zero reference.

The type of scale is just as important as the numerical scale, **Figure 13-18.** In a linear scale, the distance from one major division to another is equal. The major divisions of the graph in figure A are one-inch squares, and each small division is in tenths (1/10″). The divisions of the graph in figure B are also one inch for each

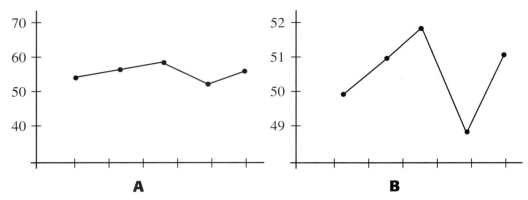

Figure 13-17. Numerical divisions affect the appearance of data on a graph. A—Graph scale with increments of ten. B—Graph scale with increments of one.

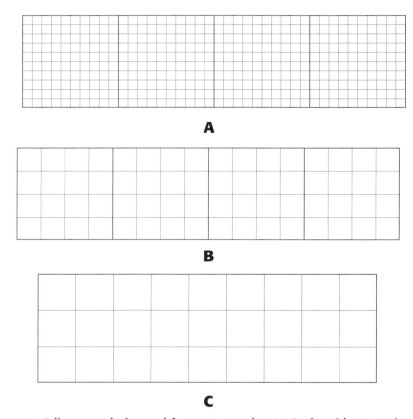

Figure 13-18. A linear scale is used for most graphs. A—Scale with ten units per inch. B—Scale with 1/4″ units. C—Scale with 1 cm units.

major division, but the smaller divisions are 1/4″. The graph in figure C shows a scale of one centimeter for each division. There are no small divisions in this scale, but the distance from one division to another is the same, indicating a linear scale.

Semilogarithmic graph paper is used for making a response curve of electronic circuits, **Figure 13-19.** A semilog graph has a linear scale across the bottom, but the vertical scale is nonlinear. The distance from point 1 to point 2 is different than the distance from point 2 to point 3. Semilog graph scale comes in 2-, 3-, 4-, 5-, and 7-cycle grids.

Full logarithmic paper is referred to as log-log scale. This type of graph scale is logarithmic in both vertical and horizontal directions; that is, there is no linear scale on the graph. See **Figure 13-20.**

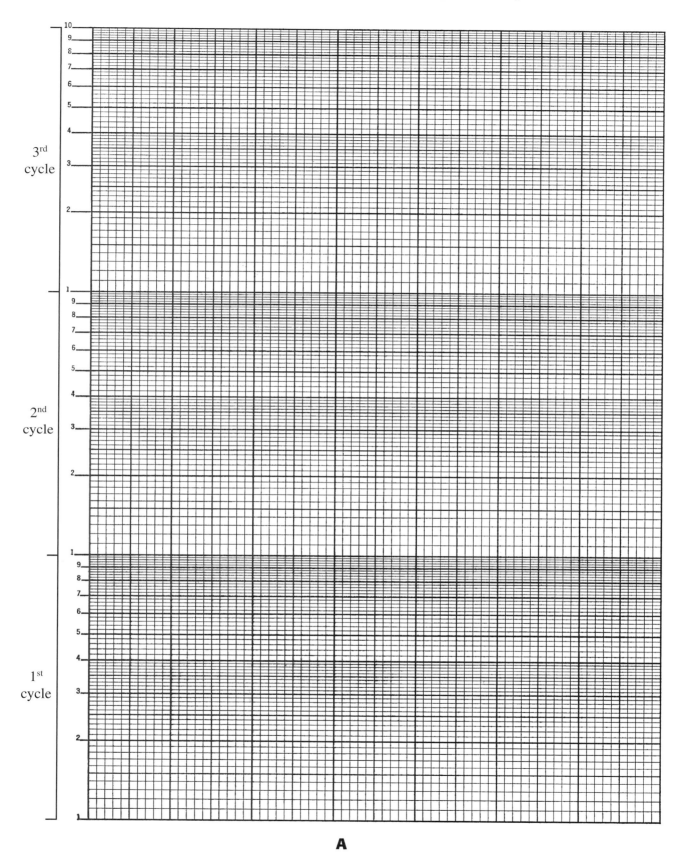

A

Figure 13-19. Semilogarithmic graph paper. A—Semilog graph scale with three cycles. *(continued)*

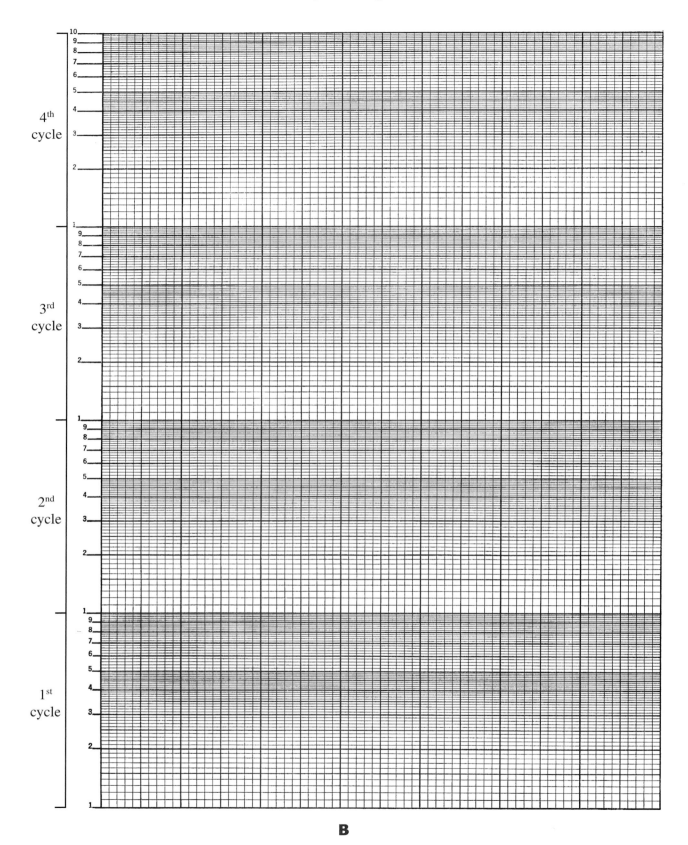

B

Figure 13-19. B—Semilog graph scale with four cycles.

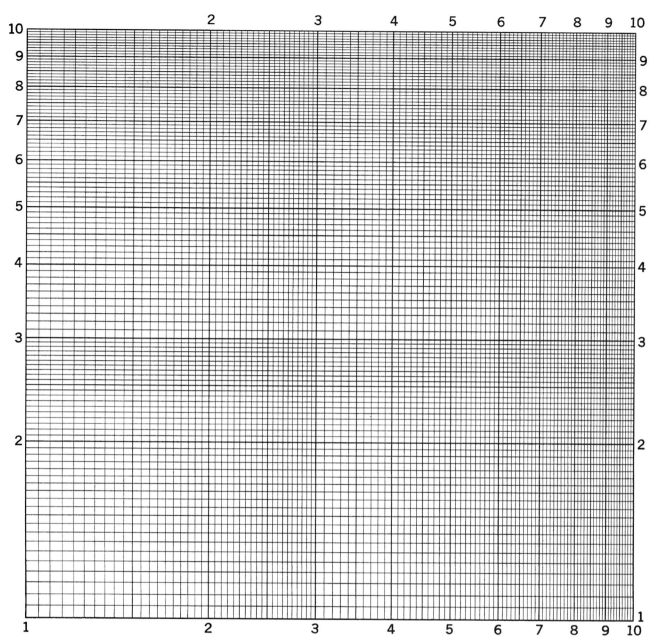

Figure 13-20. A full log or log-log graph scale has no linear portion.

Certain types of graphs, such as antenna radiation patterns, microphone pickup patterns, or vector displays, require the use of a polar coordinate graph scale. See **Figure 13-21.**

Making a Graph

Graphs are widely used in electronics. Being able to construct, draw, and read graphs is an important skill. The following steps will be helpful in making a graph:

1. Carefully consider what the graph must show. Is it supposed to convey information about voltage, resistance, current, time, frequency, decibels, or something else?
2. Once you have gathered the information, determine the type of graph scale to use: linear, semilogarithmic, log-log, or polar.

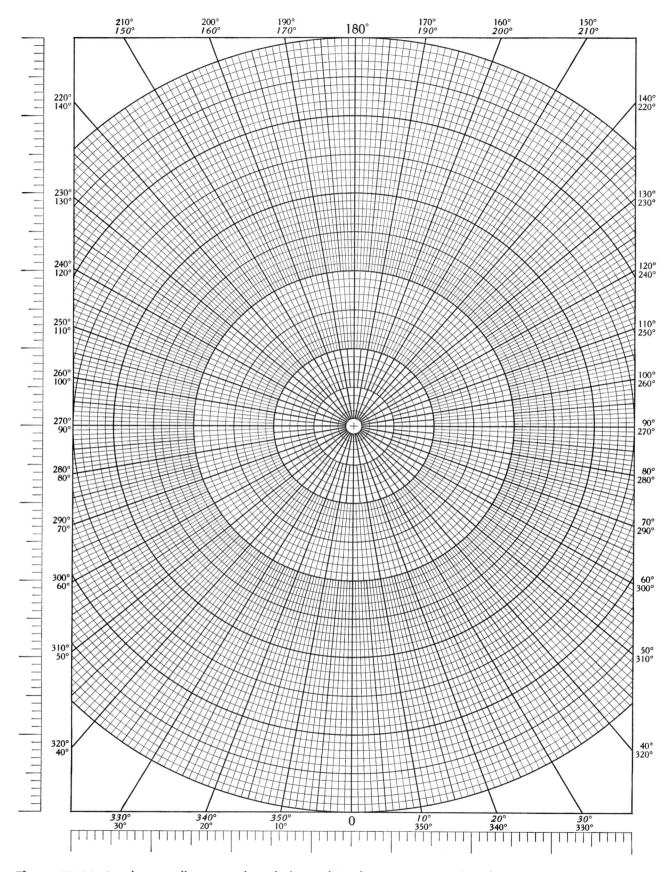

Figure 13-21. A polar coordinate graph scale is used to plot antenna or microphone patterns.

3. Determine the limits (lowest and highest numbers) of the information to be graphed. This will determine the starting and ending numbers of the graph.
4. Consider the values that will be represented by the major divisions. Consideration of the smaller divisions may be helpful.
5. Organize the data in table format and plot it on a graph, **Figure 13-22.**
6. Draw the X and Y axes for the graph, marking the required units for the data. It may take some trial and error to select the proper scale, but it becomes easier with experience.
7. Plot the data points on the graph. Be very accurate. An error will mean doing the whole graph again.
8. After plotting all the data points, connect the points with a smooth line. If the line is curved, use a French curve, not a straight ruler. Try to keep the French curve on at least three points as you draw the line, **Figure 13-23.** If there is too much blank space between the points, you may need to gather more data to place between the points to ensure the curve is correct.
9. Label the graph to indicate axes and curve points. Add other information such as your name, the date, and the project title.
10. Check the details of the graph for accuracy.

Volts	Current
2	6
4	12
6	18
12	36

Frequency	Db
500 Hz	60
1 kHz	34
5 kHz	18
10 kHz	7

A

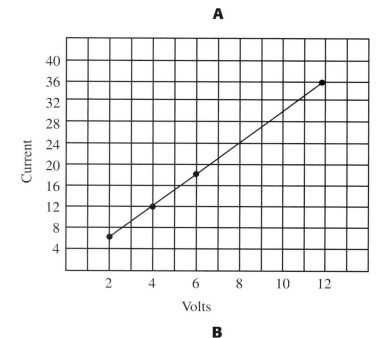

B

Figure 13-22. Making a graph. A—Sample data. B—Plotted voltage and current indicated in a linear graph. *(continued)*

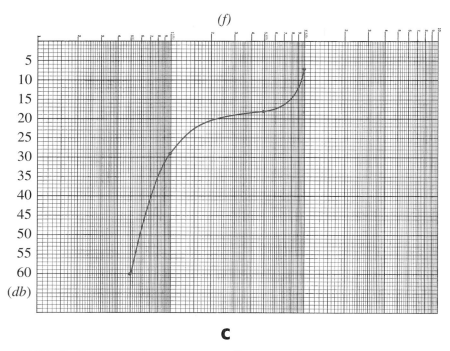

Figure 13-22. C—Frequency forms on a nonlinear graph.

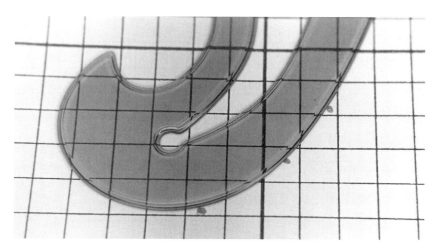

Figure 13-23. French curves allow at least three points of contact.

Summary

- A circle is divided into four quadrants: the X-axis from left to right and the Y-axis from top to bottom. The X-axis is positive on the right and negative on the left. The Y-axis is positive on the top and negative on the bottom. The circle is divided into 360 degrees.
- Points on a graph can be given by rectangular or polar notation.
- Graphs can be linear or nonlinear. The divisions of a linear graph are equal in all parts of the graph; the divisions of a nonlinear graph are not equal.
- A vector is an arrowed line indicating magnitude and direction. Vectors are used to make vector diagrams of circuit actions.
- In making a graph, the numerical scale and type of graph scale is most important.

 Important Terms

Do you know the meanings of these terms used in the chapter?

Cartesian coordinate system   rectangular notation
graticule   theta
linear   vector
nonlinear   vector diagram
origin   x-axis
polar notation   y-axis
quadrants

 Questions and Problems

Please do not write in this text. Write your answers on a separate sheet of paper.

1. Draw a circle and identify the X and Y axes. Place the positive and negative values on the axes and four quadrants.
2. Given the X-Y values 7, 4, locate the point using rectangular notation.
3. Plot a line graph for the following information:

Parts cost	Month
$300.00	July
$150.00	August
$200.00	September
$450.00	October
$ 70.00	November
$550.00	December

4. What is the difference between rectangular and polar notation?
5. Given the X-Y values –3,–4, locate the point using rectangular notation.
6. Write the polar notation when the angle is 107° and the radius is 280.
7. Indicate whether the graphs in **Figure 13-24** are linear or nonlinear.
8. Define *vector.*
9. Draw a vector diagram of the X-Y values 4, 6.
10. List the types of graph paper scales.
11. Draw two vectors having the same direction.
12. Draw two vectors indicating the same magnitude.
13. Draw a vector diagram with one vector at 45°, a second at 180°, and a third at 225° (approximate angle locations).
14. Collect samples of linear, semilogarithmic, log-log, and polar types of graph paper.

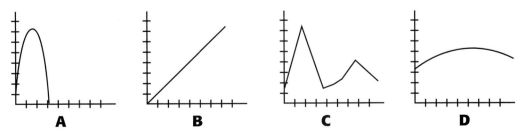

Figure 13-24. Graphs for problem 7.

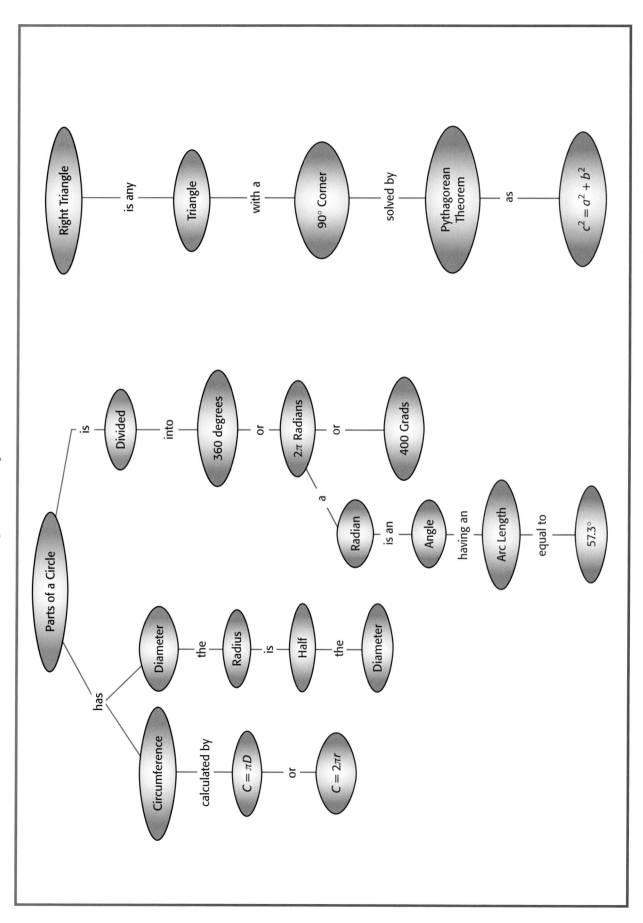

TRIGONOMETRY FOR ELECTRICITY

14

Objectives

After studying this chapter, you will be able to:
- ○ Convert between degrees, grads, and radians.
- ○ Convert between rectangular and polar notation.
- ○ Determine angles and functions using a trigonometric table and calculator.
- ○ Solve right triangles given one side and one angle.
- ○ Find the angles of a triangle given two sides.
- ○ Solve for the unknown of a triangle using the Pythagorean theorem.

Introduction

Trigonometry is a special section of the general field of mathematics. It is the study of the relationship between the sides and the angles of triangles. It is useful in the study of alternating current, true power, and phase relationships.

Parts of a Circle

A straight line through the center of a circle and touching the circle at both ends is called the *diameter*, **Figure 14-1.** From the center point to any point on the circle is the *radius.* The radius is equal to half the diameter. The equation is:

$$r = \frac{D}{2}$$

▼ *Example 14-1:*

What is the radius of a circle having a diameter of 3.5″?

$$r = \frac{D}{2}$$
$$= \frac{3.5}{2}$$
$$= 1.75″$$

The radius and diameter can be measured by any unit of length at any point around the circle.   ▲

Diameter: The distance across the middle of a circle.

Radius: The distance from the center of a circle to a point on the circumference of the circle.

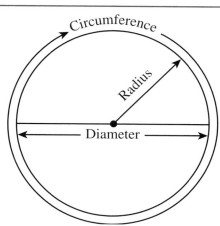

Figure 14-1. Diameter, radius, and circumference measurements of a circle.

The distance around a circle, that is, from one point on the circle, around the out-side edge of the circle and back to the beginning point, is called the **circumference.** The circumference can be calculated by the equation:

$$C = \pi D$$

or

$$C = 2\pi r$$

▼ ***Example 14-2:***

What is the circumference of a circle having a diameter of 1.5″?

$$
\begin{aligned}
C &= \pi D \\
&= 3.14 \,(1.5) \\
&= 4.71''
\end{aligned}
$$

▲

Circular Measure

A circle is divided into 360 degrees. The degree is the common unit used to measure circles and is represented by the symbol (°). Another unit for circular measure is the radian. A **radian** is an angle having an arc whose length is equal to that of the radius of the circle, **Figure 14-2.** The SI metric symbol "rad" can be used for radians although it is usually spelled out.

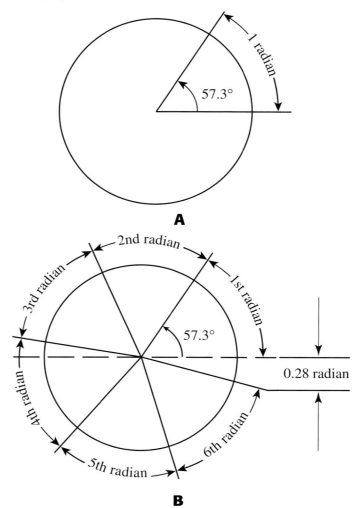

Figure 14-2. Radians. A—One radian is equal to 57.3°. B—There are 6.28 (2π) radians in a complete circle.

If both sides of the equation for circumference are divided by *r*, the equation becomes:

$$\frac{C}{r} = \frac{2\pi r}{r}$$

$$\frac{C}{r} = 2\pi$$

This means the circumference of any circle, divided by its radius, is equal to the constant 2π. Not only can a circle be divided into 360 degrees, it can also be divided into radians. The equation is:

$$2\pi \text{ radians} = 360°$$

Since π equals 3.14159, dividing both sides by 2 and then π shows:

$$\frac{\pi \text{ radians}}{\pi} = \frac{180°}{3.14159}$$

$$1 \text{ radian} = 57.3°$$

Converting Between Degrees and Radians

The conversion between degrees and radians is a common operation when working with electricity. To change degrees to radians, the equation is:

$$\text{Radians} = \text{degrees} \times \frac{\pi}{180}$$

To change radians to degrees, the equation is:

$$\text{Degrees} = \text{radians} \times \frac{180}{\pi}$$

Figure 14-3 illustrates the relationship of degrees and radians of a circle. The radians are given for each 30° increment. For example, 30° is equal to $\pi/6$ radians.

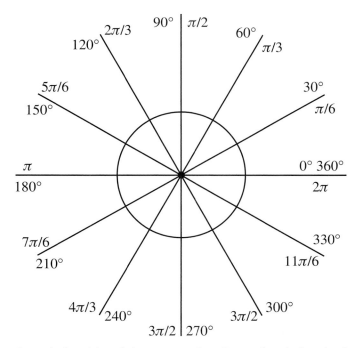

Figure 14-3. The relationship of degrees and radians of a circle. The length of half the circumference equals π radians. Therefore, the circumference equals π times the diameter.

▼ *Example 14-3:*

$$\frac{\pi}{6} = \frac{3.1415927}{6}$$

$$= 0.5235988 \text{ radians}$$

Thus:

$$\text{Degrees} = \text{radians} \times \frac{180}{\pi}$$

$$= 0.5235988 \times \frac{180}{3.1415927}$$

$$= 30°$$ ▲

Another method of conversion uses a calculator. Most calculators have a **DRG** (degrees, radians, grads) key. It may be a second or third function of a key. Press the key a few times and notice the results on the display. The display should change the labels deg, rad, and grad. Grad is another circular measure used in Europe. There are 400 grads in a circle with 100 grads in each quadrant.

Put the number 1 on the display. Press the **DRG** key several times. Notice that nothing happens to the number, only the labels in the display change. Stop the display on rad. Now press the **3rd, 2nd,** or **INV** key, then the **DRG** key. The display should change to grad with the numbers 63.661977. Press the **3rd, 2nd,** or **INV** key and the **DRG** key again. The display should change to deg with the numbers 57.29578. This key sequence has changed 1 radian to 57.3°. If the key sequence does not work on your calculator, refer to the operating manual. Try changing π radians to degrees.

The following equations summarize the mathematical conversion of angular measurement:

$$\text{Degrees} = \text{radians} \times \frac{180}{\pi}$$

$$\text{Radians} = \text{degrees} \times 0.17453$$

$$\text{Grads} = \text{degrees} \times 1.11111111$$

$$\text{Grads} = \text{radians} \times 63.662$$

Description of a Right Triangle

The rectangle in **Figure 14-4** has four corners of 90° each (360° in a complete circle around the rectangle). The rectangle is cut in half from corner to corner with a dashed line. This produces two triangles, each having one 90° corner. Any triangle with a 90° corner is called a *right triangle.* Since one of the triangles is half a rectangle, there must be a total of 180° in the triangle. The right angle is 90°, which leaves 90° for the other two corners of the triangle. Thus, the sum of the remaining two corners must be 90°.

Right triangle: A triangle that has a 90° angle.

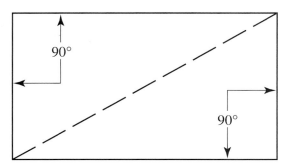

Figure 14-4. A rectangle made into two right triangles. The sum of the angles of each triangle equals 180°.

The corners of the triangle are labeled A, B, and C, **Figure 14-5.** Corner C is the 90° right angle, A is 30°, and B is 60°. The sides of the triangle (labeled a, b, and c) are opposite each of the corners. This method of identification is used when the Pythagorean theorem is used. It states the square of side c is equal to the sum of the squares of sides a and b. See **Figure 14-6.** The equation for the Pythagorean theorem is:

$$c^2 = a^2 + b^2$$

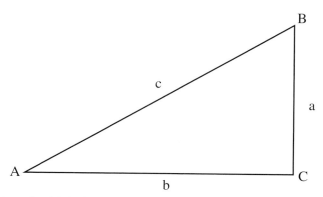

Figure 14-5. Standard labels given to the corners and sides of a right triangle. Side c is always the diagonal side.

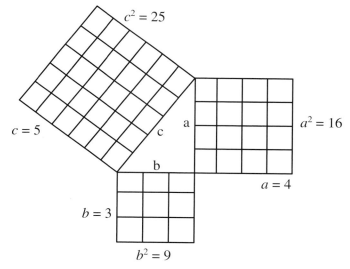

Figure 14-6. The sum of the squares of sides a and b is equal to the square of side c.

Another common method of identifying the sides of a triangle is shown in **Figure 14-7.** You must first decide which angle (A or B) is going to be used as a reference point. If angle A is the reference, then the side across from that angle is called the *opposite side* and the side connecting to that angle is the *adjacent side.* The diagonal side is called the *hypotenuse.* If angle B is the reference point, side b is the opposite and side a is the adjacent.

Trigonometric Functions

If one side of the triangle is divided into another side, the resulting number is called a *trigonometric function.* The functions are named sine, cosine, tangent, and cotangent. These functions are abbreviated in equation form as:

Opposite side: The side of a right triangle across from or opposite the angle being referenced.

Adjacent side: The side of a triangle connected to the angle.

Hypotenuse: The longest side of a right triangle or the diagonal (corner-to-corner) dimension of a rectangle.

Trigonometric function: The resulting number when one side of a triangle is divided into another side.

$$\sin \theta = \frac{opp}{Hy}$$

$$\cos \theta = \frac{adj}{Hy}$$

$$\tan \theta = \frac{opp}{adj}$$

$$\cot \theta = \frac{1}{\tan}$$

These are the trigonometric functions of the angle of a triangle. For every angle there is a sine, cosine, tangent, and cotangent. If the function is found, the angle can also be found.

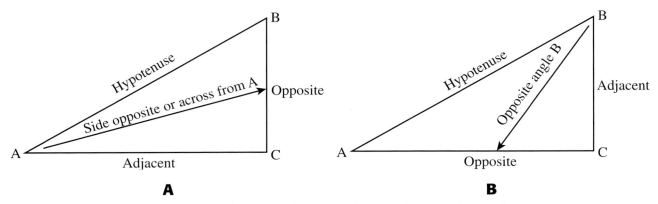

Figure 14-7. Identifying the sides of a triangle. A—Using angle A as the reference angle. B—Using angle B as the reference angle.

Finding the Angles of a Function

Finding the angles of a function can be accomplished using a chart of trigonometric functions or a scientific calculator.

▼ *Example 14-4:*
Find angle A of the triangle in **Figure 14-8.**

$$\sin \theta = \frac{opp}{Hy}$$

$$= \frac{12.4256}{16}$$

$$= 0.7766$$

▲

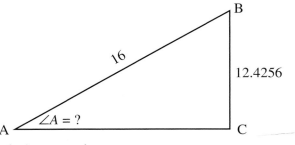

Figure 14-8. Triangle for Example 14-4.

A chart of trigonometric functions is shown in **Figure 14-9.** Look under the heading Sine for the number closest to 0.7766. To the left of the function, find angle A is 51°.

The terms arcsine (sin⁻¹), arccosine (cos⁻¹), and arctangent (tan⁻¹) are the inverse operations of the functions. For example, the equation may be written as arctan 1.429 = ? That is, if 1.429 is the tangent, what is the angle?

▼ *Example 14-5:*

Place 1.429 on the calculator display.
Press **3rd**, **2nd**, or **INV**, then **tan**. Your calculator may show a tan⁻¹ on the display. The answer is arctan 1.429 = 55°.
▲

If the angle is known, the function can be found quickly using a calculator. Place the angle 60° on the display and press the **cos** key. The number 0.5 on the display is the cosine of 60°.

Natural Trigonometric Functions							
Angle	Sine	Cosine	Tangent	Angle	Sine	Cosine	Tangent
1°	0.0175	0.9998	0.0175	46°	0.7193	0.6947	1.0355
2°	0.0349	0.9994	0.0349	47°	0.7314	0.6820	1.0724
3°	0.0523	0.9986	0.0524	48°	0.7431	0.6691	1.1106
4°	0.0698	0.9976	0.0699	49°	0.7547	0.6561	1.1504
5°	0.0872	0.9962	0.0875	50°	0.7660	0.6428	1.1918
6°	0.1045	0.9945	0.1051	51°	0.7771	0.6293	1.2349
7°	0.1219	0.9925	0.1228	52°	0.7880	0.6157	1.2799
8°	0.1392	0.9903	0.1405	53°	0.7986	0.6018	1.3270
9°	0.1564	0.9877	0.1584	54°	0.8090	0.5878	1.3764
10°	0.1736	0.9848	0.1763	55°	0.8192	0.5736	1.4281
11°	0.1908	0.9816	0.1944	56°	0.8290	0.5592	1.4826
12°	0.2079	0.9781	0.2126	57°	0.8387	0.5446	1.5399
13°	0.2250	0.9744	0.2309	58°	0.8480	0.5299	1.6003
14°	0.2419	0.9703	0.2493	59°	0.8572	0.5150	1.6643
15°	0.2588	0.9659	0.2679	60°	0.8660	0.5000	1.7321
16°	0.2756	0.9613	0.2867	61°	0.8746	0.4848	1.8040
17°	0.2924	0.9563	0.3057	62°	0.8829	0.4695	1.8807
18°	0.3090	0.9511	0.3249	63°	0.8910	0.4540	1.9626
19°	0.3256	0.9455	0.3443	64°	0.8988	0.4384	2.0503
20°	0.3420	0.9397	0.3640	65°	0.9063	0.4226	2.1445
21°	0.3584	0.9336	0.3839	66°	0.9135	0.4067	2.2460
22°	0.3746	0.9272	0.4040	67°	0.9205	0.3907	2.3559
23°	0.3907	0.9205	0.4245	68°	0.9272	0.3746	2.4751
24°	0.4067	0.9135	0.4452	69°	0.9336	0.3584	2.6051
25°	0.4226	0.9063	0.4663	70°	0.9397	0.3420	2.7475
26°	0.4384	0.8988	0.4877	71°	0.9455	0.3256	2.9042
27°	0.4540	0.8910	0.5095	72°	0.9511	0.3090	3.0777
28°	0.4695	0.8829	0.5317	73°	0.9563	0.2924	3.2709
29°	0.4848	0.8746	0.5543	74°	0.9613	0.2756	3.4874
30°	0.5000	0.8660	0.5774	75°	0.9659	0.2588	3.7321
31°	0.5150	0.8572	0.6009	76°	0.9703	0.2419	4.0108
32°	0.5229	0.8480	0.6249	77°	0.9744	0.2250	4.3315
33°	0.5446	0.8387	0.6494	78°	0.9781	0.2079	4.7046
34°	0.5592	0.8290	0.6745	79°	0.9816	0.1908	5.1446
35°	0.5736	0.8192	0.7002	80°	0.9848	0.1736	5.6713
36°	0.5878	0.8090	0.7265	81°	0.9877	0.1564	6.3138
37°	0.6018	0.7986	0.7536	82°	0.9903	0.1392	7.1154
38°	0.6157	0.7880	0.7813	83°	0.9925	0.1219	8.1443
39°	0.6293	0.7771	0.8098	84°	0.9945	0.1045	9.5144
40°	0.6428	0.7660	0.8391	85°	0.9962	0.0872	11.4301
41°	0.6561	0.7547	0.8693	86°	0.9976	0.0698	14.3006
42°	0.6691	0.7431	0.9004	87°	0.9986	0.0523	19.0811
43°	0.6820	0.7314	0.9325	88°	0.9994	0.0349	28.6363
44°	0.6947	0.7193	0.9657	89°	0.9998	0.0175	57.2900
45°	0.7071	0.7071	1.0000	90°	1.0000	0.0000	

Figure 14-9. Chart of trigonometric functions.

To find the angle of Example 14-4 using a calculator, divide the number 12.4256 by 16. The number 0.7766 should be on the display. Press **3rd**, **2nd**, or **INV**, then **sin**. The result should be 51°.

Finding the Sides of a Right Triangle

Finding the sides of a triangle is a simple process, but a step-by-step procedure must be followed. Skipping a step will make it difficult or impossible to arrive at the correct solution. Whenever possible, do not round off numbers since this will reduce accuracy. Take the time to understand the process.

The procedure for solving a right triangle is:
1. Select the angle to be used.
2. Identify the sides of the triangle.
3. Decide which side is to be found.
4. Select the side whose value is known.
5. Select the correct trigonometric equation using the sides found in steps 3 and 4.
6. Place the numbers in the trigonometric equation.
7. Solve the equation.

▼ *Example 14-6:*

Find the unknown side for the triangle in **Figure 14-10.**

With reference angle A, the sides of the triangle are labeled as shown. The side to be found is the opposite side. Since the opposite is to be found and the hypotenuse is given, the opposite and hypotenuse must be in the equation. The trigonometric function that has these two sides in the equation is the sine function. This means the sine equation must be used to solve for the unknown side.

$$\sin \theta = \frac{\text{opp}}{\text{Hy}}$$

$$\sin 64° = \frac{\text{opp}}{42}$$

$$\frac{0.8988}{1} = \frac{\text{opp}}{42} \quad \text{(Cross multiply.)}$$

$$\text{opp} = 37.8 \quad \text{(Remember to multiply by the sine of 64°, } not \text{ 64 times 42.)} \ ▲$$

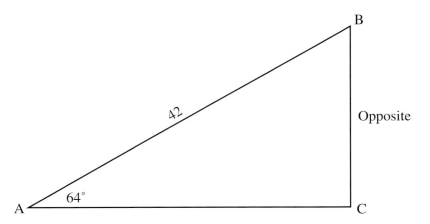

Figure 14-10. Triangle for Example 14-6.

▼ *Example 14-7:*

Find the unknown side of the triangle in **Figure 14-11.**

The reference angle is A. The side to be found is the hypotenuse; the side given is the adjacent. Therefore, the hypotenuse and adjacent must be in the equation, which is the cosine function.

$$\cos \theta = \frac{\text{adj}}{\text{Hy}}$$

$$\cos 38° = \frac{51}{\text{Hy}}$$

$$\frac{\cos 38°}{1} = \frac{51}{\text{Hy}} \quad \text{(Cross multiply.)}$$

$$(\cos 38°)\,(\text{Hy}) = 51$$

$$\frac{(\cos 38°)\,(\text{Hy})}{\cos 38°} = \frac{51}{0.7880} \quad \text{(To get the Hy by itself, divide both sides by } \cos 38°, \textit{not} \text{ by the number 38.)}$$

$$\text{Hy} = 64.72 \qquad\qquad ▲$$

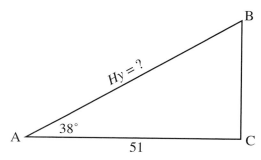

Figure 14-11. Triangle for Example 14-7.

Converting Rectangular and Polar Notation

In Chapter 13, an X, Y value was used to describe rectangular notation, and a radius and angle were used for polar notation. The conversion between these two methods can be done using either mathematics or a calculator.

First, investigate the keys of your calculator. You should find key **P<>R** or keys **P-R** and **R-P**. These keys are used with the **X<>Y** key for converting notations. Put number 3 on the display and press the **X<>Y** key. Number 3 should become 0. Next, put number 4 on the display. What has been done on the calculator is X = 3, Y = 4. Now press the proper keys for **R-P** conversion. The display should show number 5 and a small case *r* in the display. This is the radius in polar notation. Next, press the **X<>Y** key. The display should show the angle 53.13°. Polar notation would be 5 ∕53.13°. Some calculators use **A** and **B** keys in place of the **X<>Y** key. Consult the operating manual for your calculator.

Conversions can also be done by mathematical calculations.

Rectangular to polar:

$$r = \sqrt{a^2 + b^2} \text{ (Pythagorean theorem)}$$

$$\tan \theta = \frac{b}{a}$$

Polar to rectangular:

$$a = r\,(\cos \theta)$$

$$b = r\,(\sin \theta)$$

Trigonometry will be used throughout the remaining chapters. An understanding of the objectives and processes used in trigonometry will help in your study of electricity.

➤| Summary

- The parts of a circle are diameter, radius, and circumference. The circle can be divided into 360 degrees, 2π radians, or 400 grads.
- Conversion between degrees, radians, and grads is done with the DRG calculator key.
- A right triangle has 180°, one of which is 90°. The sum of the other two angles must be 90°.
- Conversion between rectangular and polar notation can be completed directly using a scientific calculator.
- A step-by-step process must be used to solve a right triangle.

➤| Important Terms

Do you know the meanings of these terms used in the chapter?

adjacent side   radian
circumference   radius
diameter   right triangle
hypotenuse   trigonometric function
opposite side

➤| Questions and Problems

Please do not write in this text. Write your answers on a separate sheet of paper.

1. What is the circumference of a circle with a diameter of 6.5″?
2. What is the radius of a circle with a circumference of 45 cm?
3. Convert the following values.
   a. 1.4 radians to degrees
   b. 328 grads to degrees
   c. 69° to radians
4. 1.7 radians + 89 grads + 41° = _____°.
5. Convert the following values from rectangular notation to polar notation.
   a. 4, 8
   b. 34, 90
   c. 29, 37
   d. 6, 9
   e. 100, 250
   f. 600, 900
6. Convert the following values from polar notation to rectangular notation.
   a. 25 ∠36°
   b. 6 ∠–18°
   c. 78 ∠127°
   d. 51 ∠14°
   e. 234 ∠45°
   f. 7 ∠–12°
7. Locate the following X, Y points on the graph in **Figure 14-12.**
   a. 4, 8
   b. –5.5, –3
   c. –3, 4.5
   d. 2, –5

8. Using a calculator or trigonometric table, find the angle for each of the following values.
   - a.   sin = 0.6428
   - b.   cos = 0.1736
   - c.   tan = 1.0724
   - d.   cot = 0.64941
   - e.   tan = 0.3249
   - f.   sin = 0.2350
   - g.   cot = 0.94999
   - h.   cos = 0.7772
9. Solve each of the following problems.
   - a.   Find AC and ∠B if AB = 20 and ∠A = 60°.
   - b.   Find BC and ∠B if AB = 30 and ∠A = 40°.
   - c.   Find BC and ∠A if AC = 400 and ∠B = 32°.
   - d.   Find AC and ∠A if AB = 94 and ∠B = 20°.
10. In **Figure 14-13,** a 6″ circle has four holes around the circle. What is the distance from the center of one hole to another? (A perfect square has 45° angles.)
11. In **Figure 14-14,** an antenna pole is 75′ long. The stake for the guy wire is placed 30′ from the base of the pole. How much wire is needed for the antenna guy wire?
12. What is the angle of the Z vector in **Figure 14-15?**

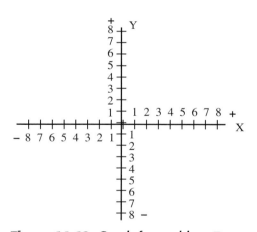

Figure 14-12. Graph for problem 7.

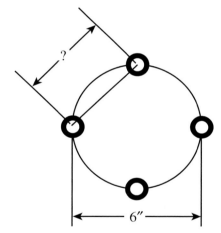

Figure 14-13. Drawing for problem 10.

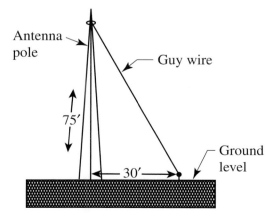

Figure 14-14. Drawing for problem 11.

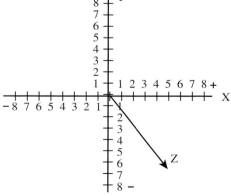

Figure 14-15. Graph for problem 12.

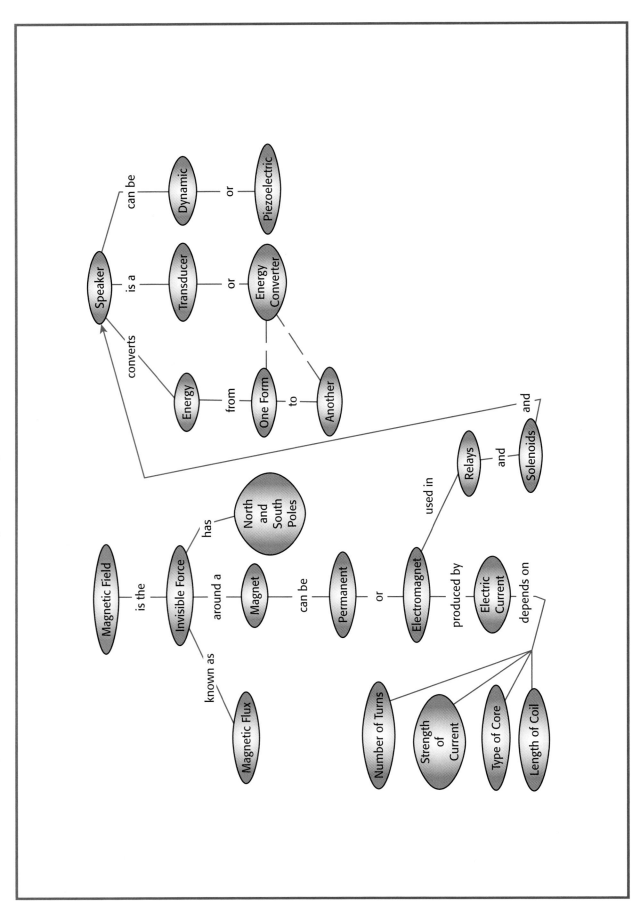

MAGNETISM — 15

Objectives

After studying this chapter, you will be able to:
○ Explain the domain theory of magnetism.
○ Calculate magnetomotive force, flux density, reluctance, permeability, and field intensity.
○ State the factors that determine the strength of a magnetic field.
○ Explain the operation of solenoids and relays.
○ Identify the Hall effect.
○ Explain how to test a speaker for proper operation and phasing.

Introduction

Magnetism is a force field that acts on some materials but not others. Magnetism is also one of the basic forms of energy. An iron compound called *lodestone* is a natural magnet. Manufactured magnets are made of iron, copper, nickel, aluminum, and cobalt and have a stronger magnetic field than lodestone. Whenever an electric current is present, a magnetic field is present as well. Magnetism is necessary to operate such devices as speakers, motors, generators, and transformers.

Lodestone: A natural magnet consisting mostly of a magnetic iron oxide called magnetite.

Magnetic Fields and Poles

A *magnetic field* is the invisible force of magnetism around a magnet. The field extends out from the magnet in all directions, **Figure 15-1**. The invisible lines of force that make up the magnetic field are known as *magnetic flux.* At the ends (poles) of the magnet, the flux lines are more concentrated and the magnetic field is stronger.

Magnetic field: The area around a natural, permanent, or electromagnet where magnetic forces can be detected.

Magnetic flux: The total magnetic lines of force in a magnetic field.

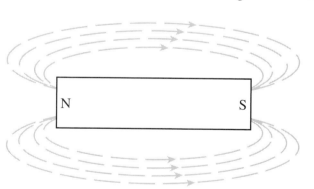

Figure 15-1. The magnetic field of a bar magnet. The dashed lines represent magnetic flux.

The poles of the magnet are labeled north (N) and south (S). The direction of the magnetic lines of force is assumed to leave the north pole and enter the south pole. If two magnets are brought close together, their fields will interact, **Figure 15-2.** When opposite poles (N and S) are close together, the fields are attracted to each other. When like poles (S and S) are together, the fields repel each other. This is the *law of magnetic poles* which states like poles repel each other and unlike poles attract each other. See **Figure 15-3.**

Law of magnetic poles: Like poles repel each other and unlike poles attract each other.

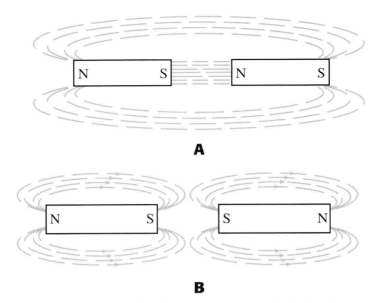

Figure 15-2. Magnetic fields and poles. A—The magnetic fields of two bar magnets with unlike poles form a single magnetic field around the magnets. B—The magnetic fields of two bar magnets with like poles cause the fields to oppose each other.

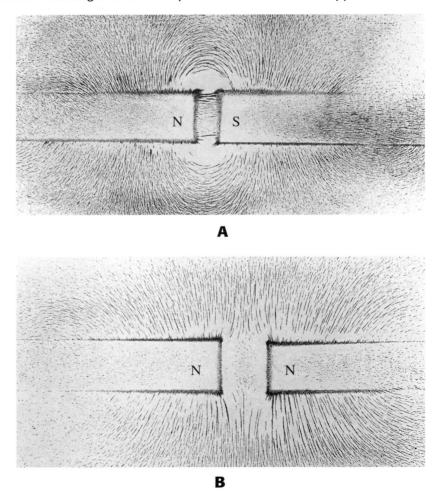

Figure 15-3. Magnetic poles. A—This photograph shows the magnetic lines of force between and around two magnets whose unlike poles are close together. The lines produce a magnetic circuit, drawing the two magnets toward each other. B—This photograph shows the lines of force between two like poles that repel each other.

Magnetic Materials

Materials are either magnetic or nonmagnetic, compared to iron, which has strong magnetic properties. Magnetic materials can be classified as ferromagnetic, paramagnetic, and diamagnetic. Each type of material has properties that are important in certain applications.

Ferromagnetic materials include iron, steel, nickel, cobalt, and alloys such as alnico and Permalloy™. These materials are easily magnetized and become strongly magnetized in the same direction as the magnetizing field.

Paramagnetic materials include aluminum, platinum, manganese, and chromium. These materials become weakly magnetized in the same direction as the magnetizing field.

Diamagnetic materials include antimony, bismuth, copper, gold, mercury, silver, and zinc. These become weakly magnetized in the opposite direction of the magnetizing field.

Ferrite is a ceramic material that acts like iron in the presence of magnetic fields. Unlike iron, ferrite is lightweight and easily formed into a desired shape. A common application is the use of a ferrite core in adjustable radio frequency (RF) transformers and coils, **Figure 15-4.** Unfortunately, ferrites are easily saturated at low magnetizing current values. This means the ferrite core cannot be used for power transformers.

Another common application is to pass a ribbon cable through a ferrite bead, **Figure 15-5.** This makes a simple, inexpensive radio frequency choke for protection against interference from unwanted signals. It can be applied to a single wire as well as a cable.

Ferromagnetic: Refers to materials that are easily magnetized.

Paramagnetic: Describes a material having a permeability slightly greater than a vacuum.

Diamagnetic: A material that is less magnetic than air or in which the intensity of magnetism is negative.

Figure 15-4. An RF transformer with an adjustable ferrite core. Specially shaped nonmetallic screwdrivers are required to adjust these types of transformers.

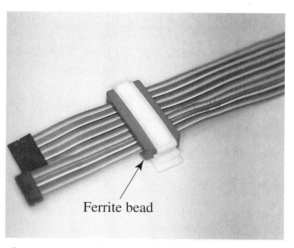

Ferrite bead

Figure 15-5. A ferrite bead on this wire protects the circuit from unwanted currents.

Why Materials Become Magnetized

The mystery of magnetism has been the subject of scientific investigation for many years. One theory of why materials become magnetized says the atomic structure and magnetic property of a material are related. Called the ***domain theory,*** it suggests each electron in an atom is spinning on its own axis as well as orbiting the nucleus of the atom like the earth spins on its axis and orbits the sun. If an equal number of electrons are spinning in opposite directions, the atom is unmagnetized. If more electrons are spinning in one direction than the other, the atom is magnetized and surrounded by a magnetic field.

Domain theory: Suggests each electron in an atom is spinning on its own axis as well as orbiting the nucleus of the atom.

Domain: Areas within a material made up of atoms with the same magnetic polarity.

Atoms within a material interact with each other and form domains. ***Domains*** are areas within the material made up of atoms with the same magnetic polarity. The domains exist in a random pattern, **Figure 15-6.** The random pattern causes the magnetic fields of the domains to cancel each other so the material is demagnetized. If a strong external magnetic field is brought close to the material, the domains become aligned and the material becomes magnetized.

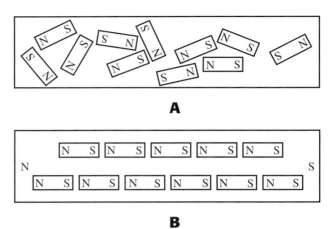

A

B

Figure 15-6. Domains. A—The domains of an unmagnetized material are scrambled and have no particular alignment. B—The domains of a magnetized material are aligned and one domain aids the other.

Magnetic Flux

As defined earlier in this chapter, magnetic flux is the quantity of magnetism or number of magnetic lines of force. Represented by the Greek letter phi (Φ), magnetic flux is measured in maxwells in the centimeter-gram-second (cgs) system or webers in the meter-kilogram-second (mks) system. One ***maxwell*** (Mx) is equal to one magnetic field line. The ***weber*** is a larger unit of measurement equal to 100 million maxwells or lines of force. It is used in the SI units of measurement.

Flux density (represented by B) is equal to the number of magnetic lines of flux per square meter. It is measured in the unit tesla (T). The equation is:

Maxwell: The centimeter-gram-second (cgs) unit of magnetic flux where 1 weber equals 10^8 maxwells.

Weber: The unit of measurement of magnetic flux.

Flux density: The number of magnetic lines of flux per square meter.

$$B = \frac{\Phi}{A}$$

Where:
B = Flux density in teslas (T)
Φ = Magnetic flux in webers (Wb)
A = Area in square meters (m^2)

▼ ***Example 15-1:***
If a magnet produces 500 mWb (milliwebers) of flux in 1 cm^2, what is the flux density?

$$
\begin{aligned}
B &= \frac{\Phi}{A} \\
&= \frac{500 \times 10^{-3}\,\text{Wb}}{1\,\text{cm}^2} \\
&= \frac{500 \times 10^{-3}}{0.0001\,\text{m}^2} \\
&= 5000\,\text{T}
\end{aligned}
$$

▲

Remember, the units for flux are maxwells and webers, which measure total lines. Flux density is a measure of the lines per unit of area.

Magnetomotive Force

A current flows in an electric circuit because electromotive force (emf) pushes the electrons through the circuit. Similarly, magnetic lines of force are pushed through a material by **magnetomotive force** (mmf), expressed as ampere-turns. The equation is:

$$mmf = N \times I$$

Where:

N = Number of turns in a wire

I = Current in amperes

Magnetomotive force: The force that produces flux in a magnetic circuit.

▼ *Example 15-2:*

If a current of 3 A flows through a coil with 25 turns, what is the magnetomotive force?

$$mmf = N \times I$$
$$= 25 \times 3$$
$$= 75 \text{ ampere-turns}$$

▲

Reluctance

Reluctance is to magnetism as resistance is to electric current. **Reluctance** (represented by $\mathcal{R}$) is the total resistance of a material to magnetic lines of force. This relationship can be seen in **Ohm's law for magnetic circuits** which states magnetic flux is equal to magnetomotive force divided by reluctance. The equation is:

$$\mathcal{R} = \frac{mmf}{\Phi}$$

Where:

Φ = Magnetic flux

mmf = Magnetomotive force

$\mathcal{R}$ = Reluctance

Reluctance: The resistance a material has to magnetic lines of force.

Ohm's law for magnetic circuits: Magnetic flux is equal to magnetomotive force divided by reluctance.

▼ *Example 15-3:*

What is the total reluctance of a material that has a mmf of 400 A and a magnetic flux of 200 μWb?

$$\mathcal{R} = \frac{mmf}{\Phi}$$
$$= \frac{400}{200 \text{ μWb}}$$
$$= 2 \times 10^6 \text{ A/Wb}$$

▲

In a magnetic circuit, mmf produces magnetic flux in a material with reluctance. This corresponds to an electrical circuit in which emf or voltage produces a current in material with resistance. There is no unit of measurement for reluctance.

Other Magnetic Terms

Three terms used to explain magnetism have just been defined — magnetic flux, magnetomotive force, and reluctance. Examples of how to find flux density, mmf, and magnetic flux were also given. The following terms are important to the study of magnetism as well.

Permeability

Permeability: Measure of how well a material conducts magnetic lines of force.

Permeability is the ability of a material to conduct magnetic lines of force. It is measured in henrys per meter (H/m). Permeability is to magnetic circuits as conductance is to electrical circuits. A good magnetic material, such as iron, has high permeability. High permeability indicates low reluctance. Represented by the Greek letter mu (μ), permeability is equal to the reciprocal of the reluctance. The equation is:

$$\mu = \frac{1}{\Re}$$

Relative permeability: The permeability of a material compared to air.

Absolute permeability: The permeability of free space.

Relative permeability (μ_r) is the permeability of a material compared to air. The permeability of air has a value of 1. The higher the number, the greater the permeability. For example, magnetic lines of force will pass through cast iron 90 times easier than through air, **Figure 15-7.**

Absolute permeability (μ_o) is the permeability of free space and is equal to $4\pi \times 10^{-7}$ (1.26×10^{-6}). Permeability is equal to relative permeability times absolute permeability. The equation is:

$$\mu = \mu_r \times \mu_o$$

Material	Permeability
Air	1
Nickel	50
Cobalt	60
Cast iron	90
Steel	450
Iron	5,500
Silicon iron	7,000
Permalloy™	100,000

Figure 15-7. Relative permeability of materials.

▼ ***Example 15-4:***

Given the relative permeability, what is the permeability of nickel?

$$\begin{aligned} \mu &= \mu_r \times \mu_o \\ &= 50\,(1.26 \times 10^{-6}) \\ &= 63 \times 10^{-6}\ \text{H/m} \end{aligned}$$

▲

Residual Magnetism

Residual magnetism: Magnetism that remains in a device after the magnetizing force is removed.

Residual magnetism is the magnetism remaining in a material after the magnetizing force is removed. The iron nail in **Figure 15-8** is being magnetized by a permanent magnet. When the magnet is taken away, a small magnetic field remains in the nail.

Retentivity

Retentivity: Property of a material that measures its ability to stay magnetized after the magnetizing force is removed.

Retentivity is the ability of a material to hold magnetism after the magnetizing force has been removed. When an iron nail is magnetized, only a small amount of magnetism remains in the nail because the nail has low retentivity. A high-quality steel retains much of its magnetic field when the magnetizing force is removed because the steel has high retentivity. A permanent magnet should have high retentivity.

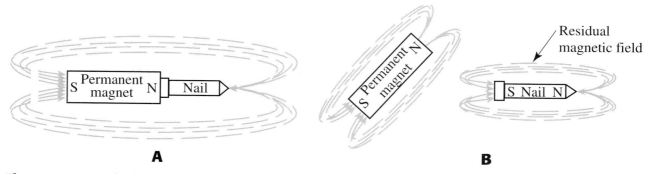

Figure 15-8. Residual magnetism. A—The nail is magnetized and becomes part of the magnet. The point of the nail becomes the north pole of the magnetic circuit. B—When the nail is removed from the magnet, it continues to have a small amount of magnetism.

Magnetic Saturation

At some point, any further attempt to increase the magnetism of a material will not increase the magnetic strength. The ***magnetic saturation*** point has been reached. This occurs when all the molecular dipoles and domains are aligned by the magnetizing force, and no additional field intensity can be produced.

Hysteresis

"Hysteresis" means to lag behind. In electricity, ***hysteresis*** is a loss in energy caused by the molecular realignment of the material. The atoms do not return exactly to their original positions when the magnetizing force is moved or changed in polarity.

Hysteresis loss occurs when a changing alternating current causes the molecules to realign at such a high speed that molecular friction becomes a resistance factor. Energy generated by the friction of the molecules rubbing together is lost as heat.

The relationship between flux density and field intensity (represented by *H*) can be illustrated by a ***hysteresis loop*** or ***B-H curve,*** **Figure 15-9.** Flux density is plotted on the Y-axis and field intensity is plotted on the X-axis. The more area enclosed by the loop, the greater the hysteresis loss.

Figure 15-10 shows an alternating current and the positions plotted on the graph. The graph starts at point A (zero) with no magnetic flux. At point B the flux density is maximum in the positive direction. The flux density cannot increase beyond saturation. At point C the magnetizing force is zero and B falls to the value of C.

Magnetic saturation: In an iron core, the point where any further increase of magnetizing force produces little or no increase in the magnetic lines of force in the core.

Hysteresis: The amount of magnetism that lags behind the magnetizing force due to molecular friction.

Hysteresis loss: The power loss caused by molecular friction.

Hysteresis loop: A graphic curve showing the relationship between a magnetizing force and the resulting magnetic flux.

B-H curve: A curve plotted on a graph to show successive states during the magnetization of a ferromagnetic material.

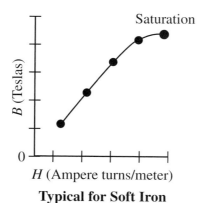

Typical for Soft Iron

Figure 15-9. B-H curve for soft iron. No values are shown near zero because the permeability can vary depending on previous magnetization.

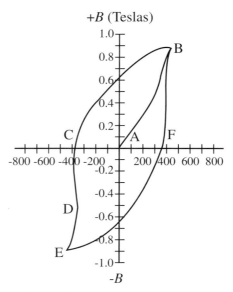

+B (Teslas)

Figure 15-10. B-H curve when using alternating current. The curve is similar to that in Figure 15-9, but the H value alternates in polarity with the alternating current.

The alternating current changes direction and causes the flux density to move to point D. The current continues to increase to the maximum negative value causing the curve to go to point E. This produces maximum flux density in the negative direction until the saturation point is reached. At point F, the magnetizing force is again zero and B falls to the value of F. The current now changes direction again and causes the flux density to go to point B. The process continues as long as the peak source voltage amplitude remains the same.

Types of Magnets

Permanent magnet: A piece of hardened steel or other magnetic material that indefinitely retains its magnetism.

Electromagnet: A temporary magnet made from a coil of wire with or without a core of iron. A magnetic field exists only when current flows in the conductor.

Magnets are classified as either permanent or electromagnet. A ***permanent magnet*** is a piece of hardened steel or other magnetic material that retains its magnetism indefinitely. Permanent magnets take the form of a bar, button, toroid (ring), or horseshoe, **Figure 15-11.** An ***electromagnet*** is a temporary magnet made by wrapping a coil of wire around a form of any shape, **Figure 15-12.** Magnetism is produced by an electrical current. The center of the coil has a core of iron or ferrite. RF coils have either air or ferrite cores. The RF ferrite core is usually adjustable.

Figure 15-11. Common types of permanent magnets are (from top to bottom) bar, toroid, and button.

Figure 15-12. An electromagnet is made of insulated wire wrapped on a form. The form is usually round but can be any shape.

Induction

The effect of one material on another without any physical contact between them is called *induction*. Magnetic induction occurs, for instance, when a magnet is brought close to an iron nail, **Figure 15-13.** The magnet causes the iron nail to become a magnet without the two objects touching. The iron of the nail is a better conductor of magnetism than the air around the nail, so the nail becomes part of a magnetic circuit. Notice the polarity of the induced magnetic poles shown in the figure. When the nail is removed from the magnetic field, its magnetism disappears except for a small amount of residual magnetism.

Shielding is used to prevent one component from affecting another. *Magnetic shielding* protects a device or system from magnetic fields. For example, the inner conductor of a coaxial cable is shielded by the outer braided conductor. Shielding materials are always made of some form of metal. A magnetic shield serves as a short circuit for the magnetic lines of flux.

Induction: The production of an electric charge or a magnetic field in a substance by the proximity of an electrified source, a magnet, or magnetic field.

Magnetic shielding: A metallic enclosure to protect devices or systems against magnetic fields.

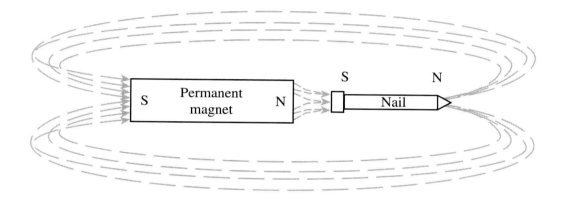

Figure 15-13. The nail is magnetized by induction. No actual physical contact is made between the nail and magnet, yet the nail becomes magnetized.

Electromagnetism

When any current flows in a conductor, a magnetic field is produced around the conductor, **Figure 15-14.** No matter how small the current or conductor, a magnetic field exists around the conductor. If a higher current flows through the conductor, the magnetic field becomes larger. A magnetic field produced by electricity is called *electromagnetism.*

The magnetic fields of conductors that are physically close can attract or repel each other, depending on the direction of the currents, **Figure 15-15.** If the currents are flowing in the same direction, the magnetic fields are attracted to each other. If the currents are flowing in opposite directions, the fields repel each other. For this reason, the placement of conductors in some electronic equipment can have an effect on the operation of the circuits. This is a major consideration with high-frequency equipment.

Electromagnetism: Magnetism produced by an electrical current.

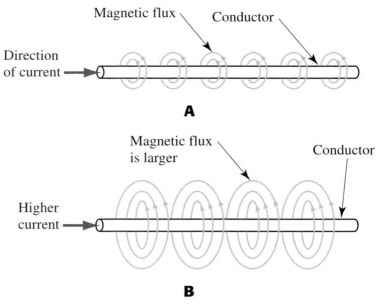

Figure 15-14. Electromagnetism. A—A conductor with a current has a magnetic field around it. B—A higher current produces a larger magnetic field around the conductor.

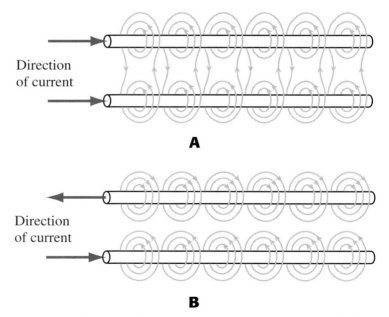

Figure 15-15. Current direction affects magnetism. A—The magnetic fields of these two conductors are attracted to each other and can interfere with the operation of the circuit. B—The magnetic fields of these conductors oppose each other.

Magnetic Field Strength

If the conductor is bent into a loop, the magnetic fields around the conductor will remain, **Figure 15-16.** More turns of the conductor will make more loops. As the loops are pushed closer together, the small magnetic fields of the individual conductors aid each other to form one large magnetic field. The strength of the magnetic field depends on four factors:

- Number of turns in the coil
- Strength of the current
- Type of core material
- Length of the coil

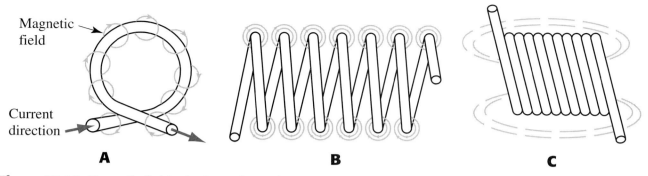

Figure 15-16. Magnetic fields. A—A conductor bent into a loop retains its magnetic fields. B—As the conductor is turned into more loops, a magnetic field remains around each loop. C—As the loops of the coil are pushed together, the individual magnetic fields become one large (stronger) magnetic field.

The more turns in the coil and the greater the current flow, the stronger the magnetic field. A magnetic core will also increase the strength of the magnetic field. Compared to an air core, for example, an iron core has a stronger magnetic field because of the high permeability of iron. A coil with a core makes an electromagnet. The strength of the magnetic field is inversely proportional to the length of the coil as shown by the equation:

$$H = \frac{\mu N I}{l}$$

Where:
H = Magnetic field intensity or strength
μ = Permeability of the core
N = Number of turns
I = Current (amps)
l = Length of the coil (meters)

▼ *Example 15-5:*
What is the magnetic field intensity of a 2 cm long electromagnet that has 400 turns, 2 A, and a nickel core?
First, find the permeability.

Permeability (μ) = Relative permeability (μ_r) × Absolute permeability (μ_o)
μ = 50 × (1.26 × 10⁻⁶) (Refer to Figure 15-7 for the relative permeability of nickel.)

$$= 6.28 \times 10^{-5}$$

Next, find the field intensity.

$$H = \frac{\mu N I}{l}$$

$$= \frac{(6.28 \times 10^{-5})(400 \text{ turns})(2 \text{ A})}{2 \times 10^{-2} \text{ m}}$$

$$= 2.5 \text{ A/m}$$ ▲

Left-hand Rule

The ***left-hand rule of conductors*** states if you hold a conductor with your left hand so your thumb points in the direction of electron flow, your fingers will show the direction of the magnetic field. The ***left-hand rule of coils*** further states if you hold a coil in your left hand so your fingers point in the direction of current flow, your thumb will point toward the north pole. See **Figure 15-17.**

Left-hand rule of conductors: If you hold a conductor with your left hand so your thumb points in the direction of electron flow, your fingers will point in the direction of the magnetic field.

Left-hand rule of coils: If you hold a coil in your left hand so your fingers point to the direction of current flow, your thumb will point toward the north pole.

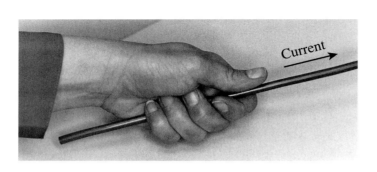
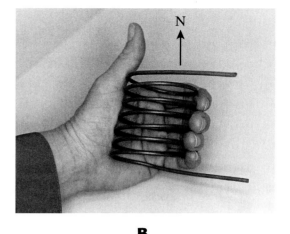

A   **B**

Figure 15-17. Left-hand rule. A—With your thumb pointed in the direction of the current, your fingers show the direction of the magnetic lines of force around the conductor. B—With your fingers pointed in the direction of the current, your thumb points in the direction of the north magnetic pole of the coil.

Solenoids

Solenoid: A device consisting of an electromagnet and a movable core used for moving valves, electric switches, and mechanical levers.

Energized: Electrically connected to a voltage source.

The attraction and repulsion characteristics of electromagnetism are used in a device called a *solenoid*. In **Figure 15-18,** the iron core is not completely inside the electromagnet. When the solenoid is energized, the magnetism of the coil pulls the movable iron core into the coil. *Energized* means a current has been caused to flow in the device. Most solenoids operate on ac power, although dc solenoids are available. Solenoids are used for moving valves, electric switches, and mechanical levers. Solenoids are installed in water valves in washing machines, gas valves in furnaces, and starters in automobiles.

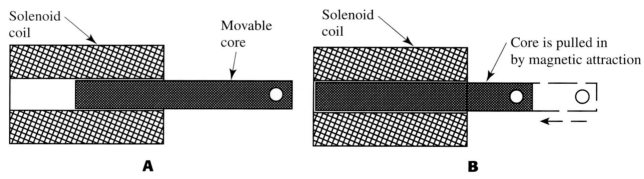

A   **B**

Figure 15-18. Solenoid. A—Without a magnetic field, the core does not move. B—The magnetic field pulls the core into the coil of the solenoid.

Relays

Relay: An electromechanical switch in which contacts are open or closed by an electromagnet.

A *relay* is a type of switch that uses a low-voltage, low-current circuit to control a high-voltage, high-current circuit. Various types of switching, such as SPST or DPDT, can be made using a relay. The two major types of relays are electromechanical and solid-state. The electromechanical relay uses sets of contacts controlled by an electromagnet. Solid-state relays are electronic devices that have no mechanical contacts.

Electromechanical Relays

Electromechanical relays fall into three main types: reed, general purpose, and industrial control. The main difference between them is design.

Reed Relays

Reed relays are small, fast-operating, SPST, normally open contacts sealed in a glass envelope. See **Figure 15-19.** Reed relays are activated by an external magnetic field, generated by a permanent magnet or dc electromagnet. When the magnetic field is brought close to the relay, the switch inside the relay is activated. The distance from the relay depends upon the strength of the magnet and the magnet pole alignment in relation to the relay. Alternating current electromagnets are not suitable for activating the reed relay because of the high switching speed capability of the reed relay.

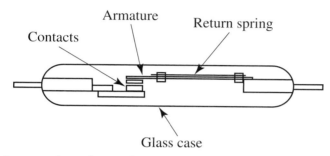

Figure 15-19. Construction of a reed relay. The contacts close when the electromagnetic coil (omitted from figure for clarity) is energized.

General Purpose Relays

A general purpose relay consists of an electromagnet and armature, **Figure 15-20.** The armature is the movable part of the relay and is held away from the electromagnet by a spring. When the relay electromagnet is energized, the armature is pulled toward the electromagnet and closes the contacts. Some relay contacts are made to "break" or "make" for switching the circuit. The circuit that controls the energizing of the relay is called the ***control circuit.*** The circuit with the relay contacts is called the ***controlled circuit.***

General purpose relays are very common in both consumer products and industrial applications. They are designed as plug-ins for quick, easy replacement and troubleshooting. General purpose relays are manufactured with a plastic cover for protection against dirt and moisture. They are used for:

- On/off control
- Limit control
- Logic operations
- Timing and sequence
- Power control
- Safety

Three terms used in describing general purpose relay contact arrangements are poles, throw, and break. ***Poles*** describe the number of separate circuits that can pass through the relay. ***Throws*** are the number of different contact positions per pole available on the relay. ***Breaks*** are the number of separate contacts the relay uses to open or close individual circuits. If the relay breaks the circuit in one place, it is a single-break switch. If it breaks the circuit in two places, it is a double-break switch.

Control circuit: A circuit that controls various motor functions such as speed.

Controlled circuit: A circuit controlled by either automatic or manual switching.

Pole: One end of a magnet.

Throws: The number of different contact positions per pole available on the relay.

Breaks: The number of separate contacts a relay uses to open or close individual circuits.

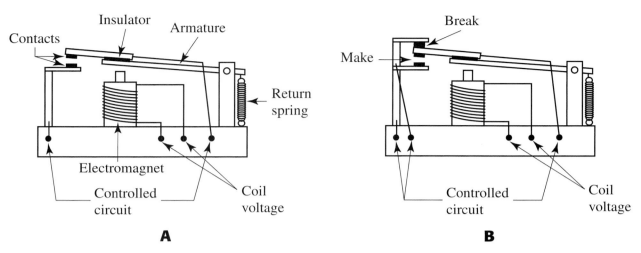

Figure 15-20. Electromagnetic relay. A—An SPST switch contact that "makes" when the coil is energized. B—An SPDT switch contact in which one set of contacts "breaks" while the other "makes."

Industrial Control Relays

Industrial control relays are like general purpose relays but are designed for heavy-duty operation, such as controlling contactors, starters, and solenoids. Industrial relays have optional accessories such as indicator lights, noise suppressors, timing controls, and special features for industrial applications. The controlling voltages can be changed quickly to adapt the relay to different circuits.

Testing and Troubleshooting Relays

Testing and troubleshooting electromechanical relays is a simple process. A common problem is a relay that does not activate. In many cases, if you place a metal screwdriver blade against the electromagnet, you can feel the presence of the magnetic field.

To test a relay, use a voltmeter and verify the controlling voltage is present. If the voltage is absent, make a systematic check of the control circuit. If the controlling voltage is present, use an ohmmeter to check the coil for continuity. Other relay problems include dirty contacts, moisture, overloading or arcing (relay contacts are burned), physical damage, and expiration of the useful life of the relay.

Identifying information on a relay has been simplified by the National Association of Relay Manufacturers (NARM). A list of codes along with the basic contact forms designated by NARM are given in Appendix D.

A relay should be replaced with one of the same make and part number. If that is not possible, the following factors should be considered when finding a replacement:

- *Type.* The proper type must be selected because of the physical requirements of the relay. If you try to replace a reed relay with an industrial relay, it will not fit in the space. Voltage-current requirements must also be considered. For example, the current requirements of an industrial relay are greater than those of a general purpose relay. A reed relay should be replaced with a reed relay. Likewise, a general purpose or industrial relay should be replaced with the same type.

- *Coil voltage and current.* The circuit was designed for a specific operating voltage and current. If the voltage rating of the new coil is low, the relay coil will operate hot or may burn up. If the voltage rating is high, the relay contacts may operate slower than required or not at all.

- *Contact current rating.* If the contacts cannot carry the required load current, the new relay contacts will quickly be destroyed by the excessive current.
- *Contact switching arrangements.* The controlled circuit must be controlled by the same switching (for example SPST, DPDT, 4PST), and the replacement relay must have the same throw and break.
- *Terminal configuration.* Even though the correct type, voltage-current ratings, and switching are found, the replacement relay terminals may not be the same as the original relay. Care must be used in connecting the new relay terminals to the circuit connections.
- *Isolation.* Although all other factors may allow a certain relay to be used as a replacement, the voltage isolation rating may not be high enough. The voltage isolation rating is determined by the insulation rating of the materials by which the relay is made. In this case, the materials that make up the relay determine whether a replacement relay can be used for another relay. If the isolation rating is not high enough, voltage insulation breakdown may occur, causing failure of the new relay and possibly fire.
- *Response time.* The response time of a relay is a measure of its ability to turn on and off. Replacing a solid-state relay (SSR) with an electromechanical relay is often impossible because the response time of the SSR may be in microseconds. Electromechanical relay response time may be as slow as a few milliseconds. Replacing an electromechanical relay with a SSR may not work because the SSR response time is too fast for the circuits being controlled.
- *Environment.* Some relays are used in an environment that contains dirt and moisture. Relays used for this service usually have a plastic cover or are completely sealed from the environment.

Solid-state Relays

Solid-state relays are highly reliable and useful, and have replaced electromagnetic relays in many applications. Even though solid-state relays that operate using semiconductor components are beyond the scope of this text, they deserve to be addressed.

The exterior construction of the SSR is very simple, **Figure 15-21.** It has mounting holes and connections for the control circuit, voltage source, and load. The SSR is sealed, protecting it from dirt and moisture. Advantages of the SSR include:
- Arcless switching of the load
- Shock and vibration resistant
- Long life
- Compatible for connecting to other electronic circuitry
- Good response time

The main disadvantages of the SSR are its sensitivity to temperature and its inability to withstand surge currents. The troubleshooting tips in **Figure 15-22** will help you diagnose and correct SSR problems.

Hall Effect

In 1879, E. H. Hall of Johns Hopkins University observed a small voltage generated across a gold conductor carrying current within an external magnetic field. The voltage was very small with common metal conductors. Little use could be made of this discovery, named the Hall effect, until semiconductors were developed and larger Hall voltage values could be generated.

Indium arsenide is one such semiconductor material used to produce a Hall effect voltage, **Figure 15-23.** Many types of devices using the Hall effect have been developed for use in the electronic industry as sensors of magnetic fields.

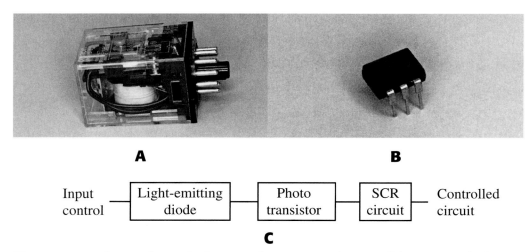

A   **B**

Input control — Light-emitting diode — Photo transistor — SCR circuit — Controlled circuit

C

Figure 15-21. Comparison of relays. A—Electromagnetic. B—Solid-state. C—Block diagram of a solid-state relay with no moving parts.

Problem	Troubleshooting
Relay fails to turn off	• Excess load current — SSR shorted. • High line voltage. • Transient voltages on the line — insufficient protection. • Insufficient heat sinking.
Relay fails to turn on	• Check input voltage. Supply the correct voltage from another source and see if the relay operates. • If the correct voltage is present, check input current with clamp-on ammeter. If no current is indicated, the SSR input is open and the relay must be replaced.
Intermittent operation	• Replace the SSR. • Check control circuit. • Check load circuit.

Figure 15-22. Troubleshooting guide for SSR problems.

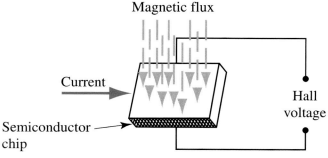

Figure 15-23. Hall voltage is generated by a magnetic field acting upon a piece of solid-state material.

Speaker: An energy converter that changes electrical energy to mechanical energy. Sound is caused by mechanical energy moving the air.

Transducer: A device that converts one form of energy to another.

Speakers

A *speaker* converts energy from one form to another. Electrical energy is converted to magnetic energy, which is converted to mechanical energy. The technical term for such a device is a *transducer.* The two types of speakers in common use are dynamic and piezoelectric.

Dynamic Speaker

The dynamic (permanent magnet) speaker is the most common type in audio stereo systems, **Figure 15-24.** A permanent magnet is mounted at the center of the frame. The *cone* is a special type of paper attached to the outside edges of the frame. The center of the cone is attached to the frame with a flexible membrane called a *spider* that acts as a hinge. Attached to the center of the cone is a circular form with a coil of small wire wound around it. This is called the *voice coil.* The position of the cone allows the permanent magnet to rest inside the voice coil.

Cone: The moving part of a speaker that causes the surrounding air to vibrate and creates audible sound.

Spider: A thin flexible piece of membrane connecting the cone of a speaker to its frame.

Voice coil: A coil of small diameter wire that produces magnetic fields corresponding to the electrical currents representing the sound. The magnetic field produced by the voice coil reacts with the permanent magnet field of the speaker in which it is installed.

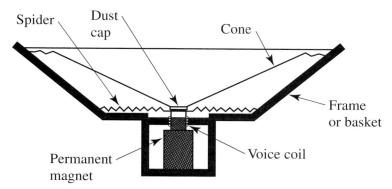

Figure 15-24. Basic construction of a dynamic speaker. The sizes of the magnet and voice coil conductor determine the power rating of the speaker.

The speaker operates because of the attraction and repulsion of magnetic fields. The permanent magnet is stationary; that is, it cannot move. As electrical current passes through the voice coil, it produces a magnetic field in the voice coil. The voice coil magnetic field interacts with the permanent magnet field, causing the voice coil to be attracted or repelled. This makes the voice coil move inward or outward from the permanent magnet. The amount of movement depends on the electrical impulses that represent the sound. With the voice coil attached to the cone, the cone vibrates, in turn moving the air as sound vibrations.

Piezoelectric Speaker

The principles of piezoelectricity, whereby voltage is produced by pressure on certain materials, was discussed in Chapter 1. In a *piezolectric speaker,* a changing electrical signal on the material causes the reverse to take place — the piezoelectric material (element) vibrates in step with the voltage change. Speakers used in buzzers, wristwatches, and computers contain a ceramic disc as the piezoelectric element.

Piezoelectric speaker: A speaker in which the mechanical movements are produced by piezoelectric action.

Testing Speakers

Most speaker problems are limited to either no sound or distorted sound. Either problem can be caused by the speaker or amplifier circuit to which the speaker is connected. If a good speaker is available, an easy test is to connect the good speaker in place of the suspected one. If the problem is still present, the amplifier is at fault.

If a good speaker is not available, check for continuity across the voice coil. A low resistance reading is normal. A slight click may be heard when the ohmmeter is first connected to the voice coil.

Also check for a rubbing cone by gently pushing the speaker cone in and allowing it to move back out. A rubbing or scraping sound indicates the voice coil is rubbing the sides of the magnet assembly. This is usually caused by excessive current through the voice coil. The cone should move freely and silently.

Speaker Phasing

When only one speaker is used in a system, phasing is not a problem. When two or more speakers are used, it is best to phase the speakers so the cones are all moving in the same direction. If the speakers are out of phase, they will move in opposite directions with the same audio signal, giving a reduced overall volume.

To connect the speakers in phase, the positive terminal of each speaker should be connected to the positive terminal of the amplifier. If the terminals are not marked, use the setup shown in **Figure 15-25** to connect the negative lead of the battery assembly to one of the speaker terminals. Momentarily touch the other terminal with the positive lead while holding your fingers on the speaker cone. Feel the cone to determine whether it moves outward. If it moves inward, reverse the test circuit leads and try again. Verify the speaker cone moves outward. The resistor must be used; otherwise, the current will be excessive and damage the voice coil.

Mark the terminals of the speaker to indicate positive or negative terminals. Then, connect the speakers so they are in phase.

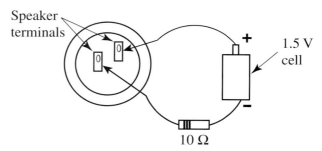

Figure 15-25. Setup for determining the phase of a speaker. The speaker voice coil can be damaged if the resistor is omitted.

Summary

- Magnetism is a form of energy.
- The law of magnetic poles states like poles repel each other and unlike poles attract each other.
- Materials are classified as magnetic or nonmagnetic. They can also be classified as ferromagnetic, paramagnetic, or diamagnetic.
- Ferrite, a ceramic material that acts like iron, is commonly used in the core of magnets.
- The domain theory explains why materials become magnetized.
- Magnets can take the form of a bar, horseshoe, toroid, or electromagnet.
- Electromagnetism is magnetism produced by an electric current through a conductor.
- The strength of the magnetic field of a coil depends upon the number of turns, current, type of core material, and length of the coil.
- The left-hand rule for conductors states if your thumb is placed in the direction of the current, your fingers will show the direction of the magnetic field. The left-hand rule for coils states if your left hand is placed in the direction of the current, your thumb will point toward the north pole of the magnet.
- A solenoid is a mechanical device operated by the attraction and repulsion of a magnet on its core.
- A relay is an electromechanical switch. Types of relays are reed, general purpose, industrial control, and solid-state.
- Shielding protects one component from another.
- A speaker (transducer) converts one form of energy to another. Speaker problems are usually evident by no sound or a distorted sound.

⏩ Important Terms

Do you know the meanings of these terms used in the chapter?

absolute permeability
B-H curve
break
cone
control circuit
controlled circuit
diamagnetic
domain
domain theory
electromagnet
electromagnetism
energized
ferrite
ferromagnetic
flux density
hysteresis
hysteresis loop
hysteresis loss
induction
law of magnetic poles
left-hand rule of a coils
left-hand rule of conductors
lodestone
magnetic field

magnetic flux
magnetic saturation
magnetic shielding
magnetomotive force
maxwell
Ohm's law for magnetic circuits
paramagnetic
permanent magnet
permeability
piezoelectric speaker
pole
relative permeability
relay
reluctance
residual magnetism
retentivity
solenoid
speaker
spider
throw
transducer
voice coil
weber

⏩ Questions and Problems

Please do not write in this text. Write your answers on a separate sheet of paper.

1. Magnetism is a form of _____.
2. The invisible magnetic lines of force are called _____.
3. What is the law of magnetic poles?
4. List the three types of magnetic materials.
5. What is a disadvantage of ferrite cores?
6. The reason materials become magnetized is explained by the _____ theory.
7. A field has a magnetic flux of 500 µWb with an area of 1 cm². What is the flux density?
8. What is the magnetomotive force of a coil that has 450 turns and 3 A of current?
9. If a coil has a permeability of 20 H/m, what is the reluctance? ($\mu = \dfrac{1}{\mathfrak{R}}$)
10. Magnets are classified as _____ or _____.
11. An electromagnet is a magnet produced by _____.
12. The strength of the magnetic field of an electromagnet is controlled by what four factors?
13. What is the magnetic field intensity of a 2 cm long coil that has an air core, 300 turns, and 550 mA current?
14. What is a solenoid?
15. What are the two major types of relays?
16. What is a relay?
17. What is the Hall effect used for?
18. Define *transducer.*
19. What are the two typical speaker problems?
20. What two steps are used to test a speaker?

Chapter 16 Graphic Overview

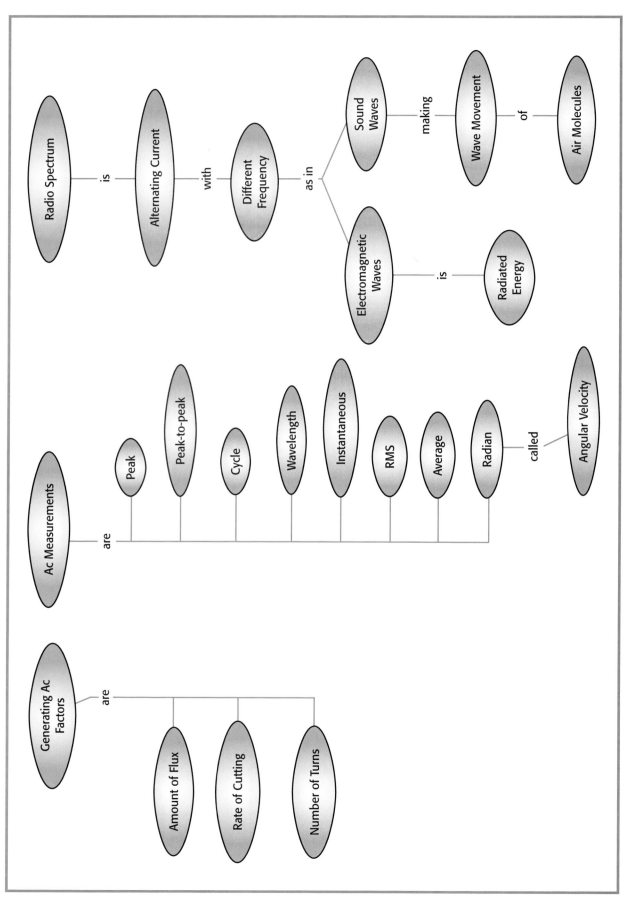

ALTERNATING CURRENT 16

Objectives

After studying this chapter, you will be able to:
○ Define the process used to generate alternating current.
○ Draw a graph of alternating current.
○ Calculate peak, peak-to-peak, instantaneous, and root-mean-square values of alternating current.
○ Identify the various frequencies on the frequency spectrum.

Introduction

Alternating current (ac) is the type of electricity most often used. Ac is used in antenna systems, amplifiers, and homes supplied with electricity by electric power companies. This chapter covers how alternating current is produced and measured.

Generating Ac

As you learned in the chapter on magnetism, when a current is flowing in a conductor, a magnetic field surrounds the conductor. In **Figure 16-1,** the opposite is taking place. As a conductor is moved through a stationary magnetic field, the action of the field causes the electrons of the conductor to move and current is produced. The process is called *induction* because a current is induced to flow in the conductor.

Induction: Producing an electric charge or magnetic field in a substance by the proximity of an electrified source, a magnet, or magnetic field.

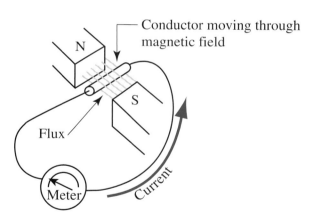

Conductor moving through magnetic field

Flux

Meter

Current

Figure 16-1. When a conductor moves through a magnetic field, a current flow is induced in the conductor.

In **Figure 16-2,** the single conductor is replaced with a loop of wire rotated within the magnetic field. As the loop is rotated, alternating current is generated. The end of the loop of wire is rotating in a circle, **Figure 16-3.** At 0° the conductor is moving parallel to the magnetic flux lines and is not cutting across any flux, Figure 16-3A. Therefore, no current is induced. This can be plotted on a graph as zero current at 0°.

As the conductor is moved to the 35° position, it cuts across the magnetic field and induces a current of some value in the conductor, Figure 16-3B. The process continues to the 90° position where the conductor cuts the maximum number of magnetic lines, Figure 16-3C. The magnetic field is also strongest at this point. The maximum or peak value of the current can be plotted at the 90° position on a graph.

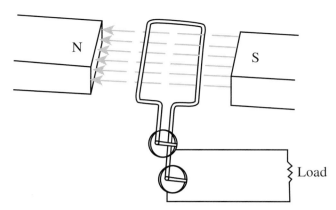

Figure 16-2. As the wire loop is rotated, ac is generated and flows through the circuit.

As the conductor continues to 120°, 135°, and 150°, the magnetic field gets weaker and the conductor cuts fewer magnetic lines, Figure 16-3D. This causes the current to weaken as the conductor rotates toward 180°. At 180° the conductor is not cutting across any magnetic lines and is traveling in the same direction as the magnetic lines. The current is again zero at 180° on a graph.

As the conductor rotates toward 270°, it is moving through the magnetic field in the opposite direction, Figure 16-3E. This is indicated below the X-axis of the graph. The current increases to a peak at 270° where the maximum magnetic field again is seen by the conductor. The conductor continues from 270° to 360° where the current is again zero, Figure 16-3F. As the conductor continues to rotate, the process repeats itself over and over.

The current flows first in one direction, then in the other direction, changing polarity every 180°, **Figure 16-4.** This is called *alternating current.*

Alternating current: A current of electrons that moves first in one direction, then the other.

Factors Determining Induced Voltage

The amount of voltage induced by a coil cutting the magnetic flux depends on three factors: number of turns, amount of flux, and rate of cutting.

The more turns in a coil, the higher the induced voltage. The induced voltage is the sum of all the individual voltages generated in each turn (connected in series). If one conductor causes a certain amount of voltage to be induced, 100 conductors would produce 100 times as much voltage.

The amount of flux is another factor determining induced voltage. The more magnetic lines of force a conductor cuts, the higher the induced voltage.

Rate of cutting is the third factor. The faster the conductor cuts the magnetic flux, the higher the induced voltage. In other words, the more lines of force the conductor cuts, the higher the induced voltage.

The angle at which a conductor cuts the magnetic field is a factor when considering the instantaneous voltage value. Angle will be discussed with radian measure and instantaneous value later in this chapter.

Cycle Frequency

One rotation of the conductor from 0° to 360° is one cycle of measured events, **Figure 16-5.** The number of times the conductor rotates each second is called the *frequency.* Standard household ac is set at 60 cycles per second. Frequency (*f*) is measured in Hertz (Hz). One Hertz is equal to one cycle per second (cps). If the frequency in the same time period (one second) is doubled, the wavelength will be half, **Figure 16-6.**

Frequency: The number of times per second (or other specified duration) that a particular action takes place.

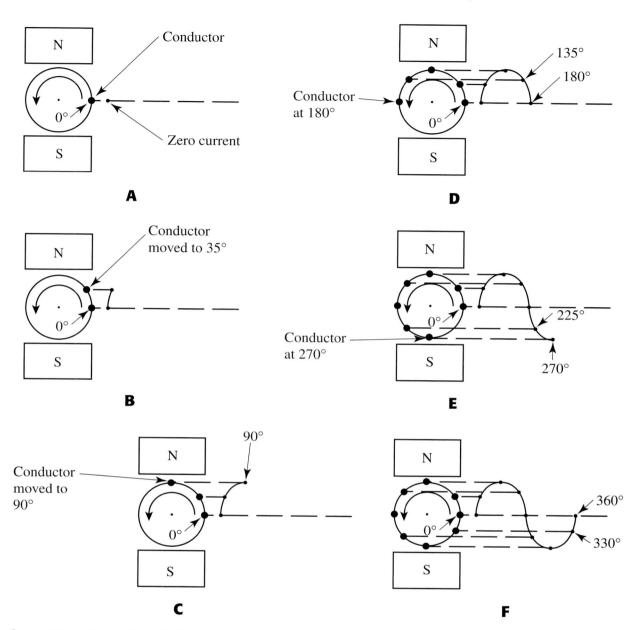

Figure 16-3. Generation of alternating current. A—At 0° the conductor has zero current in the loop. B—At 35° the loop has some current flowing. C—At 90° maximum current flows in the loop. D—At 135° the amount of current begins to decrease. At 180° the current in the loop is zero. E—At 225° the current increases in the other direction through the loop. At 270° the current is maximum in the other direction. F—At 330° the current decreases to 360° where it is again zero.

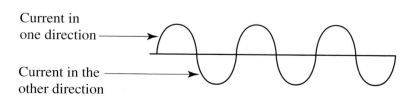

Figure 16-4. Alternating current changes direction around the X-axis every 180°.

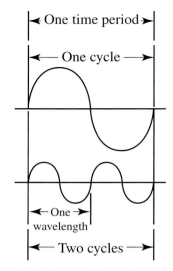

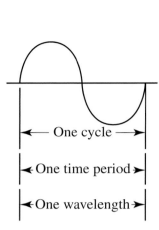

Figure 16-5. One cycle or wavelength depends on the time period of the alternating current.

Figure 16-6. Comparison of one cycle and one wavelength to two cycles and two wavelengths in the same time period.

As shown in Figure 16-5, it takes a certain period of time for the conductor to go through one rotation or cycle. One cycle can be measured by that ***time period.*** The frequency of an alternating current is equal to the reciprocal of the time period (T). The time period is equal to the reciprocal of the frequency. The equations are:

Time period: The time required for one cycle.

$$f = \frac{1}{T}$$

and

$$T = \frac{1}{f}$$

▼ *Example 16-1:*

If an alternating current has a time period of 20 μs, what is the frequency?

$$
\begin{aligned}
f &= \frac{1}{T} \\
&= \frac{1}{20 \times 10^{-6}} \\
&= 50 \text{ kHz}
\end{aligned}
$$

▲

▼ *Example 16-2:*

If the frequency of an alternating current is 1000 Hz, what is the time period (in milliseconds)?

$$
\begin{aligned}
T &= \frac{1}{f} \\
&= \frac{1}{1000} \\
&= 1 \text{ ms}
\end{aligned}
$$

▲

Wavelength

As stated earlier, it takes a period of time for the ac to complete one cycle. Radio waves move through space at the speed of light: 3×10^8 meters per second (186,000 miles per hour). By dividing the velocity (speed) of radio waves by the frequency, the specific wavelength of each frequency can be found. **Wavelength,** represented by the Greek letter lambda (λ), is a measurement of the sine wave from one point through one complete cycle. The equation for finding wavelength is:

$$\lambda = \frac{3 \times 10^8}{f}$$

Wavelength: The length in distance of one cycle.

▼ *Example 16-3:*

If an alternating current has a frequency of 5 MHz, what is the wavelength (in meters)?

$$\lambda = \frac{3 \times 10^8}{f}$$

$$= \frac{3 \times 10^8}{5 \times 10^6}$$

$$= 60 \text{ m}$$

▲

Ac Measurements

Various measurements of alternating current can be made. These include values for peak, peak-to-peak, instantaneous value, and root-mean-square (RMS), **Figure 16-7.**

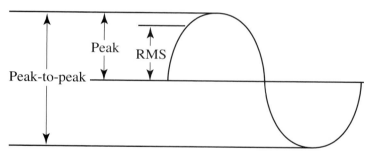

Figure 16-7. Common measurements of a sine wave are peak, peak-to-peak, and RMS.

Peak Value

Peak value is measured from the X-axis to the maximum point in the positive or negative direction. The peak value (E_P) equals 1.414 times the RMS value. The equation is:

$$E_P = 1.414 \times E_{RMS}$$

Peak value: The maximum instantaneous value of voltage, current, or power.

▼ *Example 16-4:*

Find the peak value of a 120 V RMS alternating current.

$$E_P = 1.414 \times E_{RMS}$$

$$= 1.414 \times 120$$

$$= 169.68 \text{ V}$$

▲

Peak-to-Peak Value

Peak-to-peak value of ac is measured from the maximum negative point to the maximum positive point on the ac graph. The peak-to-peak value ($E_{P\text{-}P}$) is equal to two times the peak value or 2.828 times the RMS value. The equation is:

$$E_{P\text{-}P} = 2 \times E_P$$

Peak-to-peak value: The value of alternating current from the positive peak to the negative peak.

▼ *Example 16-5:*

If the RMS value of an alternating current is 40 V, what is the peak-to-peak value?
First, find the peak value:

$$E_P = 1.414 \times E_{RMS}$$
$$= 1.414 \times 40$$
$$= 56.56 \text{ V}$$

Then, find the peak-to-peak value:

$$E_{P\text{-}P} = 2 \times E_P$$
$$= 2 \times 56.56$$
$$= 113.12 \text{ V}$$

or

$$E_{P\text{-}P} = 2.828 \times E_{RMS}$$
$$= 2.828 \times 40$$
$$= 113.13 \text{ V}$$ ▲

Instantaneous Value

The amount of induced voltage at any given time depends on the angle at which the conductor is cutting across the magnetic field. At 0° and 180° the conductor is moving parallel with the magnetic lines of force, and the induced voltage is zero. At angles between 0° and 90°, 90° and 180°, 180° and 270°, and 270° and 360°, a voltage between zero and maximum is induced.

Instantaneous value: The magnitude of a changing value at any given moment.

The *instantaneous value* (represented by lowercase *e*) is the magnitude of a changing value at any given moment. It is a measurement of a specific point (in degrees) within the cycle times the maximum voltage value (E_{MAX}). See **Figure 16-8.** The equation is:

$$e = \sin \theta \times E_{MAX}$$

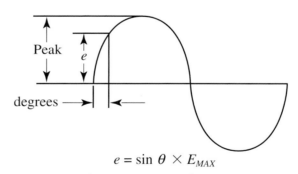

$$e = \sin \theta \times E_{MAX}$$

Figure 16-8. Instantaneous voltage is measured at one specific time or degree of the sine wave.

▼ *Example 16-6:*

Find the instantaneous value of an alternating current if the angle is 0° and the maximum voltage value is 14 V.

$$e = \sin \theta \times E_{MAX}$$
$$= \sin 0° \times 14 \text{ V}$$
$$= 0 \times 14 \text{ V}$$
$$= 0 \text{ V}$$ ▲

▼ *Example 16-7:*

Find the instantaneous value of an alternating current if the angle is 90° and the maximum voltage value is 14 V.

$$e = \sin \theta \times E_{MAX}$$
$$= \sin 90° \times 14 \text{ V}$$
$$= 1 \times 14 \text{ V}$$
$$= 14 \text{ V}$$

▲

The previous two examples show minimum induced voltage occurs at 0° and maximum induced voltage occurs at 90°.

▼ *Example 16-8:*

Find the instantaneous value of an alternating current if the angle is 60° and the maximum voltage value is 14 V.

$$e = \sin \theta \times E_{MAX}$$
$$= \sin 60° \times 14 \text{ V}$$
$$= 12.124 \text{ V}$$

This example shows an angle between 0° and 90° produces a voltage somewhere between zero and maximum induced voltage.   ▲

The instantaneous value of a current (represented by lowercase *i*) can also be found by substituting the current for the voltage value. The equation is:

$$i = \sin \theta \times I_{MAX}$$

▼ *Example 16-9:*

Find the instantaneous current value of an alternating current if the angle is 70° and the peak value is 5.5 A.

$$i = \sin \theta \times I_{MAX}$$
$$= \sin 70° \times 5.5 \text{ A}$$
$$= 5.17 \text{ A}$$

▲

RMS Value

RMS value is the electric power delivered to each residence by the power companies, commonly 120 V ac. The root-mean-square value is measured from the X-axis to a point 70.7% of the peak value. The RMS is the value of ac that will do the same work as that same value of dc. The RMS value is often called the *effective value.* The RMS voltage value equals 0.707 times the peak value. The equation is:

$$E_{RMS} = 0.707 \times E_P$$

RMS value: The value of alternating current that will do the same work as that same value in direct current.

Effective value: The amount of alternating current that will produce the same amount of work as that of direct current.

▼ *Example 16-10:*

If the peak value of an alternating current is 50 V, what is the RMS value?

$$E_{RMS} = 0.707 \times E_P$$
$$= 0.707 \times 50$$
$$= 35.35 \text{ V}$$

▲

The RMS value of a sine wave is found with the following steps, **Figure 16-9:**
1. Calculate the instantaneous values of each degree of a positive alternation.
2. Calculate the mean or average of all the instantaneous values.
3. Calculate the square root of the mean value, that is, the RMS value. Thus, RMS is the square root of the mean of the squares of an alternation.

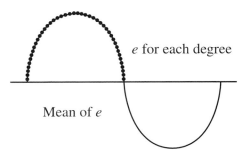

Square root of the mean (RMS)

Figure 16-9. The instantaneous voltage for each degree is squared then averaged (mean). Finally, the square root of the mean is found.

A table for peak, peak-to-peak, instantaneous, and RMS values is shown in **Figure 16-10.** Study the table to see how one value can be determined given another value.

Given this value	Multiply by this value to get		
	RMS	**Peak**	**Peak-to-peak**
Peak	0.707	–	2
Peak-to-peak	0.35361	0.5	–
RMS	–	1.414	2.828
Peak	e (instantaneous) = sin θ E		

Figure 16-10. Table for determining ac values.

Average Value

Average value: The value obtained by dividing the sum of a number of quantities by the number of quantities.

The ***average value*** (E_{AVE}) of an ac alternation is found by taking the amplitude of the current or voltage at each 1° position. The sum of the 1° positions is divided by the total number of values (averaging). For all sine waves, this comes out to 0.637 times the peak value. The equation is:

$$E_{AVE} = 0.637 \times E_P$$

▼ *Example 16-11:*
Find the average value of a sine wave with a peak value of 15.7 V.

$$E_{AVE} = 0.637 \times E_P$$
$$= 0.637 \times 15.7 \ V_P$$
$$= 10 \ V$$ ▲

Radian Measure

Angular velocity: The rate at which an angle changes.

The measurement of a circle in radians was discussed in Chapter 14. A complete circle (360°) has 2π radians. Frequency is the number of revolutions per second. The ***angular velocity,*** expressed as radians per second (rad/s), is found by combining the number of radians in a circle and the frequency. The Greek letter omega (ω) is used for $2\pi f$. The equation is:

$$\omega = 2\pi f$$

Since distance equals velocity multiplied by time (*t*), then:

$$\theta = \omega \times t$$

Notice this is the same as $\theta = (2\pi f)\ t$. This equation states the angular velocity multiplied by the time of rotation produces the angle (in radians) the conductor travels during that time period.

▼ *Example 16-12:*

What is the RMS voltage of 120 V at a frequency of 60 Hz in radian measure?

$$\begin{aligned}
\omega &= 2\pi f \\
&= 2\,(3.1416)\,(60) \\
&= 377 \text{ rad/s}
\end{aligned}$$

$$\begin{aligned}
E_P &= RMS \times 1.414 \\
&= 120 \text{ V} \times 1.414 \\
&= 170 \text{ V}
\end{aligned}$$

▲

Expressed in radian measure, it is 170 V sin 377*t*. The equations for instantaneous voltage and current in radian measure are:

$$e = E_{PEAK} \sin \omega\ t$$
$$i = I_{PEAK} \sin \omega\ t$$

▼ *Example 16-13:*

If the peak value of an alternating current is 25 V at 60 Hz, what is the instantaneous voltage at 2 milliseconds?

$$\begin{aligned}
e &= E_{PEAK} \sin \omega\ t \\
&= (25) \sin [(377)\,(2 \times 10^{-3})] \\
&= (25) \sin 0.754 \text{ (in radians)}
\end{aligned}$$

Convert to degrees:

$$\begin{aligned}
&= (25) \sin 43.2° \\
&= 17.11 \text{ V}
\end{aligned}$$

▲

Frequency Spectrum

The *frequency spectrum* is the entire range of frequencies of electromagnetic radiation, **Figure 16-11.** The ac frequency spectrum increases up the chart with audio frequencies, radio frequencies, optics, and X-rays.

Another way to look at the frequency spectrum is with audio frequencies at the bottom and cosmic rays at the top. Notice the small band labeled visible. These are the frequencies to which our eyes are sensitive. Although the cosmic rays are at the top, some experts believe the frequency spectrum keeps reaching toward infinity. This cannot be proven because the necessary technology does not yet exist. The wavelength of X-rays are often expressed in angstroms (represented by Å). An *angstrom* is equal to 10^{-10}.

Frequency spectrum: The entire range of frequencies of electromagnetic radiation.

Angstrom: A value equal to 10^{-10}.

Radio Waves

Radio waves are electromagnetic waves with a frequency less than 3 THz (terahertz) moving through space without an artificial guide such as a wire cable. See **Figure 16-12.** An electromagnetic wave is energy radiated from an antenna. The troposphere is the portion of the atmosphere 4 to 7 miles (6 to 11 km) above the earth's surface. The ionosphere is the layer above the troposphere. It extends 70 miles (112 km) above the earth's surface. These two atmospheric layers refract (bend) radio waves, providing the capability for international radio transmission.

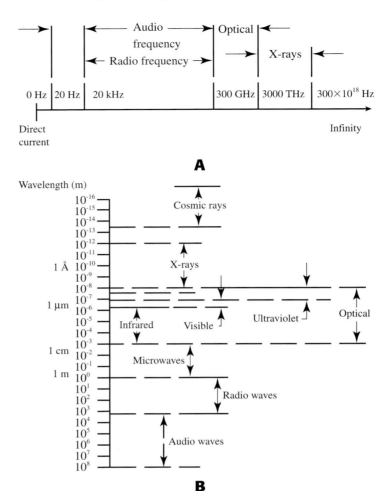

Figure 16-11. Frequency spectrum. A—The frequency spectrum begins at 0 Hz and extends to infinity. This includes all radio and television stations as well as satellite transmissions. B—The frequency spectrum with reference to wavelengths.

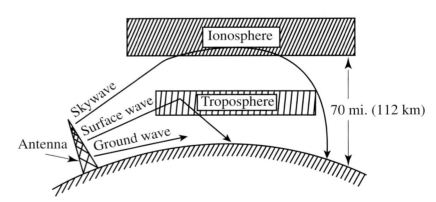

Figure 16-12. Radio waves are bent in the ionosphere, enabling them to travel around the world.

Sound Waves

Sound is a wave movement of air molecules. For example, when a tuning fork is struck, air molecules are compressed and expanded, **Figure 16-13.** The alternating compressions and *rarefactions* (expansions) produce a wave with a specific frequency. The movement of air can be detected by the human ear. There cannot be any sound in a vacuum because there are no air molecules to vibrate.

Rarefaction: To make a signal more difficult to define or distinguish.

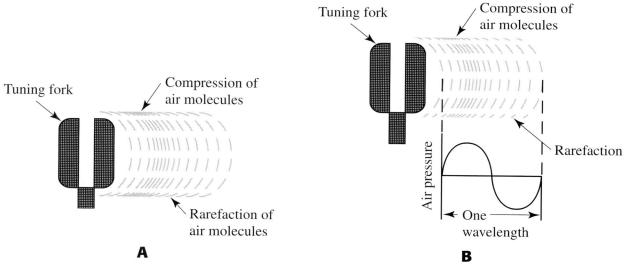

Figure 16-13. Sound waves. A—A tuning fork causes the air molecules to be compressed and expanded. B—One compression and one rarefaction make one wavelength.

The speed of sound waves through air is 343.5 meters per second. The wavelength of sound waves (in meters) is found by dividing 343.5 m/s by the frequency (in Hertz). The equation is:

$$\lambda \ (m) = \frac{343.5 \ m/s}{f \ (Hz)}$$

▼ *Example 16-14:*

What is the wavelength of a sound with a frequency of 2000 Hz?

$$\lambda = \frac{343.5 \ m}{2 \times 10^3}$$

$$= 0.17175 \ m$$
▲

The sound frequencies for notes on a scale for various musical instruments is shown in **Figure 16-14.** An oscillator circuit in an electronic organ generates a specific alternating current frequency for each note. The oscillator can be adjusted (tuned) to the correct frequency. C0 and C1 correspond to the first two octaves for the C notes. A4 (440 Hz) corresponds to the A note of the fourth octave and is the note to which all instruments in an orchestra are tuned.

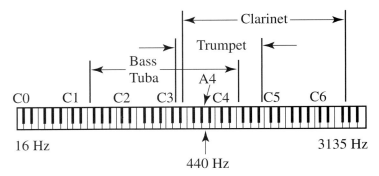

Figure 16-14. The standard musical keyboard is based on the frequency of note A in the fourth octave at 440 Hz.

Summary

- Ac is produced by rotating a loop in a magnetic field using an induction process.
- The number of times the loop is rotated each second determines the frequency of the ac. Frequency is measured in cycles per second.
- Ac voltage amplitude can be measured by peak-to-peak, peak, instantaneous, and RMS values.
- The frequency of an ac is related to its time period.
- The wavelength of an alternating current can be found by dividing its frequency into the velocity of radio waves. Sound travels slower than radio waves; therefore, sound wavelengths are calculated using a different mathematical constant.
- The frequency spectrum extends from dc to cosmic rays and beyond.

Important Terms

Do you know the meanings of these terms used in the chapter?

alternating current
angstrom
angular velocity
average value
effective value
frequency
frequency spectrum
induction

instantaneous value
peak value
peak-to-peak value
rarefaction
RMS value
time period
wavelength

⇥| Questions and Problems

Please do not write in this text. Write your answers on a separate sheet of paper.

1. What process is used to generate alternating current?
2. Draw a graph showing three cycles of alternating current.
3. Calculate the RMS value for each of the following peak values.
   a. 224 V
   b. 49 V
   c. 8.5 V
   d. 371 V
   e. 12 V
   f. 1500 V
4. Calculate the frequency for each of the following time periods.
   a. 1 ms
   b. 20 μs
   c. 4 ns
   d. 2 ms
   e. 180 μs
   f. 100 μs
5. Calculate the peak value for each of the following RMS values.
   a. 332 V
   b. 6.3 V
   c. 200 V
   d. 55 V
   e. 150 V
   f. 12.6 V
6. Calculate the time period for each of the following frequencies.
   a. 200 kHz
   b. 60 Hz
   c. 1200 GHz
   d. 10 kHz
7. Calculate the wavelength for each of the following frequencies.
   a. 20 Hz
   b. 15 MHz
   c. 100 MHz
   d. 15 kHz
8. Find the frequency for a wave of 500 meters.
9. How many degrees are in one cycle?
10. What is the frequency range of the optical spectrum?
11. The frequency of microwaves is from _____ to _____.
12. Calculate the instantaneous voltage for each of the following peak values.
    a. 45 V at 37°
    b. 12 V at 50°
    c. 80 V at 82°
    d. 137 V at 18°
13. What is the time period of an alternating current with a wavelength of 30 m?

Chapter 17 Graphic Overview

The Oscilloscope

— measures → Peak-to-peak — or — Timing — of a — Waveform

— has —

Triggering — controls are — Trigger Source, Trigger Level, Slope Switch

Vertical Amplifier — calibrated in — Volts — per — Division

Horizontal Generator (time-base) — calibrated in — Time — per — Division

Beam Controls — are — V. & H. Position, Focus, Intensity

— has —

CRT — contains — Electron Gun and Deflection Plates — and — Fluorescent Screen — with a — Graticule — that is — 8 × 10 cm

THE OSCILLOSCOPE 17

Objectives

After studying this chapter, you will be able to:
- ○ Explain the basic operation of a typical oscilloscope, including adjustment controls.
- ○ Explain the divisions on the oscilloscope graticule.
- ○ Measure peak-to-peak value and time period.
- ○ Describe the precautions to be observed when using an oscilloscope.
- ○ Measure dc voltages using an oscilloscope.
- ○ Adjust a probe for proper compensation.
- ○ Check for proper calibration of the oscilloscope.

Introduction

The oscilloscope is the electronic technician's most useful test instrument. It is also the most underused. This is unfortunate because the oscilloscope gives a visual image of the voltage present. An oscilloscope can provide information on waveform amplitude, time durations, shape, time period, and frequency. The instrument can also determine how well a circuit is processing a signal, whether the waveform has distortion, and the phase relationship between two or more waveforms.

Information gathered by the oscilloscope must be interpreted. Incorrect adjustment of the controls may make the information inaccurate or difficult to understand. Proper adjustment is a skill that is developed with practice.

Safety Precautions

- • Do not operate the oscilloscope with the cabinet removed. Oscilloscopes with metal cabinets depend on the cabinet as a shield from stray magnetic fields.

> **+ Warning:** Dangerous voltages inside the oscilloscope can cause severe shock.

- • When the beam is not in use, turn down the intensity control so the beam is not visible. Prolonged exposure of the beam on the phosphor material will shorten the life of the material. As the phosphor material ages, the trace will become more difficult to see.
- • Keep the oscilloscope away from anything generating magnetic fields. Small metal components of the oscilloscope may become slightly magnetized and affect operation of the oscilloscope display.
- • Stay within the input voltage limits of the oscilloscope. Exceeding the specified voltage will damage the insulation of the input components and the oscilloscope probe.
- • Make sure the oscilloscope is connected to the proper ground reference point.

> **⚡ Caution:** Connecting the oscilloscope ground to an improper ground reference point can damage the equipment and oscilloscope probe.

- Be aware of chassis grounding and "hot chassis" situations. In some equipment, one side of the ac power line is connected to the ground. If the power line becomes reversed, the "hot" side of the power line will be connected to the equipment chassis ground.

> **Warning:** When the ground of the oscilloscope is connected to the chassis ground, a shower of sparks and a loud noise will ensue, damaging the oscilloscope and probe. More importantly, if you touch the chassis of the equipment with one hand and pick up the bare probe ground with your other hand, you will be placing yourself directly across the ac power line (the hot chassis to ground).

Oscilloscope Operation

Your study of the oscilloscope will not involve specific oscilloscope circuits; rather, a block diagram of a typical oscilloscope will be used. See **Figure 17-1.**

Cathode Ray Tube

Cathode ray tube: A vacuum tube in which an electron beam strikes the surface of the screen and produces a particular pattern of light.

Electron gun: A mechanical assembly at the back of a cathode ray tube that produces an electron beam.

Operation of the oscilloscope centers around the *cathode ray tube* (CRT). The CRT contains an electron gun, deflection plates, and a fluorescent screen, **Figure 17-2.** The *electron gun* has many electrodes connected to pins at the base of the tube.

Figure 17-1. Block diagram of a typical oscilloscope. Although an oscilloscope may have other specific circuitry, every oscilloscope must have these basic blocks.

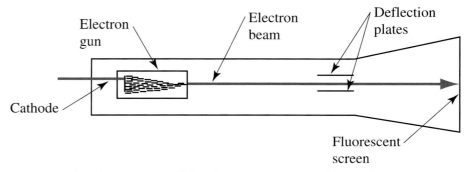

Figure 17-2. The electron gun of the CRT generates an electron beam.

The ***cathode*** of an electron gun is a piece of metal coated with a substance that makes electrons available. The cathode is heated and shoots a narrow stream of electrons called an ***electron beam.*** The electrons speed toward the oscilloscope screen and strike it. The ***fluorescent screen*** has a phosphor coating on the inside of the glass tube. The ***phosphor material*** glows when struck with the high-speed electrons. The substances in the phosphor material determine the colors on the CRT screen. The phosphor material also controls the length of time a spot will glow after the electron beam is moved to another position.

As the electron beam goes from the electron gun to the fluorescent screen, it passes between the deflection plates. The ***deflection plates*** move the spot of light to different positions on the CRT screen or faceplate, **Figure 17-3.** One pair of deflection plates is positioned to the left and right sides of the screen. Another pair is positioned toward the top and bottom. When a voltage is applied between a pair of deflection plates, the electric field between them pulls the beam toward the positive plate and pushes it away from the negative plate (electrostatic forces).

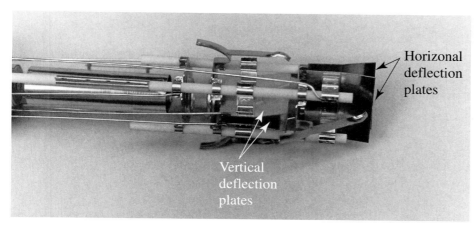

Figure 17-3. The CRT has two pairs of deflection plates — horizontal and vertical. Voltages applied to the plates cause the CRT electron beam to move, in turn making a graph of the voltage applied from the oscilloscope probe.

Figure 17-4 shows the deflection plate action with no difference in potential between the two horizontal plates. The beam travels straight, hitting the center of the CRT screen. In **Figure 17-5,** the left deflection plate is more positive, and the electron beam is attracted toward the left side of the CRT screen. If the right deflection plate were made more positive, the beam would be attracted to the right side of the screen. This movement from left to right is in a horizontal direction; therefore, the plates are called horizontal deflection plates. The other pair of deflection plates moves the beam vertically; they are called vertical deflection plates. When the proper voltage polarities are placed on the deflection plates, the electron beam can move to any position on the CRT screen.

If the polarity of the voltage on the horizontal deflection plates is changed quickly, the spot of light (beam) will move across the CRT screen quickly. If the polarity is changed fast enough, the light of the phosphor material will still be glowing when the beam gets back to the same position on the screen. The action produces a line across the CRT screen. The left to right movement of the beam is referred to as ***scanning.*** If the polarity of the vertical deflection plates is changed at the same time the horizontal polarity is changing, the line is not straight but goes vertical at the same time it goes horizontal. See **Figure 17-6.**

Cathode: The negative terminal of a semiconductor diode.

Electron beam: A stream of electrons that strikes the inner surface of a cathode ray tube.

Fluorescent screen: The screen surface of a cathode ray tube. It is coated with a material that causes light to be emitted when bombarded with electrons.

Phosphor material: A material placed upon the inner surface of a cathode ray tube.

Deflection plates: Two pairs of parallel electrodes set at right angles to each other. An electrostatic field applied to the plates causes an electron beam to move in a cathode ray tube.

Scanning: The process of moving an electron beam across a cathode ray tube.

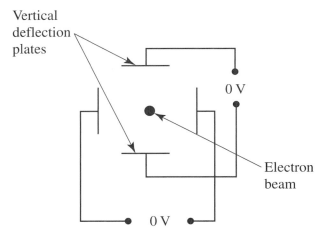

Figure 17-4. With zero potential difference on the deflection plates, the beam stays in the center of the CRT.

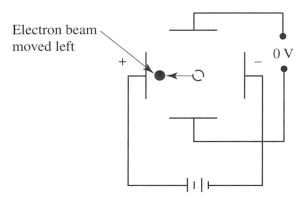

Figure 17-5. With a positive potential on the left plate, the beam is attracted toward the left side of the CRT screen.

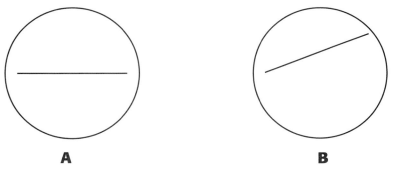

Figure 17-6. Changing polarities. A—The rapid horizontal movement of the electron beam causes a line to form on the CRT screen. B—Voltage on both the vertical and horizontal deflection plates causes the electron beam to move in an upward or downward direction while also moving from left to right.

Figure 17-7 illustrates the actual scanning process. The voltage placed on the horizontal deflection plates is not a dc voltage but a sawtooth ac voltage. The center of the screen is at the zero reference of the sawtooth waveform.

Starting at the most negative point on the sawtooth waveform, the beam is at the left side of the CRT. As the sawtooth wave continues through its cycle, the beam passes through the center of the screen and to the right side of the CRT. This is called

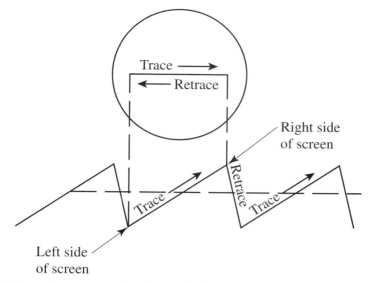

Figure 17-7. A sawtooth waveform is applied to the horizontal deflection plates to produce the trace and retrace pattern.

the *trace period* (also known as the horizontal sweep). When the sawtooth wave goes from its most positive point back to the negative point, the beam is returned very quickly to the left side of the CRT screen. This is the *retrace period.* During the retrace period, you do not want to see the beam on the screen, so the beam is turned off.

Since the sawtooth wave undergoes a linear change as it moves from negative to positive, it provides an accurate measurement of time as it goes through its cycle. This timing is used as a base for the X-axis of the oscilloscope graph.

If a voltage is placed on the vertical deflection plates at the same time the sawtooth is operating the horizontal plates, an image of the voltage is made on the CRT, **Figure 17-8.**

The CRT also has controls for focus, intensity, horizontal position, and vertical position. Refer to Figure 17-1.

Trace period: The period of time that a waveform is made on the screen of a cathode ray tube.

Retrace period: Time period during which the electron beam returns from the right side of the cathode ray tube screen to the left side to begin the next trace.

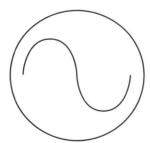

Figure 17-8. Voltage on the vertical deflection plates can form a sine wave on the oscilloscope screen. The electron beam is moved horizontally and vertically at the same time.

Horizontal Generator and Amplifier

All blocks in the diagram in Figure 17-1 come into the CRT in some way. The horizontal generator (oscillator) produces the sawtooth ac required for the horizontal deflection plates. It contains a control to adjust the frequency (time base) of the sawtooth ac.

The horizontal amplifier increases the amplitude of the sawtooth wave to a sufficient level to drive the deflection plates. The horizontal gain control may not be available on all oscilloscopes.

Vertical Amplifier

The voltage under test is connected to the vertical input, which is connected to the vertical amplifier. The vertical amplifier increases the voltage level and processes the signal to be applied to the vertical deflection plates. The vertical amplifier has two controls: the vertical range (attenuator) and the vertical gain (vertical variable control). The *attenuator* reduces the amplitude of the signal entering the vertical amplifier.

Attenuator: A resistive network used to decrease the amplitude of a signal.

Triggering: The start of the oscilloscope trace period at a predetermined point on the signal cycle.

Triggering Circuit

The *triggering* block shown in Figure 17-1 is a circuit that takes a sample from the vertical amplifier, processes the signal, and applies it to the horizontal generator. The triggering signal, which is applied to the horizontal generator, tells the generator when to begin the sawtooth ac cycle and when the electron beam should start across the oscilloscope screen.

Oscilloscope Graticule

Graticule: The graduated scale on the screen of an oscilloscope.

The oscilloscope *graticule* is a calibrated sheet of glass or plastic placed over the front of the oscilloscope faceplate or screen, **Figure 17-9.** You must understand how the graticule is divided to be able to measure voltages and time with the oscilloscope.

Each major division of the graticule is 1 cm^2. Industry has set the standard of ten divisions on the X-axis and eight divisions on the Y-axis. Each major division is divided into smaller divisions across the X and Y axes. Each small division is equal to 0.2 of a major division. The divisions progress from zero through two, **Figure 17-10.**

In **Figure 17-11,** the number of divisions counted vertically is 5.5. The number of divisions counted horizontally is 4.4. Always measure from the beginning of a major division.

Controls and Knobs

The basic controls and knobs of any oscilloscope can be classified into four groups: beam, vertical, horizontal, and triggering. Your particular oscilloscope may have additional controls or input and output terminals specific to that model.

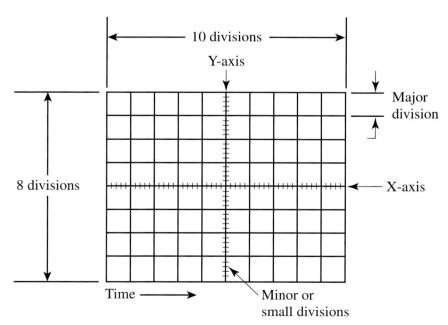

Figure 17-9. The industry standard for an oscilloscope graticule is ten horizontal and eight vertical divisions.

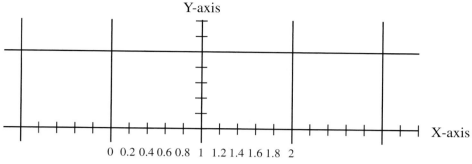

Figure 17-10. The small divisions of an oscilloscope graticule are equal to 0.2 of a major division. Each major division is 1 cm².

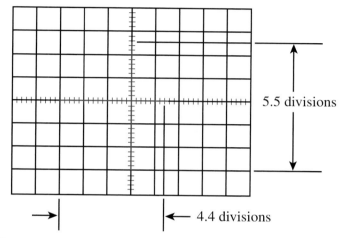

Figure 17-11. Adjusting the trace onto a major division makes it easier to count the number of smaller divisions at the other side of the waveform.

Beam Controls

The *focus control,* **Figure 17-12,** adjusts the trace to a thin line for sharpest focus. If the oscilloscope is adjusted for a single spot, the spot should be minimum size.

The *intensity control* adjusts the brightness or intensity of the trace. This control should be adjusted so the trace is just visible but not excessively bright.

 Caution: Excessive brightness shortens the life of the CRT. A single spot can cause a burn of the phosphor material.

Some oscilloscopes have a *trace tilt control* which adjusts the trace for proper alignment with the X-axis. The trace in **Figure 17-13** is going uphill from left to right instead of straight across the CRT screen. The trace tilt is adjusted with a screwdriver until the trace is aligned with the X-axis. If the oscilloscope does not have a trace control, the trace is adjusted by removing the oscilloscope from its cabinet and turning the CRT for correct trace alignment.

Vertical Controls

The *vertical position control* adjusts the trace in the vertical direction, **Figure 17-14.** The trace is usually adjusted to align with the X-axis. It can be adjusted to any point on the screen that provides the best observation of the waveform. A normal display of a waveform is stationary and focused with a readable amount of height.

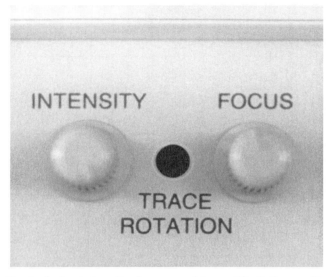

Figure 17-12. Controls for intensity, focus, and trace tilt. If the oscilloscope does not have a trace tilt control, the CRT must be turned to correct the misalignment.

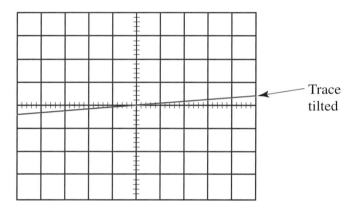

Trace tilted

Figure 17-13. An out-of-adjustment trace tilt.

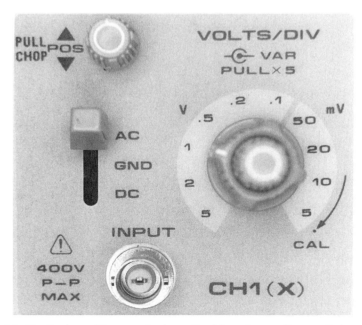

Figure 17-14. Vertical position control, vertical input coupling switch, and attenuator.

The *vertical input coupling switch* selects the type of coupling of the signal to the oscilloscope. Three typical choices are ac, ground, and dc. Refer to Figure 17-14. On ac, the input signal is capacitively coupled and the dc component is blocked. In the ground position, the switch opens the signal path and connects the vertical amplifier input to ground. This provides a zero voltage reference position for the trace. It is also helpful if you are having difficulty locating the trace. On dc, the input signal is directly coupled to the vertical amplifier.

The *vertical attenuator,* sometimes called the vertical range, adjusts the height of the trace. This control is a switch calibrated in volts per division. In the position shown in Figure 17-14, the graticule scale would be equal to 1 V for each major division. The peak-to-peak voltage is calculated using the equation:

$$E_{P\text{-}P} = \text{Number of divisions} \times \text{Attenuator voltage setting}$$

▼ *Example 17-1:*

In **Figure 17-15,** the waveform has a height of 5.8 divisions. If the attenuator is set at 1 V/Division, calculate the voltage being displayed.

$E_{P\text{-}P}$ = Number of divisions × Attenuator voltage setting

  = 5.8 × 1 V

  = 5.8 V   ▲

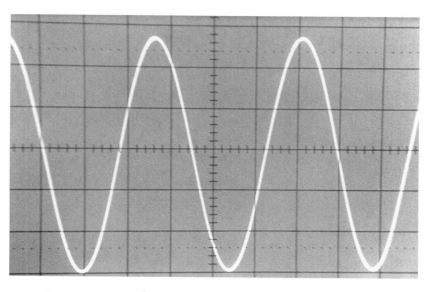

Figure 17-15. To measure peak-to-peak voltage, multiply the total divisions of the waveform from top to bottom by the voltage setting on the attenuator.

The vertical attenuator setting in **Figure 17-16** is incorrect. The amplitude is so great that the trace is beyond the limits of the CRT screen. Likewise, **Figure 17-17** shows a trace with incorrect vertical attenuator adjustment. The vertical attenuator has not been adjusted so the trace is at its maximum on the screen.

To correctly set the vertical attenuator, adjust the control until the trace goes beyond the limits of the CRT screen, as shown in Figure 17-16. Then, adjust the control until the complete trace is seen on the screen. This will give the most accurate measurement.

The *variable gain control* adjusts the height of the trace within the limits of the control. This control is often in the center of the vertical attenuator. It must be in the calibrated position for the scale of the vertical attenuator control to be correct. Making measurements out of the calibrated position is discussed later in the chapter.

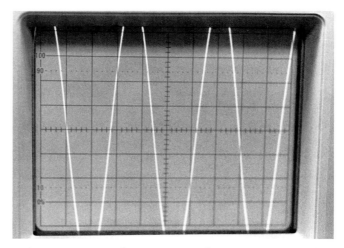

Figure 17-16. An incorrect vertical attenuator adjustment causes the waveform to go beyond the limits of the oscilloscope CRT.

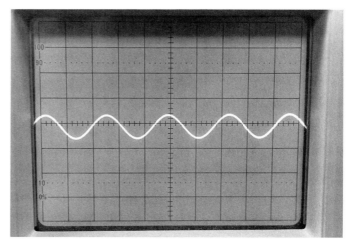

Figure 17-17. Voltage measurement can be inaccurate if the vertical attenuator is adjusted so the waveform is too small. The height should be sufficient for the waveform to be accurate and easily observed.

Horizontal Controls

The horizontal position control adjusts the trace for its position in the horizontal direction. The control is adjusted for a trace position that provides the best observation of the waveform.

The *horizontal time base control* is calibrated in milliseconds or microseconds per major division, **Figure 17-18.** For proper adjustment, the time base control should be adjusted for the best observation.

To measure time, the number of divisions on the graticule is multiplied by the time base control. The equation is:

$$T = \text{Number of divisions} \times \text{Time base setting}$$

▼ *Example 17-2:*

In **Figure 17-19,** if the number of divisions is 2.6 and the time base control is 2 ms per division, what is the time?

$$
\begin{aligned}
T &= \text{Number of divisions} \times \text{Time base setting} \\
&= 2.6 \times 2\text{ ms} \\
&= 5.2\text{ ms}
\end{aligned}
$$

▲

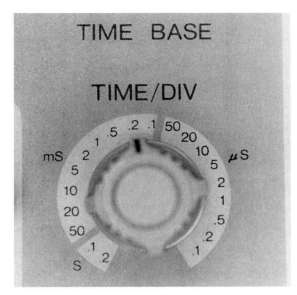

Figure 17-18. Horizontal position and time base are used to adjust the trace on the X-axis.

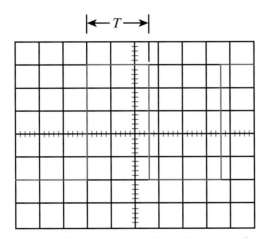

Figure 17-19. Calculating the time of a waveform for Example 17-2.

The time base steps are not valid when the *horizontal variable control* is out of the calibrated position. The main use of the horizontal variable control is to facilitate the reading of certain waveforms on the screen. Operating the oscilloscope with the horizontal variable out of the calibrated position is discussed later in this chapter along with Making Measurements.

Triggering Controls

The *trigger source switch* selects the source of the sweep trigger signal, **Figure 17-20.** The trigger signal can be selected from channel 1, channel 2, or the external jack. The jack is connected to a test point from which you want to time the horizontal time base.

The *trigger level control* determines the point at which the sweep is triggered. Points on the triggering waveform are selected by rotating the control in the negative (−) or positive (+) direction. The *slope switch* selects a positive or negative direction on the waveform as the triggering point.

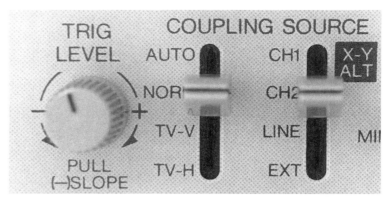

Figure 17-20. The level control on this oscilloscope can be pulled out to change from positive to negative slope triggering.

These are the usual triggering methods used when operating an oscilloscope. More advanced operations will vary with the make and model of the oscilloscope, available controls, and application of the instrument. Refer to the manual of your oscilloscope for specific operating instructions.

Making Measurements

Two basic measurements are made using an oscilloscope. The most common is the peak-to-peak measurement of a waveform. The other is the time base measurement.

Peak-to-peak Measurement

To measure the peak-to-peak voltage, multiply the number of vertical divisions by the voltage setting on the attenuator. The equation is:

$$E_{P\text{-}P} = \text{Number of divisions} \times \text{Attenuator voltage setting}$$

▼ *Example 17-3:*
In **Figure 17-21,** the number of divisions from the most negative point on the waveform to the most positive point is 5.6 divisions. With the vertical attenuator set at 0.2 V, what is the peak-to-peak voltage?

$$\begin{aligned}
E_{P\text{-}P} &= \text{Number of divisions} \times \text{Attenuator volt setting} \\
&= 5.6 \times 0.2 \text{ V} \\
&= 1.12 \text{ V}
\end{aligned}$$

▲

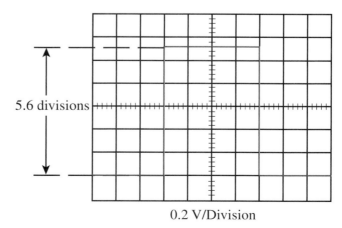

0.2 V/Division

Figure 17-21. To measure peak-to-peak voltage, set the bottom of the waveform on a major division and count the divisions upward on the screen.

If the amplitude and frequency of a certain reference voltage is known, an unknown signal may be measured for amplitude without the variable input attenuator control set to the calibrated position. Using the calibrating voltage on the front of an oscilloscope, the graticule can be calibrated for making voltage measurements. Follow this procedure:

1. Connect the probe to the calibrating voltage terminal, and adjust the oscilloscope for a normal display.
2. Adjust the vertical attenuator and variable controls so the amplitude of the calibrating voltage occupies a fixed number of divisions. After this adjustment, do not disturb the vertical variable control.
3. Calculate the vertical calibration coefficient using the equation:

$$\text{Vertical coefficient} = \frac{C}{D \times S}$$

Where:
C = Amplitude of the calibrating voltage
D = Divisions of the calibrating voltage
S = Vertical attenuator setting

4. Connect the unknown voltage to the vertical input, and adjust the vertical attenuator for best observation. *Do not move the vertical variable control.*
5. Measure the amplitude of the unknown waveform in divisions.
6. Calculate the voltage of the waveform using the equation:

Unknown voltage = Vertical division × Volts/division × Vertical coefficient

---

▼ *Example 17-4:*

If the variable control is adjusted so the amplitude of the calibrating signal is 5 divisions, the calibrating voltage is 0.2 V, the volts/division setting is 1 V, the vertical coefficient is calculated as:

$$\frac{0.2 \text{ V}}{5 \times 1 \text{ V}} = 0.04 \text{ V}$$

If the unknown waveform is measured as 3 divisions and the attenuator is set at 5 V/division, the peak-to-peak amplitude of the unknown is calculated as:

$$3 \times 5 \text{ V} \times 0.04 \text{ (vertical coefficient)} = 0.6 \text{ V}$$ ▲

Time Base Measurements

Time measurements are made between two points on a waveform, **Figure 17-22.** The beginning and ending points of one cycle may be used.

Follow this procedure to make time measurements:

1. Connect the signal to be measured to the vertical input, and adjust the oscilloscope for a normal display of the waveform.
2. Using the vertical position control, adjust one point of the waveform as a reference on the X-axis.
3. Using the horizontal position control, adjust the reference point to one of the major vertical division lines.
4. Measure the number of divisions from the reference point to the point on the waveform to be measured.
5. Multiply the number of divisions in step 4 by the time indicated on the time base control.

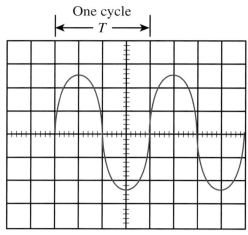

Figure 17-22. Measuring the time for one cycle.

Out of Calibrated Position

The time of an unknown waveform can be measured with the time base out of the calibrated position. Follow this procedure:

1. Connect the probe to the calibrating voltage terminal, and adjust the oscilloscope for a normal display.
2. Adjust the time base variable control so one cycle occupies a number of divisions on the X-axis.
3. Calculate the sweep coefficient using the equation:

$$\text{Sweep coefficient} = \frac{P}{D \times S}$$

Where:
P = Period of the calibrating voltage
D = Horizontal divisions of the calibrating voltage
S = Step of the time base control

4. Connect the unknown signal to the vertical input, and adjust the time base step control for a readable display. *Do not disturb the variable sweep control.*
5. Measure the width (in divisions) of one cycle of the unknown waveform or corresponding points on the waveform.
6. Calculate the time of the unknown waveform using the equation:

Unknown time = Horizontal divisions × Time/division × Sweep coefficient

---

▼ *Example 17-5:*

If the variable sweep control is adjusted so the calibrating voltage occupies four divisions, the time period of the calibrating voltage is 1 ms, and the time base is set at 0.1 ms, what is the sweep coefficient?

$$\text{Sweep coefficient} = \frac{P}{D \times S}$$

$$= \frac{1 \text{ ms}}{4 \times 0.1 \text{ ms}}$$

$$= 2.5$$

If the width of the unknown signal is 6.5 divisions and the time base is 0.2 ms, what is the time period of the unknown?

Unknown period = Horizontal divisions × Time/division × Sweep coefficient
= 6.5 × 0.2 ms × 2.5
= 3.25 ms   ▲

Calibration Voltage

Most oscilloscopes have a terminal that provides a calibration voltage, **Figure 17-23.** The voltage is a square wave of 0.2 V, 0.5 V, 1 V, or 2 V peak-to-peak. This terminal can also be used for the probe compensation adjustment. The voltage is used to calibrate the vertical and horizontal variable gain adjustments for measuring voltage and time.

Figure 17-23. Most oscilloscopes have a calibrating voltage on the front.

Circuit Loading

One advantage of measuring with an oscilloscope is its high input impedance, usually in the range of 10 megohms or higher. This means the oscilloscope will not affect the measurements and performance of the circuit under test.

Probes

Various types of probes are used with oscilloscopes. The most popular types are the one-to-one (1:1), ten-to-one (10:1), and hundred-to-one (100:1). See **Figure 17-24.**

The one-to-one probe is sometimes called a direct probe. It allows direct reading of the amplitude of the voltage being measured. The 1:1 probe has full loading effect on the circuit under measurement and also has a low bandwidth.

The ten-to-one probe increases the maximum measurable voltage by a factor of 10; that is, the voltage on the vertical attenuator is multiplied by 10. The 10:1 probe reduces the circuit loading and has an improved bandwidth. Many probes combine the 1:1 and 10:1 probe features into one probe. Many 10:1 probes have a compensation adjustment.

The hundred-to-one probe greatly reduces the circuit loading effects. The voltage per division on the attenuator must by multiplied by 100. For example, the 0.1 V per division setting would become 10 V per division. The 100:1 probe has a high bandwidth and low capacitance compared to the direct probe.

Probe Compensation

The 10:1 and 100:1 probes have a compensation adjustment for matching the probe impedance to the oscilloscope impedance, **Figure 17-25.** Since the impedance can be different for each oscilloscope and channel, the probe should be adjusted to match the input impedance of the channel with which the probe is being used. If the probe is used with a different oscilloscope, it should be adjusted again.

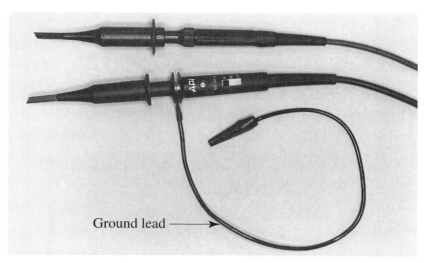

Figure 17-24. Standard oscilloscope probes. All probe ground leads should be connected to ground when measuring.

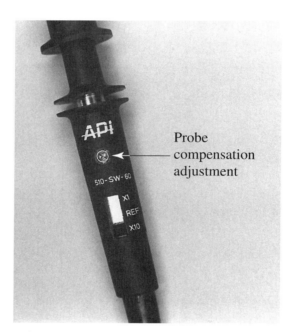

Figure 17-25. The probe compensation adjustment matches the impedance of the probe to that of the oscilloscope input.

To adjust the probe for compensation, use the following procedure:
1. Connect the probe to the calibrating terminal. Adjust the display for three or four cycles at five to six divisions of vertical amplitude.
2. Adjust the compensation trimmer on the probe for the best square wave with minimum overshoot, rounding off, and tilt. See **Figure 17-26.**

Demodulator Probe

A demodulator probe is used in circuits involving modulated radio frequency signals, **Figure 17-27.** The probe works as a simple radio detector circuit.

Oscilloscope manufacturers place many special features on their instruments. Refer to individual operating manuals for correct adjustments and applications of a particular oscilloscope.

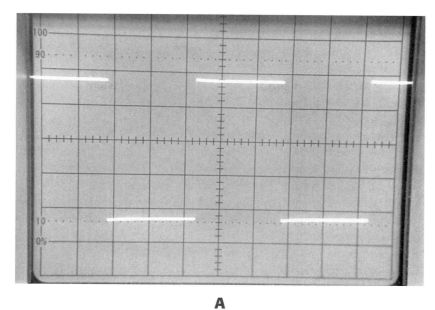

A

B

Figure 17-26. Adjusting the probe compensation. A—Correctly adjusted waveform. B—The distorted square wave is caused by an extreme misadjustment of the probe compensation.

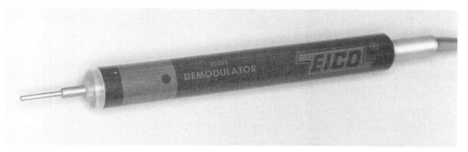

Figure 17-27. A demodulator probe is used for radio frequency signals. Use of this type of probe requires experience interpreting the output of an oscilloscope on a CRT.

Summary

- An oscilloscope consists of a CRT, vertical amplifier, horizontal generator, horizontal amplifier, and trigger circuits.
- An electron gun produces an electron beam that makes a spot of light on the fluorescent screen. Voltage on the deflection plates cause the beam to move on the screen.
- The movement of the electron beam from left to right is called scanning, horizontal sweep, or the trace period. The right to left movement of the beam is called the retrace period.
- The graticule is the calibrated sheet of glass or plastic in front of the CRT.
- Oscilloscope controls are classified as beam, vertical, horizontal, or triggering.
- To measure voltage or time, multiply the number of divisions by the setting of the vertical attenuator or time base controls.
- Oscilloscope probes may be 1:1, 10:1, or 100:1. An oscilloscope probe should be adjusted for compensation so the probe is matched to the impedance of the oscilloscope in use.

Important Terms

Do you know the meanings of these terms used in the chapter?

attenuator
cathode
cathode ray tube
deflection plates
electron beam
electron gun
fluorescent screen

graticule
phosphor material
retrace period
scanning
trace period
triggering

Questions and Problems

Please do not write in this text. Write your answers on a separate sheet of paper.

1. Draw a block diagram of an oscilloscope.
2. The electron beam is created by a(n) _____.
3. The electron beam is moved by different voltage potentials on the _____ and _____ deflection plates.
4. The left to right movement of the electron beam is called _____ or _____.
5. Explain the graduations on the oscilloscope graticule.
6. The time base is generated by the _____ generator.
7. List the four groups into which the controls and knobs of an oscilloscope are classified.
8. List three beam controls.
9. Explain what action the trigger level has on a waveform.
10. If the volts/division is 50 mV, what is the peak-to-peak value of the waveform in **Figure 17-28?**
11. If the time base is set for 0.1 ms/division, what is the time period of the waveform in Figure 17-28?
12. Write six safety precautions to observe when working with an oscilloscope.
13. List the three common types of oscilloscope probes.
14. Explain why probe compensation is necessary.
15. If the time base is set for 10 ms/division, what is the time period of **Figure 17-29?** What is the frequency?

16. If the volts/division is set for 2 V/division, what is the peak-to-peak value of **Figure 17-30?**

17. If the time base is set for 5 μs/division, what is the time period of Figure 17-30?

18. What is the frequency of Figure 17-30?

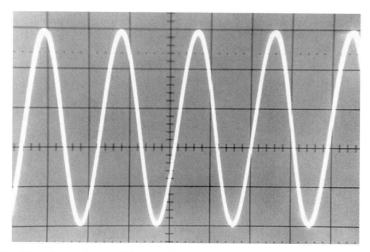

Figure 17-28. Waveform for problems 10 and 11.

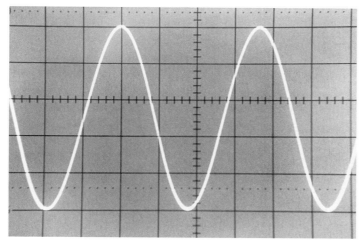

Figure 17-29. Waveform for problem 15.

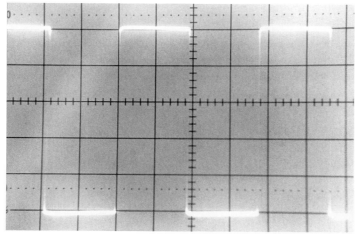

Figure 17-30. Waveform for problems 16, 17, and 18.

Chapter 18 Graphic Overview

Waveform

measured as
- Peak-to-peak
- Rise Time
- Fall Time
- Pulse Width
- Overshoot
- Preshoot
- Undershoot
- Phase Difference
- Cycles
- Tilt
- Duty Cycle

types are
- Sine
- Square
- Triangular
- Sawtooth
- Trapezoid
- Exponential
- Staircase
- Composite

a common Defect **is** Ringing

edges are
- Leading
- Trailing
- Positive Going
- Negative Going
- Positive Ramp
- Negative Ramp

WAVEFORMS AND MEASUREMENTS

Objectives

After studying this chapter, you will be able to:
○ Identify waveform edges and various types of waveforms.
○ Identify and make waveform measurements.
○ Measure the phase shift of two ac voltages.
○ Describe the differences between an ac square wave, dc pulses, and a square wave with dc offset.

Introduction

The electronics industry uses various terms to describe and evaluate waveforms. Certain words are used to identify waveform edges, measure a value, or designate the type of waveform. Successful use of an oscilloscope depends on the technician's ability to evaluate the information displayed on the oscilloscope screen. Success starts with accurately identifying and measuring waveforms.

Waveform Edges

On an oscilloscope screen, the left side of the waveform is produced first. It is called the *leading edge.* The right side of the waveform is called the *trailing edge.* See **Figure 18-1.** Other examples of leading and trailing edges are shown in **Figures 18-2** and **18-3.**

Leading edge:
1. The edge of a waveform first produced by the oscilloscope. 2. The first edge of the waveform from the left when reading a waveform from left to right.

Trailing edge: The edge toward the right to be made on a waveform.

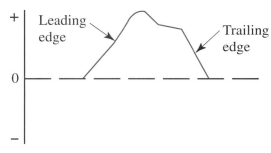

Figure 18-1. Leading and trailing edges of a positive going waveform.

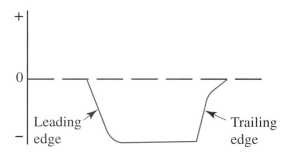

Figure 18-2. Leading and trailing edges of a negative going waveform.

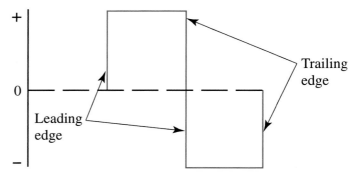

Figure 18-3. Leading and trailing edges of a square wave.

Positive going edge: The edge of a waveform that goes toward the positive direction.

Negative going edge: The edge of a waveform that decreases toward the negative direction.

Ramp: A gradual, linear, positive or negative transition in a waveform, which can be a positive ramp or negative ramp.

Other terms are used to describe the edges of a waveform, **Figure 18-4.** The leading edge is going more positive; thus, it is called the ***positive going edge.*** The other side of the waveform is going more negative; therefore, it is called the ***negative going edge.*** Notice the leading edge in Figure 18-2 is a negative going edge.

A linear positive or negative going edge is often called a ***ramp*** (negative ramp or positive ramp). See **Figure 18-5.** The steepness and length of the edge do not matter.

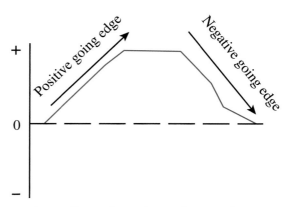

Figure 18-4. Positive and negative going edges of a waveform.

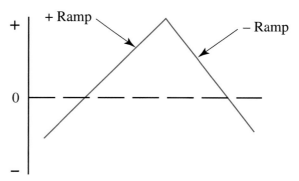

Figure 18-5. Positive and negative ramps are seen in many waveforms including square waves with tilt, sawtooth waves, and triangular waves.

Types of Waveforms

Waveforms are also identified by shape. A study of the waveform illustrations in **Figures 18-6** through **18-13** will increase your waveform vocabulary and help you recognize the various types.

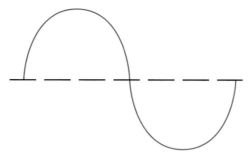

Figure 18-6. A sine wave has a smooth transition from one point to another.

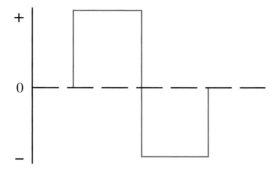

Figure 18-7. The straighter the sides, the more "ideal" the square wave.

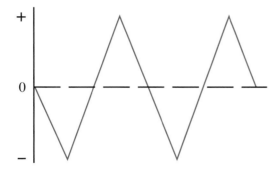

Figure 18-8. In a triangular wave, the positive and negative parts are equal in amplitude and time.

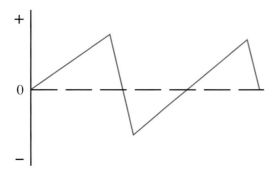

Figure 18-9. In a sawtooth wave, the positive and negative going edges have different duration times.

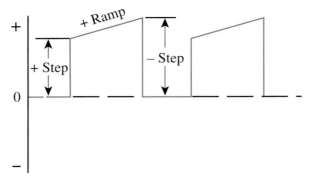

Figure 18-10. In a trapezoid wave, the waveform has a positive step and a positive going ramp followed by a negative step.

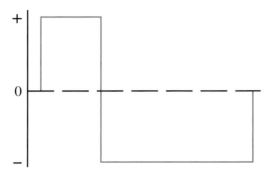

Figure 18-11. Unsymmetrical rectangular wave. Amplitudes and time durations of the negative and positive parts of the wave are unequal.

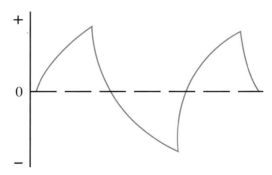

Figure 18-12. Exponential wave. The rise and fall of the waveform depends on the exponent of a mathematical equation.

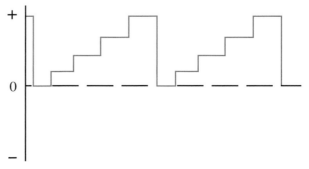

Figure 18-13. Staircase wave. The waveform is a series of positive steps followed by a negative step. It can have any number of positive steps.

Waveform Measurements

The most common waveform measurement is the **peak-to-peak value** ($E_{P\text{-}P}$). It is measured from the most positive point to the most negative point of the waveform, **Figure 18-14.**

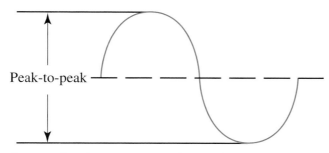

Figure 18-14. Peak-to-peak value is measured from the positive peak to the negative peak of the waveform. The measurements should not include any overshoot or preshoot.

Rise time (T_R) is the time required for the leading edge of a waveform to rise from 10% to 90% of its maximum value. See **Figure 18-15.**

Fall time (T_F) is the time required for the trailing edge of a waveform to decrease from 90% to 10% of its maximum value.

Pulse width is the time duration measured between the 50% amplitude levels of the waveform. See **Figure 18-16.**

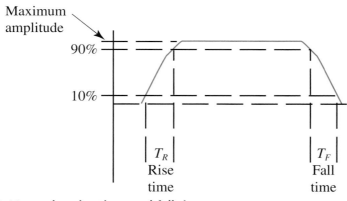

Figure 18-15. Measuring rise time and fall time.

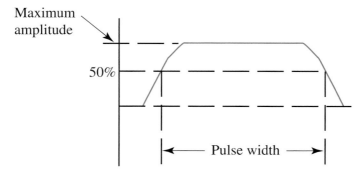

Figure 18-16. The pulse width is measured at the 50% amplitude levels.

Peak-to-peak value: The value of alternating current from the positive peak to the negative peak.

Rise time: Time required for the leading edge of a pulse to rise from 10% to 90% of its maximum value.

Fall time: Time in which a pulse drops from 90% to 10% of maximum amplitude.

Pulse width: Time duration of the pulse measured at 50% of peak amplitude of a waveform.

One cycle:
Measurement taken from one point to a corresponding point on a waveform.

Tilt: 1. A distortion of a square wave in which the horizontal flat portion of the wave becomes tilted and forms a ramp. 2. A measure of the amount of slope to the full amplitude.

One cycle is the change in a wave from zero to a positive peak, to zero, to a negative peak, and back to zero. See **Figure 18-17.** One cycle can be measured from one point on the waveform to the next corresponding point on the waveform, **Figure 18-18.**

Tilt is a percentage of a square wave, **Figure 18-19.** It is the ratio of the slope (*A*) to the full amplitude of the waveform (*B*). The equation is:

$$\text{Tilt} = \frac{A}{B}\,(100)$$

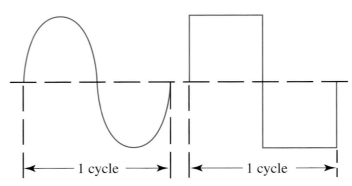

Figure 18-17. One cycle of a sine wave and a square wave.

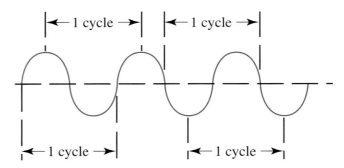

Figure 18-18. There is more than one way to measure a cycle. The measurement can be made from any point on the waveform to a corresponding point beginning the next cycle.

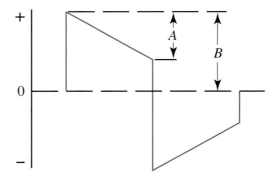

Figure 18-19. Tilt is a percentage of slope to the total amplitude.

▼ *Example 18-1:*

What is the percent of tilt in **Figure 18-20?**

$$\text{Tilt} = \frac{A}{B}\,(100)$$

$$= \frac{0.2}{6}\,(100)$$

$$= 3.3\%$$

▲

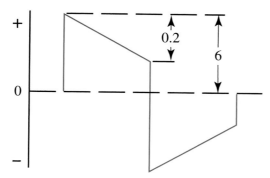

Figure 18-20. Waveform for Example 18-1. (Dimension 0.2 exaggerated for clarity.)

The ***duty cycle*** of a pulse is the fraction of time the pulse is high in relation to the time period of the waveform. See **Figure 18-21.** Duty cycle (expressed as a percentage) is found by dividing the time the pulse is high by the time period. The equation is:

$$\text{Duty cycle} = \frac{t\,\text{high}}{t\,\text{period}}\,(100)$$

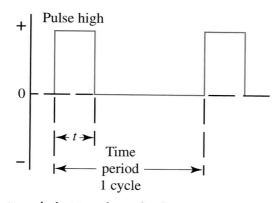

Figure 18-21. If the duty cycle is 15%, the pulse lasts 15% of the period of the waveform.

▼ ***Example 18-2:***
In **Figure 18-22,** what is the percent of duty cycle?

$$\text{Duty cycle} = \frac{t\,\text{high}}{t\,\text{period}}\,(100)$$

$$= \frac{0.3\ \text{ms}}{5\ \text{ms}}\,(100)$$

$$= 6\%$$ ▲

Distortions in Waveform Measurements

Three types of distortion are possible in a waveform. ***Overshoot*** occurs when the beginning waveform trace goes beyond the maximum amplitude levels, **Figure 18-23.**

Preshoot occurs when the waveform trace goes positive or negative before making the actual waveform trace. See **Figure 18-24.**

Undershoot is when the waveform trace changes direction after crossing the baseline before it continues to make the actual waveform trace. See **Figure 18-25.**

Duty cycle: 1. In a waveform, the ratio of the time a pulse is high to the complete time period of the waveform. 2. In equipment, the ratio of the amount of time a piece of equipment is idle to the amount of time it operates.

Overshoot: When the waveform trace goes beyond the maximum amplitude point in either the positive or negative direction.

Preshoot: A waveform distortion where the trace momentarily goes in the opposite direction before making the main portion of the trace.

Undershoot: A waveform distortion in which the trace momentarily goes in the opposite direction before making the actual waveform trace.

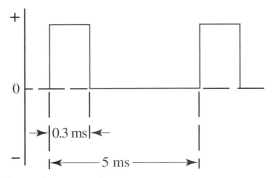

Figure 18-22. Waveform for Example 18-2.

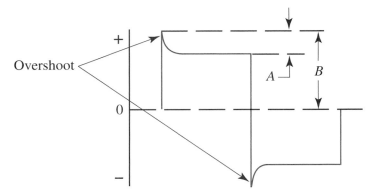

Figure 18-23. Overshoot is a common waveform distortion.

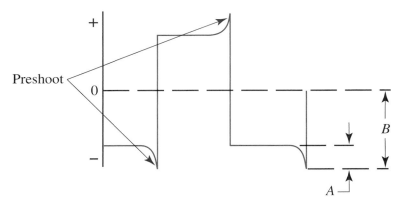

Figure 18-24. Preshoot can cause serious problems due to the sudden increase in amplitude.

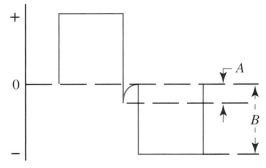

Figure 18-25. Undershoot can cause waveform timing problems.

Overshoot and preshoot are percentages calculated in the same way as tilt. The equation is:

$$\text{Percent} = \frac{A}{B} (100)$$

Measuring Phase Difference

Two common methods used to measure phase difference are the Lissajous display and the dual trace method. When measuring phase difference, one signal is compared to another of the same frequency. Although some people in the electronic field may consider the Lissajous method out of date, modern dual trace oscilloscopes with the X-Y control button make it easy to set up this type of measurement. The dual trace method, however, is considered more accurate.

Lissajous Method

The ***Lissajous method*** is used when measuring sine waves. The two signals are fed into the vertical and horizontal inputs of the deflection system and produce a 1:1 Lissajous pattern. The phase difference is displayed by the oscilloscope as an ellipse, **Figure 18-26.** Measurements *A* and *B* are needed to find the phase angle. Depending on the amount of phase shift, the oscilloscope display will appear as a straight line, circle, or ellipse. As shown in **Figures 18-27** and **18-28,** straight lines occur at 0° and 180° phase difference. Circles occur at 90° and 270° in **Figure 18-29.** Any other phase difference is displayed as an ellipse, **Figures 18-30** and **18-31.**

Lissajous method: The measurement of frequency or phase difference using Lissajous figures (patterns) produced on the screen of a cathode-ray tube.

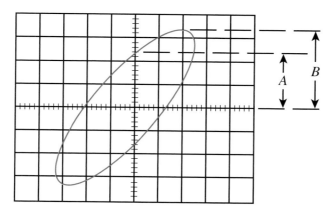

Figure 18-26. Lissajous pattern for making phase difference measurements.

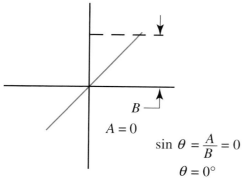

Figure 18-27. Lissajous pattern for 0° phase difference.

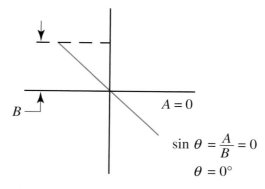

Figure 18-28. Lissajous pattern for 180° phase difference.

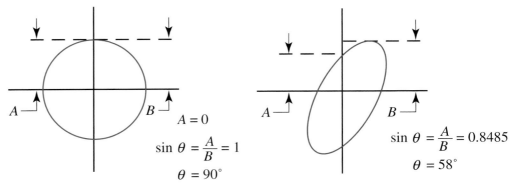

Figure 18-29. Lissajous pattern for 90° or 270° phase difference.

Figure 18-30. Lissajous pattern for 58° phase difference.

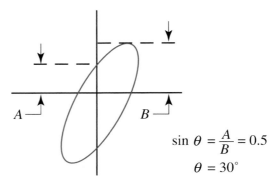

Figure 18-31. Lissajous pattern for 30° phase difference.

Lissajous Procedure

The oscilloscope adjustments that produce an elliptical display are as follows:

1. Adjust the oscilloscope for normal operation.
2. Connect the oscilloscope probes to the circuit under test. Apply the reference voltage to the X-axis or horizontal external input.
3. Ground the Y-input and adjust the X-axis gain (horizontal gain) for a line six divisions long, centered on the X-axis. See **Figure 18-32.**
4. Remove the Y-input from ground and ground the X-input. Adjust the vertical gain for a centered line on the Y-axis six divisions long. See **Figure 18-33.**
5. To check the X and Y deflections for the same amplitude and centering, repeat steps 3 and 4. This will ensure accuracy and reduce the need for adjustments later.
6. Remove both inputs from ground. The oscilloscope display should be a 1:1 Lissajous figure similar to Figure 18-26.
7. Carefully and accurately make the necessary measurements according to Figure 18-26, and calculate the phase angle (θ). The equation is:

$$\sin \theta = \frac{A}{B}$$

Dual Trace Method

The method of finding phase difference by Lissajous patterns has been replaced by the dual trace method. One reason is as the phase shift gets closer to 90°, the accuracy of the measurement diminishes. However, the availability of the dual trace triggered oscilloscope has had the greatest impact on this operation. By comparing the two

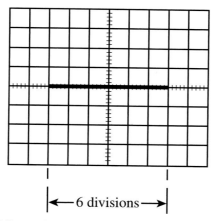

Figure 18-32. Horizontal line adjustment for a Lissajous pattern setup.

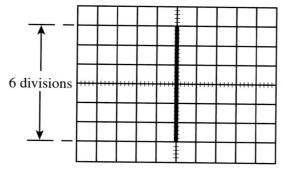

Figure 18-33. Vertical line for Lissajous pattern setup.

signals at the same time on a dual-trace screen, it can be seen instantly if any phase difference is present. Then, using some simple math, the amount of phase difference can be calculated.

The dual trace method can be applied in two ways. The first example of this method is shown in **Figure 18-34** for sine waves and **Figure 18-35** for square waves.

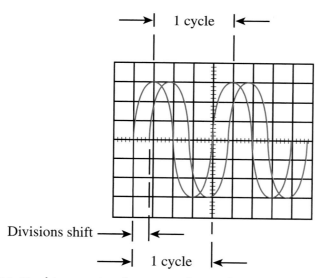

Figure 18-34. Dual trace setup for measuring a phase difference between sine waves.

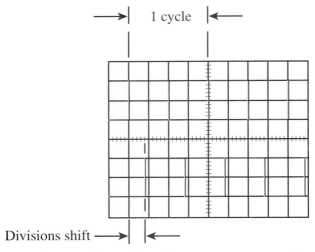

Figure 18-35. Dual trace adjusted for phase measurement of a square wave.

Dual Trace Procedure

The steps in the dual trace method are as follows:
1. Adjust the oscilloscope for normal operation.
2. Connect the oscilloscope probes to the circuit under test, applying the reference voltage to channel 1.
3. Align the channel 1 waveform as shown in **Figure 18-36.**
4. Determine the number of divisions for one cycle.
5. Determine the number of divisions of difference or horizontal phase difference. You may need to adjust the horizontal position to make it easier to read.
6. Calculate the degrees of phase difference using the following equation:

$$\text{Phase difference} = \frac{360 \times \text{Divisions difference}}{\text{Divisions for one cycle}}$$

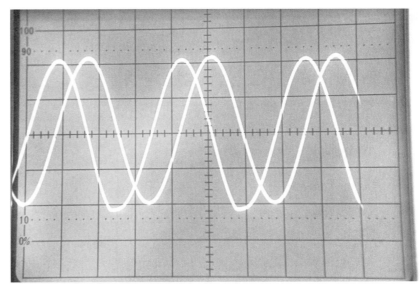

Figure 18-36. A dual trace phase measurement gives an immediate indication of any phase difference.

▼ *Example 18-3:*

What is the phase difference in Figure 18-36?

$$\text{Phase difference} = \frac{360 \times \text{Divisions difference}}{\text{Divisions for one cycle}}$$

$$\text{Phase difference} = \frac{360 \times 0.8}{3.45}$$

$$= 83.5°$$ ▲

The second dual trace method is the simplest to use and can provide measurement of smaller phase angles. Follow this procedure:

1. Set up as before for making a dual trace measurement.
2. Adjust the time base so one cycle of the left trace on the X-axis covers eight divisions. (One cycle, or 360°, divided by eight divisions equals 45°; thus, each division equals 45°.) See **Figure 18-37.**
3. Determine the horizontal divisions of phase difference between corresponding points on the two waveforms.
4. Multiply the distance (in divisions) times 45° per division to obtain the phase difference. The equation is:

> **Phase difference = Horizontal division × 45°/Division**

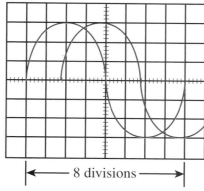

← 8 divisions →

Figure 18-37. Traces adjusted for a second dual trace method. Each horizontal division equals 45°.

▼ *Example 18-4:*

In Figure 18-37, the horizontal phase shift distance is 1.75 divisions. What is the phase difference?

$$\text{Phase difference} = \text{Horizontal divisions} \times 45°/\text{Division}$$

$$= 1.75 \times 45°$$

$$= 78.8°$$ ▲

Small Angle Measurement

The previous procedure may not give the desired accuracy for small phase differences. For greater accuracy, the sweep time/division setting may be changed to expand the waveform display, **Figure 18-38.** However, by adjusting the oscilloscope display in this manner, the relationship of one division to 45° is no longer true. Thus, the following equation must be used:

> $$\text{Phase difference} = \text{Horizontal division} \times 45° \frac{A}{B}$$

Where:

A = New time/Division setting

B = Old time/Division setting

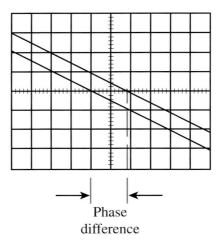

Figure 18-38. The time base is adjusted to expand the waveforms for greater accuracy in phase difference measurement.

▼ ***Example 18-5:***

A waveform is measured as 4.3 divisions with an old time/division setting of 2 ms and a new time/division setting of 10 ms. What is the phase difference?

$$\text{Phase difference} = \text{Horizontal divisions} \times 45° \frac{A}{B}$$

$$= 4.3 \times 45° \frac{2 \text{ ms}}{10 \text{ ms}}$$

$$= 38.7° \qquad \blacktriangle$$

Another method to obtain more accuracy is to use the 10× magnification to expand the horizontal display scale. Adjust the oscilloscope as in the previous step 2. Set the oscilloscope to 10× magnification. The horizontal scale is now 4.5° per division. Simply multiply the number of divisions of difference between the waveform traces by 4.5°.

In the dual trace method, notice the formula does not require the use of trigonometry as does the Lissajous display method. A single trace oscilloscope could be used with an electronic switching device. Oscilloscope displays can also have eight or 16 traces using proper switching techniques. These types of displays are seen in digital circuit analysis.

Waveform Details

The most common waveform is the sine wave. Other common waveforms are the square wave, triangular wave, and sawtooth.

Figure 18-39 illustrates the details of a ***square wave.*** Notice the time duration of the positive half (t_1) is the same as the duration of the negative half (t_2). The positive and negative peak amplitudes are also equal in value.

The ***triangular wave*** has the same type of symmetry, or the same equal parts about the axis, as the square wave. The time duration of the positive and negative ramps are equal, **Figure 18-40.**

The ***sawtooth*** waveform is illustrated in **Figure 18-41.** Notice the positive and negative ramps have unequal time durations.

Square wave: A square or rectangular periodic wave that alternately assumes two fixed values for equal lengths of time.

Triangular wave: A waveform that has rise and fall distinctly shaped like an equal triangle.

Sawtooth: A waveform shape that resembles the teeth of a saw blade.

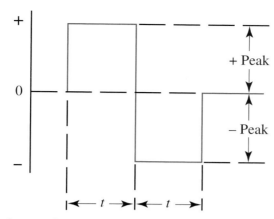

Figure 18-39. The time and amplitudes of a square wave are equal.

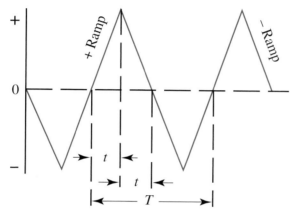

Figure 18-40. The times are equal or symmetrical in a triangular waveform.

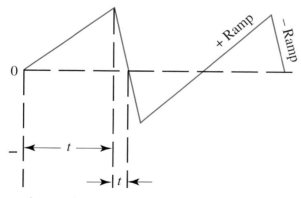

Figure 18-41. A sawtooth waveform is recognizable by the different time durations of the ramps.

Comparing Square Waves

The actual appearance of square waves may not be what is shown on the oscilloscope display. The ac square wave in **Figure 18-42** goes positive and negative on the X-axis.

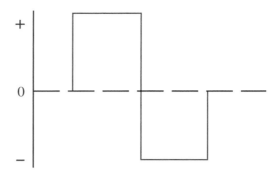

Figure 18-42. A square wave is symmetrical above and below the X-axis.

Dc pulses: Waveform pulses that do not cross the X-axis.

Dc offset: Refers to a waveform above (+) or below (–) the baseline by a given value of dc voltage.

In **Figure 18-43,** *dc pulses* stay above or below the X-axis. Moving the oscilloscope vertical input switch from ac to dc will verify which type is being displayed on the oscilloscope.

A square wave with *dc offset* is illustrated in **Figure 18-44.** The input switch of the oscilloscope must be in the dc position to correctly observe this type of waveform.

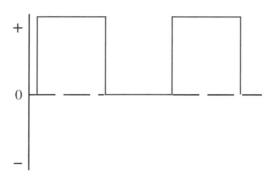

Figure 18-43. Dc pulses stay above (positive) or below (negative) the X-axis.

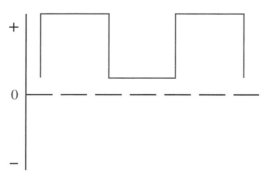

Figure 18-44. Square wave with dc offset.

Ringing

Ringing: A waveform defect that resembles a type of damped wave on a portion of a waveform..

A common defect in oscilloscopes is ringing, **Figure 18-45.** *Ringing* is a damped (decreasing) oscillation in the output signal of a system caused by a sudden change in the input signal. Often, such a reduction in magnitude of the waveform is caused by inductances in the circuitry. The problem can be very difficult to eliminate. Improper grounding is sometimes at fault.

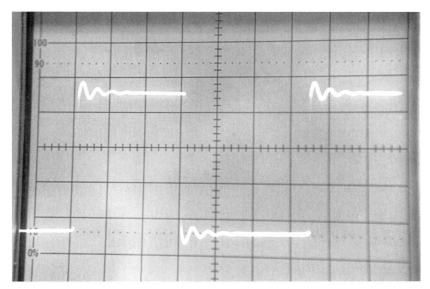

Figure 18-45. Ringing is a common problem where inductances are involved in the circuitry.

Composite Waveforms

Another term used in waveform description is ***composite.*** It simply means complex or complicated. A composite waveform is made up of many ac voltages of various amplitudes and frequencies, **Figure 18-46.** Square waves made from harmonics, music, and television video signals are examples of composite waveforms.

Composite: A waveform with many different voltages and frequencies.

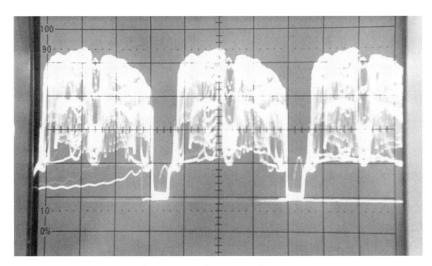

Figure 18-46. A composite television video waveform.

Summary

- Waveforms can be described using the leading edge, trailing edge, positive going edge, or negative going edge.
- Waveform types include sine, square, triangular, sawtooth, trapezoidal, unsymmetrical, exponential, staircase, or composite.
- Waveforms can be measured by peak-to-peak, rise time, fall time, pulse width, cycle, tilt, duty cycle, time period, overshoot, preshoot, and undershoot.
- The phase difference between two or more waveforms can be measured using the Lissajous or dual trace methods.
- Square waves can exist as an ac square wave, dc pulse, or a square wave with dc offset.
- Ringing is undesirable in a waveform.
- A composite waveform is one which is complex.

Important Terms

Do you know the meanings of these terms used in the chapter?

composite	preshoot
dc offset	pulse width
dc pulses	ramp
duty cycle	ringing
fall time	rise time
leading edge	sawtooth
Lissajous method	square wave
negative going edge	tilt
one cycle	trailing edge
overshoot	triangular wave
peak-to-peak value	undershoot
positive going edge	

Questions and Problems

Please do not write in this text. Write your answers on a separate sheet of paper.

1. Draw the common waveforms for a through g.
   a. sine
   b. square
   c. triangular
   d. sawtooth
   e. trapezoid
   f. exponential
   g. staircase
2. Identify the leading and trailing edges on the waveform in **Figure 18-47.**
3. What is the time period of the waveform in **Figure 18-48?** What is the peak-to-peak value?
4. What is the time period, frequency, and peak-to-peak value in **Figure 18-49?**
5. What is the time duration of the dc pulses in **Figure 18-50?** What is the duty cycle percentage?
6. Find the RMS voltage and frequency of the sine wave in **Figure 18-51.**

7. What is the peak-to-peak voltage and time period of the waveform in **Figure 18-52?**
8. In **Figure 18-53,** what is point Z called?
9. What is the time period and frequency of **Figure 18-54?**
10. What is the peak-to-peak value of **Figure 18-55?**
11. What is the pulse width and peak-to-peak value of **Figure 18-56?**
12. In **Figure 18-57,** what is the duty cycle?
13. What is the peak-to-peak value, rise time, and fall time of **Figure 18-58?**

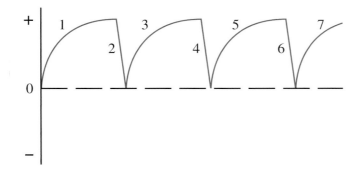

Figure 18-47. Waveform for problem 2.

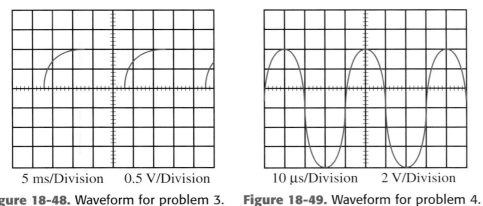

5 ms/Division   0.5 V/Division   10 µs/Division   2 V/Division

Figure 18-48. Waveform for problem 3.   **Figure 18-49.** Waveform for problem 4.

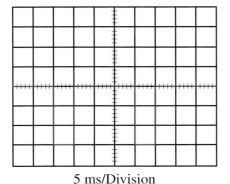
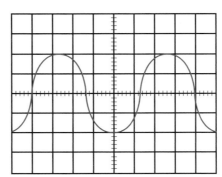

5 ms/Division   0.1 ms/Division   0.01 V/Division

Figure 18-50. Waveform for problem 5.   **Figure 18-51.** Waveform for problem 6.

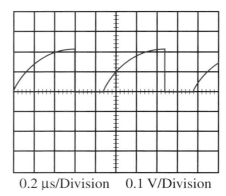

0.2 μs/Division   0.1 V/Division

Figure 18-52. Waveform for problem 7.

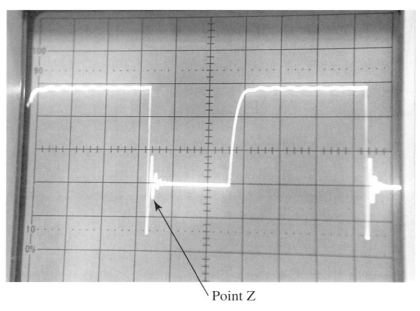

Point Z

Figure 18-53. Waveform for problem 8.

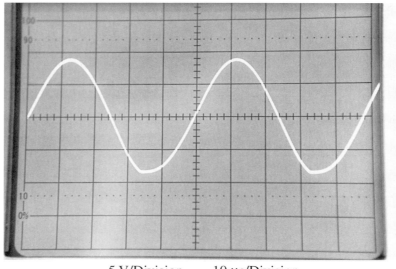

5 V/Division   10 μs/Division

Figure 18-54. Waveform for problem 9.

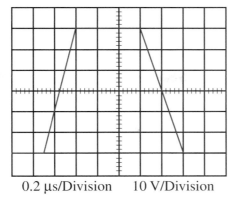

0.2 µs/Division   10 V/Division

Figure 18-55. Waveform for problem 10.

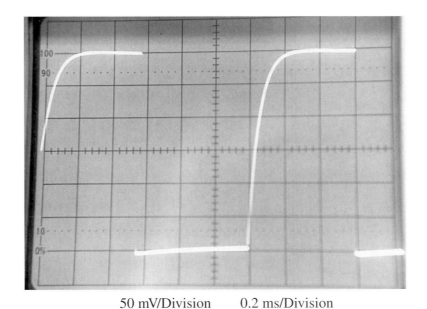

50 mV/Division   0.2 ms/Division

Figure 18-56. Waveform for problem 11.

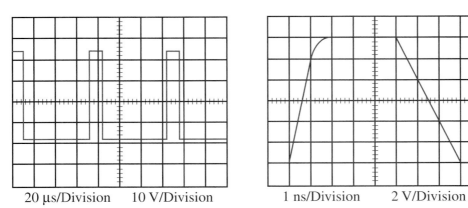

20 µs/Division   10 V/Division   1 ns/Division   2 V/Division

Figure 18-57. Waveform for problem 12.   **Figure 18-58.** Waveform for problem 13.

Chapter 19 Graphic Overview

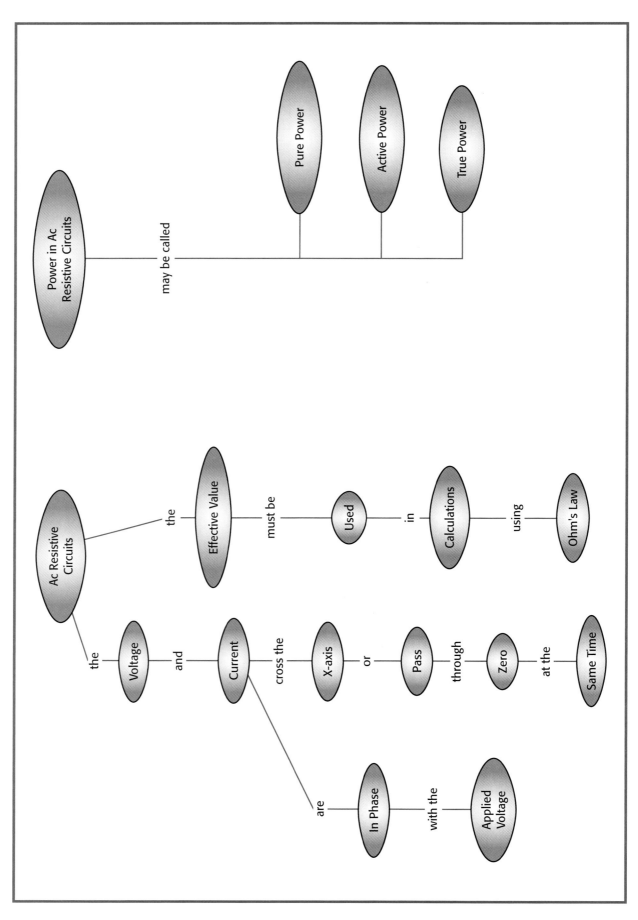

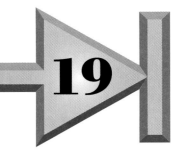

RESISTIVE AC CIRCUITS ▶ 19

 ## Objectives

After studying this chapter, you will be able to:
○ Apply Ohm's law to ac resistive circuits.
○ Describe the phase relationship between current and voltage in an ac resistive circuit.
○ Solve for voltage and currents in series and parallel ac resistive circuits.
○ Solve for power in ac resistive circuits.

Introduction

As with direct current, when alternating current flows through a resistance, the current changes in step with the voltage changes. The same method of making calculations in dc circuits can be used in ac resistive circuits.

Ac Resistive Circuit Operation

The circuit in **Figure 19-1** consists of an ac source connected to a resistive load. The voltage and current cross the X-axis, or pass through zero at the same time. When this happens, the voltage and current are ***in phase.*** Because they represent two different quantities that are measured in different units, the two waveforms do not have the same peak amplitude. An oscilloscope display of in-phase waveforms is shown.

In phase: A condition that occurs when the waveforms of two or more traces cross the X-axis at the same time and in the same direction.

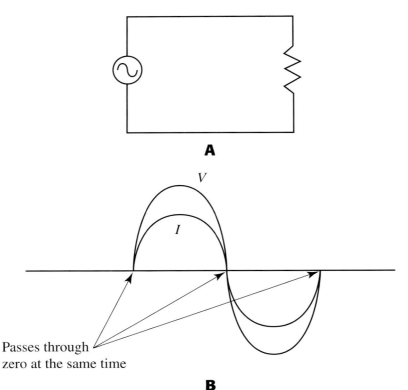

Figure 19-1. Ac resistive circuit. A—Ac source with a resistive load. B—Voltage and current are in phase. *(continued)*

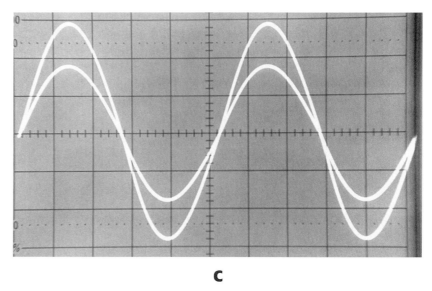

Figure 19-1. C—Oscilloscope display of two waveforms in phase.

Series Ac Resistive Circuits

The effective values of alternating current must be used for any calculations using Ohm's law. The effective value of ac will do the same work (power) as the same value of dc voltage.

▼ *Example 19-1:*
Find the voltage drops of R_1 and R_2 in **Figure 19-2.**
First, find the current:

$$I = \frac{V_T}{R_T}$$
$$= \frac{14 \text{ V}}{10 \text{ }\Omega}$$
$$= 1.4 \text{ A}$$

Then, find the voltage drops:

$$V = I \times R$$

$$V_1 = 1.4 \text{ A } (6 \text{ }\Omega)$$
$$= 8.4 \text{ V}$$

$$V_2 = 1.4 \text{ A } (4 \text{ }\Omega)$$
$$= 5.6 \text{ V}$$

A graphic representation of these calculations is shown in Figure 19-2B.   ▲

Parallel Ac Resistive Circuits

Just as in a dc parallel circuit, the voltage of a parallel ac resistive circuit is the same across all parts of the circuit. In a parallel ac resistive circuit, the total current and the individual currents are in phase with the applied voltage. To use Ohm's law for ac calculations, all the values must be in RMS or effective values.

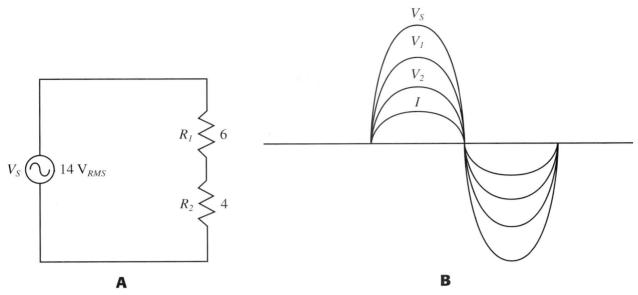

Figure 19-2. Series ac resistive circuit. A—Series ac circuit for Example 19-1. B—Graph of voltage drops.

▼ *Example 19-2:*

Find the individual currents and total current of **Figure 19-3.**

$$I = \frac{V}{R}$$

First, find the individual currents.

$$I_{R_1} = \frac{24\text{ V}}{20\ \Omega}$$
$$= 1.2\text{ A}$$

$$I_{R_2} = \frac{24\text{ V}}{30\ \Omega}$$
$$= 0.8\text{ A}$$

Then, find the total current, which equals the sum of the individual currents.

$$I_T = I_1 + I_2$$
$$= 1.2\text{ A} + 0.8\text{ A}$$
$$= 2\text{ A}$$

A graphic representation of these calculations is shown in Figure 19-3B.   ▲

Power in Ac Resistive Circuits

The power of an ac resistive circuit is calculated in the same manner as a dc circuit. It is also measured in watts and found by multiplying the current by the voltage. **Figure 19-4** illustrates the relationship of power, current, and voltage. Because the power is dissipated as heat in both the positive and negative parts of the cycle, the power curve does not go below the X-axis.

Why are both curves in the positive direction? Consider this: If the first half of the cycle is giving off heat, and the second half goes below the X-axis (power in the opposite direction), the circuit would be giving off heat during the first half of the cycle and taking in heat (returning power to the circuit) during the second half. This would mean 12 W are given off as heat and 12 W are put back, with no power consumed.

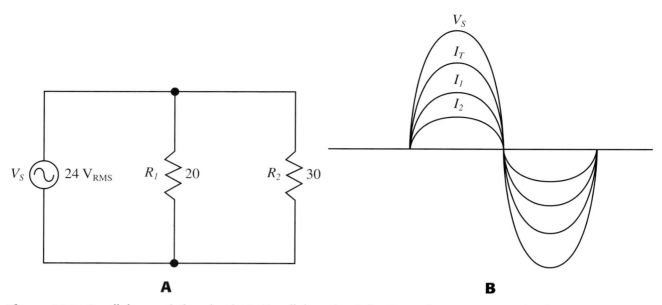

Figure 19-3. Parallel ac resistive circuit. A—Parallel ac circuit for Example 19-2. B—Graph of currents.

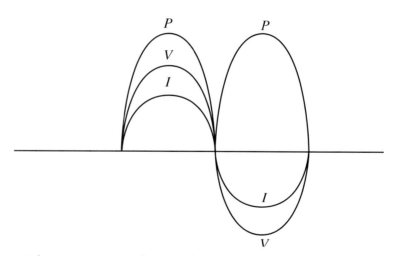

Figure 19-4. Voltage, current, and power in a resistive ac circuit produce power in the positive direction.

To further illustrate how the power is always in the positive direction in resistive circuits, see **Figure 19-5.** The figure shows a 300 Ω resistive circuit with a 150 V ac source. Using the instantaneous voltage (*e*) and resistance (*R*), the instantaneous power (*p*) is calculated using the equation:

$$p = \frac{e^2}{R}$$

The data from the calculations are compiled into table format as shown. A graphic representation of the data can then be made. The values for 90° through 180° and 270° through 360° were omitted from the graph to simplify the information.

To calculate instantaneous power, first calculate the instantaneous voltage using the equation:

$$e = \sin \theta \, E_P$$

Then, calculate the instantaneous power. Refer to Figure 19-5C.

At 30°:
$$e = \sin \theta \, E_P$$
$$= \sin 30° \, 150 \text{ V}$$
$$= 75 \text{ V}$$
$$p = \frac{e^2}{R}$$
$$= \frac{75^2}{300}$$
$$= 18.75 \text{ W}$$

At 210°:
$$e = \sin \theta \, E_P$$
$$= \sin 210° \, 150 \text{ V}$$
$$= -75 \text{ V}$$
$$p = \frac{e^2}{R}$$
$$= \frac{-75^2}{300}$$
$$= 18.75 \text{ W}$$

At 60°:
$$e = \sin \theta \, E_P$$
$$= \sin 60° \, 150 \text{ V}$$
$$= 130 \text{ V}$$
$$p = \frac{e^2}{R}$$
$$= \frac{130^2}{300}$$
$$= 56.3 \text{ W}$$

At 240°:
$$e = \sin \theta \, E_P$$
$$= \sin 240° \, 150 \text{ V}$$
$$= -130 \text{ V}$$
$$p = \frac{e^2}{R}$$
$$= \frac{-130^2}{300}$$
$$= 56.3 \text{ W}$$

At 90°:
$$e = \sin \theta \, E_P$$
$$= \sin 90° \, 150 \text{ V}$$
$$= 150 \text{ V}$$
$$p = \frac{e^2}{R}$$
$$= \frac{150^2}{300}$$
$$= 75 \text{ W}$$

At 270°:
$$e = \sin \theta \, E_P$$
$$= \sin 270° \, 150 \text{ V}$$
$$= -150 \text{ V}$$
$$p = \frac{e^2}{R}$$
$$= \frac{-150^2}{300}$$
$$= 75 \text{ W}$$

Although the negative sine function gives a negative instantaneous voltage, the negative number is squared, giving a positive power number. This causes the second alternation power curve of the graph to be above the X-axis on the graph. The equation e^2/R is used, eliminating the need to calculate the current.

Ac resistive power may also be referred to as pure power, active power, or true power. The term true power is often reserved for circuits involving inductance or capacitance with the voltage and current not in phase. A power factor is used to compensate for this out-of-phase condition. *Out of phase* refers to a situation when the voltage and current do not cross the X-axis at the same time, **Figure 19-6.**

Remember, when using Ohm's law with alternating current, RMS values must be used:

$$P = I_{RMS}E_{RMS} = I_{RMS}^2R = \frac{E_{RMS}^2}{R}$$

Peak values can be used by:

$$P = \left(\frac{1}{2}\right) I_P E_P$$

Out of phase: When two or more traces do not cross the X-axis of the oscilloscope graticule at the same time or in the same direction.

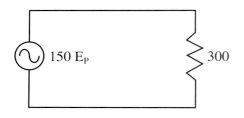

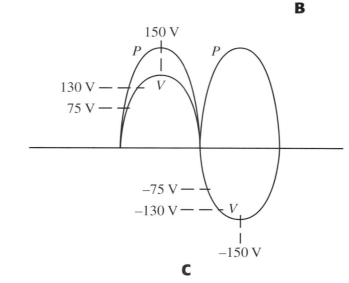

Degrees	30	60	90	210	240	270
Voltage	75	130	150	−75	−130	−150
Power	18.75	56.3	75	18.75	56.3	75

B

Figure 19-5. Calculating instantaneous power. A—Ac resistive circuit. B—Compilation of data. C—Graphic representation of calculated data.

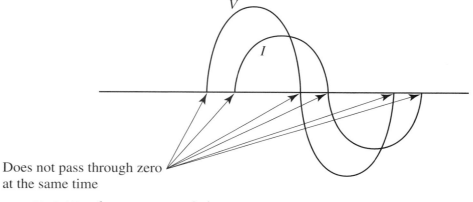

Figure 19-6. Waveforms are out of phase.

Summary

- The voltage and current are in phase in an ac resistive circuit.
- The calculations of voltage, current, resistance, and power for ac resistive circuits are the same as those for dc. However, RMS values must be used in the calculations.

Important Terms

Do you know the meanings of these terms used in the chapter?

in phase
out of phase

Questions and Problems

Please do not write in this text. Write your answers on a separate sheet of paper.

1. What is the phase relationship of current and voltage in a resistive ac circuit?
2. Find the voltage drops and power dissipated for each of the resistors in **Figure 19-7.**
3. Calculate the total power for the circuit in problem 2.
4. What determines whether the current and voltage are in phase or out of phase?
5. Draw a line graph of the voltage and current for problem 2.
6. Find the individual currents and power for **Figure 19-8.**
7. What are the different terms used for ac resistive power?
8. What is the instantaneous power at 68° of the circuit in Figure 19-8? (*Note:* The 10 V source is an RMS value.)
9. Draw a graph of voltage and current out of phase.
10. Calculate the voltage drops for each of the resistors, the power of the individual resistors, the currents, and the total power in **Figure 19-9.**

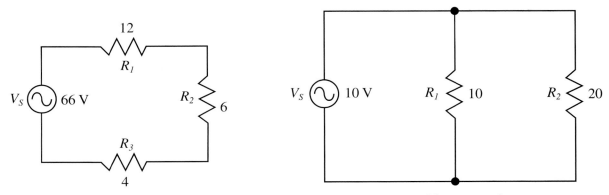

Figure 19-7. Circuit for problem 2.   **Figure 19-8.** Circuit for problems 6 and 8.

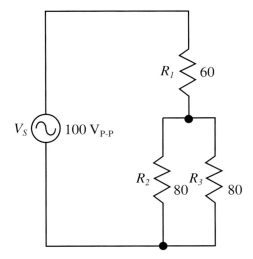

Figure 19-9. Circuit for problem 10.

Section IV

Inductance, Capacitance, and RCL Circuits

Section IV Activity

Testing Inductors

Testing an inductor consists of checking for continuity. If the inductor has continuity, it is usually good. Always remember to disconnect the inductor from the circuit when checking for resistance or continuity. This will prevent any parallel circuit paths from entering into the meter results. An inductance bridge can be used to test an inductor; however, the inductance will seldom change and is most critical at high frequencies.

Objective

In this activity, you will test whether an inductor is good, shorted, or open.

Materials and Equipment

4–Inductors (see instructor)
1–VOM
1–Inductance bridge (optional)

Procedures

1. Using an ohmmeter, test each inductor for continuity and record the results in the table in **Figure A.**
2. Check the inductance of each of the inductors; then, record the inductance in the table.

Inductor	Good	Open	Shorted	Inductance
1				
2				
3				
4				
5				
6				

Figure A.

Chapter 20 Graphic Overview

Inductors

- types are:
  - Air Core
  - Iron Core
  - Ferrite Core
  - Toroid
- ratings are:
  - Inductance Value
  - Tolerance
  - Maximum Current
  - Dc Resistance
  - Size and Package
- have: Reactance
- may be a: Shock Hazard — called — Inductive Kick
- has: **Inductance**

Mutual Inductance
- letter symbol: L_M
- depends on: Coefficient of Coupling — letter symbol — k

Inductance
- factors are:
  - Size and Shape
  - Number of Turns
  - Number of Layers
  - Kind of Core
- letter symbol: L
- produces a: Voltage — across the — Coil — called — Counter Emf
- is measured in: Henrys
- is the: Property — that — Opposes — any — Change — in the — Current

INDUCTANCE 20

Objectives

After studying this chapter, you will be able to:
- ○ Explain the process of electromagnetic induction.
- ○ State the factors that determine inductance.
- ○ Identify the different types of inductors.
- ○ State the ratings used for inductors.
- ○ Calculate the total inductance of series and parallel inductances.
- ○ Explain the operation of inductors in dc and ac circuits.
- ○ Calculate the inductive reactance of an inductor.
- ○ Calculate the Q-factor of an inductor.

Introduction

Inductance is the ability of a coil to store magnetic energy. The higher the inductance, the greater the magnetic energy stored. At high frequencies, the inductance quality of a single wire conductor must be considered. Coils manufactured for a specific inductance value are called inductors. Many electronic devices operate using the principles of inductance. An understanding of inductance is important to the study of electricity and electronics.

Electromagnetism

The principles of electromagnetism discussed in Chapter 15 will be reviewed here. When a current flows in a conductor, a magnetic field is produced around the conductor. No matter how small the current or the size of the conductor, a magnetic field exists around the conductor. If a larger current flows, the magnetic field becomes larger. See **Figure 20-1.** A magnetic field produced by electricity is called *electromagnetism.*

Electromagnetism:
Magnetism produced by an electrical current.

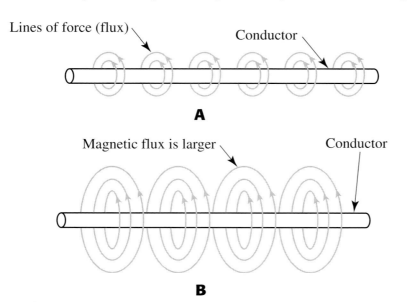

Figure 20-1. Electromagnetism. A—Any conductor carrying a current has a magnetic field around it. B—A larger current causes a larger magnetic field.

When the switch in a circuit is closed, a current begins to flow. As the electrical current increases, it produces a magnetic field. The current does not immediately go to its maximum value because some of the electrical energy is being converted to magnetic energy. When the current reaches its maximum value, no further increase in the magnetic field takes place. See **Figure 20-2.** The magnetic field produced around the coil is electrical energy converted to magnetic energy.

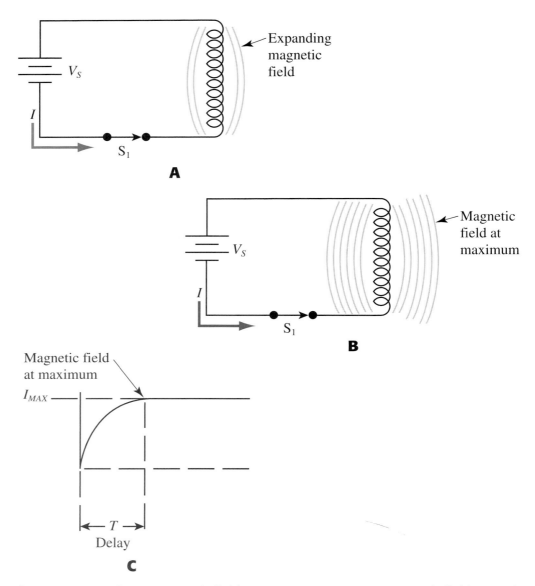

Figure 20-2. Maximum magnetic field. A–Dc current causes a magnetic field around an inductor. B–A larger current causes a larger magnetic field around the inductor. C–The current is delayed in getting to maximum value because the electrical energy is being converted to magnetic energy.

Meaning of Inductance

In your study of alternating current, you learned electromagnetic induction is the process of a magnetic field inducing a current in a conductor. Any time magnetic lines of force cut across a conductor or a conductor cuts across a magnetic field, a current is induced in the conductor.

When the switch in a circuit is opened, the current stops flowing. The magnetic energy stored as a magnetic field about the coil collapses into the coil. As the field collapses, the magnetic lines of force cut across the conductor and produce a current. See **Figure 20-3.** This current is in the direction of the original current and tends to keep it flowing. Therefore, as the current in a coil tries to increase, the electric energy is converted to magnetic energy. As the current stops flowing, the magnetic energy returns to the circuit as electric energy. The action of the coil in the circuit tends to oppose any change in the current. That property of a circuit or device opposing any change in the current is called *inductance* (represented by *L*).

Inductance: That property of a coil that tends to oppose any change in the current.

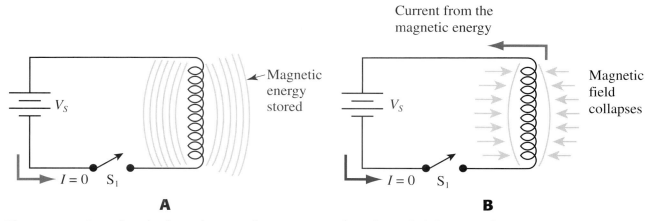

Figure 20-3. Stored and released magnetic energy. A—When the switch is open, the magnetic energy remains stored in the inductor. B—The magnetic field collapses, returning the magnetic energy to the circuit.

As just discussed, the action of the magnetic field cutting across the wire conductor induces a current. This process of inducing a current in a coil is called *self-induction.* The magnetic lines of force that cut across the conductor also produce a voltage across the coil called a *counter emf* (cemf).

The unit of measurement for inductance is the *henry* (H). The henry is a large unit. Inductors with values between 1 H and 20 H are used for low-frequency applications. High-frequency circuits have inductors in millihenrys or microhenrys.

Self-induction: The property that causes a counter-electromotive force to be produced in a conductor or coil when the magnetic field collapses or expands with a change in the amplitude of the current.

Factors Determining Inductance

All circuits have some inductance. A straight conductor has inductance distributed along the length of the conductor. Coils and transformers have concentrated or "lumped" inductance. The technical name for a coil is *inductor.*

A conductor has a very small inductance. An inductor, on the other hand, has a large inductance. The magnitude of inductance depends on several factors:

* Size and shape of the coil
* Number of turns
* Number of layers of turns
* Kind of core material

The inductance of an inductor will depend upon its size and shape. Increasing the size of an inductor will increase the area of the coil and produce more magnetic lines. The toroid core inductor is the most efficient type, while the least efficient is the simple rod core inductor. Calculating the size and shape of inductors is part of the field of magnetic engineering and beyond the scope of this text.

Counter electromotive force (cemf): The voltage generated by a magnetic field around a coil.

Henry: The unit of measure for inductance.

Inductor: A conductor wound in a spiral or coil to increase its inductance.

An inductor can be a single- or multiple-layer type, **Figure 20-4.** The turns can also be tightly or loosely wound. The inductance of a single-layer, tightly wound inductor is determined by the equation:

$$L = \frac{(N \times r)^2}{9r + 10l}$$

Where:

L = Inductance (in microhenrys)
N = Number of turns
r = Radius of the inductor (in inches)
l = Length of the inductor (in inches)

For tightly wound multilayer inductors, the equation becomes:

$$L = \frac{0.8 \, (N \times r)^2}{6r + 9l + 10D}$$

Where:

D = Depth of the inductor in inches

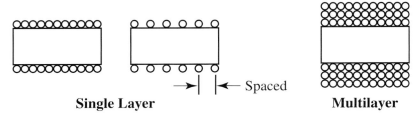

Single Layer   ←| |← Spaced   **Multilayer**

Figure 20-4. Inductors can have a single layer or multiple layers of turns.

The more turns in an inductor, the stronger the magnetic field. Increasing the number of turns is limited by the possibility of magnetic saturation of the core.

An inductor is an electromagnet. Core materials include air, iron, and ferrite. If the inductor is wound on a form made of such nonmagnetic materials as paper, ceramic, porcelain, plastic, or glass, the inductor is still called an air core inductor. An inductor wound on a magnetic material such as iron or ferrite has a much larger inductance. The permeability of the core material causes an increase in the magnetic flux. Calculations of inductance using various types of core materials and loosely wound inductors are beyond the scope of this textbook.

Types and Ratings

Inductors come in various sizes and shapes, **Figure 20-5.** They are specified as air core, ferrite core, iron core, or some special winding. See **Figure 20-6.** Another common type is called the *toroid* wound inductor, **Figure 20-7.** Stray magnetic fields are minimized because no air gap exists. Other advantages of the toroid wound inductor are smaller size and easier mounting. The cores of toroids are usually made of formed powdered iron or silicon steel.

Toroid: A very efficient type of coil wound upon a "doughnut-shaped" core.

Ratings of Inductors

The key ratings or specifications of inductors are:

- Inductance value
- Tolerance specification
- Maximum current
- Dc resistance
- Size
- Package

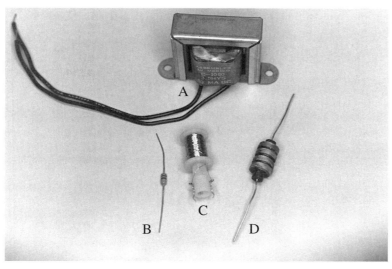

Figure 20-5. Types of conductors. A—Large iron core inductor used in power supplies. B—Small single-winding air core inductor. C—Adjustable ferrite core inductor. D—Multiple-winding air core inductor.

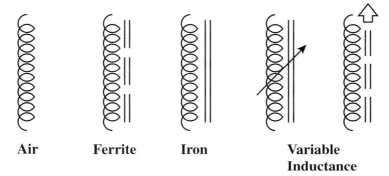

Figure 20-6. Schematic symbols of various inductors.

Figure 20-7. A toroid wound inductor keeps the magnetic fields inside the inductor.

Inductance value and tolerance specifications are important in high-frequency circuits. An incorrect inductance value or tolerance specification can cause circuits to operate at undesired frequencies or not at all.

The maximum current must not be exceeded; otherwise, the inductor will overheat and its useful life will be shortened. The dc resistance can also cause overheating or improper operation of the circuit.

When replacing an inductor, the replacement must fit within the space of the defective inductor. Although the replacement meets all other specifications, the size and type of package may make it impossible to use.

Inductances in Series and Parallel

As with other components, inductors can be connected in series or parallel. The equations for total inductance follow the resistor equations. For example, in **Figure 20-8,** the inductors are connected in series. The total inductance is equal to the sum of the individual inductances.

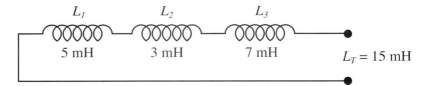

Figure 20-8. The total inductance of inductors connected in series is equal to the sum of the inductances.

▼ *Example 20-1:*

What is the total inductance in Figure 20-8?

$$L_T = L_1 + L_2 + L_3$$
$$= 5 \text{ mH} + 3 \text{ mH} + 7 \text{ mH}$$
$$= 15 \text{ mH}$$ ▲

In **Figure 20-9,** the inductors are connected in parallel. The total inductance is calculated using the reciprocal equation:

$$L_T = \cfrac{1}{\cfrac{1}{L_1} + \cfrac{1}{L_2} + \cfrac{1}{L_3}}$$

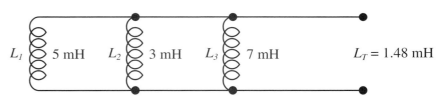

Figure 20-9. Inductors connected in parallel. The total inductance is always less than the smallest inductance.

▼ *Example 20-2:*

What is the total inductance in Figure 20-9?

$$L_T = \cfrac{1}{\cfrac{1}{5} \text{ mH} + \cfrac{1}{3} \text{ mH} + \cfrac{1}{7} \text{ mH}}$$

$$= 1.48 \text{ mH}$$ ▲

Mutual Inductance

The previous equations assume no mutual inductance exists between the inductors. When two inductors or conductors are close together, their magnetic fields will react by either aiding or opposing each other, **Figure 20-10.** The two inductors are said to be "coupled" by means of their magnetic fields or to have ***mutual inductance.***

Mutual inductance: The inductance between two or more inductors. It is represented by L_M.

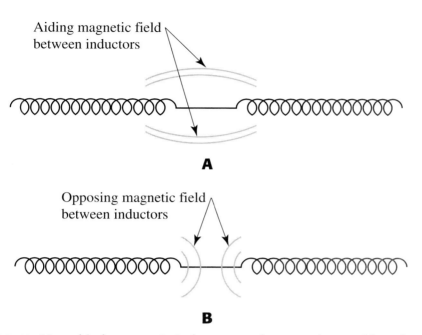

Aiding magnetic field between inductors

A

Opposing magnetic field between inductors

B

Figure 20-10. Mutual inductance. A—Inductors can be wound or positioned so their magnetic fields aid each other. B—Inductors can also be wound or positioned so their magnetic fields oppose each other.

To calculate the total inductance of two inductors with mutual inductance, the equation is:

$$L_T = L_1 + L_2 \pm 2L_M$$

When the inductors are in series *aiding,* the mutual inductance (L_M) is plus, which increases the total inductance. When the inductors are in series *opposing,* the mutual inductance is minus, which decreases the total inductance. Inductors can also be connected in parallel with mutual inductance coupling. However, such calculations are beyond the scope of this text.

Coefficient of Coupling

Mutual inductance depends on the amount of coupling between the two inductors and is determined by the coupling factor or ***coefficient of coupling.*** If the two inductors have the maximum coupling between them, the coefficient of coupling is 1. The coefficient of coupling (represented by k) cannot be greater than 1. The equation is:

Coefficient of coupling: The amount of coupling between two components or circuits.

$$k = \frac{L_M}{\sqrt{L_1 L_2}}$$

The mutual inductance is equal to:

$$L_M = \frac{k}{\sqrt{L_1 L_2}}$$

Stray Inductance

As stated earlier, all conductors have a certain amount of inductance. The inductance of any wiring other than the inductors can be considered ***stray inductance.*** In most instances, the stray inductance is not a factor. For example, a piece of copper wire with a diameter of 0.032″ and length of 4″ has an inductance of approximately 0.1 μH. At low frequencies this inductance can be ignored. However, at high frequencies this amount of inductance can have a major effect on circuit operation. The leads of resistors, capacitors, and other components can become inductors at high frequencies. Therefore, the connecting leads must be kept short in RF circuits. Components should not be moved from their positions. Doing so can cause a change in circuit characteristics.

Inductors in Dc Circuits

The action of an inductor in a dc circuit was shown in Figure 20-2. Once the current reaches its maximum value, nothing more takes place in the circuit until the current value changes. Inductors play a greater role in an ac or dc circuit whose voltage level is changing, such as in dc power supply circuits.

In certain circuits, the quick return of the magnetic energy to electrical energy in an inductor can cause the cemf voltage to be so great that the voltage damages switch or relay contacts. It can also be a shock hazard. Such a voltage is called an ***inductive kick.*** To prevent inductive kick on relays and other coil-operated devices, a solid-state diode is placed across the electromagnet coil, **Figure 20-11.** The diode allows current in only one direction, providing a current path in the opposite direction of the line current for the cemf.

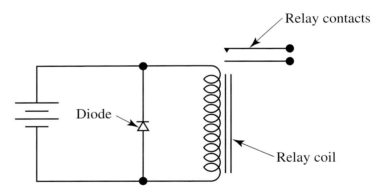

Figure 20-11. A semiconductor diode connected across the coil of a relay prevents cemf problems.

Inductors in Ac Circuits – Inductive Reactance

When an ac source is connected to an inductor, the current is changing constantly, **Figure 20-12.** This changing current is opposed by the inductance property of the inductor. Anything that opposes the current flow is considered a resistance. The resistance of an inductor to the flow of ac is ***reactance*** (represented by X). Since the reactance is that of an inductor, it is called ***inductive reactance*** (X_L). It is measured in ohms.

The equation for inductive reactance is:

$$X_L = 2\pi f L$$

Figure 20-12. An inductor connected to an ac source has a reactance toward the current.

▼ *Example 20-3:*

Find the inductive reactance when an inductance of 2 H is operating at 500 Hz.

$$X_L = 2\pi fL$$
$$= 2\,(3.14)\,(500)\,(2)$$
$$= 6280\;\Omega$$ ▲

Q-factor

The **Q-factor** of an inductor, or the **figure of merit**, is the measure of how well an inductor does its job. If an inductor has a high Q-factor, the losses in the inductor are low. The Q-factor is only a ratio and is not given a unit of measurement. The Q-factor can be found by the equation:

$$Q = \frac{X_L}{R}$$

Where:

X_L = Inductive reactance of the coil at the specified frequency

R = Resistance of the inductor

The power loss of an inductor is the power dissipated in the resistance of the copper winding of the coil. The energy stored in the magnetic field is returned to the circuit.

Q-factor: In an inductor or capacitor, the ratio of reactance to effective series resistance at a given frequency. Also called the **figure of merit.**

Skin Effect

At high frequencies, the electrons tend to travel at the outer portion of the conductor, **Figure 20-13.** This is called the **skin effect** and can be measured as a higher resistance at high frequencies. The higher the frequency, the more the skin effect becomes a factor in circuit operation. Its effect is minimized using *litz wire,* which has small diameter strands insulated from each other. Litz wire is actually a multiple-conductor cable. By reducing the resistance at high frequencies, less power is dissipated in the conductors.

Skin effect: The tendency of radio frequency currents to flow near the surface of a conductor.

Litz wire: A conductor made of fine, separately insulated strands woven together to make one conductor. Litz wire reduces the skin effect at high frequencies.

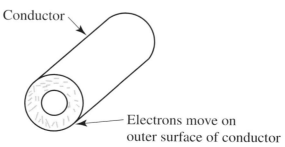

Conductor

Electrons move on outer surface of conductor

Figure 20-13. At high frequencies, current tends to flow at the outer portion of a conductor.

Testing Inductors

Typical problems with inductors include:

- Burned inductor (open) due to excess current
- Open connection where the coil conductor connects to the leads of the inductor
- Open inside the inductor
- Shorted turns inside the inductor

Continuity testing is the main test to determine an inductor problem. If the inductor has continuity, generally it is good. When checking for resistance or continuity, the inductor should be disconnected from the circuit to prevent parallel paths from interfering with the measurements. If a short exists between windings inside the inductor, it may physically appear to be normal; however, it will not work in the circuit as designed. If other verification of the inductor's condition is needed, use an inductance bridge or analyzer to test for the correct inductance. Proper type, inductance, and current capacity should be considered when selecting a replacement.

Summary

- Inductance is the property of a coil (inductor) that opposes a change in current.
- Inductance is determined by the size and shape of the coil, number of turns, number of layers of turns, and kind of core material.
- The unit of inductance (L) is the henry (H).
- Inductors may be air, iron, ferrite, or some special core.
- The key ratings of inductors are inductance, tolerance, maximum current, dc resistance, size, and package.
- Mutual inductance (L_M) is the inductive coupling between two or more conductors or inductors. The amount of mutual inductance is determined by the coefficient of coupling.
- The Q-factor represents how well an inductor is doing its job.
- Inductive reactance is the resistance an inductor has toward alternating current.
- The skin effect must be considered at high frequencies.
- Troubleshooting mainly consists of testing inductors for continuity.

Important Terms

Do you know the meanings of these terms used in the chapter?

coefficient of coupling	litz wire
counter emf	mutual inductance
electromagnetism	Q-factor
figure of merit	reactance
henry	self-induction
inductance	skin effect
inductive kick	stray inductance
inductive reactance	toroid
inductor	

Questions and Problems

Please do not write in this text. Write your answers on a separate sheet of paper.

1. Define *inductance.*
2. Self-induction produces a voltage called _____.
3. List the four factors that determine inductance.
4. The letter symbol for inductance is _____, and the unit of measurement is the _____.
5. The cores of inductors may be _____, _____, or _____.
6. List the key specifications of inductors.
7. Define *mutual inductance.*
8. Find the total inductance (the mutual inductance is aiding) in **Figure 20-14.**
9. Find the total inductance in **Figure 20-15.**
10. Explain why the connecting leads must be kept short in high-frequency circuits.
11. Calculate the inductive reactance of an inductor with an inductance of 1.5 mH operating at 500 Hz.
12. Calculate the Q-factor of an inductor with an inductance of 30 µH operating at 35 KHz with a resistance of 0.2 Ω.
13. The skin effect is a major factor at _____ frequencies.
14. What are the possible problems with inductors?
15. What instrument(s) is (are) used for testing inductors?

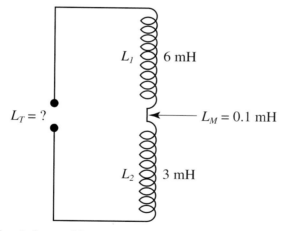

Figure 20-14. Series circuit for problem 8.

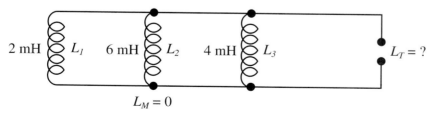

Figure 20-15. Parallel circuit for problem 9.

Chapter 21 Graphic Overview

Impedance
- measured in → **Ohms** → found by → **Vector Sum** → of the → **Resistance** — and — **Reactance**
- is the → **Total Resistance** → to → **Alternating Current**
- letter symbol → **Z**

RL Circuits
- power → **Calculations** → must use → **RMS Values** → for → **Apparent Power** — and — **Power Factor** → for → **True Power**
- can be → **Parallel** → use → **Current Calculations** → using → **Vector Sum**
- can be → **Series** → use → **Voltage Calculations**
- have → **Time Delay** → produces a → **Transient Response Curve**
- **Time Delay** → of → **63.2%** → of → **Maximum Current** → is → **One Time Constant**

RL CIRCUITS 21

Objectives

After studying this chapter, you will be able to:
- ○ Calculate the time constant of a resistor-inductor circuit.
- ○ Explain the circuit action of an inductor and cemf.
- ○ Calculate the impedance, current, and voltage drops of series RL circuits.
- ○ Draw a vector diagram of an RL circuit.
- ○ Calculate the currents, impedance, and phase angle of a parallel RL circuit.
- ○ Calculate the power of series and parallel RL circuits.

Introduction

This chapter analyzes circuits that combine resistance and inductive reactance. In many cases, Ohm's law still applies. In others, the values cannot be found so easily. The placement of an inductance in a circuit causes the voltage and current to be out of phase and changes the method of calculating power.

L/R Time Constant

If no inductor were in the circuit shown in **Figure 21-1,** the current would immediately go to maximum. Because the inductor *is* in the circuit, some of the electrical energy is converted to magnetic energy. Thus, the current has a ***time delay*** and rises gradually to the maximum current. The time required for the current to rise to 63.2% of its maximum value is known as one ***time constant.***

With the switch closed for a period of time so maximum current is flowing, then opened, one time constant would be required for the current in the inductor to fall to 36.8% of the maximum value. During the next time constant, the current would drop to a value 36.8% of the value remaining.

Time delay: The amount of time required for a signal to travel between two points.

Time constant: The time required for an exponential quantity to change by an amount equal to 63.2% of the total maximum value.

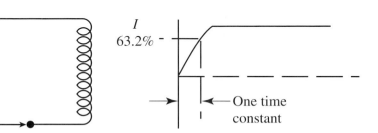

Figure 21-1. One time constant is 63.2% of the maximum current.

When a resistor is added to the circuit, the time constant changes, **Figure 21-2.** With a resistor in the circuit, the time for one time constant is found by the equation:

$$t = \frac{L}{R}$$

Where:
t = Time constant (in seconds)
L = Inductance (in henrys)
R = Resistance (in ohms)

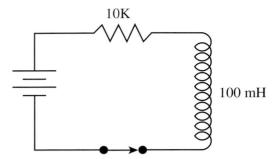

Figure 21-2. The time constant does not depend on the circuit voltage.

▼ *Example 21-1:*

Calculate the time constant in Figure 21-2.

$$t = \frac{L}{R}$$

$$= \frac{100 \times 10^{-3}}{10 \times 10^{3}}$$

$$= 10 \ \mu S$$

It would take 10 μS for the current to increase or decrease to 63.2% of its maximum value.   ▲

The universal time constant chart gives the percentages of rise and fall (decay) of the current of an RL circuit, **Figure 21-3.** After five time constants, the circuit is in a balanced state and no further change will occur. Such a graph may also be called a *transient response* curve. Transient refers to how something responds in a short period of time.

Transient response: Refers to how a circuit responds in a short period of time.

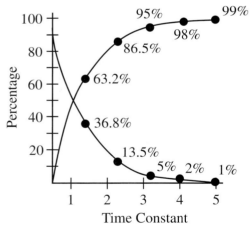

Figure 21-3. Universal time constant chart. The curves show the rise and fall of the current in an inductor.

Ac and an Inductor

An inductor is connected to an ac source (no resistance) in **Figure 21-4.** The source and the inductor are in parallel. At any instant, the two voltages must be the same. Because the voltage across a parallel is the same, no lead or lag time can exist between these two voltages.

The relationship between the circuit current and voltages is complex. Since the current is the same in all parts of the circuit, it is best to look at the circuit with the current as a reference point. The current cycle is from a maximum positive current to a maximum negative current, and back to the maximum positive value again, **Figure 21-5.** The time intervals t_1 through t_5 are for reference only during this discussion.

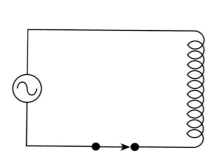

Figure 21-4. An ac source connected to an inductor has the voltage and current out of phase.

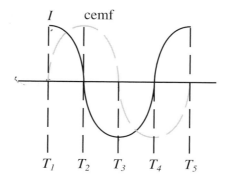

Figure 21-5. Relationship between the current and the cemf. The current leads the cemf by 90°.

The magnetic field increases and decreases in step with the current. When the current reaches its maximum positive value (instantaneous, t_1), its rate of change is zero. This means the field strength is maximum, but the magnetic field movement is zero. With no field movement, no cemf is induced across the inductor. Therefore, maximum current and zero cemf occur at the same instant (t_1).

As the current drops toward zero, the magnetic field collapses toward the center of the coil, cutting across the conductor of the coil and producing the cemf. As the current approaches zero (t_2), its rate of change is very rapid, producing the maximum induced cemf as it goes through zero.

The current increases in the negative direction, producing an expanding magnetic field in the opposite polarity. This reverse polarity expanding field produces a cemf of the same polarity as the original collapsing field. Reversing the current direction and changing the collapsing field to an expanding field is a double reversal. As the current reverses direction, the induced cemf is still in the same direction and continues as long as the current is changing. When the current reaches its maximum negative value (t_3), its rate of change is zero. The induced cemf is again zero at t_3.

As the current drops from the maximum negative value toward zero, its rate of change increases. An induced cemf is developed again, now in the negative direction. As the current approaches zero (t_4), its rate of change is very rapid, which produces the maximum cemf.

When the current goes from zero to a maximum positive value (t_5), the induced cemf drops from maximum negative to zero. This results in one complete cycle of current and induced cemf. Figure 21-5 illustrates the induced cemf and current are always 90° out of phase.

Figure 21-6 shows the phase relationship of the current, source voltage, and induced cemf in a purely inductive circuit. The source voltage is 180° out of phase with the cemf. In any pure inductance, the current always lags the source voltage by 90°.

Figure 21-7 is a vector diagram of the graph in Figure 21-6. With the current at 0°, the source leads the current by 90° and the cemf lags the current by 90°. The source voltage (E_S) and the cemf are 180° out of phase.

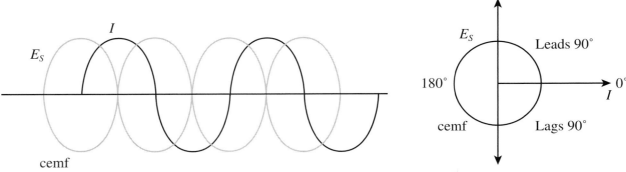

Figure 21-6. The source voltage and cemf are 180° out of phase, with the source voltage leading the current by 90°.

Figure 21-7. Vector diagram of the current, source voltage, and cemf.

Impedance

Impedance: The total resistance toward alternating current.

The *impedance* (represented by Z) of a circuit or component is the total opposition or resistance to alternating current. Impedance is measured in ohms. The impedance of a circuit can be found by replacing the resistance (R) in the Ohm's law formula with Z for impedance. The equation is:

$$Z = \frac{E}{I}$$

Series RL Circuits

Some resistance is always present in a practical circuit; therefore, the amount out of phase will never reach 90°. The circuit in **Figure 21-8A** has an inductive reactance and a resistance. It is called a series RL circuit. The total resistance to the ac is a combination of the reactance and resistance, which is the impedance. Unlike series resistance circuits, impedance cannot be found by simply adding the two quantities together. The impedance is the *vector sum* of the resistance and reactance.

Vector sum: In alternating current circuit applications, the addition of quantities using vectors. Mathematically, the square root of the sum of the squares.

Since the current is always in phase through a pure resistance, the resistance of $4\ \Omega$ is placed at 0°, **Figure 21-8B**. The voltage leads the current through an inductor by 90°. Thus, the vector for the inductive reactance is placed at 90°. To complete the vector diagram, the rectangle is completed using dashed lines. The impedance vector is drawn from the origin to a point where the dashed lines cross.

The vector diagram in **Figure 21-8C** is shown from a different viewpoint. The vectors make a triangle with R as the adjacent side and Z as the hypotenuse. You can find the out-of-phase angle by calculating the cosine of the angle using the equation:

$$\cos \theta = \frac{R}{Z}$$

Thus:
$$\cos \theta = \frac{R}{Z}$$
$$= \frac{4}{5}$$
$$= 0.8$$
$$\theta = 36.9°$$

Since the vectors make a triangle, the Pythagorean theorem can also be used to find impedance. In this case, the Pythagorean theorem changes from:

$$c = \sqrt{a^2 + b^2}$$

to

$$Z = \sqrt{R^2 + X_L^2}$$

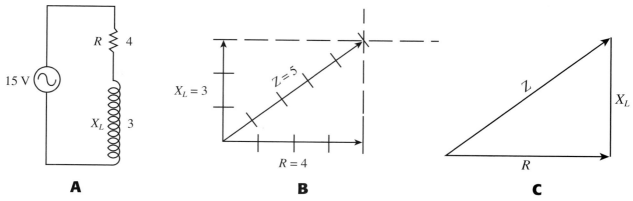

Figure 21-8. Series RL circuit. A—4 Ω of resistance in series with 3 Ω of inductive reactance combines by vector sum for 5 Ω of impedance. B— Vector diagram. C— Vector diagram as related to a triangle.

▼ *Example 21-2:*

If the resistance is 4 Ω and the inductive reactance is 3 Ω, what is the impedence?
The impedance is found by:

$$Z = \sqrt{R^2 + X_L^2}$$
$$= \sqrt{4^2 + 3^2}$$
$$= 5\ \Omega$$

The circuit current is found by:

$$I = \frac{E}{Z}$$
$$= \frac{15\ \text{V}}{5\ \Omega}$$
$$= 3\ \text{A}$$

The voltage drops (series circuit) are found by:

$$V_R = IR$$
$$= 3\ (4)$$
$$= 12\ \text{V}$$

$$V_{X_L} = IX_L$$
$$= 3\ (3)$$
$$= 9\ \text{V}$$

▲

At first glance, this violates the series voltage rule, which states the sum of the voltage drops equals the source voltage. Remember, though, you are dealing with ac. Always think in terms of the "vector sum." The vectors in **Figure 21-9** represent the respective voltage drops, and the voltage across the impedance (V_Z) is the source voltage.

Thus:

$$V_Z = \sqrt{V_R^2 + V_{X_L}^2}$$
$$= \sqrt{12^2 + 9^2}$$
$$= 15\ \text{V (source voltage)}$$

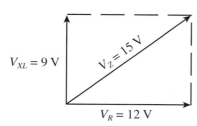

Figure 21-9. Voltage vector diagram. The voltages are substituted for the resistance and reactance. The vector sum is the source voltage.

Parallel RL Circuits

Calculations for series and parallel RL circuits differ mainly in one way. In the series circuit, the voltages are calculated. In the parallel RL circuit, the currents are calculated, **Figure 21-10.**

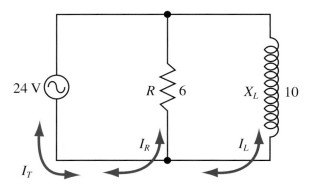

Figure 21-10. The currents must be used when dealing with a parallel RL circuit.

▼ *Example 21-3:*

Find the currents in Figure 21-10.

$$I_R = \frac{E}{R}$$

$$= \frac{24}{6}$$

$$= 4 \text{ A}$$

$$I_{X_L} = \frac{E}{X_L}$$

$$= \frac{24}{10}$$

$$= 2.4 \text{ A}$$

The total current is:

$$I_T = \sqrt{I_R^2 + I_{X_L}^2} \text{ (vector sum)}$$

$$= \sqrt{4^2 + 2.4^2}$$

$$= 4.66 \text{ A}$$

The impedance is calculated as:

$$Z = \frac{E}{I}$$

$$= \frac{24}{4.66}$$

$$= 5.15 \ \Omega$$

Or, the impedance can be calculated as:

$$Z = \frac{R \times X_L}{\sqrt{R^2 + X_L^2}}$$

$$= \frac{6 \times 10}{\sqrt{6^2 + 10^2}}$$

$$= \frac{60}{\sqrt{136}}$$

$$= \frac{60}{11.66}$$

$$= 5.15 \ \Omega$$

Figure 21-11 shows a vector diagram of the parallel RL circuit. The source voltage is the reference since it is the same across a parallel circuit. I_R is in phase with the source voltage at 0°. The current of an inductor lags the source voltage by 90°; therefore, the current (I_L), goes downward, giving a negative angle. The total current vector is drawn from the origin where the dashed lines cross. The Pythagorean impedance equation applies using the current values.

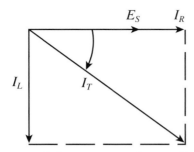

Figure 21-11. Vector diagram of the currents of a parallel RL circuit. The voltage drops are substituted for the resistance and reactance. The vector sum is the source voltage across the impedance.

Power in RL Circuits

The voltage and current in **Figure 21-12** are in phase. Using only RMS values, the power is equal to the voltage times the current. This is called the *apparent power.* When the voltage and current are out of phase, power is no longer equal to the product of the voltage and current. In **Figure 21-13,** the current is lagging the voltage by 50°. Between 0° and 50°, no current exists. Thus, zero current times any voltage is zero power. A *power factor* (*pf*) must be used to account for the out-of-phase condition. By multiplying the apparent power by the power factor, the *true power* of the circuit is found. The power factor is simply the cosine of the out-of-phase angle.

Apparent power: Electrical power equal to the voltage times the current.

Power factor: The cosine of the phase angle used when calculating power.

True power: The actual power consumed by a circuit or device which considers the voltage-current phase angle in the calculation.

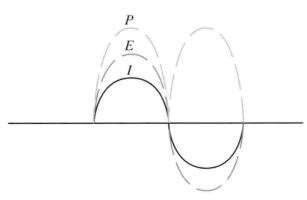

Figure 21-12. Sine waves in phase in a resistive ac circuit produce power in the positive direction for both halves of the ac cycle.

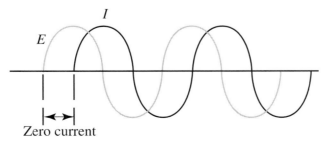

Figure 21-13. In a pure inductive circuit, the voltage and current are 90° out of phase.

▼ *Example 21-4:*

Find the power in **Figure 21-14**:

$$X_L = 2\pi fL$$
$$= 2\,(3.14)\,(3 \times 10^3)\,(35 \times 10^{-3})$$
$$= 660\ \Omega$$

$$Z = \sqrt{R^2 + X_L^2}$$
$$= \sqrt{400^2 + 660^2}$$
$$= 772\ \Omega$$

$$I = \frac{E}{Z}$$
$$= \frac{40}{772}$$
$$= 51.8\ \text{mA}$$

$$pf = \cos\theta$$
$$= \frac{R}{Z}$$
$$= \frac{400}{772}$$
$$= 0.5181$$

$$P_{APPARENT} = IE$$
$$= 51.8 \times 10^{-3}\,(40)$$
$$= 2.072\ \text{W (expressed in volt-ampere units)}$$

$$P_{TRUE} = IE\,pf$$
$$= 51.8 \times 10^{-3}\,(40)\,(0.5181)$$
$$= 1.075\ \text{W}$$

Remember, RMS values must be used in these calculations.   ▲

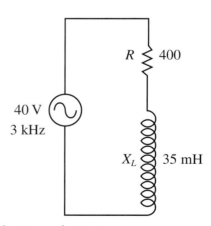

Figure 21-14. RL circuit for Example 21-4.

▷| Summary

- In an inductive circuit, one time constant is the time in which the current increases to a value 63.2% of its maximum value. After five time constants, the circuit is in a balanced state and no further change occurs.
- In a pure inductance, the current lags the source voltage by 90°. The source voltage and the cemf are 180° out of phase.

- Impedance is the total resistance toward alternating current. Impedance is represented by the letter Z and measured in ohms.
- In working calculations of parallel circuits, the voltage across the parallel is the same and calculations are made for the currents. Always consider the vector sum when working with ac.
- The power of RL circuits must be calculated for true power by using the power factor that is the cosine of the angle.

Important Terms

Do you know the meanings of these terms used in the chapter?

apparent power   time delay
impedance   transient response
power factor   true power
time constant   vector sum

Questions and Problems

Please do not write in this text. Write your answers on a separate sheet of paper.

1. What is the time constant in **Figure 21-15?**
2. What is the current in Figure 21-15 after three time constants?
3. Draw a schematic diagram of a series RL circuit using a 40 Ω resistor, a 30 Ω inductive reactance, and a 50 V source. Find impedance, current, and the voltage drops across the resistance and reactance.
4. Draw a vector diagram of problem 3.
5. What is the true power of the series RL circuit in problem 3? What is the phase angle?
6. What is the impedance of a coil with 6 V across and a current of 100 mA?
7. Draw a schematic diagram of a parallel RL circuit using an 18 V source at 200 Hz, 9 Ω resistance, and 1.4 mH inductor. What is the current through the resistor, the inductor, and the total current?
8. What is the impedance in problem 7? What is the phase angle?
9. Draw a vector diagram of problem 7.
10. Calculate the true power for problem 7.
11. An inductance has 16 V across it with a current of 40 mA. Operating at a frequency of 4 kHz, what is X_L in ohms?
12. If the frequency in problem 3 is increased, will the current increase or decrease? Explain.

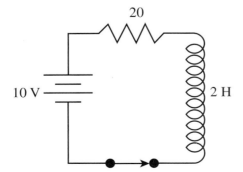

Figure 21-15. Circuit for problems 1 and 2.

Chapter 22 Graphic Overview

Transformers

- **have**
  - **Core**
    - **can be**
      - Air
      - Iron
      - Ferrite
  - **Primary** — **and** — **Secondary**
    - **is the** — Input
    - **is the** — Output
      - **depends on** — Turns Ratio
    - **may have** — More — **than** — One Secondary
- **types are**
  - Audio
  - RF — **can be** — Adjustable
  - Power
- **losses are**
  - I^2R
  - Eddy Currents
  - Magnetic Saturation
  - Hysteresis

TRANSFORMERS 22

Objectives

After studying this chapter, you will be able to:
- ○ Explain the operation of a transformer.
- ○ State transformer losses.
- ○ Calculate the efficiency, turns ratio, voltage, and current of a transformer.
- ○ Explain the difference between loaded and unloaded transformers.
- ○ Explain how to test for common transformer problems.

Introduction

A transformer changes alternating current to some other form of ac. Transformers do not change ac to dc as some people believe. Transformers do not work on direct current. A transformer must have a moving magnetic field to operate. Transformers change, or transform, ac to a higher or lower voltage. A transformer also controls the value of the current involved in the circuit. Sometimes a transformer is used to isolate a load from a power source. Transformers come in many sizes; they can be as small as a sugar cube or as large as a house.

Transformer Construction

If you take a small transformer apart, you will find two coils of wire wound on an iron core. All transformers have a primary winding, secondary winding, and core. The input to the transformer is connected to the **_primary_** and the output is taken from the **_secondary_** and connected to the load. See **Figure 22-1.** The only connection between the primary and secondary is the magnetic field of the primary.

A coil is an electromagnet and may consist of various types of core materials, such as air, iron, or ferrite, **Figure 22-2.** Air core transformers are used in high-frequency circuits. Iron cores are used in power transformers.

Primary: The input winding of a transformer.

Secondary: The output winding of a transformer.

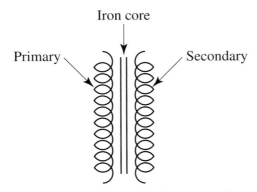

Figure 22-1. Three main parts of a transformer. If air is used as the core, the primary and secondary windings are wound on some type of form.

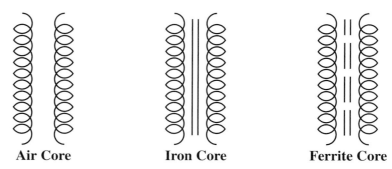

Air Core   **Iron Core**   **Ferrite Core**

Figure 22-2. Schematic symbols for the common types of transformers.

Types and Insulation

Transformers take many forms. Small transformers are used in the electronics field, **Figure 22-3. Figure 22-4** shows a large power transformer. Huge transformers are used by electrical power companies, **Figure 22-5.** A transformer can have more than one secondary, **Figure 22-6.**

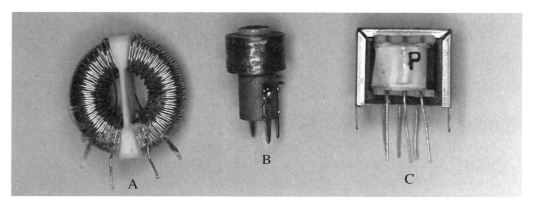

Figure 22-3. Transformers used in electronics. A—Toroid transformer used in a special industrial control circuit. B—Small, adjustable, ferrite-core transformer used in high-frequency FM radio circuits. C—Small transistor circuit transformer.

Figure 22-4. Large power transformers are used in dc power supplies. Notice the ventilation holes for cooling the transformer.

Figure 22-5. Huge transformers weighing several tons are constructed by carefully placing the coils over the cores. (McGraw-Edison Co.)

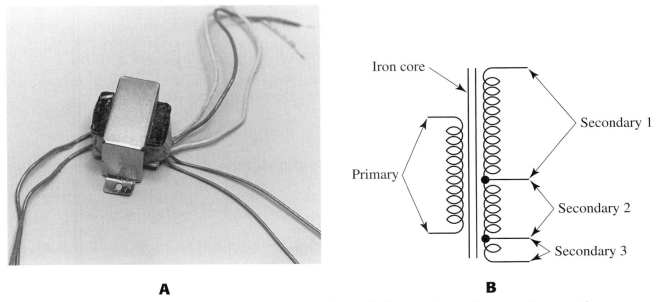

A   **B**

Figure 22-6. Transformers with more than one secondary winding. A—A transformer with more than one secondary has four or more wires. B—Schematic of a transformer with three secondaries.

Small transformers are the dry type. The only insulation on the winding conductors is varnish, so no special consideration is needed. Transformers operating above 35,000 V require a special facility for housing. Some transformers contain a nonflammable liquid that insulates and cools the windings. Other transformers are oil-insulated to protect the windings and cool the transformer. Such transformers are cooled by metal fins that radiate heat into the atmosphere. They may also have cooling fans for proper air circulation.

Transformer Operation

The transformer has an alternating current flowing in the primary which causes a magnetic field around the primary coil. The strength of this magnetic field is always changing in step with the primary current. The magnetic field of the primary cuts across the conductor of the secondary, **Figure 22-7.**

Remember, when a magnetic field cuts across a conductor, a current is induced in that conductor. Therefore, with the primary magnetic field cutting across the secondary, a current is induced in the secondary. Energy is transferred, or coupled, from the primary to the secondary only by the magnetic field. Another way to look at it is the electrical energy in the primary is converted to magnetic energy. The magnetic energy is converted back to electrical energy in the secondary. Due to losses between the input and output, the output energy is always less than the input energy.

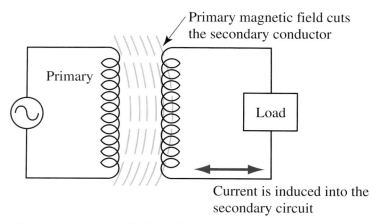

Figure 22-7. When the magnetic field of the primary cuts the conductor of the secondary, it causes current to flow in the secondary.

Transformer Losses

If a transformer were perfect, no losses would occur. Transformers are imperfect for several reasons:

- The conductor material of the windings always has some resistance. Thus, there is always an I^2R power loss.
- The core material, such as iron, is a conductor. The magnetic field cuts across the core conductor causing currents in the core called **eddy currents.** Eddy currents are a loss in energy. The energy is dissipated in the core as heat, and never recovered and delivered to the secondary. To reduce eddy current loss, the iron core is made of thin sheets of iron (laminations) insulated from each other.
- All magnetic lines of flux produced in the windings should pass through the core and on to the secondary. A certain amount of flux follows other paths, such as the surrounding air, and produces **magnetic leakage.**
- Most magnetic materials do not immediately demagnetize then magnetize in the other polarity. It takes time and energy to get the domains to change polarity and position within a material. The resulting loss of energy is called **hysteresis loss.**

Transformer manufacturers are able to make high-quality transformers with very small losses. The windings are made of conductors large enough to reduce the I^2R loss. Using different alloys for the core reduces hysteresis loss and keeps the core reluctance low. This reduces magnetic leakage.

Eddy currents: Currents induced in the core of a transformer.

Magnetic leakage: Passage of magnetic flux outside the path along which it can do useful work, causing a loss of energy.

Hysteresis loss: The power loss caused by molecular friction.

Transformer Efficiency

The output power of a transformer is equal to the input power minus the sum of all losses. If the output power of a transformer were equal to the input power, the transformer would be 100% efficient. With the losses involved, however, the output is never equal to the input. Transformer efficiency (expressed as a percentage) is equal to:

$$\text{Efficiency} = \frac{\text{Output power}}{\text{Input power}} \ (100)$$

▼ *Example 22-1:*

Find the efficiency of a transformer with an input power of 220 W and an output power of 186 W.

$$\text{Efficiency} = \frac{\text{Output power}}{\text{Input power}} \ (100)$$

$$= \frac{186}{220} \ (100)$$

$$= 85\%$$ ▲

Turns Ratio

The voltage at the secondary depends on the number of turns in the primary and secondary, and is called the ***turns ratio*** (expressed as N_{PRI}/N_{SEC}). The voltage ratio and current ratio depend on the turns ratio. The ratio proportions are:

Turns ratio: The ratio of the primary to secondary turns in a transformer.

$$\frac{E_{PRI}}{E_{SEC}} = \frac{N_{PRI}}{N_{SEC}} = \frac{I_{SEC}}{I_{PRI}}$$

The turns ratio can be found by finding a common denominator and reducing the fraction.

▼ *Example 22-2:*

A transformer has 5000 turns on the primary and 2250 turns on the secondary. What is the turns ratio?

$$\frac{N_{PRI}}{N_{SEC}} = \frac{5000}{2250} \ \text{(Divide both by 50.)}$$

$$= \frac{100}{45} \ \text{(Divide both by 5.)}$$

$$= \frac{20}{9}$$

$$= 20:9$$ ▲

▼ *Example 22-3:*

A transformer has 120 V on the primary and 12 V on the secondary. What is the turns ratio?

$$\frac{E_{PRI}}{E_{SEC}} = \frac{120\text{ V}}{12\text{ V}} \ \text{(Divide both by 12.)}$$

$$= \frac{10}{1}$$

$$= 10:1$$ ▲

▼ *Example 22-4:*

The transformer in **Figure 22-8** has a turns ratio of 20:1 and a primary voltage of 120 V. What is the secondary voltage?

$$\frac{E_{PRI}}{E_{SEC}} = \frac{N_{PRI}}{N_{SEC}}$$

$$\frac{120}{E_{SEC}} = \frac{20}{1} \quad \text{(Put the numbers into the equation.)}$$

$$20\,E_{SEC} = 120 \quad \text{(Cross multiply and divide.)}$$

$$E_{SEC} = 6\text{ V} \quad ▲$$

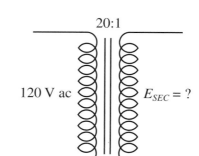

Figure 22-8. Transformer for Example 22-4.

▼ *Example 22-5:*

A transformer has a primary voltage of 120 V and primary current of 2 A. If the secondary voltage is 6 V, what is the secondary current?

$$\frac{E_{PRI}}{E_{SEC}} = \frac{I_{SEC}}{I_{PRI}}$$

$$\frac{120}{6} = \frac{I_{SEC}}{2}$$

$$6\,I_{SEC} = 240$$

$$I_{SEC} = 40\text{ A} \quad ▲$$

If the voltage is increased, the transformer is called a ***step-up transformer.*** If the voltage is reduced, it is called a ***step-down transformer.***

Autotransformer

An ***autotransformer*** consists of one continuous winding with a tapped connection somewhere in the length of the coil winding, **Figure 22-9.** A single winding makes the autotransformer smaller, more efficient, and inexpensive. The tap of a step-up transformer is at 50 turns of the complete winding, and the secondary has all 1200 turns. A step-down transformer uses all the transformer turns for the primary. See **Figure 22-10.**

Unloaded and Loaded Actions

In **Figure 22-11,** the load on the secondary is disconnected or open. No power is dissipated by the secondary. The no-load primary current, called the ***excitation current,*** is very small. It is usually less than 3% of the rated current. The excitation current that supplies the power to the losses in the transformer, and the power stored in the magnetizing field, is returned to the circuit (minus any loss).

Step-up transformer: A transformer that has more turns in the secondary than in the primary.

Step-down transformer: A transformer that has fewer turns in the secondary than in the primary.

Autotransformer: A single winding transformer with a tapped point that operates as both a primary and secondary.

Excitation current: The current that flows in the primary of a transformer when the secondary is open.

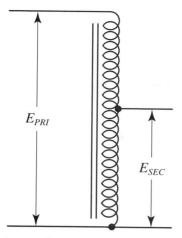

Figure 22-9. An autotransformer has one winding that acts as the primary and secondary.

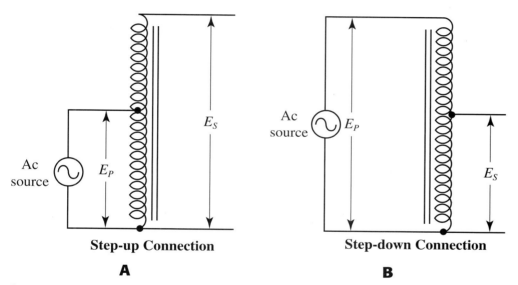

Figure 22-10. Autotransformers. A—A step-up autotransformer has fewer turns in the primary. B—A step-down autotransformer has fewer turns in the secondary. Both step-up and step-down have only three connections.

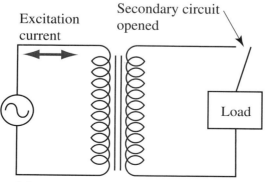

Figure 22-11. Because of the inductive characteristics of the primary, only an excitation current flows in an unloaded transformer.

If the secondary load is a resistance, the secondary current can be found by Ohm's law, **Figure 22-12.** The secondary current is dependent on the resistance of the load. If the secondary current increases, the primary current increases to deliver the power required by the secondary. As the power requirements of the transformer increase, the temperature of the transformer also increases. An overheated transformer usually indicates overloading.

If the load includes an inductive load, the current is placed out of phase. The true power depends on the value of the phase angle.

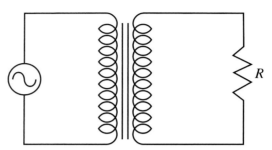

Figure 22-12. Transformer with a resistive load.

Impedance Matching

Sometimes transformers are used to provide impedance matching between two circuits or isolation of one circuit from another. Remember, the maximum power is transferred when the output impedance is equal to the load impedance. These types of transformers may be called *coupling transformers.* They connect one circuit to another circuit. The impedance ratio of transformers is related to the turns ratio by the equation:

Coupling transformer: A transformer used to connect one circuit to another.

$$\frac{N_{PRI}}{N_{SEC}} = \sqrt{\frac{Z_{PRI}}{Z_{SEC}}}$$

▼ *Example 22-6:*

If a transformer has a turns ratio of 5:2 and the primary impedance is 1000 Ω, what is the secondary impedance?

$$\frac{N_{PRI}}{N_{SEC}} = \sqrt{\frac{Z_{PRI}}{Z_{SEC}}}$$

$$\frac{5}{2} = \sqrt{\frac{1000}{Z_{SEC}}}$$

$$2.5 = \sqrt{\frac{1000}{Z_{SEC}}}$$

$$6.25 = \frac{1000}{Z_{SEC}}$$

$$6.25\, Z_{SEC} = 1000$$

$$Z_{SEC} = 160\ \Omega$$ ▲

Isolation

A transformer secondary is not electrically connected to the primary. This provides isolation between the two circuits and reduces the shock hazard. Another advantage of the isolated secondary is blockage of any dc on the primary. Sometimes a transformer with a 1:1 turns ratio is used for isolation from the 60 Hz power. The autotransformer does not provide this circuit isolation.

Magnetic Shielding

Preventing the magnetic field of one component from affecting another is called *magnetic shielding.* In a coaxial cable, the outer braided conductor shields the inner conductor. In radio frequency circuits or delicate instruments, the circuitry often needs shielding from magnetic fields. High-frequency transformers are often placed in a metal container to provide magnetic shielding to the surrounding circuits, **Figure 22-13.**

Magnetic shielding: A metallic enclosure to protect devices or systems against magnetic fields.

Figure 22-13. RF transformer with an aluminum cover for shielding magnetic fields. The metal shield acts as a short circuit to the lines of magnetic flux.

Troubleshooting Transformer Problems

The main problems with transformers are open windings and overheating. Open windings cause complete failure of the circuits. Sometimes a conductor will break at the transformer connections. This is usually caused by technicians subjecting the terminal to physical abuse. Overheating can be caused by an overload condition, a shorted winding, or improper ventilation.

Testing

If the transformer is inoperative, a voltage test of the primary should be made first. A voltage test at the secondaries can also be made to determine if the transformer has the correct output voltage. If it is more convenient or safe, a continuity test of the primary and secondary can be performed. When making a continuity test, it may be necessary to disconnect all connections from the transformer winding. Disconnecting will prevent reading circuits that are in parallel with the transformer. Depending on the number of turns and wire size, a resistance check made on RF-type transformers may have resistance values of less than 1 to 20 ohms. Transformers used in 60 Hz power and audio may have a resistance from 10 Ω to 500 Ω. In the case of an auto-transformer, check from one of the transformer leads to the other two leads.

 Warning: Use extreme caution when testing transformers. By applying a small test voltage to the primary, you may find a dangerously high voltage present at the secondary. Even a low ohms/volt VOM can deliver enough current to a primary to make sufficient voltage on the secondary to produce a dangerous shock.

Shorted winding problems can be troublesome to analyze. Repairing an internal short in a winding is not economical, so the transformer should be replaced. A replacement transformer should have the same part number as the used one. Often this is not possible and a substitute must be located. Factors to consider when replacing a transformer include:

- Primary voltage and current
- Secondary voltage and current
- Turns ratio
- Power capability
- Core material
- Primary and secondary impedance
- Package and style
- Operating frequency
- Weight
- Environmental considerations

A "ringing" test can be used for a shorted winding. For this test, the transformer must be disconnected from the circuits and connected to a test circuit, **Figure 22-14.** Depending on the type of transformer and frequency, the ringing test produces waveforms similar to those in **Figures 22-15** and **22-16.**

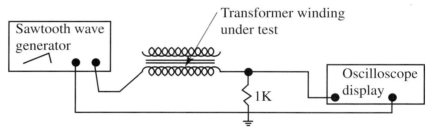

Figure 22-14. Ringing test circuit.

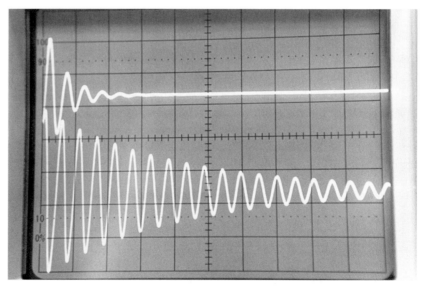

Figure 22-15. Ringing test waveforms. The bottom waveform indicates a good transformer, while the top waveform indicates a defective transformer.

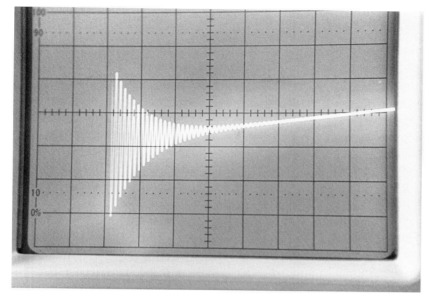

Figure 22-16. The ringing test waveform can have many variations in appearance.

Another possible short is a leakage path occurring from one of the windings to the core. With the chassis ungrounded, a leakage path can pose great danger. To check this condition, disconnect all the secondaries. Take a voltage measurement from the core or frame of the transformer to each of the primary connections. Then, take a voltage measurement from the core to each of the secondaries. A resistance measurement from the windings to the core should indicate infinity. However, this does not test under the actual working voltage and a leakage may occur.

> **Warning:** A leakage path can cause strange, unexplainable, even illogical symptoms in equipment and deliver a dangerous electrical shock.

Summary

- The basic parts of a transformer are the primary, secondary, and core. The core material may be air, iron, or ferrite.
- A transformer operates by means of mutual inductance.
- Transformer losses consist of I2R power loss, eddy currents, magnetic leakage, and hysteresis loss.
- Output voltage, current, and impedance are determined by the turns ratio of the transformer.
- A transformer can step up or step down the voltage applied to the primary.
- The autotransformer consists of a single-tapped winding.
- Transformer characteristics include primary voltage and current, secondary voltage and current, turns ratio, power capability, impedance, package, and style.
- Coupling transformers connect one circuit to another by matching their impedances.
- The main problems with transformers are open windings and overheating.
- Dangerous voltages can be present when working with transformers. Always use caution.

⇥ Important Terms

Do you know the meanings of these terms used in the chapter?

autotransformer
coupling transformer
eddy currents
excitation current
hysteresis loss
magnetic leakage
magnetic shielding

primary
primary step-up transformer
secondary
secondary step-down transformer
step-down transformer
step-up transformer
turns ratio

⇥ Questions and Problems

Please do not write in this text. Write your answers on a separate sheet of paper.

1. Transformers do *not* work on what type of current?
2. The input winding of a transformer is called the _____.
3. Transformer cores may be _____, _____, or _____.
4. List four types of transformer losses.
5. Why are the cores of transformers constructed of laminated sheets of iron?
6. What is the efficiency of the transformer in **Figure 22-17?**
7. If a transformer has 250 turns in the primary and 75 turns in the secondary, what is the turns ratio?
8. A transformer has a turns ratio of 3:1 and the secondary has 205 turns. How many turns are on the primary?
9. A transformer with a turns ratio of 3:2 has 120 V on the primary. What is the secondary voltage?
10. In problem 9, the primary current in the transformer is 1.5 A. What is the secondary current?
11. Calculate the secondary voltage of the autotransformer of **Figure 22-18.** If the secondary current is 0.6 A, what is the efficiency of the transformer?
12. The excitation current of a transformer is usually less than _____ of the rated current.
13. If a transformer has a turns ratio of 3:2 and the primary impedance is 500 Ω, what is the secondary impedance?
14. Explain the purpose of an isolation transformer.
15. Why is a coil or transformer placed inside an aluminum cover?
16. The main problems with transformers are _____ and _____.
17. Why must extreme caution be used when testing transformers?
18. What is the resistance of an open primary winding?

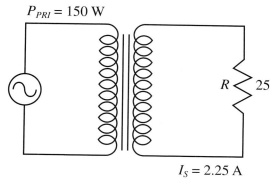

$P_{PRI} = 150$ W

R ⟩ 25

$I_S = 2.25$ A

Figure 22-17. Circuit for problem 6.

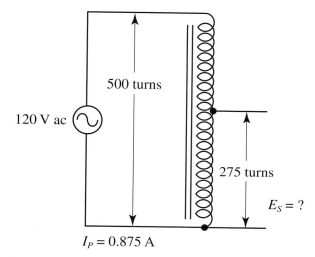

500 turns

120 V ac

275 turns

$E_S = ?$

$I_P = 0.875$ A

Figure 22-18. Circuit for problem 11.

Chapter 23 Graphic Overview

Capacitors
- problems are: Shorted, Leakage, Open, Changed Value
- specifications are: Type, Capacitance, Working Voltage, Tolerance, Physical Size, Package, Lead Arrangement
- have: Capacitive Reactance
- types are:
  - Polarized → are: Electrolytic, Tantalum
  - Unpolarized → are: Ceramic, Mica, Glass, Paper, Plastic, Air (usually Variable)
- have: Capacitance → depends on: Area of Plates, Distance between Plates, Type of Dielectric
- have: Two Metal Plates and Insulator called a Dielectric
  - with: Charge has Electrostatic Field

CAPACITANCE

➤❙ Objectives

After studying this chapter, you will be able to:
- ○ Calculate the charge on a capacitor.
- ○ Indicate the factors that determine the amount of capacitance.
- ○ Calculate the total capacitance of capacitors in parallel and in series.
- ○ List the various types of polarized and unpolarized capacitors.
- ○ Calculate capacitive reactance.
- ○ Identify capacitors by code markings.
- ○ Describe common problems found with capacitors.

➤❙ Introduction

Just as an inductance has characteristics for current variations, capacitance has similar but opposite characteristics for voltage variations across an insulator. A capacitor stores energy in the form of a charge in coulombs of electrons. Capacitors are used in circuits as RC filters, waveform shaping circuits, timing circuits, power supply filters, coupling, and bypass capacitors in amplifiers and power supplies. Capacitors used to be called condensers. The term is still used in some applications, such as automotive electrical systems.

Capacitor Operation

In a capacitor, two metal plates are separated by an insulator called a *dielectric,* **Figure 23-1.** The plates are usually made of metal foil. The dielectric can be ceramic, paper, mica, glass, Mylar™, air, or some other insulating material.

Dielectric: The insulation between the plates of a capacitor.

As you know, an inductor stores energy in the form of magnetism. A capacitor stores energy in the form of electrons. The plates of the capacitor are connected to a battery. The voltage pushes electrons onto one plate, making it negative, and takes electrons away from the other plate, making it positive (charged). If the voltage of the source is increased, more electrons are pushed onto the capacitor, and the amount of charge is increased. See **Figure 23-2.**

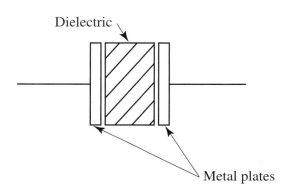

Figure 23-1. Basic construction of a capacitor.

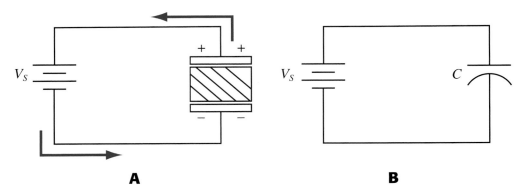

Figure 23-2. A–Dc source charging of a capacitor. B–Schematic diagram of a capacitor circuit.

Capacitance is the ability of a device to store an electrical charge. It is measured in farads (represented by *F*). A capacitance of one *farad* of charge is stored when one coulomb with a potential difference of one volt is on the capacitor.

The amount of charge on the capacitor is a measure of the number of electrons in coulombs (represented by *C*). The amount of charge is directly proportional to the voltage and capacitance. The charge in coulombs is calculated by the equation:

$$Q = CV$$

Where:
Q = Charge (in coulombs)
C = Capacitance (in farads)
V = Voltage across the capacitor

▼ *Example 23-1:*

How much charge is stored in a 10 µF capacitor with 50 V across?

$Q = CV$
$\quad = (10 \times 10^{-6})\, 50$
$\quad = 500$ µC (microcoulombs)

Note values for common capacitors are measured in microfarads or picofarads.   ▲

Discharging a Capacitor

In **Figure 23-3,** if the capacitor is disconnected from the source by opening switch 1, the charge will remain on the capacitor. If switch 2 is closed, it will make a path for the electrons stored on the negative plate to flow to the positive plate. The capacitor then discharges. When an equal number of electrons exists on both plates, the capacitor is discharged and the potential difference across the capacitor is zero.

Electrostatic Fields

When a voltage exists across a capacitor, the opposite polarity plates of the capacitor are attracted to each other. This attraction is seen as dielectric lines of force, **Figure 23-4.** Like magnetic lines of force, the dielectric lines of force may go through almost any type of material. Dielectric lines of force are also known as electrostatic lines of force. Just as an electrical current produces a magnetic field, an electrical field is produced by a voltage. The electrostatic lines, which begin at one plate and end at the other plate, are straight. This group of electrostatic lines makes up an *electrostatic field.*

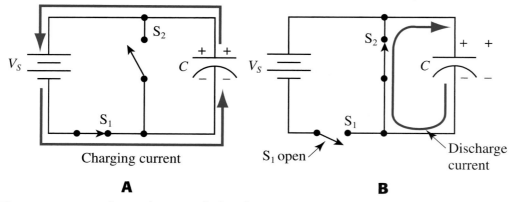

Figure 23-3. A—When S₁ is opened, the charge remains on the capacitor. B—Closing S₂ discharges the capacitor.

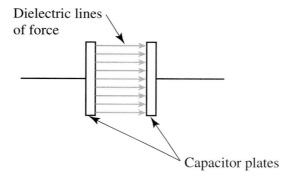

Figure 23-4. A charged capacitor has an electrostatic field between the plates.

A material that easily conducts the dielectric lines of force has a high *dielectric constant.* The dielectric constant measures a dielectric's ability to establish an electrostatic field. A vacuum is the least effective dielectric and has a dielectric constant of 1. This is the reference point for other materials. See **Figure 23-5.**

The dielectric must be an insulator. It must not conduct electrons. If electrons do pass through the dielectric, an undesirable *leakage current* is created.

Factors That Determine Capacitance

Three physical factors determine the capacitance of a capacitor. See **Figure 23-6.**
- *Area of the plates.* The larger the plate area, the greater the capacitance.
- *Distance between the plates.* The closer the plates are to each other, the greater the capacitance.
- *Type of dielectric material.* The amount of electrostatic energy stored in a dielectric is determined by the dielectric material. For example, glass has a dielectric constant of 7.5; therefore, glass can cause an electrostatic field 7.5 times as strong as a vacuum. Since capacitance is proportional to the dielectric constant, the glass capacitor will have a capacitance 7.5 times the same size vacuum dielectric capacitor.

Dielectric constant: In a capacitor, the ratio of capacitance with a given dielectric to one with a dielectric of air.

Leakage current: An undesirable small current that flows between two points.

Material	Dielectric Constant
Vacuum	1.0
Air	1.0006
Teflon™	2.0
Polyethylene	2.3
Paper, waxed	2.5
Polystyrene	2.6
Polyester	4.0
Polycarbonate	4.5
Mica	5.0
Ceramic (low *K*)	6.0 (varies)
Porcelain	6.5
Glass	7.5
Aluminum oxide	8.4
Tantalum oxide	26
Ceramic (high *K)*	7500 (varies widely)

Figure 23-5. Table of dielectric constants.

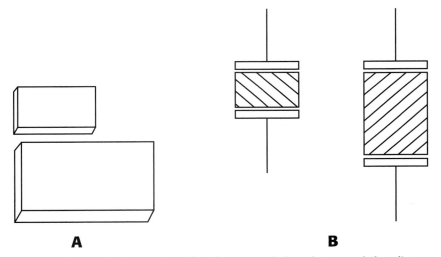

A   **B**

Figure 23-6. Capacitance is determined by the area of the plates and the distance between them. A—A larger plate area has a higher capacitance than a smaller plate. B—Increasing the distance between the plates of a capacitor decreases capacitance.

Capacitance is equal to:

$$C = \frac{(8.85 \times 10^{-12}) \times K \times A}{d}$$

Where:
C = Capacitance (in farads)
8.85×10^{-12} = Absolute permitivity constant
K = Dielectric constant
A = Plate areas (in square meters)
d = Distance (in meters) between plates

▼ ***Example 23-2:***
Calculate the capacitance for two plates, each with an area of 4 m² separated by 1.5 cm and a glass dielectric.

$$C = \frac{(8.85 \times 10^{-12}) \times K \times A}{d}$$

$$= \frac{(8.85 \times 10^{-12}) \times (7.5) \times (4)}{1.5 \times 10^{-2}}$$

$$= 17.7 \ \mu F \text{ (microfarads)}$$ ▲

Total Capacitance

The equations for total capacitance are the *reverse* of the equations for total resistance. The equation for series resistance is the same as for parallel capacitors.

Capacitors in Parallel

The total capacitance of capacitors connected in parallel is the sum of all the individual capacitors. The equation is:

$$C_T = C_1 + C_2 + C_3$$

▼ ***Example 23-3:***
In **Figure 23-7,** find the total capacitance of the 5 μF, 10 μF, and 50 μF capacitors connected in parallel.

$$C_T = C_1 + C_2 + C_3$$
$$= 5 + 10 + 50$$
$$= 65 \ \mu F$$ ▲

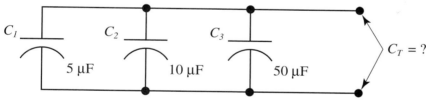

Figure 23-7. The total capacitance of capacitors in parallel is equal to the sum of the individual capacitors.

Capacitors in Series

The equations for total capacitance in series are the opposite of those for resistors in parallel. For two capacitors connected in series, the total capacitance is found by using the product over the sum equation:

$$C_T = \frac{C_1 C_2}{C_1 + C_2}$$

▼ ***Example 23-4:***
In **Figure 23-8,** find the total capacitance of a 20 μF and a 30 μF capacitor connected in series.

$$C_T = \frac{C_1 C_2}{C_1 + C_2}$$

$$= \frac{(20) \ 30}{20 + 30}$$

$$= 12 \ \mu F$$

Since all values are in microfarads, the 20 + 30 powers of ten can be ignored and the answer is expressed in microfarads.   ▲

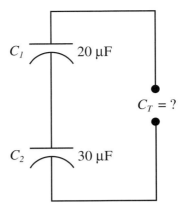

Figure 23-8. Example 23-4. Calculating the total capacitance of two capacitors in series.

For three or more capacitors in series, use the reciprocal of the sum of the reciprocals equation:

$$C_T = \frac{1}{\dfrac{1}{C_1} + \dfrac{1}{C_2} + \dfrac{1}{C_3}}$$

▼ *Example 23-5:*

In **Figure 23-9,** find the total capacitance of three capacitors with values of 0.1 µF, 0.01 µF, and 0.2 µF connected in series.

$$C_T = \frac{1}{\dfrac{1}{C_1} + \dfrac{1}{C_2} + \dfrac{1}{C_3}}$$

$$= \frac{1}{\dfrac{1}{(0.1 \times 10^{-6})} + \dfrac{1}{(0.01 \times 10^{-6})} + \dfrac{1}{(0.2 \times 10^{-6})}}$$

$$= 8.7 \text{ nF or } 8700 \text{ pF} \qquad \blacktriangle$$

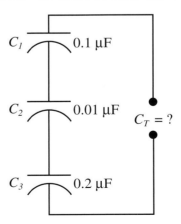

Figure 23-9. Example 23-5. Calculating the total capacitance of three capacitors in series.

Types of Capacitors

Capacitors are classified by the type of dielectric in the capacitor. They are also classified as unpolarized or polarized. As with resistors, capacitors can be a fixed or variable in value.

Unpolarized Capacitors

Various types of unpolarized capacitors are available, **Figure 23-10.** In molded ceramic capacitors, the foil plates are separated by ceramic material, stacked together, and covered with a molded plastic case. See **Figure 23-11.**

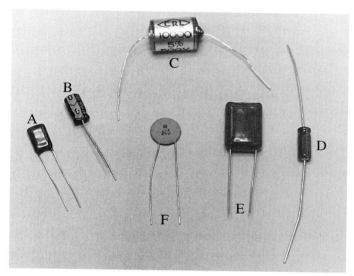

Figure 23-10. Various types of capacitors. A—Mylar™. B—Electrolytic (radial leads). C—Mylar™. D—Electrolytic (axial leads). E—Polyester. F—Ceramic disk.

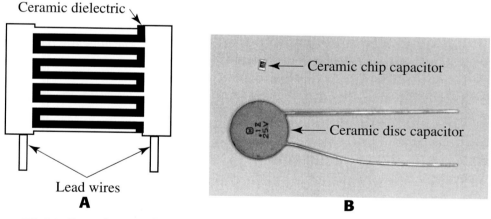

Figure 23-11. Ceramic capacitors. A—Construction of a molded ceramic capacitor. B—Surface mounted ceramic chip capacitor (top). A common ceramic disc capacitor (bottom) is approximately the size of a dime.

In mica capacitors, the foil plates are stacked and separated by sheets of mica, **Figure 23-12.** The capacitor package can be a dipped or molded plastic.

Glass capacitors are constructed in the same way as ceramic or mica capacitors. However, glass capacitors have a higher dielectric constant.

In a paper capacitor, two long strips of foil are separated by a wax paper dielectric, **Figure 23-13.** The two strips are rolled into a cylinder shape and given a molded or dipped plastic case.

Plastic capacitors are constructed like paper capacitors. A plastic film of Mylar™, polycarbonate, Teflon™, or polypropylene is used as the dielectric, **Figure 23-14.** The plastic capacitor has nearly replaced the paper capacitor in common use.

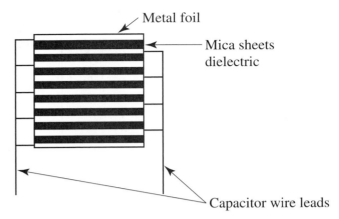

Figure 23-12. Construction of mica capacitors. Layers of metal foil are separated by sheets of mica.

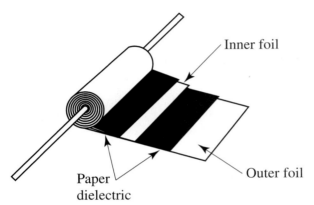

Figure 23-13. Construction of paper capacitors. Layers of metal foil are separated by paper and wound to form a tubular component.

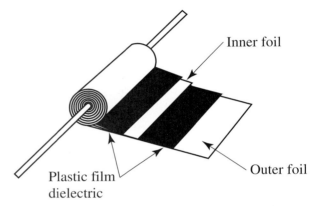

Figure 23-14. Construction of plastic capacitors. Constructed like paper capacitors, the paper is replaced with plastic.

Air dielectric capacitors are usually the variable type, **Figure 23-15.** The movable set of plates is called the rotor, and the stationary set of plates is called the stator. The capacitance is varied by rotating the rotor, which adjusts the effective plate area and the amount of capacitance. When the rotor plates are completely out, the capacitance is minimum. Likewise, when the rotor plates are completely in, the capacitance is maximum.

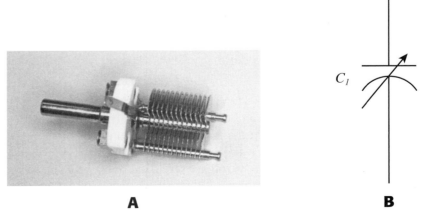

A   **B**

Figure 23-15. Air variable capacitor. A—These types of capacitors use a ceramic frame for mounting the capacitor plate assembly. B—Schematic symbol for this type of capacitor.

Trimmer capacitors are a type of variable capacitor used to adjust (trim) the capacitance of a circuit for proper circuit operation, **Figure 23-16.** The dielectric is often mica, ceramic, or plastic. A plate is moved by a simple screw adjustment.

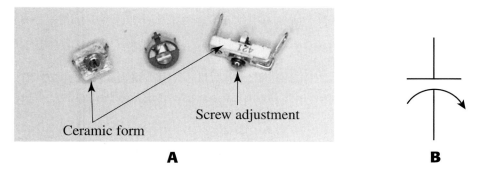

A   **B**

Figure 23-16. Trimmer capacitors. A—These types of capacitors are used in high-frequency circuits. Adjustment usually requires a nonmetallic screwdriver. B—Schematic symbol for trimmer capacitor.

Polarized Capacitors

Electrolytic and tantalum are two common types of polarized capacitors, **Figure 23-17.** Electrolytic capacitors have various packages, **Figure 23-18.** The basic types are axial lead, radial lead, and metal can.

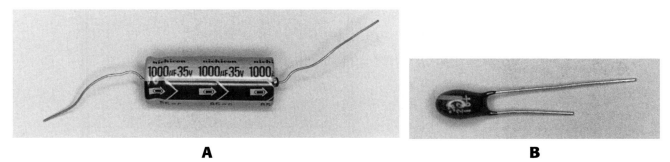

A   **B**

Figure 23-17. Polarized capacitors. A—Electrolytic capacitor. B—Tantalum capacitor.

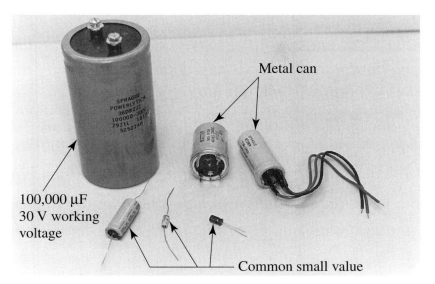

Figure 23-18. Various electrolytic capacitor packages. High-capacitance capacitors (top row) are used in dc power supplies and can produce a dangerous shock. Discharge these types of capacitors before working on equipment.

A liquid-soaked (electrolyte) gauze separates two aluminum foils. The electrolyte reacts with the aluminum to form a thin oxide layer, which becomes the dielectric. Positive and negative polarity is the most important consideration when connecting an electrolytic capacitor.

 Warning: If the capacitor is connected in the wrong polarity, it will heat up, moisture inside the capacitor will turn to steam, and the capacitor will explode.

The tantalum capacitor uses the metal tantalum instead of aluminum. Although the tantalum capacitor is more expensive than the aluminum type, it has some advantages:
- Better leak-resistance
- Better temperature stability
- Smaller size
- Longer life

Capacitor Symbols

Figure 23-19 shows the various schematic symbols for capacitors. Some industries use their own symbols. The industrial symbol for a normally open relay contact can be mistaken for a capacitor, **Figure 23-20.**

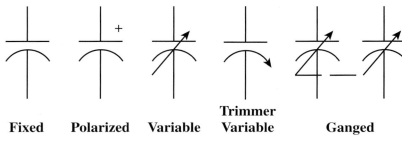

| Fixed | Polarized | Variable | Trimmer Variable | Ganged |

Figure 23-19. Capacitor schematic symbols. Ganged means two or more capacitors are adjusted by the same shaft at the same time.

Figure 23-20. Industrial symbol for an open relay contact.

Capacitive Reactance

A capacitor causes a delay of the voltage in a circuit. As in inductance, this is a reactance to the flow of current-voltage. It is called ***capacitive reactance*** (represented by X_C). Capacitive reactance is measured in ohms and found by the equation:

Capacitive reactance: The opposition or resistance to alternating current by a capacitor or component that has a capacitive quality.

$$X_C = \frac{1}{2\pi fC}$$

▼ *Example 23-6:*

Find the capacitive reactance of a capacitance of 0.01 μF and a frequency of 1500 Hz.

$$X_C = \frac{1}{2\pi fC}$$

$$= \frac{1}{2 \,(3.14) \,(1500) \,(0.01 \times 10^{-6})}$$

$$= 10,616 \ \Omega$$ ▲

Capacitive Ac Circuit Operation

In a purely capacitive ac circuit, the current leads the voltage by 90°, **Figure 23-21A.** In the inductive circuit, the current lagged the voltage by 90°. The circuit action is opposite for the two quantities, inductance and capacitance.

In **Figure 23-21B,** the ac current is flowing, making the capacitor plates charge first in one direction, then in the other. An ammeter placed in the circuit would suggest a current flows through the capacitor. The capacitor *blocks dc* and *passes ac.* With a dielectric between the plates of a capacitor, a current flowing through the capacitor is not possible under normal working conditions. However, to the external circuit it appears an ac current does flow through the capacitor.

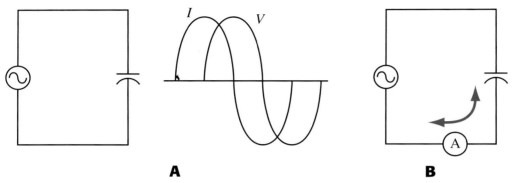

A   **B**

Figure 23-21. Capacitive ac circuit. A—Capacitor connected to an ac source. Note the voltage-to-current phase relationship. B—An ammeter indicates current flowing in the capacitor circuit.

Stray Capacitance

A capacitor is a lumped capacitance purposely placed in the circuit. Capacitance is also present in a circuit as *stray capacitance* due to the physical construction of the circuit, **Figure 23-22.** Since a capacitor consists of two pieces of metal with an insulator between them, two wires running beside each other form capacitors along the length of the wire.

Stray capacitance: Capacitance introduced into a circuit by the leads and wires connecting the circuit components.

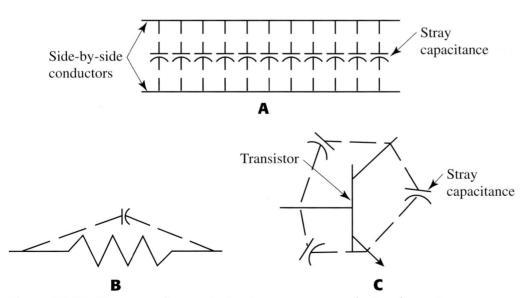

Figure 23-22. Stray capacitance. A—Any two or more conductors have stray capacitance between them. B—Even resistors have a stray capacitance between their leads. C—Stray capacitance of a transistor circuit.

Stray capacitance is also present from the wire to the chassis and is found between the leads of the components. Stray capacitance is more of a problem in high-frequency circuits.

Capacitor Specifications

The capacitor manufacturing industry lacks standardization in marking capacitors. However, certain specifications are required for using capacitors in electronic circuits.

- *Type of dielectric*. The type of dielectric should be the same for any replacement capacitor. Characteristics such as voltage rating, tolerance, temperature coefficient, and temperature range are unique to each dielectric.
- *Capacitance value*. The capacitance of a capacitor is specified in microfarads (μF) or picofarads (pF). A replacement capacitor should have the same capacitance value as the original.
- *Tolerance*. The tolerance of the capacitor is plus or minus a percentage of the rated value at a particular temperature. This specification is not important at low frequencies; however, at high frequencies it is a major factor in proper operation of the circuit.
- *Working voltage*. This rating specifies the maximum voltage that can be applied across the plates without puncturing the dielectric. The voltage given is for dc and shown as DCWV. An ac voltage must be lower due to the internal heat produced by the continuous charging and discharging of the capacitor. A replacement capacitor should have a voltage rating equal or greater than the original for the safety and long life of the capacitor.
- *Lead arrangement*. The typical lead arrangements are axial, radial, or surface mount device (SMD).
- *Temperature characteristics*. Specifications for operating temperature and temperature coefficient are a concern at high frequencies and when the equipment is to be operated in extremely low or high temperatures.

Code Markings

Code markings are used to indicate the capacitance and tolerance of capacitors. The code markings on the capacitor in **Figure 23-23** provide the following information:

- Electronics Industries Association (EIA) style number Z5U (used to identify a replacement).
- Capacitance of 5000 pF.
- Tolerance of ±20% as indicated by the letter M, **Figure 23-24.**
- Voltage rating of 1 KV.

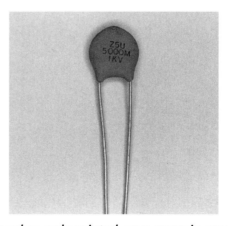

Figure 23-23. Capacitor value code printed on a ceramic capacitor.

Letter Code	Tolerance
F	1
G	2
H	3
J	5
K	10
M	20
Z	+80, −20
P	+100, −0

Note: Tolerance of +100, −0 is referred to as Guaranteed Minimum Value (GMV).

Figure 23-24. Table of tolerance codes by the Electronics Industries Association (EIA).

Code markings are not standardized. The capacitance and tolerance values of the ceramic capacitor in **Figure 23-25** follow EIA standards:
- Capacitance value is indicated by the three-digit number 104 indicates the capacitor is 100,000 picofarads (meaning one, zero, and four more zeros). The first number is the first digit value, the second number is the second digit value, and the third number is the multiplier (number of zeros). If the first number is a zero, the capacitance is given in microfarads. If the first number is not zero, capacitance is given in picofarads.
- Tolerance is indicated by the letter Z. (Refer to the table in Figure 23-24.)

Other specifications found by referencing EIA manuals include:
- Lowest usable temperature indicated by a single letter.
- Highest usable temperature indicated by a single letter.
- The highest possible percentage change in capacitance over the usable temperature range, indicated by a single letter.

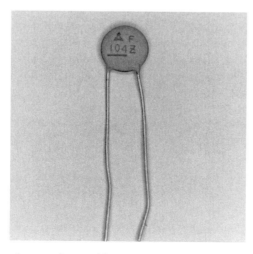

Figure 23-25. EIA capacitor code markings.

Capacitor Problems

Four types of capacitor problems may occur: shorted, leaking, open, or changed value. A capacitor can develop a direct short circuit between the plates in which the dielectric has been damaged. This defect can be verified by an ohmmeter test.

Leakage in a capacitor is caused by a high-resistance path from one plate to the other through the dielectric. (Low resistance is considered a short). This problem is most common in electrolytic capacitors. An ohmmeter test may not show this defect because the ohmmeter does not provide the working voltage necessary to push the electrons through the dielectric. In that case, a capacitor tester or a test circuit that provides the proper high-voltage test for the capacitor should be used.

An open capacitor is not uncommon. This problem is usually caused when the capacitor leads disconnect from the plates. A high-capacitance capacitor can be tested by placing an ohmmeter across the capacitor leads. If the capacitor is good, a constant increase in resistance will be seen as the capacitor charges. A low-capacitance capacitor can be tested by connecting the capacitor in series with a signal generator. A meter or oscilloscope will indicate if the signal is able to pass through the capacitor.

A capacitor can change value in capacitance, especially over time. A change in value is not a problem in most circuits. However, in high frequency circuits a change in capacitance value can be a major problem. Because of the relaxation of the dielectric material, ceramic capacitors can change value as much as 15% during the first 12 to 18 months in a circuit. Electrolytic capacitors change value as their electrolytic solution dries out. A capacitor tester or analyzer can be used to check a capacitor for the correct value, or a known good capacitor can be substituted for one suspected of having changed value.

Some capacitor problems occur when the capacitor temperature changes. To isolate this type of problem, a heat gun can be used to increase the temperature of the capacitor, or a freeze spray can be used to lower the temperature. However, excessive use of heat or cold can take the capacitor beyond normal limits.

Selecting a Replacement Capacitor

Any replacement capacitor should be the same type and value as the used one. Other factors to consider are:
- Working voltage
- Tolerance
- Physical size
- Package and lead arrangement

 Warning: Electrolytic capacitors can store a dangerous charge. Use a shorting wire to safely discharge the capacitors before working on the circuits. A shorting wire has a resistor to limit the discharge current. Allow sufficient time for the capacitor to fully discharge before beginning to work on the equipment.

 # Summary

- A capacitor consists of two metal plates separated by an insulator called a dielectric. An electrostatic field exists between the metal plates when a charge is present on the capacitor.
- The capacitance of a capacitor is determined by the area of the plates, distance between the plates, and type of dielectric material.

- Capacitors are named for their type of dielectric material: ceramic, mica, glass, paper, air, or Mylar.
- Electrolytic and tantalum capacitors are polarized and must be connected in the circuit with the proper polarity.
- The resistance of a capacitor toward ac is called capacitive reactance and is measured in ohms.
- In a purely capacitive circuit, the current leads the voltage by 90°.
- Conductors separated by an insulation have some stray capacitance.
- Capacitor problems may be described as shorted, open, leaky, or changed value.
- When selecting a replacement capacitor, consider the type, capacitance value, and working voltage.

Important Terms

Do you know the meanings of these terms used in the chapter?

capacitance
capacitive reactance
dielectric
dielectric constant

electrostatic field
farad
leakage current
stray capacitance

Questions and Problems

Please do not write in this text. Write your answers on a separate sheet of paper.

1. The nonconducting material between the plates of a capacitor is called a(n)_____.
2. What is the maximum charge in coulombs of a capacitor having a capacitance of 4.7 µF with a voltage of 50 V?
3. When electrons go through the dielectric, it is called a(n) _____ current.
4. What factors determine the capacitance of a capacitor?
5. If a 60 µF and 40 µF capacitor are connected in series, what is the total capacitance?
6. What is the total capacitance of **Figure 23-26?**
7. A glass capacitor has a plate area of 2 m² and is separated by 0.75 cm. What is its capacitance?
8. List the various types of capacitors.
9. What types of capacitors are polarized?
10. A capacitor _____ dc and _____ ac.
11. A capacitor has a capacitance of 0.001 µF and operated at a frequency of 1500 Hz. What is the capacitive reactance?
12. What are the various capacitor problems?

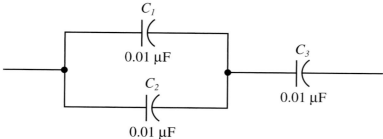

C_1

0.01 µF

C_2

0.01 µF

C_3

0.01 µF

Figure 23-26. Circuit for problem 6.

Chapter 24 Graphic Overview

RC Circuits

- **is** → Differentiator — **or** — Integrator
- **power** → Calculation — **must use** → RMS Values — **for** → Apparent Power — **and** — Power Factor — **for** — True Power
- **can be** →
  - Parallel — **use** → Current Calculations
  - Series — **use** → Voltage Calculations
  - Current Calculations and Voltage Calculations — **using** → Vector Sum
- **has** → Time Delay — **of** — 63.2% — **of** — Maximum Current — **is** → One Time Constant — **equals** → $T = RC$

RC CIRCUITS — 24

Objectives

After studying this chapter, you will be able to:
- ○ Calculate the time constant of a resistor-capacitor circuit.
- ○ Calculate the current, impedance, phase angle, and voltage drops of a series RC circuit.
- ○ Draw a vector diagram of an RC circuit.
- ○ Calculate the currents, impedance, and phase angle of a parallel RC circuit.
- ○ Calculate the power of series and parallel RC circuits.

Introduction

As with RL circuits, resistor-capacitor (RC) circuits also cause the voltage and current to be out of phase. This chapter analyzes circuits that combine resistance and capacitive reactance. Capacitors are combined with resistors and inductors to form electronic circuits used in various types of equipment.

RC Time Constant

One time constant is the time required for the voltage or current to obtain a value 63.2% of the maximum value. A capacitor fully charges or discharges during five time constants. The actual time length depends on the resistance and capacitance in the circuit.

An **RC time constant** is calculated by the equation:

$$t = RC$$

The time constant of the line graph in **Figure 24-1** can be for any value of resistance and capacitance. The voltage or current will charge or discharge to the same percentage, but the time it takes will change depending on the RC product.

RC time constant:
The product of resistance times the capacitance of a resistor-capacitor circuit ($t = RC$). The time required for the current or voltage to decrease or increase 63.2% of maximum value.

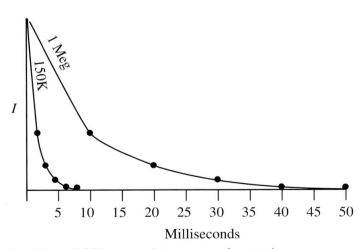

Figure 24-1. The effect of different resistances on the RC time constant curve.

For 150 K resistance:
t = (15×10^4) 0.01 µF
    = 1.5 ms
For 1 meg resistance:
t = (1×10^6) 0.01 µF
    = 10 ms

If the capacitance is changed to 10 µF, the time constants become 1.5 seconds and 10 seconds.

The percentage of charge or discharge of current or voltage for a given time constant can be determined using the universal time constant chart, **Figure 24-2.**

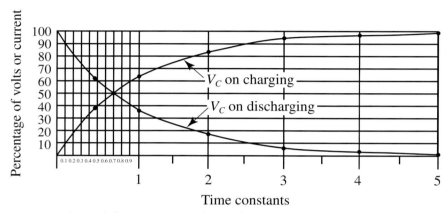

Figure 24-2. Universal time constant chart. The amount of voltage or current values for any time constant can be determined.

Series RC Circuit Operation

The circuit in **Figure 24-3** has a capacitive reactance and resistance. This type of circuit is called an **RC circuit.** The impedance is the vector sum of the resistance and capacitive reactance.

RC circuit: A series, parallel, or series-parallel circuit made up of a resistance and capacitance.

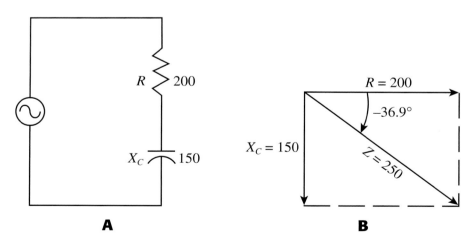

Figure 24-3. Series RC circuit. A— The circuit has a capacitive reactance and resistance. B—Vector diagram. Since X_L and X_C are 180° apart, the X_C gives a negative angle.

The current is the same in all parts of a series circuit and in phase with the voltage in a resistance. The resistor vector is placed at 0°. The voltage lags the current in a capacitor; thus, the X_C vector is placed at −90°. The rectangle is completed using dashed lines. The impedance vector is drawn from the origin to the point where these lines cross.

The impedance can be calculated using the equation:

$$Z = \sqrt{R^2 + X_C^2}$$

▼ *Example 24-1:*

Find the impedance and voltage drops in **Figure 24-4.**

$$\begin{aligned} Z &= \sqrt{R^2 + X_C^2} \\ &= \sqrt{300^2 + 700^2} \\ &= 762 \ \Omega \end{aligned}$$

The circuit current is:

$$\begin{aligned} I &= \frac{E}{Z} \\ &= \frac{65}{762} \\ &= 85.3 \ \text{mA} \end{aligned}$$

$$\begin{aligned} V_R &= IR \\ &= (85 \times 10^{-3}) \ 300 \\ &= 25.59 \ \text{V} \end{aligned}$$

$$\begin{aligned} VX_C &= IX_C \\ &= (85 \times 10^{-3}) \ 700 \\ &= 59.71 \ \text{V} \end{aligned}$$

▲

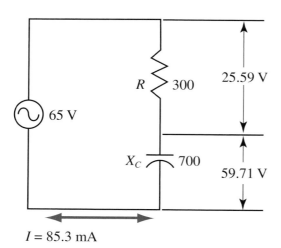

$$I = 85.3 \ \text{mA}$$

Figure 24-4. Series RC circuit for Example 24-1.

Parallel RC Circuit

As in RL parallel circuits, the currents are calculated instead of the voltages.

▼ ***Example 24-2:***

Find the currents and impedance of **Figure 24-5A.**

$$I_R = \frac{E}{R}$$

$$= \frac{15}{30}$$

$$= 0.5 \text{ A}$$

$$IX_C = \frac{E}{X_C}$$

$$= \frac{15}{50}$$

$$= 0.3 \text{ A}$$

The total current is:

$$I_T = \sqrt{I_R^2 + I_{X_C}^2}$$

$$= \sqrt{0.5^2 + 0.3^2}$$

$$= 0.583 \text{ A}$$

The impedance is:

$$Z = \frac{E}{I}$$

$$= \frac{15}{0.583}$$

$$= 25.73 \ \Omega$$ ▲

Figure 24-5B is a vector diagram of the parallel RC circuit. The source voltage is the same across all parts of the circuit. I_R is in phase with the source voltage at 0°.

The current of a capacitor leads the source voltage by 90°; therefore, the capacitor current (I_C) goes upward, giving a positive angle. The total current vector (0.583 A) is drawn from the origin to the point where the dashed lines cross. The phase difference (θ) is shown as 30.9°.

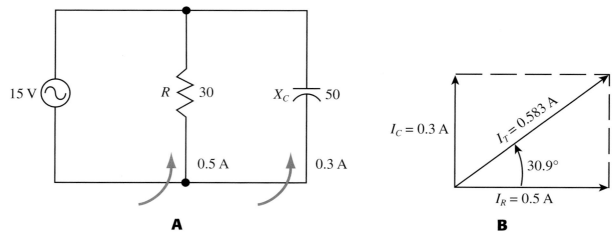

Figure 24-5. Parallel RC circuits. A—Circuit for Example 24-2. B—Vector diagram. The capacitor current leads by 90°, and the total current is the vector sum of the two currents.

Power in RC Circuits

As in RL circuits, when the voltage and current are out of phase, the impedance is equal to the vector sum of the resistance and capacitive reactance. The power can be calculated using the power factor.

▼ *Example 24-3:*

Find the true power of Example 24-1.

First, find the power factor (*pf*):

$$pf = \frac{R}{Z}$$
$$= \frac{300}{762}$$
$$= 0.3937$$

Then, find the true power:

$$P = IEpf$$
$$= (85 \times 10^{-3})\,(65)\,(0.3937)$$
$$= 2.18 \text{ W}$$

▲

RC Integrator

The term ***integrator*** comes from a mathematical function in calculus. An integrator circuit is recognized by the series RC connections and output taken across the capacitor, **Figure 24-6.** If a 12 V, 100 Hz square wave is applied to the integrator circuit with a time constant of 0.5 ms, the capacitor is fully charged in 3 ms.

The positive alternation of the square wave is 5 ms. Thus, the capacitor is fully charged 2 ms before the positive alternation ends. At the end of the positive alternation, the input voltage goes to 0 V and the charged capacitor now discharges in 3 ms.

If the input square wave frequency is changed to 1000 Hz, the positive alternation only lasts one time constant and the capacitor is unable to fully charge. During one time constant (0.5 ms), the capacitor charge will reach only 63.2% of the input voltage, charging the capacitor to 7.584 V. See **Figure 24-7.** During the 0 V portion of the input, the capacitor discharges to 63.2% of 7.584 V, or 2.79 V.

During the next alternation of the input voltage, the capacitor charges to a level of 8.61 V (from 2.79 V). Then, it discharges to 3.17 V. On the next alternation, the capacitor charges to 8.75 V. This action continues until the output voltage builds to an average value of approximately 6 V. The same effect is produced if the time period of the square wave is decreased or the time constant is increased.

Integrator: A device or circuit in which the output is proportional to the integral of the input signal.

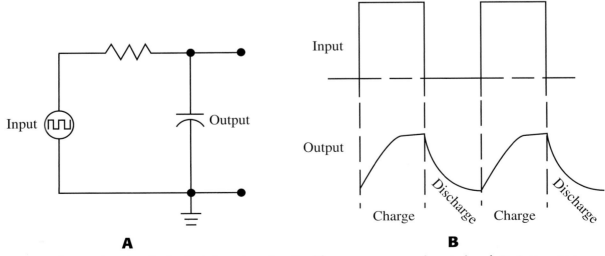

Figure 24-6. Integrator circuit. A—An integrator circuit with a square wave input signal. B—Integrator response to a square wave.

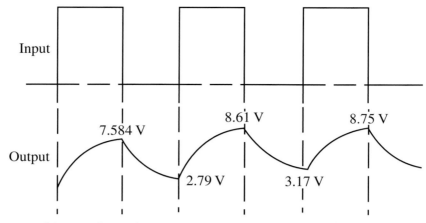

Figure 24-7. This waveform shows incomplete charging of the capacitor due to a change in applied frequency.

RC Differentiator

Differentiator: A circuit in which the output voltage is proportional to the rate of change of the input voltage.

In the RC *differentiator* circuit, the output is taken across the resistor, **Figure 24-8.** The output waveform shows the result of the charging and discharging of the capacitor. When the 12 V square wave is applied to the circuit, the maximum voltage is across the resistor and the capacitor begins to charge. As more voltage is developed across the capacitor, less voltage appears across the resistor.

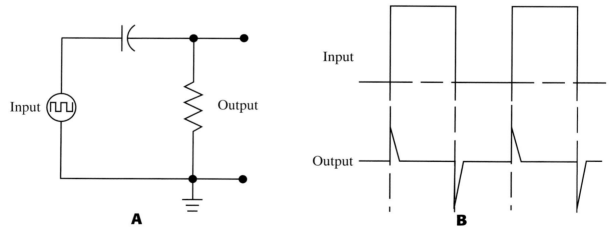

Figure 24-8. RC differentiator circuit. A—Output is taken across the resistor. B—Waveform of the differentiated square wave. The time constant is always short with respect to the input time.

 Summary

- In a capacitive circuit, the time constant is equal to the product of the resistance and capacitance.
- In a pure capacitance, the current leads the source voltage by 90°.
- The impedance is the vector sum of the resistance and capacitive reactance.
- The power of an RC circuit must be calculated for true power by using the power factor, which is the cosine of the phase angle.
- Integrator and differentiator circuits are used for waveshaping and filter circuits throughout the electronics industry.

Important Terms

Do you know the meanings of these terms used in the chapter?

differentiator	RC circuit
integrator	RC time constant

Questions and Problems

Please do not write in this text. Write your answers on a separate sheet of paper.

1. What is the time constant in **Figure 24-9**?
2. What is the current in Figure 24-9 after two time constants?
3. Draw a schematic of a series RC circuit using a 500 Ω resistor, a 1200 Ω capacitive reactance, and a 60 V source. Find impedance, current, voltage drops across the resistor and capacitor, and true power.
4. Draw a vector diagram of problem 3.
5. What is the impedance of a capacitor with 18 V across and a current of 2 mA?
6. Draw a parallel RC circuit using a 9 V source operating at 200 Hz, 30 Ω resistance, and a 10 μF capacitor. Find impedance, currents, phase angle, and true power.

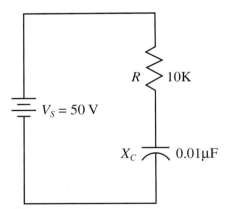

Figure 24-9. Circuit for problems 1 and 2.

Chapter 25 Graphic Overview

RCL Circuits

has
- Bandwidth — is the — Frequency Range — will — Pass — determined by — Half-power Points
- Resonance — when — X_L — and — X_C — are — Equal — is — Resonant Frequency

for
- Parallel — called a — Tank Circuit
  - has — Maximum Impedance — and — Minimum Line Current
- Series — has — Minimum Impedance — and — Maximum Line Current

Maximum Line Current — replaces — Energy Dissipated — by — Resistances

the
- Impedance — is the — Vector Sum — of the — Resistance — and — Net Reactance

can have
- Net Reactance — is the — Difference — between — X_L and X_C

Objectives

After studying this chapter, you will be able to:
- ○ Calculate the net reactance of a circuit.
- ○ Calculate the impedance of series and parallel RCL circuits.
- ○ Calculate the resonant frequency of a circuit.
- ○ Identify the currents of a resonant tank circuit.
- ○ Identify a damped wave.
- ○ Calculate the Q of a resonant circuit.
- ○ Calculate the bandwidth of a circuit.
- ○ State the relationship between circuit Q and selectivity.

Introduction

Previous chapters covered circuits involving resistance, inductance, and capacitance. This chapter will cover circuits that combine resistors, capacitors, and inductors (RCL). RCL circuits are used in RF circuits in all types of radio/television receivers and transmitters.

Net Reactance

Figure 25-1 shows an inductor and capacitor connected in series with zero resistance. An inductance causes the current to lag the voltage, while the capacitance causes the current to lead the voltage. As a result, the reactances cancel each other, and the voltage and current are in phase. To be in phase, the two reactances must be equal in value ($X_L = X_C$). If the two reactances are not equal, the **net reactance** (X) is the difference between the two reactances. The equation is:

$$X = X_L - X_C$$

The circuit will have the characteristics of the remaining reactance. If X_L is greater than X_C, the circuit will be inductive. If X_C is greater than X_L, the circuit will be capacitive.

Net reactance: The difference between the inductive reactance X_L and the capacitive reactance X_C.

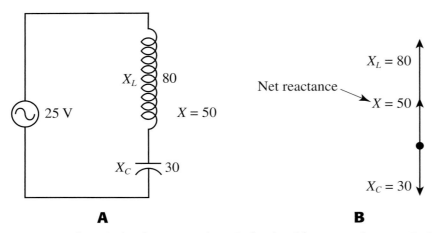

Figure 25-1. Series LC circuit. A—A series LC circuit with zero resistance. B—Vector diagram showing the vector sum net reactance of the inductive reactance and capacitive reactance.

The circuit in Figure 25-1A has an X_L of 80 Ω and an X_C of 30 Ω. This gives a net reactance of 50 Ω, which is inductive.

Thus:

$$X = X_L - X_C$$
$$= 80 - 30$$
$$= 50 \; \Omega$$

To calculate the current and voltage drops:

$$I = \frac{E}{X}$$
$$= \frac{25}{50}$$
$$= 0.5 \text{ A}$$

$$V_L = IX_L$$
$$= 0.5 \, (80)$$
$$= 40 \text{ V}$$

$$V_C = IX_C$$
$$= 0.5 \, (30)$$
$$= 15 \text{ V}$$

If these voltages are substituted for the reactances, the source voltage is equal to the difference between the reactance voltages.

Thus:

$$V_S = V_L - V_C$$
$$= 40 - 15$$
$$= 25 \text{ V}$$

In **Figure 25-2**, X_L and X_C are in parallel across the 25 V source. The currents are used in a parallel circuit. The X_L current is 0.3125 A, and the X_C current is 0.8333 A. Since the two currents are 180° out of phase, the net current is 0.5208, the difference between I_L and I_C.

Thus:

$$I_{NET} = I_C - I_L$$
$$= 0.8333 - 0.3125$$
$$= 0.5208 \text{ A}$$

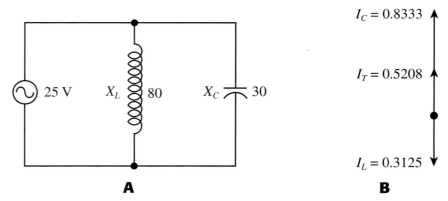

Figure 25-2. Parallel LC circuit. A— X_L and X_C are in parallel across the voltage source. B—Vector diagram.

Series RCL Circuit

When a resistance is placed in series with the two reactances, the impedance equation becomes:

$$Z = \sqrt{R^2 + (X_L - X_C)^2}$$

The impedance is the vector sum of the resistance and the net reactance.

▼ ***Example 25-1:***

Find the impedance of the series RCL circuit in **Figure 25-3.**

$$\begin{aligned} Z &= \sqrt{R^2 + (X_L - X_C)^2} \\ &= \sqrt{400^2 + (600 - 400)^2} \\ &= 447\ \Omega \end{aligned}$$

Find the current.

$$\begin{aligned} I &= \frac{E}{Z} \\ &= \frac{80}{447} \\ &= 0.179\ A \end{aligned}$$

Find the voltage across each of the components.

$$\begin{aligned} V_R &= IR \\ &= 0.179\ (400) \\ &= 71.6\ V \end{aligned}$$

$$\begin{aligned} V_L &= IX_L \\ &= 0.179\ (600) \\ &= 107.4\ V \end{aligned}$$

$$\begin{aligned} V_C &= IX_C \\ &= 0.179\ (400) \\ &= 71.6\ V \end{aligned}$$

If the voltages are placed in the impedance equation, the result is the source voltage.

$$\begin{aligned} V_S &= \sqrt{V_R^2 + (V_L - V_C)^2} \\ &= \sqrt{71.6^2 + (107.4 - 71.6)^2} \\ &= 80\ V \end{aligned}$$

Using the cosine or tangent, the phase angle can be found as before.   ▲

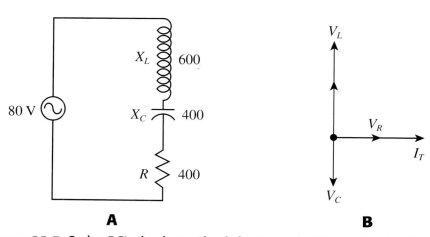

Figure 25-3. Series RCL circuit. A—Circuit for Example 25-1. B—Vector diagram.

Parallel RCL Circuit

As in parallel reactive circuits, the voltages are the same across all parts of a parallel circuit, and the calculations are made for the currents. First, each of the branch currents must be found, **Figure 25-4.**

$$I_R = \frac{E}{R}$$

$$= \frac{120}{60}$$

$$= 2 \text{ A}$$

$$I_L = \frac{E}{X_L}$$

$$= \frac{120}{30}$$

$$= 4 \text{ A}$$

$$I_C = \frac{E}{X_C}$$

$$= \frac{120}{40}$$

$$= 3 \text{ A}$$

The total current is found by:

$$I_T = \sqrt{I_R^2 + (I_L - I_C)^2}$$

$$= \sqrt{2^2 + (4 - 3)^2}$$

$$= \sqrt{4 + 1}$$

$$= 2.236 \text{ A}$$

The phase angle is found by:

$$\theta = \text{arc tan } \frac{I_L - I_C}{I_R}$$

$$= \text{arc tan } \frac{4 - 3}{2}$$

$$= 0.5$$

$$= 26.56°$$

or

$$\theta = \text{arc cos } \frac{I_R}{I_T}$$

$$= \text{arc cos } \frac{2}{2.236}$$

$$= 0.8944544$$

$$= 26.56°$$

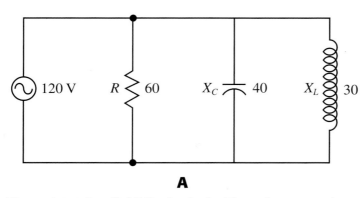

A

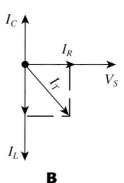

B

Figure 25-4. Parallel RCL circuit. A— The voltages are the same across all parts of a parallel circuit. B—Vector diagram. The total vector replaces the impedance vector.

Resonance

When the X_L and X_C are equal (net reactance equals 0), the circuit is at **resonance** and called a **resonant circuit.** Since frequency is in each of the reactance equations, the frequency at which resonance is obtained is called the **resonant frequency.** Resonant frequency is calculated by the equation:

$$f_r = \frac{1}{2\pi\sqrt{LC}}$$

Resonance: In a tuned circuit, the condition when X_L and X_C are equal in value.

Resonant circuit: A circuit containing both inductance and capacitance and tuned to resonance.

Resonant frequency: The frequency at which inductive reactance is equal to capacitive reactance.

▼ **Example 25-2:**

Calculate the resonant frequency in **Figure 25-5.**

$$f_r = \frac{1}{2\pi\sqrt{LC}}$$

$$= \frac{1}{2(3.14)\sqrt{(4 \times 10^{-3})(200 \times 10^{-12})}}$$

$$= 178 \text{ kHz}$$ ▲

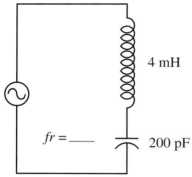

Figure 25-5. LC circuit for calculating resonant frequency in Example 25-2.

Series Resonant Circuit

In **Figure 25-6,** X_L and X_C are equal in a resonant circuit; therefore, the impedance is equal to the resistance. The circuit current is:

$$I = \frac{E}{Z}$$

$$= \frac{30}{600}$$

$$= 50 \text{ mA}$$

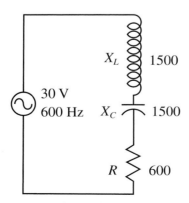

Figure 25-6. The reactances are equal in value in a resonant series RCL circuit.

In **Figure 25-7,** X_L and X_C are not equal, and the impedance is calculated as:

$$Z = \sqrt{R^2 + (X_L - X_C)^2}$$
$$= \sqrt{600^2 + (1500 - 1200)^2}$$
$$= 671 \ \Omega$$

The current is:

$$I = \frac{E}{Z}$$
$$= \frac{30}{671}$$
$$= 45 \text{ mA}$$

Any increase in a reactance increases the impedance, which reduces the current. Since the net reactance is zero at resonance, a series RCL circuit has *minimum imped-ance* and *maximum line current* at resonance.

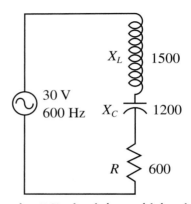

Figure 25-7. A nonresonant series RCL circuit has a higher impedance than a resonant circuit.

In **Figure 25-8,** if an output voltage is taken from across one of the reactances, the voltage is higher than the voltage applied to the circuit. In this case, 30 V has been increased to 75 V.

Thus:

$$V_{X_C} = IX_C$$
$$= 50 \times 10^{-3} (1500)$$
$$= 75 \text{ V}$$

Although the voltages across X_L and X_C are reactive, they are still actual, measur-able voltages. Some circuit designers make use of the reactive step-up characteristic.

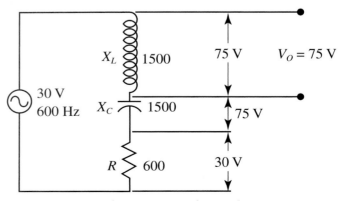

Figure 25-8. Reactances can produce a stepped-up voltage.

Parallel Resonant Circuit

With X_L and X_C in parallel, two currents are involved in a parallel LC circuit: the line current and the current within the LC parallel network, **Figure 25-9.** The parallel LC network is often called a **tank circuit.** The current is called tank, flywheel, circulating, or oscillating.

Tank circuit: A parallel LC or RCL circuit.

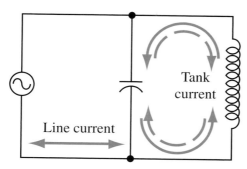

Figure 25-9. A parallel LC circuit is also called a tank circuit. It has both a line current and tank current.

To fully understand the circuit action of a tank circuit, study the process in **Figure 25-10**. The LC tank circuit consists of equal XL and XC connected to a dc source. The circuit action is as follows:

1. With the switch closed, the capacitor charges to the source voltage level, Figure 25-10A. The current through the inductor stores the maximum energy in its magnetic field (the arrows indicate an expanded magnetic field).
2. The switch is opened, Figure 25-10B. The line current is zero, and energy has been stored in the capacitor and inductor. The graph shows the voltage across the capacitor.
3. The capacitor discharges through the inductor. At the same time, the magnetic field collapses, creating current flow in the same direction as the capacitor current. This charges the capacitor in the opposite direction, Figure 25-10C. Again, the graph shows the voltage across the capacitor charged in the opposite direction. The amount of charge voltage is less since some energy was dissipated in the resistance of the inductor.
4. The capacitor again discharges through the inductor in the other direction, Figure 25-10D. The arrows on the inductor show the magnetic field expands then collapses, again charging the capacitor.
5. Charging-discharging of the capacitor continues until all energy stored in the capacitor and inductor is dissipated in the resistances of the circuit.

Figure 25-11A shows a completed graph of the circuit action. The wave drops in voltage on each cycle of charging. This type of wave is called a **damped wave.**

In **Figure 25-11B,** the same tank circuit action occurs, except the tank current does not become reduced on each cycle. The energy dissipated in the circuit resistances is replaced on each cycle by the line current. The higher the inductor resistance, the greater the loss, and more line current must flow from the source to replace the tank loss.

Damped wave: A wave in which each cycle constantly decreases in amplitude.

When the tank circuit is not at resonance, the I_L or I_C is greater and the reactive currents do not cancel. The line current is the difference between the two reactive currents, plus the current necessary to replace the energy dissipated in the component resistances. When the tank circuit is off resonance, the current is higher or the impedance is less.

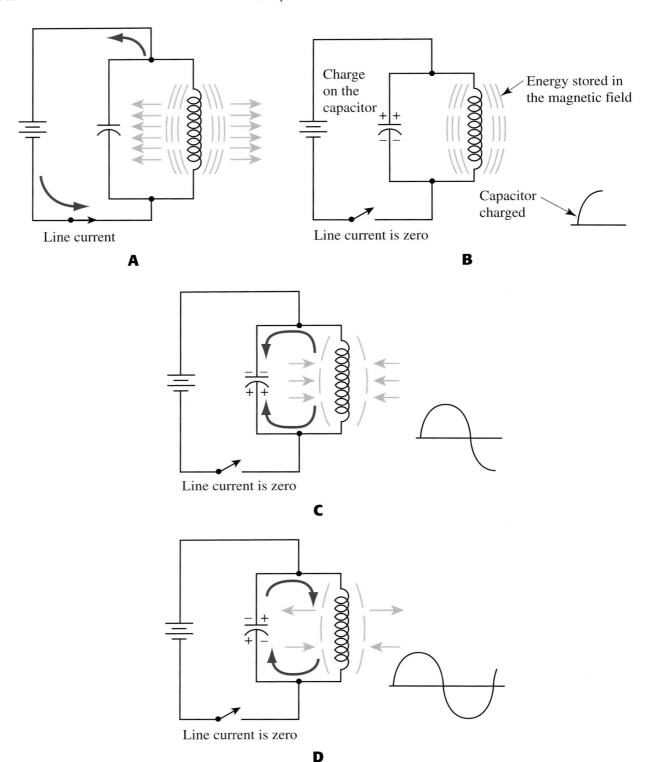

Figure 25-10. Tank circuit action. A—The initial line current causes electrons to be stored in the capacitor and a magnetic field to be created around the inductor. B—When the switch is opened, maximum energy is stored in the capacitor and inductor. C—The stored energy creates the tank current. During this time period, the capacitor is charged in the opposite direction. D—The energy stored in the capacitor reverses the direction of the current, and continues the tank current action.

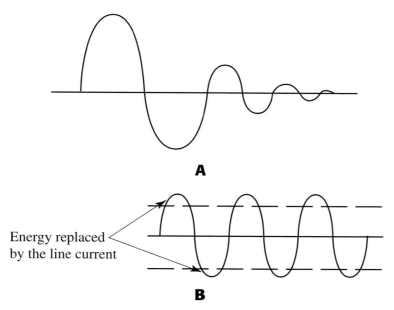

Figure 25-11. Charging-discharging of the capacitor. A—The tank current energy creates a damped wave. B—The energy dissipated in the tank circuit is replaced by the line current.

At resonance, the tank currents I_L and I_C are equal, but 180° out of phase. Since the currents are opposite, they cancel each other in the main line. The net line current flowing is only the current necessary to replace the energy dissipated in the resistances. A parallel RCL circuit has *maximum impedance* and *minimum line current* at resonance.

Q-factor of Resonant Circuits

The **Q-factor** (quality), or *figure of merit* (Q), of a resonant circuit measures the sharpness of resonance. At resonance, the higher the ratio of the reactance to the series resistance, the higher the Q and the sharper the resonance effect. Q is a numerical factor without any units.

In **Figure 25-12,** the Q of a series circuit is equal to:

$$Q = \frac{X_L}{R_S}$$

Where:
R_S = Series resistance

Thus:
$$Q = \frac{1200}{6}$$
$$= 200$$

Since X_L and X_C are equal at resonance, the Q has the same value if calculated with X_C. The inductor is usually used in calculations because it contains the series resistance; therefore, the inductor contains the circuit resistance. If an additional series resistor is added to the circuit, the Q of the circuit will be less. The highest possible circuit Q is the Q of the inductor. Because the series circuit limits the amount of current at resonance, the lower the resistance, the higher the current and the Q.

Q-factor: A measure of the sharpness of resonance or frequency selectivity.

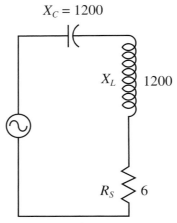

$X_C = 1200$

X_L 1200

R_S 6

Figure 25-12. Circuit used for calculating the Q of a series RCL circuit.

The Q of a parallel circuit is equal to the Q of the inductor. The resistance of the inductor is still R_S, which is in series with the inductive reactance. A resistance in parallel with the tank circuit lowers the circuit Q.

Shunt damping resistor: A resistor in parallel with a parallel RCL circuit to increase the bandwidth.

A resistance placed in parallel is often used in high-frequency circuits for increasing the bandwidth, **Figure 25-13.** It is called a *shunt damping resistor* and is said to be loading or damping the tank circuit.

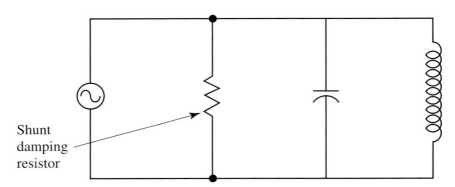

Shunt damping resistor

Figure 25-13. A damping resistor increases the bandwidth.

Bandwidth

Bandwidth: The given band of frequencies effectively passed by a circuit or device and measured at the half-power points of maximum amplitude.

Bandwidth (represented by *BW*) refers to the range of frequencies a circuit will pass. A resonant circuit will pass the resonant frequency with maximum output from the circuit. The frequencies above and below resonance through which the circuit will pass determine the circuits bandwidth.

The bandwidth is measured at points on the response curve that are 70.7% of the maximum voltage or current, **Figure 25-14.** The bandwidth is the difference between the frequencies f_1 and f_2, in this case, 30 kHz.

Thus:

$BW = f_2 - f_1$
$= 65 \text{ kHz} - 35 \text{ kHz}$
$= 30 \text{ kHz}$

If the circuit resonant frequency and Q are known, the bandwidth is equal to:

$$\Delta f = \frac{f_r}{Q}$$

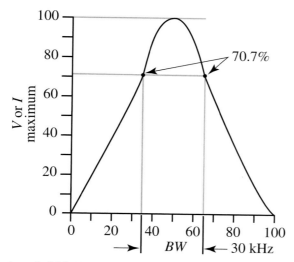

Figure 25-14. The bandwidth is measured at the 70.7% point of the response curve.

Half-power point: The frequency limits at which the maximum voltage or current drops 70.7% (3 dB).

Cutoff frequency: The frequency at which the gain of an amplifier or circuit falls below 0.707 times the maximum gain (3 dB).

The width of the bandwidth depends on the Q of the resonant circuit, **Figure 25-15.** The 70.7% points on the curve of **Figure 25-16** are not halfway between 0 and 100, but are the *half-power points.* Do not confuse these with half-voltage or current points. Frequencies f_1 and f_2 are called the ***cutoff frequency*** because the tuned circuit rejects frequencies outside these half-power points.

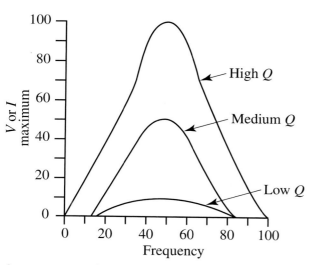

Figure 25-15. These curves show the effect on the response and bandwidth as the Q changes.

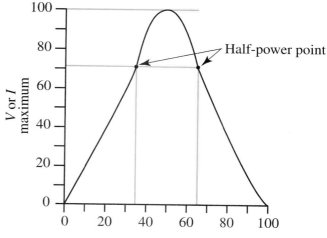

Figure 25-16. The half-power points are located at 70.7% of maximum amplitude.

▼ *Example 25-3:*

If the maximum current in a circuit is 16 mA, and the resistance is 80 Ω, what is the half-power?

$$P_{MAX} = I^2R$$
$$= (16 \times 10^{-3})^2 (80)$$
$$= 20.48 \text{ mW}$$

$$P_{HALF} = (I \times 70.7\%)^2 (R)$$
$$= [(16 \times 10^{-3}) (0.707)]^2 (80)$$
$$= 10.24 \text{ mW}$$

▲

Selectivity: Measure of how well a circuit or piece of equipment can distinguish one frequency from another.

The ability of a circuit to choose certain frequencies over others is called *selectivity.* A circuit with a narrow bandwidth has high selectivity, while a circuit with a wide bandwidth has low selectivity. In **Figure 25-17,** the high Q-response curve has a bandwidth of 20 kHz; the medium Q has a bandwidth of 40 kHz; and the low Q-curve has a bandwidth of 70 kHz. A higher Q means a narrow bandwidth and higher selectivity.

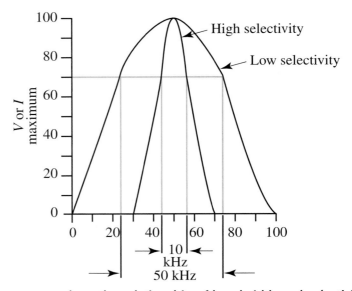

Figure 25-17. These curves show the relationship of bandwidth and selectivity.

▶ Summary

- Net reactance is equal to the difference between X_L and X_C.
- The impedance of a circuit is the vector sum of the values.
- Resonance occurs when X_L is equal to X_C.
- In a series resonant circuit, the circuit has minimum impedance and maximum line current. In a parallel resonant circuit, the circuit has maximum impedance and minimum line current.
- A tank circuit has a tank current and a line current. The line current replaces the energy dissipated in the tank circuit.
- The Q-factor of a circuit determines the ability of a circuit to select a band of frequencies.
- The bandwidth of a resonant circuit is the frequencies a circuit will pass at half-power.
- A higher circuit Q has a narrow bandwidth and higher selectivity.

▶ Important Terms

Do you know the meanings of these terms used in the chapter?

bandwidth	resonance
cutoff frequency	resonant circuit
damped wave	resonant frequency
figure of merit	selectivity
half-power point	shunt damping resistor
net reactance	tank circuit
Q-factor	

Questions and Problems

Please do not write in this text. Write your answers on a separate sheet of paper.

1. A circuit has a 2 mH inductance and 0.001 µF capacitance. What is the resonant frequency of the circuit?
2. If the inductive reactance is 500 Ω and the capacitive reactance is 300 Ω, what is the net reactance?
3. A circuit has 2000 Ω resistance, 5000 Ω X_L, and 4000 Ω of X_C. What is the impedance?
4. State the impedance and current characteristics of series and parallel resonant circuits.
5. A circuit has greater selectivity when the circuit Q is _____ (high, low).
6. In what type of circuit does a flywheel current flow?
7. What is the bandwidth in **Figure 25-18?**
8. What effect does a damping resistor have on circuit Q?
9. In **Figure 25-19,** what is the resonant frequency, circuit Q, and bandwidth?
10. Does the circuit in **Figure 25-20** have high or low selectivity? Explain.
11. Find the true power in **Figure 25-21.** What is the phase angle?
12. Draw a waveform of a damped wave.

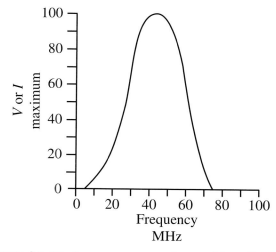

Figure 25-18. Frequency curve for problem 7.

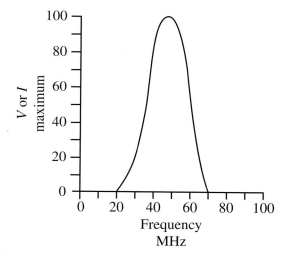

Figure 25-19. Frequency curve for problem 9.

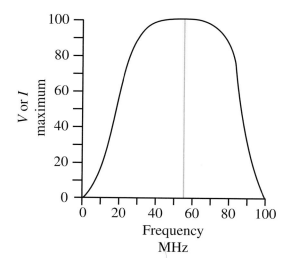

Figure 25-20. Frequency curve for problem 10.

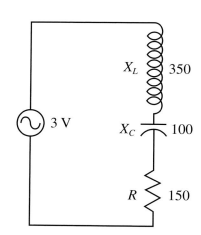

Figure 25-21. RCL circuit for problem 11.

Chapter 26 Graphic Overview

Attenuator Circuit

- used to — **Reduce** — a — **Signal Strength** — without — **Distortion**
- types are — **T, equal Z** / **T, unequal Z** / **Pi, equal Z** / **Pi, unequal Z** / **0, balanced** / **0, unbalanced**

Filter Circuits

- can make — **Speaker Crossover Network** — types are — **Tweeter** / **Midrange** / **Woofer**

- types are:
  - **M-derived** — can control — **Impedance** — or — **Attenuation**
  - **Constant-k** — makes an — **Impedance Match** — at — **One Frequency**
  - design requires — **Decibels** — and — **Logarithims**
  - can be — **Low-pass** / **High-pass** / **Band-pass** / **Band-reject**

- separate — **One Group** — of — **Frequencies** — from — **Another** — indicated by — **Response Curve** — determined by — **Cutoff Frequency**

FILTER CIRCUITS **26**

Objectives

After studying this chapter, you will be able to:
○ Find the logarithm and antilogarithm of a number.
○ Calculate voltage or power gain in decibels.
○ Identify the different types of filter networks.
○ Explain the uses of filter circuits.
○ Explain the process of making a response curve.
○ Explain the process of designing an attenuator.

Introduction

A filter circuit separates mixed frequencies. General filters use inductors, capacitors, and resistors to make a filter network. Filter circuits are used to separate audio from radio frequencies, to separate ac variations from an average dc voltage, to match impedances, or to separate the frequencies in a speaker system.

Decibels

The unit of measurement for audio and other signal-related values is the bel, named for Alexander Graham Bell. The bel unit is so large the metric prefix "deci" is used for normal measurements. The *decibel* (dB) is equal to $\frac{1}{10}$ of a bel. It is based on a mathematical equation using the power or voltage ratio.

The ear is more sensitive to a change in sound intensity at low volume rather than high volume. A change of 2 W in power output, from 4 W to 6 W, will sound much louder than a 2 W increase from 20 W to 22 W, which is only slightly noticeable. Loudness to the human ear depends on the ratio of the two powers. The ratio of 4 W to 6 W is 1.5, a 50% increase in loudness. The ratio of 20 W to 22 W is 1.1, an increase of only 10%. The ratio of the powers or voltages determines the change in loudness. In the calculation of decibels, the small and large numbers must be compressed for the perception of the human ear. That is why logarithms are used in the equations. A decibel table for gain or loss of voltage and power is given in Appendix E.

Decibel: The standard unit of measurement for gain or loss in voltage or power levels.

Logarithms

The two types of *logarithms* are common and natural. Natural, or Naperian, logarithms are based on the number 2.7182818. They are indicated by the letter e and found with the **ln** calculator key. Natural logs are used in various math operations, such as the instantaneous voltage across a charging capacitor.

Common logarithms use the base ten number system and can be related to the powers of ten. The power of ten of 100 is 10^2, and the power of ten of 1000 is 10^3. The logarithm of 100 is 2 and the log of 1000 is 3. The log is the value of the exponents. What is the power of ten of 492? Think of logarithms as the power of ten between the whole numbers of the exponents; therefore, the power of ten for 492 is somewhere between 10^2 and 10^3, or $10^{2????}$. In this case, the log of 492 is equal to $10^{2.6919651}$. This may seem strange if you assume the halfway point between 10^2 and 10^3 is $10^{2.5}$, or 500.

Logarithm: The exponent to which a base number is raised to obtain the number.

Logarithms are based on a nonlinear scale. Think of semilog graph paper. If the exponent is considered a logarithm, the numbers can be worked on the calculator using the log key.

▼ *Example 26-1:*

Find the logarithm of 492.

Place 492 on the calculator display and press the log key. The display should show 2.6919651.   ▲

Antilogarithms

Antilogarithm: The inverse of logarithm.

All math operations have an opposite function. The opposite of logarithm is *antilogarithm,* or antilog. It is the **INV, 2nd,** or **3rd** function of the log key on a calculator.

Power Gain-Loss

The equation for calculating power gain or loss is:

$$dB = 10 \log \frac{P_2}{P_1}$$

The equation actually looks like this:

$$dB = 10 \left[\log \left(\frac{P_2}{P_1} \right) \right]$$

The order of operations rules allow the brackets and parentheses to be left out of the equation, but you must work the equation from the inside out. First, do the work inside the parentheses, then log, and finally times ten.

▼ *Example 26-2:*

What is the decibel gain for an increase of power from 15 W to 30 W?

$$dB = 10 \log \frac{30}{15}$$
$$= 10 \log (2)$$
$$= 10 (0.3)$$
$$= 3 \text{ dB}$$

This indicates a power gain of 3 decibels.   ▲

▼ *Example 26-3:*

What is the power gain for a decrease in power from 12 W to 3 W?

$$dB = 10 \log \frac{3}{12}$$
$$= 10 \log 0.25$$
$$= 10 (-0.6)$$
$$= -6 \text{ dB}$$

This indicates a power loss of 6 decibels.   ▲

Voltage Gain-Loss

Voltage gain or loss can be calculated the same way. The voltage ratio for calculating decibels uses the equation:

$$dB = 20 \log \frac{V_2}{V_1}$$

▼ *Example 26-4:*

What is the decibel gain for an increase of voltage from 30 mV to 6 V?

$$dB = 20 \log \frac{6}{30}$$
$$= 20 \log 200$$
$$= 20 \ (2.3)$$
$$= 46 \ dB$$

This indicates a 46 dB gain for a voltage 200 times greater than the input voltage.   ▲

The above equation is commonly used; however, it applies only to V_2 and V_1 across the same value of impedance. If different impedances are involved, the equation becomes:

$$dB = 20 \log \left(\frac{V_2}{V_1} \right) \times \left(\sqrt{\frac{Z_1}{Z_2}} \right)$$

▼ *Example 26-5:*

What is the dB gain with a V_2 of 18 mV across 75 Ω and a V_1 of 3 mV across 300 Ω?

$$dB = 20 \log \frac{18 \ mV}{3 \ mV} \times \sqrt{\frac{300 \ \Omega}{75 \ \Omega}}$$
$$= 20 \log (6) \ (2)$$
$$= 20 \log 12$$
$$= 20 \ (1.0791812)$$
$$= 21.58 \ dB$$ ▲

Reference Levels

Two values are necessary for a decibel comparison. When only one value of power or voltage is given, the other is assumed. The following references are in common use:

$dB = 6$ mW (0.006 W) referenced in 500 Ω

$dBm = 1$ mW (0.001 W) referenced in 600 Ω

$dBmV = 1$ mV (0.001 V) referenced in 75 Ω

6 mW Reference

Any power level can be compared with the 6 mW reference using the equation:

$$dB = 10 \log \frac{P}{6 \ mW}$$

▼ *Example 26-6:*

An amplifier has an output of 18 W. What is the output in decibels?

$$dB = 10 \log \frac{18 \ W}{6 \ mW}$$
$$= 10 \log 3000$$
$$= 10 \ (3.48)$$
$$= 34.8 \ dB$$

The output is 34.8 dB above the standard reference of 6 mW.   ▲

With 0 dB at 1.73 V, the 6 mW in 500 Ω is so common it is often indicated on the scales of analog ac voltmeters. See **Figure 26-1.** This is found by transposing the equation:

$$P = \frac{V^2}{Z}$$

to

$$V = \sqrt{PZ}$$

Thus:

$$V = \sqrt{PZ}$$
$$= \sqrt{(0.006)\,(500)}$$
$$= \sqrt{3}$$
$$= 1.73 \text{ V}$$

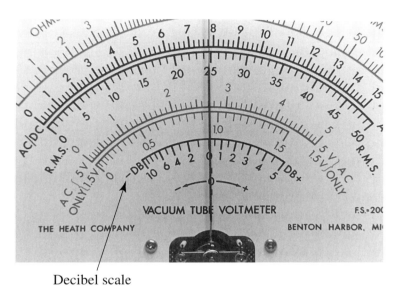

Decibel scale

Figure 26-1. Many analog meters have a decibel scale.

1 mW Reference

The 1 mW reference is used for telephone systems and audio equipment in radio broadcast stations. The equation is:

$$dBm = 10 \log \frac{P}{1 \text{ mW}}$$

▼ ***Example 26-7:***

Calculate the dBm level for a 25 mW audio signal.

$$dBm = 10 \log \frac{25 \text{ mW}}{1 \text{ mW}}$$
$$= 10 \log 25$$
$$= (10)\,(1.4)$$
$$= 14 \text{ dBm}$$ ▲

Volume Units

Volume unit: The unit of measurement for the power level of voice waves.

The radio broadcasting industry uses the VU unit for measuring audio levels. In **Figure 26-2,** the 1 mW in 600 Ω reference is used to define the *volume unit* (VU) with an ac voltmeter calibrated in VU to monitor the audio modulation. The VU meter circuit is designed for operating at 1000 Hz and measuring the complex waveforms in voice and music signals.

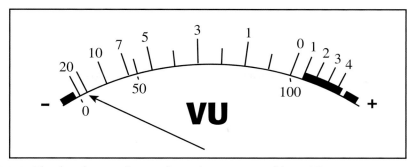

Figure 26-2. The loudness of sound can be measured by a VU meter.

Solving Problems with Decibels

Given a block diagram of a signal system, **Figure 26-3,** the signal gain can be calculated by adding the decibels of each of the amplifier blocks.

Thus:

$$dB = 8 + (-20) + 35 + (-5) + 40$$
$$= 58 \text{ dB}$$

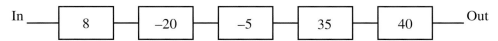

Figure 26-3. A signal system represented by a block diagram.

▼ *Example 26-8:*

In **Figure 26-4,** two amplifiers are connected by a coaxial cable. What is the output voltage of amplifier 2?

The RG59 coax has a loss of 4.9 dB per 100′.

$$dB = -4.9 \text{ dB (6)   (600′ of coaxial cable)}$$
$$= -29.4 \text{ dB   (Loss in the coaxial cable)}$$

Add the decibels to find the total decibels.

$$dB = 25 + (-29.4) + 30$$
$$= 25.6 \text{ dB}$$

Place the numbers into the equation and solve.

$$dB = 20 \log \frac{V_2}{V_1}$$

$$25.6 = 20 \log \frac{V_2}{60 \ \mu V} \quad \text{(Divide both sides by 20.)}$$

$$1.28 = \log \frac{V_2}{60 \ \mu V} \quad \text{(Take the antilog of both sides.)}$$

$$19.1 = \frac{V_2}{60 \ \mu V} \quad \text{(Place the left side over 1 and cross multiply.)}$$

$$= 1.14 \text{ mV}$$

The output of the second amplifier is 1.1 mV.   ▲

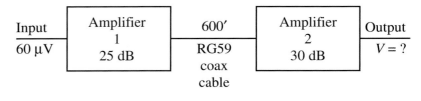

$$dB = -29.4$$

Figure 26-4. Block diagram for Example 26-8. The coaxial cable between the amplifiers results in a loss in decibels.

Sound Power and Loudness

The loudness of sound to the human ear depends on the amount of air pressure striking the eardrum, the frequency of the sound, the physical structure of the person's ear and head, and the age of the individual. If the cone of a speaker disturbs the air, vibrations will cause variations in the air pressure. The greater the variation in air pressure, the louder the sound.

Since the ear is most sensitive to the lower and middle ranges of the audio spectrum (250 Hz to 8 kHz), a 1000 Hz signal is standard for testing circuits and equipment. A graph of sound levels in decibels and frequency is called a ***response curve,*** **Figure 26-5.** The graph indicates how the ear responds to a given set of ac frequencies. Most graphs of audio frequencies use semilog scales, which relate the graph to the logarithmic characteristic of the ear.

Response curve: A graph that plots the output in relation to the frequency of a circuit or device.

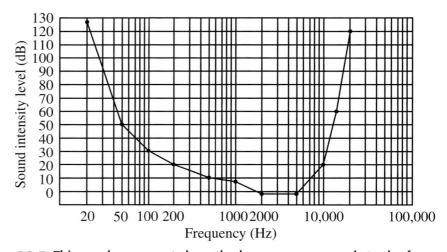

Figure 26-5. This graph represents how the human ear responds to the frequency of sound. The ear is most sensitive to sound frequencies between 500 Hz and 5,000 Hz.

Another factor determining the loudness of sound to the human ear is the structure of the ear and head. Every person has different resonant cavities.

Finally, the age of a person is a factor in the loudness of sound. As old age sets in, a loss of hearing is often found at the high end of the normal speech spectrum (200 kHz to 8 kHz). The amount of hearing loss also depends on the amount of abuse to a person's ears over a lifetime.

The common reference for decibels is the threshold of hearing (0 dB) to the threshold of pain (140 dB). See **Figure 26-6.**

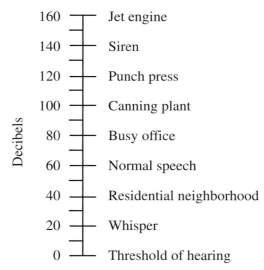

Figure 26-6. Common sounds and corresponding decibels.

Passive-Active Devices

Devices are classified as either passive or active. ***Passive devices*** include resistors, inductors, and capacitors. These components allow a signal to pass through them. Due to the internal losses of the device, the output is usually less in amplitude than the input. The output characteristics depend on whatever change the device causes.

Active devices include transistors, integrated circuits, op-amps, and optocouplers. Active devices provide amplification or some control over the output. Opinions differ as to which class diodes belong. Some people put diodes in the active class because of their solid-state nature. Others place diodes in the passive class because they only pass a changed and reduced input to the output.

Filters

Filtering is the process of separating one group of frequencies from another. The four general types of passive filter circuits are low-pass, high-pass, band-pass, and bandstop. These circuits can take many different forms using resistors, inductors, and capacitors.

Filters can be designed using two methods. The ***constant-k*** filter design presents an impedance match to the line at only one frequency and a mismatch at all the other frequencies. With the ***m-derived*** filter design, either the impedance or the ***attenuation*** characteristics of the filter circuit can be controlled. Attenuation means to decrease or reduce.

Low-pass Filter

A low-pass filter passes the lower frequencies and attenuates the higher frequencies. The point of pass or reject is determined by the ***cutoff frequency,*** the frequency at which the output is reduced by 3 dB of the maximum amplitude. See **Figure 26-7.**

High-pass Filter

A high-pass filter passes the frequencies higher than the cutoff frequency and attenuates the frequencies lower than the cutoff frequency. See **Figure 26-8.**

Passive device: An idle component that may control or change, but does not create or amplify, energy.

Active device: An electronic component whose output responds to its input.

Constant-k: Refers to a filter in which the values of inductance and capacitance are designed so the product of X_L and X_C is constant at all frequencies.

M-derived (filter): A modified form of the constant-k filter, based upon the ratio of the filter cutoff frequency to the frequency of infinite attenuation.

Attenuation: The decrease in amplitude or intensity of a signal.

Cutoff frequency: The frequency at which the gain of an amplifier or circuit falls below 0.707 times the maximum gain (3 dB).

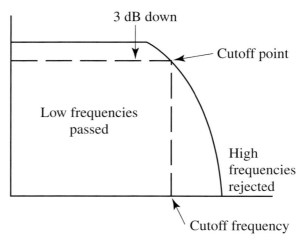

Figure 26-7. The 3 dB down cutoff point determines the cutoff frequency. The lower frequencies are passed while the high frequencies are blocked.

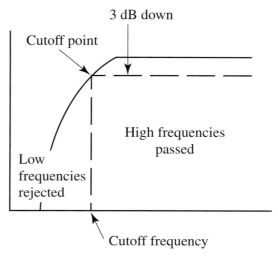

Figure 26-8. The lower frequencies are blocked in a high-pass response curve.

Types of Filter Circuits

Constant-k filters are designed for general use filtering and are of three types: L, T, or pi. The filters are combinations of resistors, capacitors, and inductors. The attenuation of the L section is equal to half the T or pi sections. **Figures 26-9, 26-10,** and **26-11** illustrate the low-pass filter circuits.

The values for *L, C, Z,* and f_C can be calculated using the equations:

$$L = \frac{Z}{\pi f_C}$$

$$C = \frac{1}{\pi f_C Z}$$

$$Z = \sqrt{\frac{L}{C}}$$

$$f_C = \frac{1}{\pi \sqrt{LC}}$$

Notice the values of *L* and *C* must be divided in half.

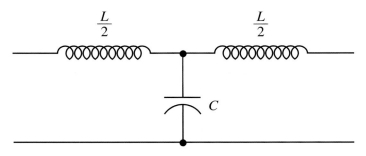

Figure 26-9. T low-pass filter.

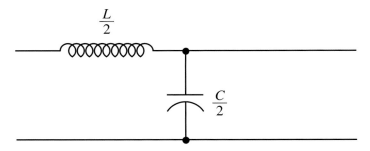

Figure 26-10. L low-pass filter.

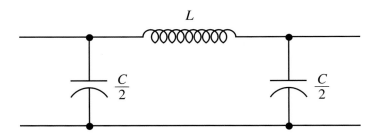

Figure 26-11. Pi low-pass filter.

Figures 26-12, 26-13 and **26-14** illustrate high-pass filter circuits. The values for *L*, *C*, *Z*, and f_C can be calculated using the equations:

$$L = \frac{Z}{4\pi f_C}$$

$$C = \frac{1}{4\pi f_C Z}$$

$$Z = \sqrt{\frac{L}{C}}$$

$$f_C = \frac{1}{4\pi\sqrt{LC}}$$

The value for *C* must be doubled in the T and L sections. Likewise, the value of *L* must be doubled in the L and pi sections.

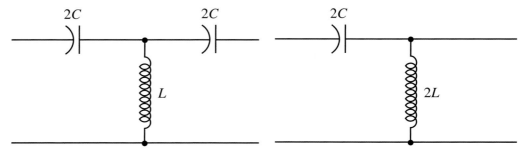

Figure 26-12. T high-pass filter.   **Figure 26-13.** L high-pass filter.

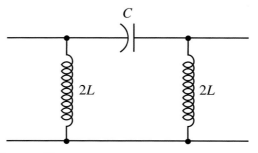

Figure 26-14. Pi high-pass filter.

Bandpass Filter

Bandpass filter circuits are designed to pass a certain band of frequencies and reject all others, **Figure 26-15.** The calculations for *L* and *C* are:

$$L_1 = \frac{Z}{\pi (f_2 - f_1)}$$

$$L_2 = \frac{(f_2 - f_1)}{4\pi f_1 f_2}$$

$$C_1 = \frac{(f_2 - f_1)}{4\pi f_1 f_2 Z}$$

$$C_2 = \frac{1}{\pi (f_2 - f_1) Z}$$

$$f_m = \frac{1}{2\pi \sqrt{L_1 C_1}}$$

$$Z = \sqrt{\frac{L_1}{C_2}} = \sqrt{\frac{L_2}{C_1}}$$

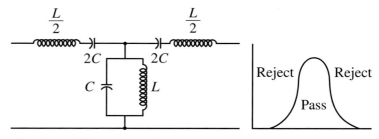

Figure 26-15. Bandpass filter network.

Bandstop Filter

Bandstop filter circuits are designed to stop or reject a certain band of frequencies, **Figure 26-16.** The output consists of a certain band of the input frequencies. The calculations for L and C are:

$$L_1 = \frac{(f_2 - f_1)}{\pi f_1 f_2}$$

$$L_2 = \frac{Z}{4\pi\,(f_2 - f_1)}$$

$$C_1 = \frac{1}{4\pi\,(f_2 - f_1)\,Z}$$

$$C_2 = \frac{(f_2 - f_1)}{\pi f_1 f_2 Z}$$

$$f_m = \frac{1}{2\pi\sqrt{L_1 C_1}}$$

$$Z = \sqrt{\frac{L_1}{C_2}} = \sqrt{\frac{L_2}{C_1}}$$

Where:
f_1 and f_2 = Bandpass frequencies
f_m = Center frequency

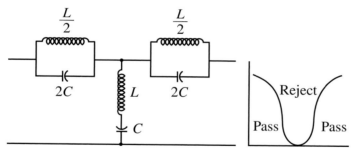

Figure 26-16. Bandstop (also called band-reject) filter network.

By using the m-derived filter design, you can control either the impedance or attenuation characteristics of the filter. The values are calculated as for a constant-k filter, then modified by an algebraic expression containing the constant m. The value of m is determined by an involved formula. The best impedance match is obtained when m is equal to 0.6; therefore, this value is usually used for m. **Figure 26-17** illustrates the m-derived low-pass filter circuits. The calculations for L and C are:

$$L_1 = m\left(\frac{Z}{Z\pi f_C}\right)$$

$$L_2 = \left(\frac{1-m^2}{4m}\right) \times \left(\frac{Z}{2\pi f_C}\right)$$

$$C_2 = m\left(\frac{1}{\pi f_C Z}\right)$$

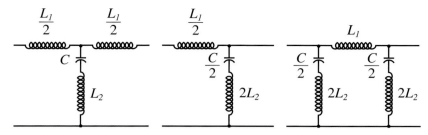

Figure 26-17. M-derived low-pass filter networks.

Figure 26-18 illustrates the m-derived high-pass filter circuits. The calculations for *L* and *C* are:

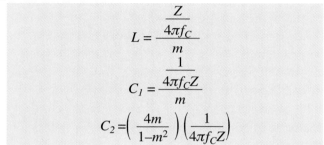

$$L = \frac{\dfrac{Z}{4\pi f_C}}{m}$$

$$C_1 = \frac{\dfrac{1}{4\pi f_C Z}}{m}$$

$$C_2 = \left(\frac{4m}{1-m^2}\right)\left(\frac{1}{4\pi f_C Z}\right)$$

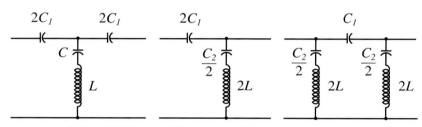

Figure 26-18. M-derived high-pass filter networks.

Response Curves

As discussed earlier, a response curve indicates how the ear responds to a given set of ac frequencies. Response curves are also used to graphically display how filter circuits, amplifiers, and other electronic circuitry respond to a given set of ac frequencies. The curve may indicate decibels-to-frequency, voltage-to-frequency, or any other desired quantities. For example, a graph of an RC circuit can be considered a response curve of voltage or current and time.

Making a Response Curve

With all the equipment available to show circuit response, making a response curve may seem to have no value. However, not everyone has the luxury of such sophisticated equipment. To make a response curve, apply what you learned in Chapter 13 about making a graph. The information to be illustrated is usually voltage or decibels as related to frequency. The graph is often made on semilog graph paper. Follow this procedure:

1. Set up the circuit or device and equipment for making the test,
   Figure 26-19. Allow at least five minutes for the equipment to warm up and stabilize. A wide-band oscilloscope is a good device for measuring the input and output voltages.

2. Determine the low- and high-frequency limits desired, and mark the graph for the frequency scale.
3. Adjust the output voltage of the generator to get a workable input voltage for the circuit or device under test. The output should not be distorted. *The input voltage level must be maintained at this level.*
4. Scan the generator frequencies and note the low and high limits of the peak-to-peak voltage. This will also ensure the test is set up correctly and working properly.
5. Make a chart of the different frequencies, including a column for voltage and decibels.
6. Starting with the lowest frequency, measure the peak-to-peak voltage. Record it on the chart.
7. Adjust the generator to the next frequency. Check the input voltage and adjust the level to maintain the voltage determined in step 3.
8. Measure the output voltage and record its value in the chart.
9. Continue this process with each frequency until the chart is completed. Do *not* turn off any equipment. You may need to return to check a measurement or make a voltage measurement at a different frequency.
10. If desired, calculate the decibels and complete the chart.
11. Transfer the voltage measurements, or decibels, to the graph paper.
12. Using a French curve, connect the dots to make a smooth, curved line.
13. Label the chart.

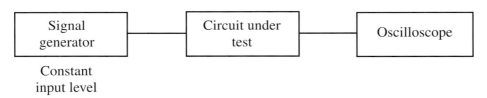

Constant
input level

Figure 26-19. Block diagram of equipment for making a response curve.

Speaker Crossover Networks

A high-quality speaker system will be able to produce sound at three classes of frequencies. The speakers must be the proper size for the band of frequencies desired. Ordinary 4″ and 6″ speakers are used for *midrange frequencies.* Midrange frequencies span approximately 1 kHz to 4 kHz. A small speaker called a *tweeter* is required for reproducing the high frequencies (above 4 kHz). Low frequencies (below 1 kHz) are reproduced by a large speaker called a *woofer.*

With three ranges of frequencies, circuitry based on the X_L and X_C characteristics are used to direct the specified frequencies to the desired speaker. With higher frequencies, X_L increases so the inductor will pass the lower frequencies. With higher frequencies, X_C decreases to allow a capacitor to pass the higher frequencies.

An LC network, called a *crossover network,* is used to direct the desired frequencies to the proper speaker, **Figure 26-20.** The better crossover systems use three-way circuitry to produce better sound, **Figure 26-21.**

Attenuator Circuits

An *attenuator* is a network of noninductive resistors used to reduce the strength of a signal without introducing distortion. Attenuators can be designed to work between equal or unequal impedances and are often used as impedance matching networks. The resistor used may be fixed or variable.

Midrange frequencies: Audio frequencies that span approximately 1 kHz to 4 kHz.

Tweeter: A small speaker designed to reproduce high frequencies.

Woofer: A speaker used to reproduce the lower frequencies.

Crossover network: A filter for a multispeaker system that separates the output signal of an amplifier into two or more frequency bands.

Attenuator: A resistive network used to decrease the amplitude of a signal.

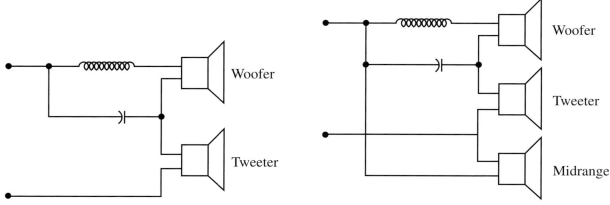

Figure 26-20. A two-way speaker crossover network only has provisions for the woofer and tweeter. The midrange frequencies are not considered.

Figure 26-21. A three-way speaker crossover network has provisions for high, low, and mid-range frequencies.

An attenuator placed between unequal impedances must introduce a certain minimum insertion loss, **Figure 26-22.** The impedance ratio equals the input impedance divided by the output impedance, or the reciprocal of the impedance ratio, whichever is greater than 1.

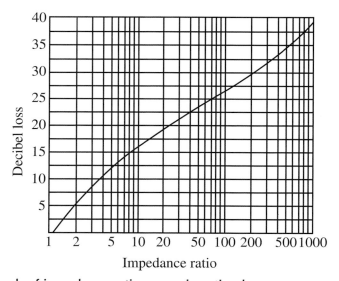

Figure 26-22. Graph of impedance ratio versus insertion loss.

The *K*-factor is used in the calculation of the resistor values of attenuator networks. It is the ratio of voltage, current, or power corresponding to a value of attenuation in decibels. **Figure 26-23** gives the value of *K* for the more common voltage loss values.

Three common types of attenuator circuits are presented in the following equations. (Refer to a design manual for a complete listing of networks.)

- *T, equal impedance,* **Figure 26-24.**
- *T, unequal impedance,* **Figure 26-25.**
- *Pi, equal impedance,* **Figure 26-26.**
- *Pi, unequal impedance,* **Figure 26-27.**
- *O, balanced,* **Figure 26-28.**
- *O, unbalanced,* **Figure 26-29.**

dB	K	dB	K	dB	K	dB	K
0.05	1.0058	9.5	2.9854	29	28.184	49	281.84
0.1	1.0116	10	3.1623	30	31.623	50	316.23
0.5	1.0593	11	3.5481	31	35.481	51	354.81
1.0	1.1220	12	3.9811	32	39.811	52	398.11
1.5	1.1885	13	4.4668	33	44.668	53	446.68
2.0	1.2589	14	5.0119	34	50.119	54	501.19
2.5	1.3335	15	5.6234	35	56.234	55	562.34
3.0	1.4125	16	6.3096	36	63.096	56	630.96
3.5	1.4962	17	7.0795	37	70.795	57	707.95
4.0	1.5849	18	7.9433	38	79.433	58	794.33
4.5	1.6788	19	8.9125	39	89.125	59	891.25
5.0	1.7783	20	10.000	40	100.0	60	1000.0
5.5	1.8837	21	11.2202	41	112.202	65	1778.3
6.0	1.9953	22	12.589	42	125.89	70	3162.3
6.5	2.1135	23	14.125	43	141.25	75	5623.4
7.0	2.2387	24	15.849	44	158.49	80	10,000
7.5	2.3714	25	17.783	45	177.83	85	17,783
8.0	2.5119	26	19.953	46	199.53	90	31,623
8.5	2.6607	27	22.387	47	223.87	95	56,234
9.0	2.8184	28	25.119	48	251.19	100	100,000

Figure 26-23. Table of *K*-factors.

$$R_1 \text{ and } R_2 = \left(\frac{K-1}{K+1}\right) Z$$
$$R_3 = \left(\frac{K}{K^2-1}\right) 2Z$$

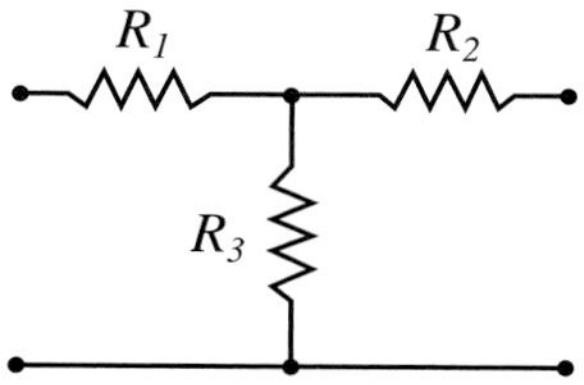

Figure 26-24. Equal impedance T-attenuator.

$$R_1 = Z_1\left(\frac{K^2+1}{K^2-1}\right) - 2\sqrt{Z_1 Z_2}\left(\frac{K}{K^2-1}\right)$$
$$R_2 = Z_2\left(\frac{K^2+1}{K^2-1}\right) - 2\sqrt{Z_1 Z_2}\left(\frac{K}{K^2-1}\right)$$
$$R_3 = 2\sqrt{Z_1 Z_2}\left(\frac{K}{K^2-1}\right)$$

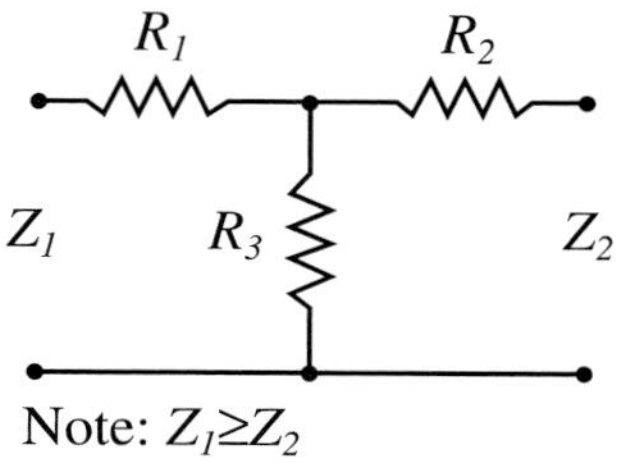

Note: $Z_1 \geq Z_2$

Figure 26-25. Unequal impedance T-attenuator.

$$R_1 = Z\left(\frac{K+1}{K-1}\right)$$
$$R_2 = \left(\frac{Z}{2}\right)\left(\frac{K^2-1}{K}\right)$$

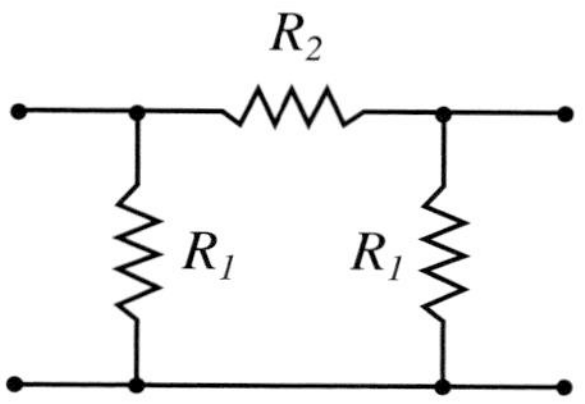

Figure 26-26. Equal impedance pi attenuator.

$$R_1 = Z_1\left(\frac{K^2-1}{K^2-2KS+1}\right)$$
$$R_2 = \left(\frac{\sqrt{Z_1 Z_2}}{2}\right)\left(\frac{K^2-1}{K}\right)$$
$$R_3 = Z_2\left(\frac{K^2-1}{K^2-2\frac{K}{S}+1}\right)$$ 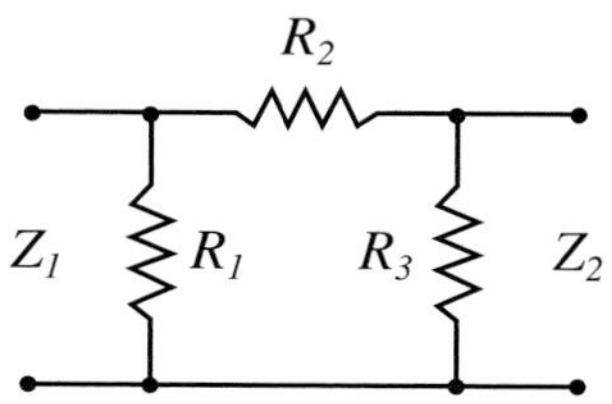
$$S = \sqrt{\frac{Z_1}{Z_2}}$$

Figure 26-27. Unequal impedance pi attenuator.

$$R_1 = Z\left(\frac{K+1}{K-1}\right)$$

$$R_2 = \frac{\left(\frac{Z}{2}\right)\left(\frac{K^2-1}{K}\right)}{2}$$

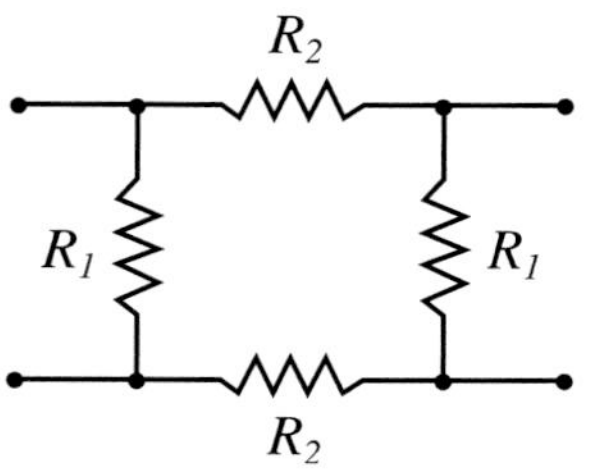

Figure 26-28. Balanced O attenuator.

$$R_1 = Z_1\left(\frac{K^2-1}{K^2-2KS+1}\right)$$

$$R_2 = \left(\frac{\sqrt{Z_1 Z_2}}{2}\right)\left(\frac{K^2-1}{K}\right)$$

$$R_3 = Z_2\left(\frac{K^2-1}{K^2-2\frac{K}{S}+1}\right)$$

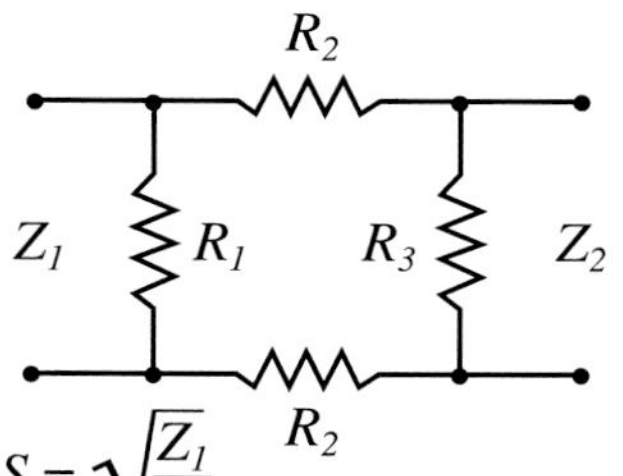

$$S = \sqrt{\frac{Z_1}{Z_2}}$$

Figure 26-29. Unbalanced O attenuator.

The four steps in designing attenuator networks are:
1. Determine the type of network required.
2. If the impedances are unequal, calculate the impedance ratio and refer to Figure 26-21 for the minimum loss value.
3. Find the value of K for the loss value. Refer to Figure 26-23.
4. Using the formulas in Figures 26-24 through 26-29, calculate the resistor values.

Summary

- The decibel unit is used to measure the gain or loss in sound, audio, or radio signals.
- Two types of logarithms are common and natural.
- When only one value of power or voltage is known, the other is assumed.
- The broadcasting industry uses the VU unit of measurement for audio levels.
- Passive devices include resistors, capacitors, and inductors. Active devices include transistors, integrated circuits, op-amps, and optocouplers.
- Filtering is the process of separating one group of frequencies from another group. Filters can be low-pass, high-pass, bandpass, or bandstop.
- Speaker crossover networks direct a certain band of frequencies to a specific speaker.
- Attenuators are used to reduce the strength of a signal.

Important Terms

Do you know the meanings of these terms used in the chapter?

active device	logarithm
antilogarithm	m-derived
attenuation	midrange frequencies
attenuator	passive device
constant-k	response curve
crossover network	tweeter
cutoff frequency	volume unit
decibel	woofer

Questions and Problems

Please do not write in this text. Write your answers on a separate sheet of paper.

1. Find the natural logarithm of the following numbers.
   a. 600
   b. 25
   c. 470,000
   d. 133
2. Find the common logarithm of the following numbers.
   a. 1500
   b. 26
   c. 450
   d. 350,000
   e. 1,465,000
   f. 12
3. Find the antilog of the following numbers.
   a. 1.69897
   b. 3.0791812
   c. 3.748188
   d. 6.27
   e. 2.33
   f. 0.60206
4. What is the dB gain for an increase of voltage from 3 mV to 15 V?
5. What is the dB loss for an input power of 35 W and an output of 26 W?
6. What do the following represent?
   a. dBm
   b. dBmV
   c. dB
7. An amplifier has a 500 Ω impedance and an output power of 45 W. What is the dB gain of the amplifier?
8. The human ear is most sensitive to what range of frequencies?
9. Explain the difference between passive and active devices.
10. Calculate the cutoff frequency for the Pi filter in **Figure 26-30.**
11. If the impedance is 600 Ω and the cutoff frequency is 2 kHz, what is the required inductance and capacitance for a low-pass T filter?
12. If the impedance is 800 Ω and the cutoff frequency is 5 kHz, calculate the inductance and capacitance for an m-derived high-pass pi filter.
13. What are the three types of speakers?
14. What is a crossover network?
15. Design a T attenuator having equal impedances of 800 Ω and 25 dB attenuation.
16. Design a type O balanced attenuator for 8 dB attenuation with equal impedances of 600 Ω.

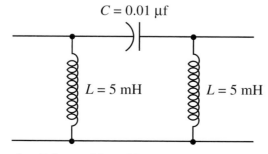

$C = 0.01 \mu f$

$L = 5$ mH $L = 5$ mH

Figure 26-30. Pi filter circuit for problem 10.

Chapter 27 Graphic Overview

Complex Circuit Analysis

for → **Parallel Circuits** — change → **Reactances** — to → **Susceptances** — and → **Impedances** — to → **Admittances**

Susceptances and **Admittances** — are → **Reciprocals**

Parallel Circuits / **Series Circuits** — use → **Product** — over → **Sum**

Complex Plane — makes a → **X-axis** / **J-axis**

Resistance Values — on the → **X-axis**

Reactance Values — on the → **X-axis** or **J-axis**

uses → **Complex Numbers** — that is → **Real Number** — and → **Imaginary Number**

Resistance Values used for **Real Number**

Reactance Values used for **Imaginary Number**

Imaginary Number — is the → **Square Root** — of → **−1** — is called → **J-operator**

464

COMPLEX CIRCUIT ANALYSIS 27

Objectives

After studying this chapter, you will be able to:
- ○ Locate points on a complex plane.
- ○ Express a power of j in its simplest form.
- ○ Add, subtract, multiply, and divide complex numbers.
- ○ Apply complex numbers to RCL circuit analysis.

Introduction

Complex numbers form a number system that includes magnitude and phase angle. Since the phase angle is included, the use of complex numbers is the best way to analyze ac circuits. The phase angle is never in question.

Imaginary Numbers and the *J*-Operator

The square root of 36 is easily found on your calculator, but try to find the square root of −36. Your calculator will show ERROR. A negative number can be regarded as the product of −1 and a positive number. For example, −16 can be thought of as 16 times −1.

The square root of −16 can be written:

$$\sqrt{-16} = \sqrt{16}\,(-1)$$
$$\sqrt{16\,(-1)} = \sqrt{16} \times \sqrt{(-1)}$$
$$= 4\sqrt{-1}$$

▼ **Example 27-1:**

$$\sqrt{-36} = \sqrt{36} \times \sqrt{-1}$$
$$= 6\sqrt{-1}$$
$$= j6$$

▲

▼ **Example 27-2:**

$$= \sqrt{-16X^2}$$
$$= \sqrt{16X^2} \times \sqrt{-1}$$
$$= 4X \times \sqrt{-1}$$
$$= 4Xj$$

▲

The square root of −1 is called an ***imaginary number.*** The letter i is used in place of $\sqrt{-1}$; however, since i represents current in electricity, j is used in electronics. It is called the j-operator.

Complex Numbers

A ***complex number*** is a real number and imaginary number united by a plus or minus sign. Imaginary numbers are used to describe vectors or phasors; therefore, imaginary numbers can be used to represent reactance values.

Imaginary number: A number pertaining to the square root of a negative number and used to represent a reactance.

Complex number: A two-part number containing a real number and imaginary number (also known as a j-number).

In the following example, the real numbers represent resistances, while the imaginary numbers represent reactances.

▼ *Example 27-3:*

$5 + j7$

$16 - j14$

$18.6 + j4$

$1400 - j427$

$900 + j80$

$7.2 - j27.54$ ▲

Complex numbers can be expressed in rectangular, polar, or exponential form:

- Rectangular: $15 + j6$
- Polar: $12 \angle 35°$
- Exponential: $13e^{jX}$ (The j exponent is expressed in radians.)

In **Figure 27-1,** the Y-axis is changed to the imaginary or j-axis, and the graph figure is called a ***complex plane.*** The location of points on the complex plane is the same as when using the X and Y axis labels. For rectangular notation, the real number is placed on the X-axis and the imaginary number on the j-axis. Polar notation can be used for complex numbers, **Figure 27-2.** *Be sure to mark the correct sign of each number.*

Complex plane: A rectangular method of graphing using an imaginary axis and real axis.

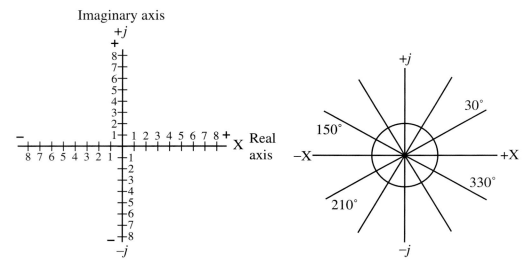

Figure 27-1. The complex plane. The j-number replaces the Y-axis.

Figure 27-2. Polar notation of complex numbers.

Powers of *J*

The powers of j sequence repeats after j is raised to the fourth power. The sequence is $j, -1, -j, 1$, and the cycle repeats. This makes it possible to reduce a power of j to its simplest form.

The powers of j are:

$j^1 = \sqrt{-1} = j$

$j^2 = j \times j = \sqrt{-1} \times \sqrt{-1} = -1$

$j^3 = j^2 \times j^1 = -1 \times j = -j$

$j^4 = j^2 \times j^2 = -1 \times -1 = 1$

$j^5 = j^3 \times j^2 = -j \times -1 = j$

$j^6 = j^4 \times j^2 = 1 \times -1 = -1$

To express a power of j in its simplest form, divide the power by 4 and consider only the *fraction* of 4 remaining. The quantity of fours, no matter how many, is equal to 1 and can be ignored.

▼ ***Example 27-4:***

$$j^7 = j^4 \times j^3$$
$$= 1 \times j^3$$
$$= -j$$

▲

▼ ***Example 27-5:***

$$j^{41} = j^4 \times j^4 \times j^4 \times j^4 \times j^4 \times j^4 \times j^4 \times j^4 \times j^4 \times j^4 \times j^1$$
$$= 1 \times 1 \times 1 \times 1 \times 1 \times 1 \times 1 \times 1 \times 1 \times 1 \times j^1$$
$$= 1 \times j^1$$
$$= j$$

▲

▼ ***Example 27-6:***

$$j^{18} = j^4 \times j^4 \times j^4 \times j^4 \times j^2$$
$$= 1 \times 1 \times 1 \times 1 \times -1$$
$$= -1$$

▲

Using the *J*-Operator in Math

The *j*-operator can be used to add, subtract, multiply, and divide complex numbers. These mathematical processes are often necessary in electrical calculations.

Adding Complex Numbers

The addition of complex numbers follows the rules for adding algebraic numbers.

▼ ***Example 27-7:***

Add $5 + j6$, $3 + j5$, and $4 - j3$.

$$5 + j6$$
$$3 + j5$$
$$+ \ 4 - j3$$
$$\overline{12 + j8}$$

▲

Subtracting Complex Numbers

To subtract complex numbers, change the sign of the subtrahend (the number to be subtracted), and add the two complex numbers.

▼ ***Example 27-8:***

Subtract $2 - j6$ (the subtrahend) from $8 + j9$.

$$8 + j9 - (2 - j6)$$

Change the signs of the subtrahend.

$$8 + j9 - 2 + j6$$

Add.

$$8 + j9$$
$$+ \ -2 + j6$$
$$\overline{6 + j15}$$

▲

▼ ***Example 27-9:***
Subtract $4 + j12$ from $12 + j6$.
$$12 + j6 - (4 + j12)$$
Change the signs of the subtrahend and add.

$$\begin{array}{r} 12 + j6 \\ + \; {-4} - j12 \\ \hline 8 - j6 \end{array}$$

▲

Multiplying Complex Numbers

Although multiplication of complex numbers is possible using algebraic multiplication, it may be more easily accomplished by converting the rectangular notation to polar notation, then multiplying. Multiplying polar notation is done by multiplying the radii and adding the angles.

▼ ***Example 27-10:***
Multiply $8 \angle{-60°}$ by $2 \angle{20°}$.

$$\begin{array}{r} 8 \angle{-60°} \\ \times \; 2 \angle{20°} \\ \hline 16 \angle{-40°} \end{array}$$

▲

Dividing Complex Numbers

As in multiplication of complex numbers, division is possible using algebraic multiplication. It too may be more easily accomplished by converting the rectangular notation to polar notation, then dividing. Dividing polar notation is done by dividing the radii and subtracting the angles.

▼ ***Example 27-11:***
Divide $18 \angle{80°}$ by $3 \angle{20°}$.

$$\frac{18 \angle{80°}}{3 \angle{20°}}$$
$$= 6 \angle{60°}$$

▲

RCL Circuit Analysis

Complex numbers can be used for RCL circuit analysis. Solving problems for series RCL circuits, parallel RCL circuits, parallel complex impedances, parallel complex currents, and three or more parallel complex impedances are presented next.

Applying Complex Numbers to Series RCL Circuits

By using complex numbers, the solution of RCL circuits is an easy process. The inductive reactance is a positive j value, and the capacitive reactance is a negative j value. The resistance is the real number in the complex number.

▼ ***Example 27-12:***

Find the current in **Figure 27-3.**

Add.

$$X_L = 0 + j600$$
$$X_C = 0 - j300$$
$$+ \quad R \ = 400 + j0$$
$$Z \ = 400 + j300$$

Change the sum to polar notation.

$$= 500 \angle 36.9°$$

Find the current.

$$I \ = \frac{V}{Z}$$

$$= \frac{15 \angle 0°}{500 \angle 36.9°}$$

$$= 30 \angle{-36.9°} \text{ mA}$$

The negative current angle indicates the current is lagging the voltage by 36.9°. ▲

▼ ***Example 27-13:***

Find the current in **Figure 27-4.**

Add.

$$X_L = 0 + j200$$
$$X_C = 0 - j800$$
$$+ \quad R \ = 200 + j0$$
$$Z \ = 200 - j600$$

Change the sum to polar notation.

$$= 632 \angle{-71.6°}$$

Find the current.

$$I \ = \frac{V}{Z}$$

$$= \frac{50 \angle 0°}{632 \angle{-71.6°}}$$

$$= 79 \angle 71.6° \text{ mA}$$

▲

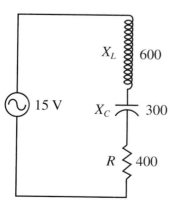

Figure 27-3. Series RCL circuit for Example 27-12.

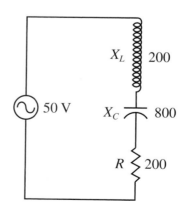

Figure 27-4. Series RCL circuit for Example 27-13.

Applying Complex Numbers to Parallel RCL Circuits

Complex numbers can be applied to parallel circuits to calculate circuit imped-ance. The circuit is changed to an equivalent series circuit.

▼   *Example 27-14:*

In **Figure 27-5,** find the circuit impedance using the product over the sum equation.
Add.

$$30 + j0$$
$$\underline{\quad 0 + j40}$$
$$+ \ 30 + j40 \text{ (sum)}$$

Multiply.

$$30 \underline{/0°}$$
$$\underline{\times 40 \underline{/90°}}$$
$$1200 \underline{/90°} \text{ (product)}$$

Convert the sum to polar notation and divide.

$$Z = \frac{1200 \underline{/90°}}{50 \underline{/53.13°}}$$
$$= 24 \underline{/36.86°} \ \Omega$$

▲

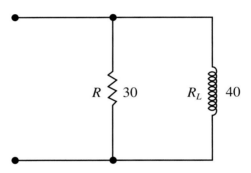

Figure 27-5. Parallel RL circuit.

Just as conductance was used in the calculation of parallel resistive circuits, parallel complex circuits can make use of the reciprocal of impedance and reactance. The reciprocal of impedance is called ***admittance*** (represented by Y). The reciprocal of reactance is called ***susceptance*** (represented by B). The equations are:

Admittance: The reciprocal of impedance.

Susceptance: The reciprocal of reactance.

$$Y = \frac{1}{Z}$$

$$B = \frac{1}{\pm X}$$

With parallel branches of conductance and susceptance, the total admittance (Y_T) is calculated as:

$$Y_T = G \pm jB$$

▼   *Example 27-15:*

Find the impedance in **Figure 27-6.**

First, change the values to conductance and susceptance.

$$25 + j0 = 0.04 + jB0$$
$$0 + j40 = 0 + jB0.025$$
$$0 - j50 = 0 - jB0.02$$

Add the rectangular values.
$$Y_T = 0.04 + jB0.005$$

Change to polar notation.
$$Y_T = 0.04031 \angle 7.13°$$

Take the reciprocal of Y_T to find the impedance.
$$Z = \frac{1}{Y_T}$$
$$= 24.8 \angle{-7.13°}\ \Omega$$
or
$$Z = \frac{1}{\sqrt{G^2 + B^2}} \quad \text{(The phase angle is lost using this equation.)}$$

▲

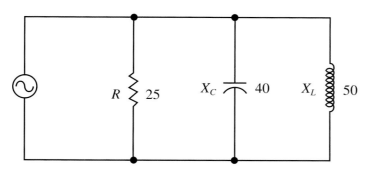

Figure 27-6. Parallel RCL circuit for Example 27-15.

Parallel Complex Impedances

To find the total impedance of two complex impedances, use the product over the sum of the impedances.

▼   ***Example 27-16:***
Find the total impedance in **Figure 27-7.** Write the impedances in rectangular notation and convert them to polar notation.

Add.
$$\begin{aligned} Z_1 &= 12 - j15 \\ + Z_2 &= 8 + j4 \\ \hline &= 20 - j11 \end{aligned}$$

Change the sum to polar notation.
$$22.83 \angle{-28.8°}$$

Convert each to polar notation and multiply.
$$\begin{aligned} &\ 19.2 \angle{-51.3°} \\ \times\ &8.94 \angle{26.56°} \\ \hline &171.7 \angle{-24.73°} \end{aligned}$$

Divide the product by the sum.
$$Z_T = \frac{171.7 \angle{-24.73°}}{22.83 \angle{-28.8°}}$$
$$= 7.52 \angle{4.07°}\ \Omega$$

▲

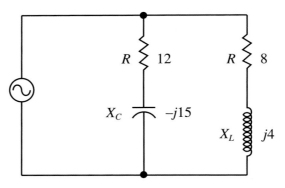

Figure 27-7. Series-parallel circuit for Example 27-16.

Parallel Complex Currents

With two parallel circuit currents, the total current is the sum of the complex branch currents.

▼ *Example 27-17:*

Find the total current and impedance in **Figure 27-8.**

Add the branch currents.

$$I_T = (8 + j8 \text{ A}) + (4 - j6 \text{ A})$$
$$= 12 + j2 \text{ A}$$

Change to polar notation.

$$I_T = 12.17 \angle 9.46° \text{ A}$$

The impedance is the voltage divided by the current.

$$Z = \frac{V}{I}$$
$$= \frac{60 \angle 0°}{12.17 \angle 9.46°}$$
$$= 4.93 \angle -9.46° \text{ } \Omega$$

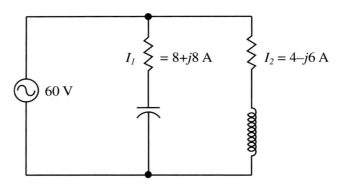

Figure 27-8. Series-parallel circuit for Example 27-17.

Three or More Parallel Complex Impedances

When more than two impedances are in parallel, the branch currents are used to calculate the total impedance. The process is quite involved, with considerable conversion between rectangular and polar notation.

The steps in the calculations are:
1. Convert each branch impedance to polar notation for dividing each impedance into the source voltage. This gives each of the branch currents.
2. Convert the branch currents to rectangular notation so they may be added to find the total current.
3. Convert the total current to polar notation for dividing into the source voltage to find the impedance.

▼   ***Example 27-18:***

Find the total current and impedance in **Figure 27-9.**
Convert the branch impedances to polar notation.

$$Z_1 = 25 + j30 = 39.05 \angle 50.2°$$
$$Z_2 = 30 + j40 = 50 \angle 53.13°$$
$$Z_3 = 60 - j60 = 84.85 \angle{-45°}$$

Calculate each of the branch currents.

$$I_1 = \frac{V}{Z_1}$$
$$= \frac{120}{39.05 \angle 50.2°}$$
$$= 3.07 \angle{-50.2°}$$

$$I_2 = \frac{V}{Z_2}$$
$$= \frac{120}{50 \angle 53.13°}$$
$$= 2.4 \angle{-53.13°}$$

$$I_3 = \frac{V}{Z_3}$$
$$= \frac{120}{84.85 \angle{-45°}}$$
$$= 1.414 \angle 45°$$

Convert each of the branch currents to rectangular notation. Add them together to find the total current.

$$I_1 = 1.97 - j2.36$$
$$I_2 = 1.44 - j1.92$$
$$+ \quad I_3 = 0.999 + j0.999$$
$$\overline{\quad I_T = 4.409 - j3.281 \quad}$$

Convert the total current to polar notation and divide into the source voltage to find the impedance.

$$I_T = 4.409 - j3.281$$
$$= 5.5 \angle{-36.66°}$$

$$Z_T = \frac{V}{I_T}$$
$$= \frac{120 \angle 0°}{5.5 \angle{-36.66°}}$$
$$= 21.82 \angle 36.66° \ \Omega$$

▲

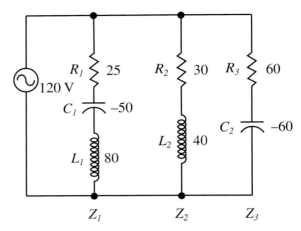

Figure 27-9. Series-parallel circuit for Example 27-18.

Summary

- A complex number is a real number combined with an imaginary number.
- A graph using complex numbers is called a complex plane.
- Powers of j can be simplified to the lowest term of j.
- The process for the solution of RCL circuits is to add the complex numbers, change the total to polar notation, and perform the mathematical operations using polar numbers.

Important Terms

Do you know the meanings of these terms used in the chapter?

admittance
complex number
complex plane

imaginary number
susceptance

Questions and Problems

Please do not write in this text. Write your answers on a separate sheet of paper.

1. Convert the following rectangular notations to polar notation.
   a. $4 + j15$
   b. $130 - j60$
   c. $45 + j45$
   d. $6 - j25$

2. Express the following numbers in terms of j.
   a. $\sqrt{-9}$
   b. $\sqrt{-49}$
   c. $\sqrt{-144}$
   d. $\sqrt{-32}$
   e. $\sqrt{-8X^2}$
   f. $-\sqrt{-169}$
   g. $\sqrt{-81a^4b^2}$
   h. $\sqrt{-64X^4Y^2}$

3. Simplify the following numbers in terms of j.
   a. $j7$
   b. $j16$
   c. $j31$

   d. *j*36

   e. *j*21

   f. *j*54

   g. *j*27

   h. *j*93

   i. *j*13

   j. *j*73

4. Convert the following polar notation to rectangular notation. (*Note:* Polar-rectangular notation conversions were discussed in Chapter 13.)

   a. $15 \angle 100°$

   b. $6 \angle 90°$

   c. $22 \angle -60°$

   d. $100 \angle -46°$

   e. $18.5 \angle 3°$

   f. $40 \angle 53°$

   g. $12 \angle -32°$

   h. $40 \angle -116°$

5. Solve **Figure 27-10** for the total impedance.

6. Solve **Figure 27-11** for the total current and impedance.

7. Solve **Figure 27-12** for the total current and impedance.

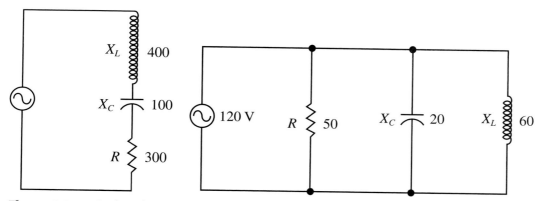

Figure 27-10. Series circuit for problem 5.

Figure 27-11. Parallel circuit for problem 6.

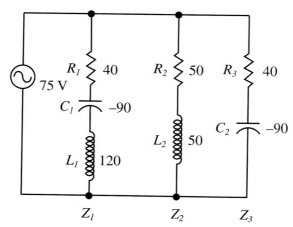

Figure 27-12. Series-parallel circuit for problem 7.

Section V

Ac Power and Motors

Section V Activity

Ladder Diagram Project

A ladder diagram helps you understand how any 60 Hz power system is wired. This information is necessary to determine if the power supply system is sufficient. By drawing a ladder diagram of the wiring, the power and current to be used by the electrical equipment can be calculated.

Objective

In this activity, you will make a ladder diagram of a wired room.

Materials and Equipment

1–Graph paper
1–Ruler
1–Pencil

Procedures

1. Using proper electrical symbols, make a ladder diagram of the electrical system assigned by your instructor.
2. Identify the fuse or circuit breaker for each part of the diagram.
3. Identify the watts of power for each load on the diagram.
4. Calculate the current for each part of the diagram.
5. Calculate the total maximum power required by each part of the diagram.
6. How do the calculated powers and currents compare with the fused circuit values?

_____

_____

_____

7. Is the fuse or circuit breaker system adequate?

_____

_____

_____

Chapter 28 Graphic Overview

Three-phase Power — is — **Voltages** — that are — **120° Apart** — generated by — **Revolving-field Alternator** — or — **Revolving-armature Alternator**

Three-phase Power — is — **More Efficient**

Three-phase Power — can be — **Wye** / **Delta** / **Four-wire Delta** — provides — **Single-phase Power**

60 Hz Power — uses — **Ladder Diagrams**

Ladder Diagrams — the — **Rules** — are — **Each Line Numbered** / **One Load per Line** / **Loads Connected to L_2**

Ladder Diagrams — are — **Read** — from — **Left** — to — **Right**

Ladder Diagrams — with — **L_1** — the — **Hot Side** — should be — **Switched**

60 Hz Power — is — **Distributed** — using — **Step-down Transformers**

Distributed — to — **Cities** / **Factories** / **Residential Homes**

60 Hz Power — generated by — **Generator** — having — **Stator** — and — **Rotor**

CAUTION — only the — **Hot Side** — should be — **Switched**

60 HERTZ AND THREE-PHASE AC POWER

28

Objectives

After studying this chapter, you will be able to:
- State the main parts of a generator.
- Describe the generation of three-phase power.
- Describe the distribution of residential power.
- Draw and read ladder diagrams.
- Diagram various three-phase transformers and a four-wire wye system.
- Describe a typical manufacturing distribution system.
- State the advantages of three-phase power.

Introduction

Nearly all homes in the United States are supplied 115 V to 125 V RMS alternating current with a frequency of 60 Hz. Electrical power is produced by large kilowatt generators and distributed by power lines across the country to homes and industries. The large motors used in industry are more efficient than those supplied to residential homes because they utilize three-phase power with higher voltages.

Generating Single-phase Power

All generators have two main parts. The *stator* is the stationary outside part and the movable inside part is the *rotor*. The rotor is mounted on a shaft that extends outside the generator case.

To generate electricity, a main magnetic field is needed. This magnetic field is placed in the stator. Commonly, the magnetic field is a permanent magnet and the armature, a coil, is rotated within the stator magnetic field. See **Figure 28-1.** Generators without a permanent magnet main field have an electromagnetic main field.

Stator: The stationary part of a motor.

Rotor: The moving part of a motor.

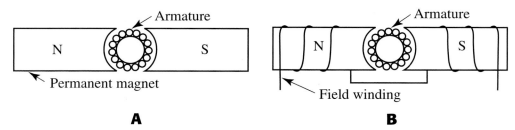

A   **B**

Figure 28-1. Generating single-phase power. A—The basic parts of a generator are a magnetic field and armature. B—A magnetic field is created by a field winding.

Generators have many uses. Large generators are found in electrical power companies, and small generators are used in homes or businesses, **Figure 28-2.**

By mounting the generator on wheels, it can be moved to a construction site and operated by means of a gasoline or diesel motor. A motor home, for instance, can stop anywhere and use the generator to run an air conditioner, microwave, and other equipment, allowing the motor home to be completely self-contained. See **Figure 28-3.**

Figure 28-2. Portable generator used for electric welding. (The Lincoln Electric Co.)

Figure 28-3. A generator mounted in a recreational vehicle, such as a motor home, provides electricity anywhere the vehicle is parked.

Power Distribution

Electrical power produced at a generating plant must be distributed to many factories, businesses, and homes. The output of the generator is connected to a transformer that steps up voltages between 12,000 V and 25,000 V to voltages between 100,000 V and 1,000,000 V. This high-voltage ac is connected to transmission (distribution) lines that take the power across the country to various substations, **Figure 28-4.** At each substation, step-down transformers reduce the ac power to between 30,000 V and 12,000 V. These voltages are again transformed to normal industrial and residential voltages.

Residential Distribution

Figure 28-5 illustrates the typical distribution of ac power in a residential home. The high voltage at the power company pole is stepped down by a transformer and brought into the residence by the service entrance cable. The service entrance cable is connected to the electric meter and distribution panel. The distribution panel can be either a fuse or circuit breaker-type control box, **Figure 28-6.**

Figure 28-4. Distribution lines carry electrical energy across the country for use in homes and factories.

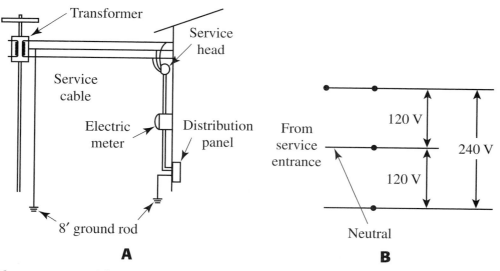

Figure 28-5. Residential distribution of ac power. A—A common method for supplying electrical power to a residence. Some entrance lines are buried underground. B—Common voltages for residential service.

 Warning: The neutral wire in a 60 Hz electrical system should never have its continuity interrupted by switches, relays, fuses, or circuit breakers. Such an interruption can create an electrical shock hazard and make it impossible to use the neutral wire as a ground return to the distribution panel. Only the hot side of the circuits should be interrupted.

The electrical system is grounded at the pole and distribution panel by a grounding rod at least 8′ long. The rod is driven completely into the ground except for a few inches. For additional information and rules on grounding, refer to the National Electrical Code.

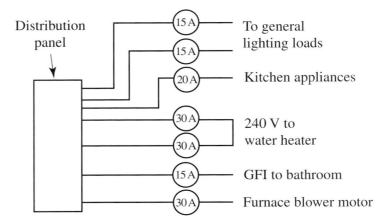

Figure 28-6. A distribution panel (circuit breaker) supplies power to different residential circuits.

Ladder Diagrams

Since all loads are connected in parallel, ladder diagrams (or line diagrams) with a 60 Hz power source are convenient to use. See **Figure 28-7.** Ladder diagrams use industrial electrical symbols instead of the common electronic schematic symbols, **Figure 28-8.**

Ladder diagrams are always read from left to right. L_1 is the hot line and L_2 is the neutral line. The lines resemble the side rails of a ladder. Each device that requires power is connected across the two lines like steps on a ladder. Although it may appear simple, the ladder diagram can become complicated when devices such as relays or multiple pole switches control more than one step on the ladder.

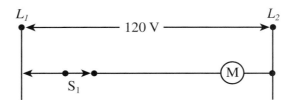

Figure 28-7. Ladder diagrams are used for schematic diagrams of 60 Hz and three-phase power.

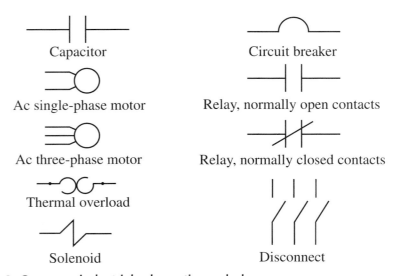

Figure 28-8. Common industrial schematic symbols.

In the ladder diagram in **Figure 28-9,** a motor is controlled by a float switch. The float switch is controlling a sump pump in the basement of a home. As water rises into the sump reservoir, the pump motor is turned on and empties the reservoir. A manual override switch is connected in parallel with the float switch and allows the sump pump to be run manually. This design is typical of a simple industrial control system. The ability to read such a diagram is essential when working with 60 Hz and three-phase power.

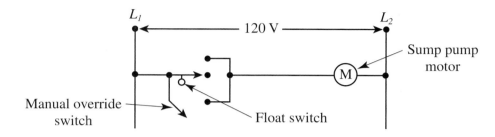

Figure 28-9. Sump pump motor control by a float switch.

The electrical industry has developed a universal set of symbols and rules for drawing ladder diagrams. Understanding how to read, interpret, and apply these symbols and rules is necessary for compliance with electrical industry standards.

Each line in a ladder diagram is numbered, starting at the top with 1 and proceeding down the diagram, **Figure 28-10.** A line is a complete path from L_1 to L_2 and contains no load. All ladder diagrams are numbered no matter how simple they are. As ladder diagrams become more complicated, the importance of numbering becomes obvious.

Only one load is connected to a line. When more than one load must be connected, additional loads are connected in parallel, **Figure 28-11.** A pilot light and solenoid are connected from L_1 to L_2 with a push-button switch. Notice the solenoid retains the line 2 identification.

As the previous diagrams have illustrated, loads are always connected to L_2. The controlling device is connected between L_1 and the load.

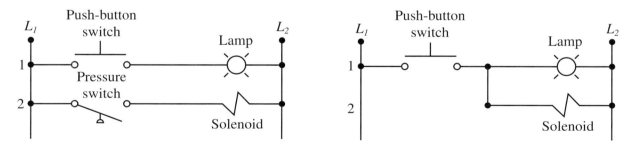

Figure 28-10. Each line of a ladder diagram should be numbered 1, 2, etc.

Figure 28-11. A circuit properly connecting two loads to one control switch.

Generating Three-phase Power

As you know, residential power is single-phase 120 V/240 V power. For more efficient use of electrical power, manufacturing and commercial industries use three-phase power.

Three-phase:
Related to a power
circuit or device
powered by a three-
terminal source 120°
apart in phase.

**Revolving-field
alternator:**
Generates an output
voltage by rotating
the magnetic field
while the output is
taken from the
stationary conductors.

Three-phase ac has three different voltages 120 degrees apart in timing, **Figure 28-12.** An oscilloscope display of three-phase alternating current is shown in **Figure 28-13.** In the most common method of generating the three phases, each phase coil in the alternator is 120 degrees apart mechanically and rotating the dc magnetic field. This is called a *revolving-field alternator.* The high power needed in many three-phase systems requires the use of large conductors and iron cores in the armature. For this reason, it is easier to make the armature stationary and rotate the magnetic field winding.

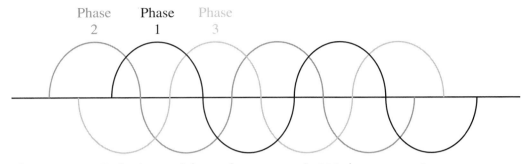

Phase 2   Phase 1   Phase 3

Figure 28-12. Each phase of three-phase power is 120 degrees apart.

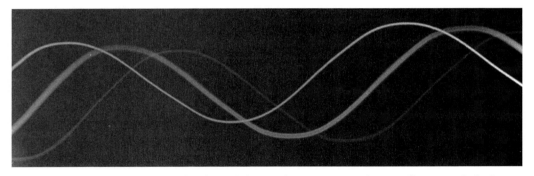

Figure 28-13. Oscilloscope display of three-phase power. The oscilloscope is being triggered on the brightest trace.

The rotor of a revolving-field alternator is made up of field poles mounted on a shaft, driven by a motor or some type of turbine. The magnetic flux produced by the rotating field cuts across the conductors of the stator winding to produce the induced current on the stator output terminals.

**Revolving-armature
alternator:**
Generates an output
voltage by turning
the armature
conductors in a
magnetic field.

The other type is the *revolving-armature alternator* which operates like the single-phase alternator. The only difference is three separate windings in the three-phase alternator instead of one set for the single-phase.

The motionless conductors of the stator windings of the revolving-field alternator are cut by the flux of the rotating field poles. In the revolving-armature alternator, the armature conductors cut the flux of the stationary field poles. With both types of alternators, the field excitation current must be dc to provide a constant magnetic field for the field poles. The field terminals of a generator or alternator are standardized by F_1 for the positive connection and F_2 for the negative field dc excitation voltage.

Three-phase Wiring

Wye: Type of
transformer
connection used in
three-phase ac
power.

Three-phase alternator output windings are usually in a *wye* (or star) connection. The three transformer windings used for either the primary or secondary form a Y shape, **Figure 28-14.** All three coils are connected at one end, with the other ends used for the output terminals. Any pair of output terminals is across two coils in series. The output voltage of this type of connection is equal to $1.73 \times 120 \text{ V} = 208 \text{ V}$.

Delta Connection

The three windings in **Figure 28-15** are in a *delta* connection. Any pair of output terminals is across one of the windings. The other windings are in a parallel connection across that output terminal. Therefore, the total current capability of the alternator is increased by the 1.73 factor.

Delta: 1. Letter of the Greek alphabet used for indicating "a change in." 2. Type of transformer connection used in three-phase ac power.

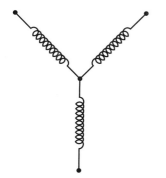

Figure 28-14. Wye transformer connection.

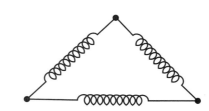

Figure 28-15. Delta transformer connection.

Three-phase Transformers

Most three-phase installations require a lower voltage to operate motors, lighting and other equipment than the voltage delivered to a manufacturing plant or commercial industrial site. Transformers used in three-phase installations can be either a single-core, three-phase transformer or three separate single-phase transformers interconnected to form the three-phase configuration. Four methods of connecting the transformers include wye-wye, wye-delta, delta-wye, and delta-delta. See **Figures 28-16** through **28-19.**

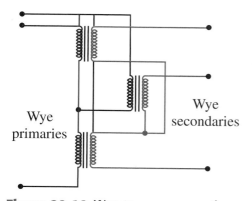

Figure 28-16. Wye-to-wye connection.

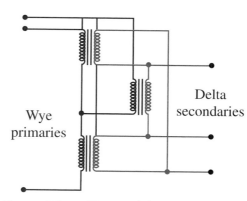

Figure 28-17. Wye-to-delta connection.

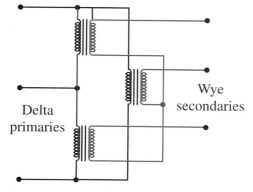

Figure 28-18. Delta-to-wye connection.

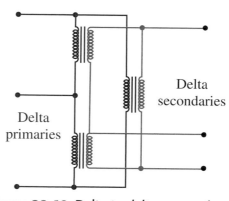

Figure 28-19. Delta-to-delta connection.

In each of the three-phase transformer methods, three transformers are required for the necessary connections. For example, in **Figure 28-20,** three single-phase transformers are connected in a wye-to-wye step-down configuration to supply the necessary voltages. This method of connection supplies three different voltages: a 208 V three-phase, a 120 V single-phase, and a 208 V single-phase. This type of service is commonly used in commercial stores, business offices, and educational facilities.

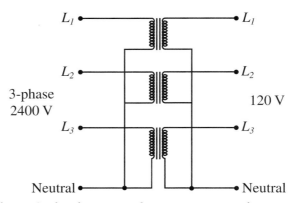

Figure 28-20. Three single-phase transformers connected wye-to-wye.

Four-wire Wye System

The center point of the wye can be used for a fourth line, the neutral conductor of this three-phase system. See **Figure 28-21.** In addition to the three-phase power, 120 V single-phase power is readily available by connecting across the neutral to any one of the three phases. Using the four-conductor system, both 208 V three-phase and 120 V single-phase power are available without special transformers or complicated wiring.

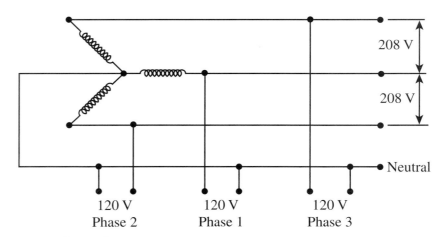

Figure 28-21. Four-wire power system.

Power Distribution Systems

Many different loads are connected to a power distribution system. Loads may include a single light circuit with one control switch, high current motor circuits, or complicated control switching and protection circuits. When the distribution system is installed, its design must be carefully considered, not only for present electrical needs, but future ones as well. A poor design could cause major electrical changes with costly redesign work a few years down the road.

Industrial manufacturing is the most common user of three-phase power. The power must be distributed over a large area with various power needs. The distribution system must provide a way to change the available power for different production machinery or assembly line requirements. The ability to make changes is accomplished with power busways in which connections are made by using a plug-in power panel. Busways allow machinery to be added or subtracted from the system as needed.

Three-phase Advantages

Manufacturing plants and commercial industry require large quantities of electrical power. The demand would make it expensive to operate with single-phase power. Three-phase power is less expensive to transmit from the power generating plants because three-phase motors used in manufacturing are simpler, more powerful, and less expensive to build and operate than similar motors of single-phase operation. Also, ac induction motors are self-starting when three-phase alternating current is used.

⏩ Summary

- Generators have two main parts: the stator (or the stationary part) and the turning rotor.
- The neutral wire of a 60 Hz electrical system should never be interrupted by switches, fuses, relays, or circuit breakers.
- Loads and controls are connected across the two lines in a ladder diagram. Loads are connected to L_2 and controls are connected to L_1. Ladder diagrams are always read from left to right.
- Always follow the rules of ladder diagrams for compliance with the electrical industry standards.
- Alternators may be either the revolving-armature or revolving-field type.
- Three-phase connections are a combination of wye and delta connections.
- The main advantages of three-phase power are reduced cost and more powerful electrical energy than single-phase alternating current.

⏩ Important Terms

Do you know the meanings of these terms used in the chapter?

delta	stator
revolving-armature alternator	three-phase
revolving-field alternator	wye
rotor	

⏩ Questions and Problems

Please do not write in this text. Write your answers on a separate sheet of paper.

1. The two main parts of an alternator are the _____ and _____.
2. List several types of ac generators.
3. What type of electricity is used for the excitation field?
4. What type of control panel is in your residence, and where is it located?
5. Why must the neutral wire never be switched or fused?
6. Draw a ladder diagram of your kitchen and workshop.
7. What are the two types of alternators?
8. Draw a schematic diagram of the wye and delta connections.
9. Draw a schematic diagram of the four three-phase transformer configurations.
10. Draw a schematic diagram of the four-wire wye system.
11. Why is the design of a distribution system important?
12. What are advantages of three-phase electrical power?

Chapter 29 Graphic Overview

Electric Motor

can be
- **Ac** — types are: Induction, Squirrel-cage, Split-phase, Shaded-pole, Synchronous, Three-phase
- **Dc** — types are: Series, Shunt, Compound, Permanent Magnet, Stepper

problems are: Fails to Start, Runs Hot, Runs Slower than Normal, Runs Noisily

has — Armature — called a — Rotor — and — Commutator — for — Brushes

operates by — Magnetic Fields — producing — Torque

has — Identification Plate — that specifies: Model and Serial Number, Frame Size, Type, Amperes, Volts, Phase, kVA, RPM, Duty

ELECTRIC MOTORS 29

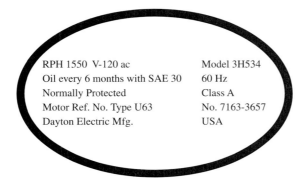

Objectives

After studying this chapter, you will be able to:
○ Identify the parts of an electric motor.
○ Explain the basic operation of a dc motor.
○ Explain the basic operation of a squirrel-cage motor.
○ Explain the operation of a shaded-pole motor.
○ Identify the problems associated with motors.
○ Develop a motor maintenance schedule.

Introduction

Electric motors play an important part in our daily lives, from starting automobiles to washing and drying clothes to manufacturing products. An in-depth study of motors would require a complete textbook on the subject. This chapter presents only basic motor principles, common problems, and motor maintenance.

Motor Identification Plates

Motor nameplates identify important information in the selection and installation of a motor. With this information, an electrician can use the National Electric Code to determine the conductors, conduits, safety devices, and control devices needed for an installation. The information also is important in selecting a replacement motor. See **Figures 29-1** and **29-2**. The nameplate for the three-phase motor in **Figure 29-3** specifies the following information:
• Model and serial number
• Frame size
• Type of enclosure
• Current (in amperes) drawn from the line when the motor is operating at the rated voltage and frequency at the fully rated nameplate horsepower
• Voltage value measured at the motor terminals

RPH 1550 V-120 ac	Model 3H534
Oil every 6 months with SAE 30	60 Hz
Normally Protected	Class A
Motor Ref. No. Type U63	No. 7163-3657
Dayton Electric Mfg.	USA

Figure 29-1. A motor name plate gives important information about the motor.

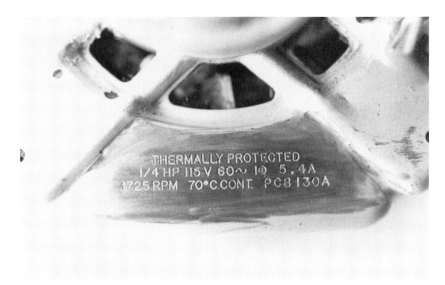

Figure 29-2. Motor name plate of a single-phase motor.

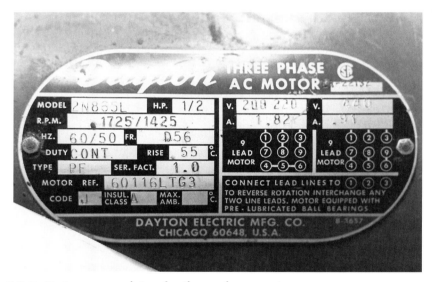

Figure 29-3. Motor name plate of a three-phase motor.

- Number of voltage phases for which the motor is designed
- Locked-rotor kVA per horsepower. This is used to determine the starting and protection devices for the motor. A table of codes can be found in the National Electric Code.
- Revolutions per minute (rpm) when all other conditions on the nameplate exist.
- Duty, which indicates the motor can be operated 24 hours a day (continuously) at a full load. If intermediate, a time will be specified indicating the specified time the motor can operate under full load. Then, it must be stopped and cooled before starting again.

Motor Principles

When electricity is generated, magnetism causes current to flow in a conductor. In a motor the reverse process takes place. The electrical current delivered to a motor is changed to magnetic fields that cause the shaft of the motor to turn.

The basic parts of the motor are the field winding, armature, commutator, and brushes, **Figure 29-4**. The *field* provides a stationary magnetic field. The *armature* is the shaft part of the motor. It rotates and is sometimes called the *rotor*. The *commutator* is a conductive ring. It is divided into separate parts that are insulated from each other, and it rotates with the rotor. The *brushes* ride against the rotating commutator and make the electrical connection to the rotor winding.

All motors operate because of the attraction and repulsion effects of magnetic fields, **Figure 29-5**. The armature loop is attracted and repelled (depending on which side of the loop is considered) by the interaction of the field and armature magnetic fields. The effect is the armature turns (rotates) in a clockwise direction.

Field (winding): The winding in a motor that produces the main magnetic field.

Armature: The movable part of a generator, alternator, or relay.

Rotor: The moving part of a motor.

Commutator: Part of an armature to which the coils of a motor are connected.

Brush: A piece of conductive material that rides on the commutator of an electric motor.

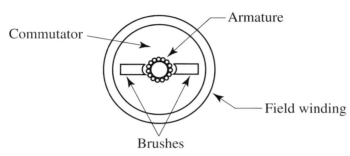

Figure 29-4. Basic construction of a motor.

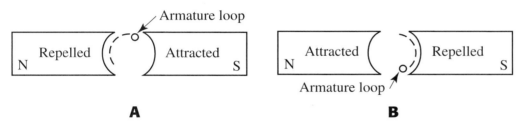

Figure 29-5. Principles of motor operation. A—The loops of the armature are repelled and attracted by the magnetic fields. B—The repelling-attracting action on the armature causes it to rotate.

The magnetic fields of the stator and armature interact, turning the armature to a certain point in its rotation. This causes the brushes to move to the next segment on the commutator and changes the direction of the current through the armature loop, in turn changing the polarity of the armature field. The armature turns in the same direction until it turns again to a certain point. The commutator segment again changes, repeating the process over and over, turning the motor shaft.

Torque

Torque is the twisting action that causes a device to rotate, **Figure 29-6**. Torque (T) is the product of the force (F) times the distance (d) between the axis of rotation and the point of the applied force. The equation is:

$$T = F \times d$$

The unit of measurement for torque can be in newton-meters (Nm) for metric or in pound-feet (lb.-ft.) for conventional measurement. Conversion from metric is made by the factor 0.73756 lb.-ft. = 1 Nm.

Torque: A measure of the circular twisting motion of a shaft.

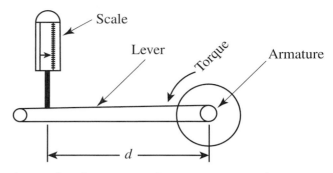

Figure 29-6. Torque is equal to force times distance ($T = F \times d$).

Locked Rotor

If a rotor could not turn, the motor would require maximum current. This condition exists when power is first applied to a motor and the armature is not turning. For a brief moment, the armature acts as if in a ***locked-rotor*** state. This condition produces the maximum armature current and torque with the opposing counter emf at zero.

Dc Motors

The following discussion of dc motors covers these types: series, shunt, compound, permanent magnet, and stepper.

Series Dc Motor

A series dc motor is shown in **Figure 29-7.** The motor windings have low resistance values, approximately 50 Ω for the field and 10 Ω for the armature. This low resistance causes high current in the windings and produces high starting torque.

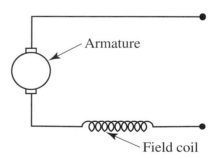

Figure 29-7. Series dc motor. The field coil is connected in series with the armature coil.

When the armature begins to rotate, the movement of the armature conductors through the magnetic flux of the field causes a counter emf to be produced across the armature. If the value of the counter emf could reach the source emf, the motor would not increase in speed. However, the cemf never reaches the source emf value. Thus, the speed of the series dc motor may continue to increase and simply fly apart due to centrifugal force. This condition can easily take place if the load is disconnected from the motor. The direction of the series dc motor can be reversed by reversing the direction of the current through the armature or the field winding.

Shunt Motor

In the shunt dc motor, the field coil and armature are connected in parallel, **Figure 29-8.** The shunt dc motor does not have the starting torque of a series motor but has much better speed regulation.

If a constant voltage is applied across the motor, the magnetic flux of the field coil and armature remains constant. If the load on the motor is increased, the speed of the armature is reduced and the armature current increases. This, in turn, increases the torque of the motor. When the torque equals the load, the motor remains at a constant speed.

The armature of ⅛ hp motor armature may have a resistance of 1000 Ω. With a load on the motor, loss of the field current would stop the motor. The armature current would rise to a point where the armature would burn out if no safety device were used. The shunt dc motor is usually reversed by changing the polarity of the armature winding.

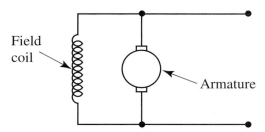

Figure 29-8. Shunt dc motor. The cemf provides an automatic load-no-load speed control.

Compound Dc Motor

A compound dc motor has both a series and shunt field, **Figure 29-9.** It usually has interpoles also. The speed of the motor is controlled by a resistance in series with the shunt field. To change the direction of rotation of a compound motor, the armature and interpoles are reversed.

A compound dc motor can be wired with the series magnetic field polarity reversed. Then it is called a cumulative compound motor or differential compound motor, **Figure 29-10.** Of the two types, the cumulative is the preferred because of good starting torque and speed regulation. The differential motor has poor speed regulation under heavy loads, and the starting torque is low even though the armature current is high.

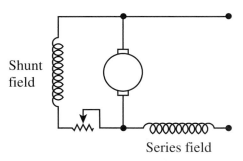

Figure 29-9. A compound dc motor has the advantages of both series and shunt-wound motors.

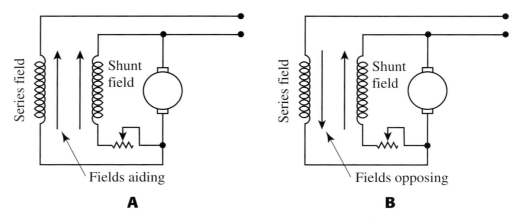

Figure 29-10. Compound dc motors. A—The fields of a cumulative compound motor aid each other. B—The fields of a differential compound motor oppose each other.

Permanent Magnet Motor

The field of a permanent magnet (PM) motor uses a permanent magnet instead of a field winding, **Figure 29-11.** The motor needs electrical connections for the armature only, simplifying the wiring of the motor and making the PM motor less expensive to manufacture. PM motors are used in electric vehicles such as forklifts and wheelchairs, and in power seats, power windows, and automobile windshield wipers. Small PM motors are used in consumer products such as audio and video tape players.

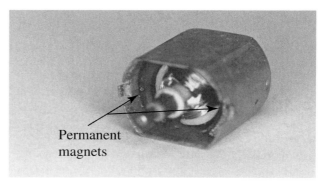

Figure 29-11. A permanent magnet motor is an inexpensive motor used extensively in the consumer electronics industry.

Stepper Motor

An electric motor armature turns at a specific speed and torque, depending on the voltage or frequency applied to the motor. In the stepper motor, the armature turns only a certain number of degrees, **Figure 29-12.** The number of degrees in rotation (called steps) the motor turns depends on the design of the individual motor. The motor receives a pulse of voltage, turns to a given point, and stops until it receives another pulse. Stepper motors are used for applications in which the position of the motor shaft must be precise in either direction.

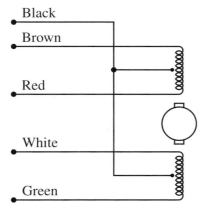

Figure 29-12. Wiring of a five-wire stepper motor. These motors have a wiring color code standard.

Ac Motors

Dc motors are used where direct current is the only available power source. Alternating current, with its wide voltage selection and application of frequency in motor design and motor control, makes ac motors more popular. The following discussion of ac motors covers these types: resistance-start, capacitor-start, shaded-pole, synchronous, and three-phase.

Induction Motor

Figure 29-13 illustrates the rotor construction of an induction motor. Rotor conductors are usually made of heavy conductors resembling bars of copper. The conductors are connected at each end by a copper or brass ring. This makes a short-circuit on the ends of the conductors. The core of the rotor is made of laminated iron, which reduces the conductor air gap reluctance and concentrates the flux between the rotor conductors. Because of the appearance of the rotor, it is called a squirrel-cage rotor and the motor is called a *squirrel-cage motor.* Many single-phase ac motors are of this type.

Squirrel-cage motor: A motor that uses a rotor resembling a cage.

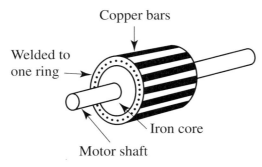

Figure 29-13. Rotor of an induction motor. The rotor acts as a transformer secondary 180° out of phase with the field winding.

Each conductor of the rotor forms a shorted turn with the conductor on the opposite side of the cage. When the cage is between the poles of an electromagnetic field, a strong current is induced into the rotor conductors. The rotor and stator magnetic fields are stationary, causing the rotor to stay in one position. The field polarity must rotate for the rotor to move.

A *split-phase motor* uses additional field pole pieces that are fed out-of-phase currents. The two sets of pole pieces develop maximum magnetic fields at different times, causing the rotor to turn. Two methods used to create the out-of-phase currents on the two sets of field poles are resistance-start and capacitor-start.

Split-phase motor:
A single-phase induction motor with two field windings. Starting torque is developed by phase displacement between the field windings.

Resistance-start Motor

In a resistance-start motor, the start winding is wound with a smaller conductor, giving it a higher resistance, **Figure 29-14.** With a higher resistance, the current lags the voltage less than the main winding. As the ac current flows in the start winding, the magnetic field polarity changes. This causes the rotor to move and makes the motor self-starting. As the rotor reaches 75% of full running speed, a centrifugal switch disconnects the start winding from the circuit, **Figure 29-15.**

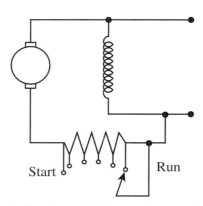

Figure 29-14. The start winding has a higher resistance than the run winding.

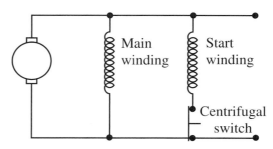

Figure 29-15. The centrifugal switch disconnects the start winding when the motor reaches a certain speed.

Capacitor-start Motor

In the capacitor-start motor, a capacitor is placed in series with the start winding, **Figure 29-16.** This causes the current to lead the applied voltage by 40°. The capacitor-start winding is disconnected from the motor circuit using the same centrifugal switch method.

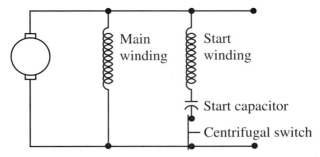

Figure 29-16. A charged capacitor provides the extra current surge needed to get the rotor turning.

Shaded-pole Motor

Another method of producing a rotary field is to shade the pole pieces. Such a motor is called a **shaded-pole motor**. The field poles are slotted and a copper ring is placed around one part of the slotted pole, **Figure 29-17.**

As the current increases in the field coil, the magnetic field expands and induces a current in the copper ring. This produces a magnetic field around the ring, which opposes the magnetism in the pole. Maximum magnetic field is produced in the unshaded part of the pole and minimum magnetic field results in the shaded part of the pole.

When the field current reaches maximum, the magnetic field no longer moves and no current is induced in the copper ring. The ring has no effect on the pole piece, and the maximum magnetic field is produced across the whole pole piece. As the field current decreases, the magnetic field collapses and induces a current in the copper ring in the opposite direction. This produces the maximum magnetic field in the shaded part of the pole. The process continues as the current in the field changes. Thus, the maximum magnetic field moves from the unshaded to the whole pole piece to the shaded part, as the ac cycle continues in the field winding. This maximum magnetic field movement produces the rotating field necessary to cause the squirrel-cage motor to be self-starting.

Shaded-pole motor: A single-phase induction motor with one or more auxiliary short-circuited stator windings magnetically displaced from the main winding.

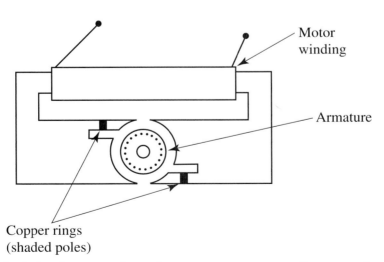

Figure 29-17. Shaded-pole motor. Sometimes one or two turns of copper wire are used in place of the rings.

Synchronous Motor

The synchronous motor is one of the most efficient motors in use. Its rotor contains a winding powered from a dc source called the dc exciter. The motor rotation is synchronized with the 60 Hz ac; therefore, it rotates 60 times per second or 3600 revolutions per minute. If the load on a synchronous motor becomes too large, the motor slows down, loses synchronism, and stops. An electric clock is an example of use of a synchronous motor. As long as the ac remains at the correct frequency, the clock will keep time. Only the frequency is important, not the voltage value.

Three-phase Motor

For industrial applications, a three-phase motor is the most desirable squirrel-cage rotor type motor. Each of the three phases are connected to separate field coils wound on separate field poles. Since the phases are 120° apart, the action gives a true rotating field.

Motor Starters

Motors, especially large ones, can require high electrical current. A motor starter is used to prevent excessive starting current. A ***motor starter*** is a switching device that starts the motor, accelerates its speed, then puts the motor circuit in a run condition. A ***motor controller*** is a device that controls the power to a motor and may have speed control. Motor control circuits will be discussed in Chapter 30.

Motor starter: A device or circuit that enables a motor to start.

Motor controller: A circuit that controls the on/off and speed of a motor.

Motor Maintenance

A planned inspection and maintenance program will prevent serious damage to electric motors. Motor maintenance begins with a good record-keeping system. The frequency of inspections and extent of the items to be checked depend on the:
- Type of service the motor provides
- Percentage of the day the motor operates
- Environment in which the motor operates

Any inspection schedule must be designed for a particular motor and its application. Always follow the maintenance specifications and procedures of the motor manufacturer.

Maintenance of small motors used in home consumer products consists of:
- Checking for loose connections.
- Cleaning the dust from the ventilating slots of the motor and cabinet.
- Placing a small drop of turbine oil on each side of the armature shaft or several drops on the oiler felt, **Figure 29-18.**

 Caution: Never overoil a motor. Too much oil can get on the commutator and damage the brushes. Too much oil also aids in the collection of dirt.

Figure 29-18. Oiler holes have pieces of felt to supply oil over a long period of time.

The following schedule can be used as a guide for industrial ac or dc motor maintenance:
- On a *daily* basis, check that the motor speed is reached in a normal period of time.

On a *weekly* basis:
- Examine the commutator and brushes for wear.
- Check for loose connections on the motor, switches, fuses, and other control equipment.
- Check the motor speed with a metering device.

Every *six months:*
- Clean the motor thoroughly. Blow any dirt from the windings and air vents.
- Clean the commutator and brushes.
- Check for the correct brush pressure and position.
- Check and clean the brush holders. The brush should be free to move in the holder.
- Replace the brushes if they are worn down more than halfway.
- Check the grease in the ball or roller bearings. Many motors have sealed bearings that do not require maintenance. When bearing problems occur, the bearings should be replaced or a new motor installed.
- Check the motor mount bolts, pulleys, gear set-screws, and keys.
- Check the coupling mechanisms.
- Check the shaft end play.
- Check that all covers are tightly in place.
- Inspect and tighten the connections on the motor and control circuits.
- Check for worn gears, chains, and belts.
- Check for vibration.
- Check the input current and compare it with normal current.

Troubleshooting Electric Motors

Most electric motor problems occur when a motor:

- Fails to start
- Runs hot
- Runs slower than normal speed
- Runs noisily

No matter the type of problem, certain steps should be taken to troubleshoot the problem.

If the motor fails to start:

1. Check the incoming power voltage. It should be within 10% of the voltage listed on the motor nameplate. If the power is absent, the power source and motor control circuits should be investigated.
2. If the incoming power voltage is correct, turn off the power and lock out the source. Try to rotate the motor shaft by hand. If the motor is connected to a load that makes shaft rotation impossible, disconnect the load from the motor.
   a. If the motor does not turn freely, inspect it to determine why the motor shaft is stalling.
   b. If the motor turns freely without the load, turn the power source back on and try to run the motor without the load. If the motor runs, use a clamp-on ammeter to check the run current. Make sure it is within the nameplate specifications. If the motor runs correctly, the load should be suspected as the cause of the problem.

If the motor runs hot:

1. The first test of a hot motor should be a clamp-on ammeter reading of the run current. The current should not be higher than specified on the motor nameplate. The closer the current is to the specified current, the hotter the motor will be. Remember, the current specification on the nameplate is the maximum current, not the normal working current. An increase in motor current can be caused by an increase in the load.
2. If the motor current is a normal value, make a test of the supply voltage. The motor voltage should be within 10% of the voltage specified on the motor nameplate.
   a. A low voltage will cause a lower starting torque and an increase in full-load motor temperature.
   b. A higher voltage will cause a higher starting torque and an increase in motor current with higher motor temperature.

Unbalanced voltages on a three-phase motor can cause a motor to run hot. Such unbalanced voltages can cause more damage than high or low voltage over the same short period of time. The measurement of the three-phase voltages should be approximately equal from each phase to ground.

Loss of a phase on a three-phase motor will make the motor run hot and is caused by an open switch contact or fuse. The loss of a phase is often called *single-phasing*. If single-phasing occurs before the motor is started, the motor usually will hum and not start. If single-phasing occurs while the motor is running, the motor will continue to run as long as the load torque requirement does not exceed the torque delivered by the motor. The motor will run hot, vibration will be increased, and the condition will quickly damage the motor. Furthermore, with single-phasing, a motor can start in any direction, which can cause damage to the load.

A motor that runs slower than normal should be accompanied by an increase in motor current and is associated with low power voltage or increased load. (Remember, the bearings are a form of load on the motor). Often the motor runs hotter than normal.

Single-phasing: The tendency of a rotor to continue to rotate when one winding is open and the other remains excited.

If the motor runs noisily, the problem may be bad bearings, insufficient lubrication, loose mounting bolts, or loss of a phase on a three-phase motor.

The cause of any type of motor failure must be found to prevent failure of a new motor. For instance, a motor that runs hot with maximum current may be too small, and a larger horsepower motor may have to be installed. Consideration of the design of the motor and proper maintenance will ensure long motor life.

Summary

- Motor identification plates provide important information for the design and replacement of a motor.
- Motors operate by the attraction and repulsion of magnetic fields.
- Torque is the twisting action of a device.
- Dc motors may be of the series, shunt, compound, or stepper type.
- A motor starter is a switching device that starts and accelerates a motor. A motor controller regulates the power and speed of a motor.
- Ac motors may be of the resistance-start, capacitor start, shaded-pole, synchronous, or three-phase type.
- A good motor maintenance program prevents breakdowns.
- Proper motor maintenance utilizes a planned set of checklists on each motor.
- Problems occur when a motor fails to start, runs hot, runs slower than normal, or runs noisily.

Important Terms

Do you know the meanings of these terms used in the chapter?

armature	rotor
brush	shaded-pole motor
commutator	single-phasing
field	split-phase motor
locked-rotor	squirrel-cage motor
motor controller	torque
motor starter	

Questions and Problems

Please do not write in this text. Write your answers on a separate sheet of paper.

1. List the types of information found on a motor nameplate.
2. What are the four main parts of a dc motor?
3. The instant a motor is started the rotor acts as if it is in the _____ state.
4. Draw a schematic diagram of a series and a shunt dc motor.
5. What happens when the load is removed from a series dc motor?
6. What type of motor is often used in consumer products?
7. Explain the difference between a motor starter and a motor controller.
8. List the different types of ac motors.
9. If a motor produces 45 Nm of torque, what is the torque in ft.-lbs.?
10. Why is a good motor maintenance program important?
11. Design a maintenance schedule for a motor. Make a chart on a separate piece of paper.
12. What are the four possible problems found with motors?

Chapter 30 Graphic Overview

Motor Control

— uses — Control System — can be —
- Open-loop — a — Series Circuit
- Closed-loop — has — Feedback — a sample — Output — back to — Controller

— is — Manual — or — Electronic

— uses — Logic Functions — which are —
- AND
- OR
- NOT
- NAND
- NOR
- Memory — called a — Latching Circuit

— and — Troubleshoot
- the — Power Circuit / Control Circuit / Motor Circuit
- follow — Safety Rules

— has — Input — and — Decisions — and — Output

MOTOR CONTROLS ▶ **30** ◀

Objectives

After studying this chapter, you will be able to:
○ Explain the logic functions AND, OR, NOT, NAND, NOR, and memory.
○ State the types of motor controllers.
○ Explain the difference between open-loop and closed-loop control circuits.
○ List the general rules for troubleshooting motor circuits.
○ Explain the process for troubleshooting a motor circuit.

Introduction

In this chapter, you will continue your study of ladder diagrams. You will learn about logic functions, which show how switching devices control circuits. Types of motor controllers, principles of control circuits, and steps for troubleshooting motor control circuits will also be introduced.

Principles of Control Circuits

The purpose of a control circuit is to do a specific job in a predesigned way. A control circuit has three parts: input, decision, and output. See **Figure 30-1.**

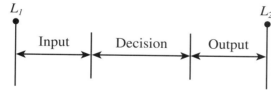

Figure 30-1. Any control circuit contains an input, a decision, and an output.

Inputs

Inputs (signals) start or stop the flow of current by opening or closing the control device contacts. If the contacts are open, current does not flow through the control device. If the contacts are closed, current flows through the control device.

The input signal may be manual, mechanical, or automatic. Inputs tell the control circuit a certain condition has been met. An input signal can be given manually by a manual switch, mechanically by a limit switch, or automatically by a level switch or signal from an electronic transducer. See **Figure 30-2.**

Logic Functions

The decision part of the control circuit determines how the control devices are connected for output, **Figure 30-3.** The decisions give the circuit some "logical" function.

To understand ladder diagrams and circuit design, you must understand the *logic* of the circuits. Control devices cause the circuit to operate in a logical manner; thus, *logic functions* are simply a method of controlling how a circuit operates. Logic functions are used in many areas, such as electricity, electronics, math, hydraulics, pneumatics, and digital electronics.

Logic: A science dealing with the basic principles and applications of truth tables, switching, and gating.

Logic function: A method of controlling how a circuit operates, such as AND, OR, and NOT gates, or a combination of these.

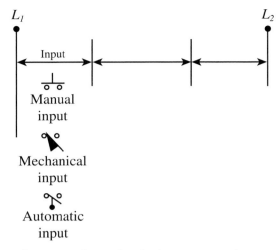

Figure 30-2. Input can be manual, mechanical, or automatic.

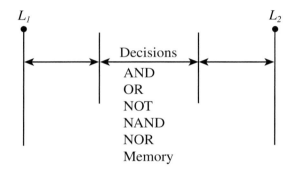

Figure 30-3. The six basic decisions.

AND Logic

The AND function circuit contains a pilot lamp and two push-button switches in series, **Figure 30-4.** If switch 1 is turned on and switch 2 is off, the lamp will remain off; that is, no output will come from the lamp. If switch 1 is off and switch 2 is on, the lamp will be off. If switch 1 and switch 2 are on, the lamp will be on. The only time the lamp will be on is when both switch 1 *and* switch 2 are in the on position. Therefore, it is called the *AND* function.

AND: A logic gate that produces an output only when all of its inputs are at a high level.

OR: The logic gate function OR.

L_1 Fuse

PB$_1$ PB$_2$

L_2

L

Figure 30-4. AND decision. Both push-button switches must be closed before the lamp will light.

OR Logic

The second logic function is called the OR function, **Figure 30-5.** With both switch 1 and switch 2 in the off positions, the lamp is off. If switch 1 is on, the lamp is on. If switch 2 is on, the lamp is on. Therefore, the lamp is on when either switch 1 *or* switch 2 is on. This is called the *OR* function.

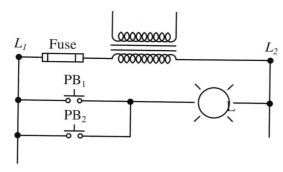

Figure 30-5. OR decision. If either switch is closed, the lamp will light.

NOT Logic

The AND and OR logic functions use switch contacts that are normally open (relay contacts are also switch contacts). If normally closed contacts are used, the logic function is different.

In **Figure 30-6,** the switch contacts are normally closed, causing the lamp to be on. In other words, an output is present if the control signal is off. Therefore, if the load is to be energized, there must *not* be an input signal. This is called the **NOT** function.

NOT: The logic function of inverse.

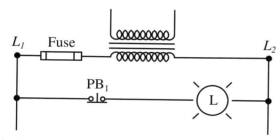

Figure 30-6. NOT decision. If the switch is pressed, the light will go out.

NAND: The logic gate function NOT AND where the output is an inverted AND.

NAND Logic

NAND logic is a combination of NOT logic and AND logic. The switch arrangement is two (or more) normally closed switches in parallel with a lamp circuit, **Figure 30-7.** If switch 1 or switch 2 contacts are opened, the lamp remains on. Only when both switches are open will the lamp be de-energized. Switches 1 *and* 2 must be opened to have *no* output; thus, NOT (no output) 1 AND 2. This is called the **NAND** function.

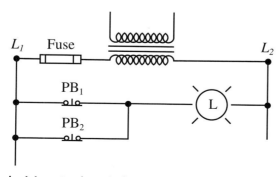

Figure 30-7. NAND decision. Both switches must be pressed before the lamp will go out.

NOR Logic

Like the NAND function, NOR logic is a combination of NOT logic and OR logic. The switch arrangement is two (or more) normally closed switches in series with a lamp circuit, **Figure 30-8.** If switch 1 or switch 2 contacts are opened, the lamp will be de-energized. Opening either switch 1 *or* switch 2 will cause *no* output; thus, NOT (no output) 1 OR 2. This is called the *NOR* function.

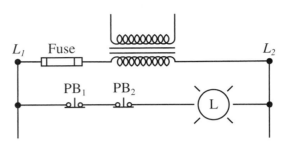

Figure 30-8. NOR decision. Pressing either switch will cause the lamp to go out.

Memory Logic

Memory logic is used when a circuit must remember the input signal after the signal is removed. When a circuit is on, it remains on until it is turned off by another signal even though the on signal was removed. Memory logic is accomplished by mechanical relay contacts or a solid-state relay, which keeps the circuit in the on state until an off signal is received.

The start switch is in parallel with a relay contact, **Figure 30-9.** When the start button is pressed, the control relay (CR) contacts close and remain closed when the start button is released. The lamp stays on until the stop button is pressed, which removes power from the CR contact coil. This is sometimes called a latching circuit, and it is a *memory* function.

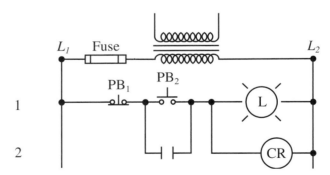

Figure 30-9. The relay contact provides memory.

Control Systems

Control circuits may be of an open-loop or closed-loop system. A *system* is any organized assembly of parts, operations, or procedures.

Open-loop System

A basic open-loop system consists of an input, controller, load, and output, **Figure 30-10.** The open-loop system is a series circuit. It is loopless, meaning there is no loop from the output to the controller. The power source is the 60 Hz ac or three-phase power supply. Any electrical devices or appliances with an on/off control are open-loop.

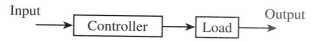

Figure 30-10. Block diagram of an open-loop system.

In the open-loop system, a change in load, voltage, motor speed, or operating environment causes a change in the operating characteristics of the system. Some manual compensation to the changes may be provided, but it is not necessary to accomplish the desired outcome. An electric motor circuit is a good example of this type of system.

Closed-loop System

A closed-loop system has a feedback loop consisting of a circuit or transducer that samples the output and delivers a correction signal to the controller. Many closed-loop systems use an electronic controller.

The feedback circuit samples (or measures) the output voltage, current, speed, temperature, or pressure and develops a correction voltage. The correction voltage is compared to the reference voltage and produces an error correction voltage. The error correction voltage is applied to the controller, which changes the load to compensate for the incorrect output. The process continues until the error correction voltage is zero and the system is stable. Any future change in the load is corrected in the same way so the system operates at maximum performance. In some systems, the reference may be part of the feedback block, **Figure 30-11.** The feedback delivers the error correction voltage directly to the controller.

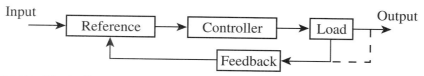

Figure 30-11. Block diagram of a closed-loop system.

Types of Motor Controllers

So many types and applications of motor controllers exist that a complete discussion is not within the scope of this book. Two broad types of motor controllers will be discussed: manual and electronic.

Manual Motor Controller

Figure 30-12 shows a contactor that controls the voltage applied to a motor. The contactor requires a low voltage (24 V) circuit, manually controlled by a start/stop switch. Such a circuit can also be controlled by a relay, but the relay would still be controlled by a manual switch. This type of motor control is common in home and industry. Simple motor control systems are found in such devices as a drill press, grinder, table saw, conveyor, air compressor, and air conditioner blower.

Electronic Motor Controller

Electronic programmable controllers are used to control the speed, on/off, and timing of motors, **Figure 30-13.** Electronic controllers use more memory and sensing devices than do manual controllers. The electronic controller does not directly control the motor. Contactors and relays do the high electrical power switching. Manually controlled motor speed has become obsolete with the use of solid-state electronic speed control.

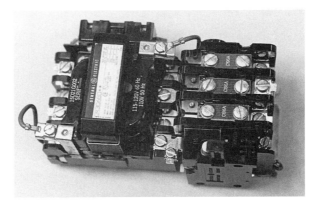

Figure 30-12. Manually activated contactor.

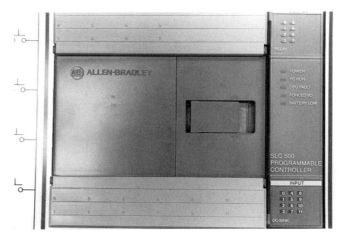

Figure 30-13. Programmable logic controller.

Troubleshooting Motor Systems

The first step in troubleshooting any motor circuit is to understand the circuit and its normal operation. Without this understanding, troubleshooting the circuit may take more time than necessary.

General rules for troubleshooting a motor circuit include:
- Find out if someone else has worked on the unit.
- Listen to the concerns of people who operate the motor circuit.
- Read the manufacturer's instructions if available.
- Check obvious trouble areas, such as fuses, circuit breakers, and overload resets.
- Look for unusual heat, odors, and noises.
- Apply the basic rules of electricity.
- Isolate the trouble to the power circuit, control circuit, or motor.

 Warning: When working with electricity, never work alone. Always lock-out the control box so someone else cannot turn on the power while you are working on the circuits. Never use a test instrument beyond its rated capacity.

Power Circuit

To isolate the problem in a single-phase motor system, check the power supply with a voltmeter reading across the L_1 and L_2 power lines. See **Figure 30-14.** Absence of power indicates blown fuses, circuit breakers, or loss of power distribution elsewhere in the system. Do not assume anything; test and verify power does exist.

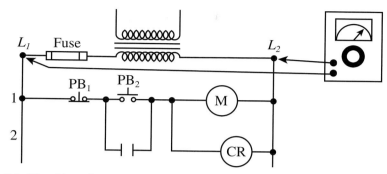

Figure 30-14. Checking the power source and fuse.

Control Circuit

Connect a voltmeter across each control switch, one at a time, **Figure 30-15.** Then, press the start button. If the control switch is open, the meter will indicate voltage is present, and the control circuit is at fault.

A fused jumper can be connected across each of the controls to determine which is defective. However, this can damage equipment because other motors and equipment may depend on the synchronization of each device in the system.

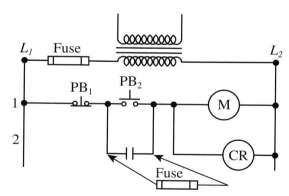

Figure 30-15. Control circuit checks should be made with a fused jumper.

Motor Circuit

Check for voltage at the motor terminals. If voltage is present, check the resistance of the motor windings. If voltage is not present at the motor, use the fused jumper across the control circuits so power is applied to the motor circuit contactor. Check for voltage at the output of the contactor. If voltage is absent, the contactor is at fault. If voltage is present at the output of the contactor, check each overload contact with a fused jumper to make sure the overload contacts are not causing the problem.

Summary

- Logic functions include AND, OR, NOT, NAND, NOR, and memory.
- Motors are controlled by either manual or electronic controls.
- Control circuits may be open-loop or closed-loop. The closed-loop circuit contains a transducer that samples the output and produces a correction signal to the controller.
- Good troubleshooting entails understanding the circuits being tested and following a systematic plan to locate the problem in the least amount of time.

Important Terms

Do you know the meanings of these terms used in the chapter?

AND	NOR
logic	NOT
logic functions	OR
memory	system
NAND	

Questions and Problems

Please do not write in this text. Write your answers on a separate sheet of paper.

1. Why is an understanding of logic functions important?
2. What are the logic functions in **Figures 30-16** through **30-20**?
3. By what two methods are motors controlled?
4. What type of control system is in **Figure 30-21?**
5. What are the three parts of an open-loop control system?
6. What is the purpose of the feedback system in a closed-loop control system?
7. List the general rules for troubleshooting a motor circuit.
8. What are the possible causes of a nonrunning motor in **Figure 30-22?**
9. What are the possible causes of a nonrunning motor in **Figure 30-23?**
10. Motor problems should be isolated to what circuits?

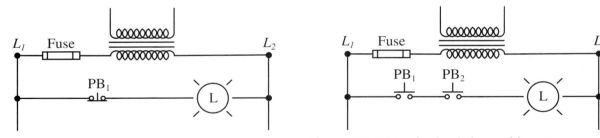

Figure 30-16. Logic circuit for problem 2.   **Figure 30-17.** Logic circuit for problem 2.

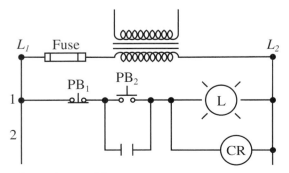

Figure 30-18. Logic circuit for problem 2.

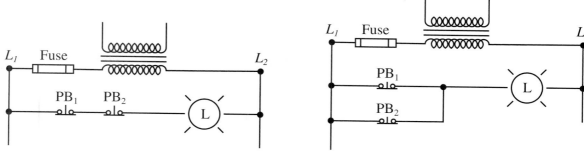

Figure 30-19. Logic circuit for problem 2.

Figure 30-20. Logic circuit for problem 2.

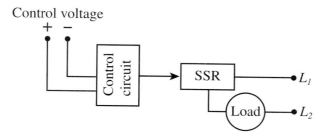

Figure 30-21. Control system for problem 4.

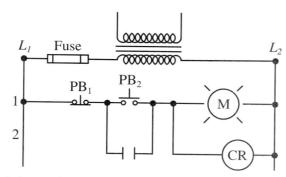

Figure 30-22. Circuit for problem 8.

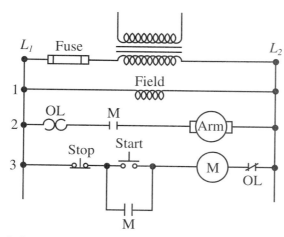

Figure 30-23. Circuit for problem 9.

Section VI

Electronic Principles

Section VI Activity

Testing Transistors

Testing a transistor is a common task for a technician. While treating the transistor as two diodes back to back, the simplest test uses an ohmmeter to check for a short across the junctions of the transistor. With the base used as the common connection, each of the junctions (e-b and c-b) are tested for low resistance in one direction and high resistance in the other. A zero resistance indicates a short at that particular junction.

Objective

In this activity, you will test transistors for good, open, or shorted condition.

Materials and Equipment

Set of transistors to be tested (see instructor)
1–VOM

Procedures

1. Ask your instructor for a set of transistors to be tested.
2. Using an ohmmeter or the diode test position on a digital meter, test each of the transistors for good, open, or shorted condition. Record each of the results in the table in **Figure A.**
3. If possible, note in the table which junction failed.

Transistor	NPN or PNP	Good	Shorted or Open	Junction Failed
2N 3904				
2N 3053				
TIP 31C				
Power				

Figure A.

Chapter 31 Graphic Overview

Diodes

consider replacement
- Forward Current
- Surge Current
- PRV
- Type
- Power Dissipation

types are
- Junction
- Zener
- Schottky
- PIN
- IMPATT
- Tunnel
- Varactor

Junction *has*
- Cathode — *is the* — Banded End
- Knee Voltage — *of* — 0.65 V SI, 0.3 V Ge
- Anode
- Breakdown Voltage — *called* — PRV or PIV

PN Junction *has applied*
- Forward Bias — *causes a smaller* — Depletion Zone
- Forward Bias — *causes* — Forward Current
- Reverse Bias — *causes a larger* — Depletion Zone
- Reverse Bias — *has a* — Small Leakage Current

PN Junction *which has a* — Forward Current

SEMICONDUCTOR FUNDAMENTALS

Objectives

After studying this chapter, you will be able to:
- ○ Explain the operation of a PN junction with forward and reverse bias.
- ○ Determine how much current is flowing in a circuit.
- ○ Identify special-purpose diodes.
- ○ Explain the process of testing a diode using an ohmmeter.
- ○ List the factors to consider when replacing a diode.

Introduction

You will recall from Chapter 1 that conductors have free electrons for current conduction, while insulators do not. This chapter will discuss semiconductors in which the material can become a conductor under certain conditions. Semiconductors are used in diodes, transistors, integrated circuits, thyristors, and other special devices. Materials used in semiconductors and the semiconductor diode will also be covered.

Semiconductor Materials

Devices made from semiconductor materials are referred to as **solid-state** because they are made from solid materials. Silicon, germanium, and selenium are three types of materials used to make semiconductor devices. The element gallium is used for certain applications. Silicon is used most often because of its abundance, refining ease, and crystal structure.

A **covalent bond** occurs when atoms are bound together in such a way that they share valence electrons. See **Figure 31-1.** Semiconductor materials are crystals made up of atoms with four valence electrons held together in covalent bonds. They share their four outer electrons with the atoms next to them. In pure form, semiconductor materials do not have enough free electrons to allow current flow. They are neither good conductors nor insulators. However, because of a change in their crystal structure, semiconductor materials have become the very heart of the modern electronic industry.

Solid-state: Any component, device, or study based on semiconductor theory.

Covalent bond: A type of linkage between atoms in which each atom contributes one electron to be shared with the others.

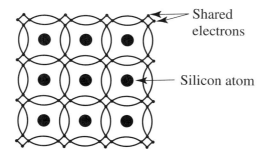

Shared electrons

Silicon atom

Figure 31-1. Nine atoms of silicon bound together and sharing valence electrons.

Doping Semiconductor Crystals

Doping: Adding P-type or N-type materials to a semiconductor material.

N-type (material): A material in which electrons have been added so the electrons serve as the majority current carrier.

Hole: A vacancy caused by the absence of an electron in the electronic valence structure of a semiconductor.

P-type (material): A material in which doping causes an excess of holes.

The crystal structure of a solid-state atom is changed by placing other materials in the structure. This process is called *doping*, **Figure 31-2.** A silicon semiconductor can be doped with phosphorus, **Figure 31-3.** Silicon with extra electrons is called *N-type* material. Placing boron in the silicon crystal structure produces a position in the crystal that needs an electron. The position is called a *hole*, **Figure 31-4.** Silicon with an absence of electrons is called *P-type* material.

Material	Symbol	Valence Electrons
For P-type material		
Aluminum	Al	3
Boron	B	3
Gallium	Ga	3
Indium	In	3
For N-type material		
Antimony	Sb	5
Arsenic	As	5
Bismuth	Bi	5
Phosphorus	P	5

Figure 31-2. Various materials used in doping.

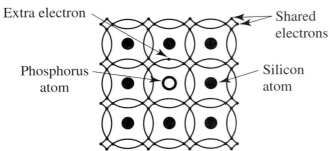

Figure 31-3. Silicon atom doped with phosphorus provides extra electrons in the crystal structure.

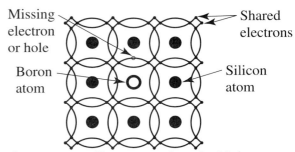

Figure 31-4. Silicon atom doped with boron produces a missing electron or hole.

PN Junction

PN junction: The region of transition between P-type and N-type material in a single semiconductor crystal.

Depletion zone: The area extending on both sides of the junction in which the current carriers have been removed by a voltage.

When a piece of silicon is doped, a solid-state semiconductor is formed with a P-type region on one side and an N-type region on the other side. The P and N regions are made within a single piece of silicon. The *PN junction* is the place where the P and N regions come together, **Figure 31-5.**

The excess electrons of the N-type material are attracted to the P-type material closest to the junction. This creates a nearly neutral area, which is the actual junction of the two types of material. The slightly positive and negative charges on the edges of the junction form the *depletion zone.* The depletion zone separates the N-type and P-type material so no electrons can move across the zone from the N to the P material. See **Figure 31-6.**

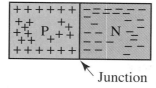

Figure 31-5. Different doping within the silicon piece produces a PN junction. The plus (+) signs in the P-type material are holes. The minus (–) signs in the N-type material are electrons.

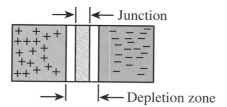

Figure 31-6. Depletion zone of a PN junction.

Applying a Voltage to the PN Junction

When a voltage is applied to the PN junction, the depletion zone becomes narrow and electrons can move across it. Electrons move from the negative side of the voltage source, across the depletion zone, across the P-type material, to the positive side of the voltage source. The positive side of the source constantly removes electrons from the P-type material, while the negative side of the source replaces the electrons on the N-type material. A current flows from the source negative terminal, through the PN junction, to the source positive terminal. When a voltage is connected so a current flows through a PN junction, the PN junction is *forward biased.* See **Figure 31-7.** The voltage connected to cause the current is called the **bias voltage,** and the current is the forward current (I_F).

In **Figure 31-8,** the source voltage is connected in the other direction. The positive is connected to the N-type material, and the negative is connected to the P-type material. This causes the depletion zone to become so large that electrons cannot move across it. When the voltage is connected so current does not flow in the PN junction, the junction is *reverse biased.* The voltage connected in the reverse direction is the reverse bias voltage.

Forward biased: A voltage in such polarity across a PN junction that current flows through the junction.

Bias voltage: The voltage across a semiconductor P-N junction.

Reverse biased: A voltage applied to a semiconductor junction in such a way that no current (except leakage current) flows.

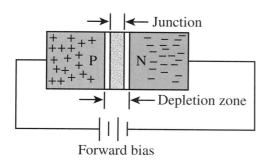

Figure 31-7. Forward bias causes the depletion zone to become narrow.

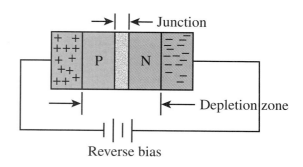

Figure 31-8. Reverse bias causes the depletion zone to become wider.

Solid-state Diode

Diode: A semiconductor device that conducts current in only one direction.

Cathode: The negative terminal of a semiconductor diode.

Anode: The positive terminal of a device.

The PN junction makes a device called a ***diode***. The term diode is a holdover from the vacuum tube era. A vacuum tube with two electrodes is called a diode. (The prefix "di" means two, and "ode" is taken from the end of the word electrode.) Both the tube and solid-state diode have two electrodes. The symbol for the solid-state diode has a ***cathode*** and ***anode*, Figure 31-9.** The PN junction is shown only for comparison with the symbol. Also shown is the direction of the electron current (I_F) through the diode from cathode to anode.

Since the diode is made up of a PN junction, it allows current to flow in the forward direction but not the reverse direction. Current flows in only one direction.

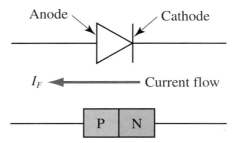

Figure 31-9. Schematic symbol for a diode. Notice the cathode is on the N-type part of the PN junction.

Forward Direction

In the forward bias direction, a certain amount of bias voltage (knee voltage) is needed to overcome the depletion zone and allow conduction. See **Figure 31-10.** The knee voltage is 0.3 V for germanium and 0.65 V for silicon. The knee voltage is a constant voltage drop across the PN junction.

The knee voltage for selenium diodes can range from 0.6 V to 1.8 V, depending on the load current. As the load current increases, so does the voltage drop across the diode. Selenium diodes have all but disappeared from use in electronic assemblies. They are still in use in the motorcycle industry and other special applications.

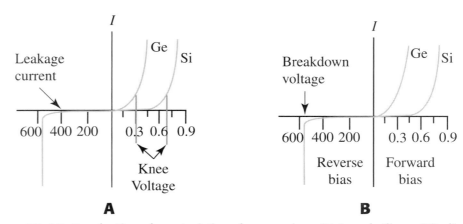

Figure 31-10. Conduction characteristics of germanium (Ge) and silicon (Si) diodes in the forward bias direction. A—Voltage-current curves. B—Knee voltage curves.

▼ *Example 31-1:*

How much current is flowing in the circuit in **Figure 31-11?**

Because this is a series circuit, the sum of the voltage drops V_R and V_D must equal the source voltage. Because the diode is silicon, it has a constant voltage drop of 0.65 V. The voltage across the resistor is:

$$V_R = V_S - V_D$$
$$= 18 - 0.65$$
$$= 17.35 \text{ V}$$

The circuit current is:

$$I = \frac{V_R}{R}$$
$$= \frac{17.35 \text{ V}}{600 \ \Omega}$$
$$= 28.92 \text{ mA}$$

▲

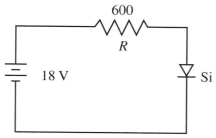

Figure 31-11. Series circuit for Example 31-1.

Reverse Direction

Reverse voltage is usually shown in hundreds of volts. As the reverse voltage is increased, the reverse current increases slightly. This small reverse current is called *leakage current,* refer to Figure 31-10B. At a given reverse voltage value, the voltage overcomes the depletion zone and causes a high current to flow in the reverse direction through the diode. This is the *breakdown voltage* or *avalanche point* of the diode. The breakdown voltage is given as the *peak reverse* (inverse) *voltage* (PRV or PIV) rating of the diode. It is the absolute maximum voltage that can be applied to a diode in the reverse direction. When reverse current occurs, the diode usually becomes shorted and must be replaced.

Diode Applications

PN junction diodes are mainly used for demodulation or rectification. A radio transmitter uses a modulator to put information onto radio waves that carry the information to the radio receiver. A PN junction diode circuit inside the receiver detects, or demodulates, the information from the carrier.

The process of changing alternating current to direct current is called *rectification.* The process is possible due to the ability of the PN junction diode to allow current in only one direction. Diodes used in power supply circuits to change ac to dc are called rectifiers.

Leakage current: In a capacitor, a small value stray current that flows through or across the surface of an insulator or the dielectric of a capacitor.

Breakdown voltage: The voltage across a semiconductor that causes maximum current in the reverse direction through the semiconductor material. Also called **avalanche point.**

Peak reverse voltage (PRV): The peak ac voltage a PN junction can withstand in the reverse direction.

Rectification: The process of changing alternating current to direct current.

Diode Packages

General purpose diodes are found in many dc power supplies. A stud-mounted diode is used for higher current. See **Figure 31-12.** The diode in **Figure 31-13** is used in high-current industrial power applications. Notice the anode and cathode (band) identifications.

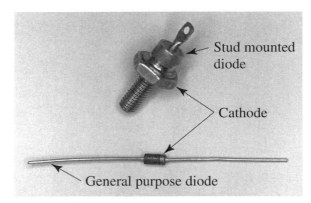

Figure 31-12. General purpose diodes are used in dc power (bottom). This higher current diode is mounted to a heat sink to carry away excess heat (top).

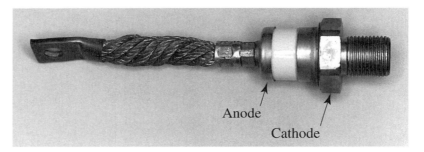

Figure 31-13. This diode can carry 200 A.

Special-purpose Diodes

Besides the PN junction diode, several special types of diodes have been developed. These include the zener, varactor, Schottky, tunnel, PIN, and IMPATT.

Zener Diode

Zener diodes look like ordinary junction diodes but operate differently in a circuit. The cathode takes the form of a "Z," **Figure 31-14.** In the forward direction, a zener diode operates essentially the same as a junction diode, but in the reverse direction it operates very differently. The zener diode silicon chip is heavily doped in manufacturing. By controlling the amount of doping, the breakdown voltage is regulated from under 4 V to over 120 V.

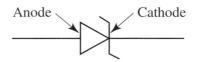

Figure 31-14. Schematic symbol for a zener diode.

The characteristic curve in **Figure 31-15** shows that as the reverse bias is increased, the PRV breakdown point is reached, and the diode conducts in the reverse direction. Unlike the PN junction diode, the zener diode normally operates in the breakdown region. The breakdown voltage is called the zener voltage point. Although the current through the zener diode may vary greatly, the zener voltage remains fairly constant, **Figure 31-16.** For this reason the zener diode is used as a voltage regulator. Resistor R_1 limits the current through the zener diode.

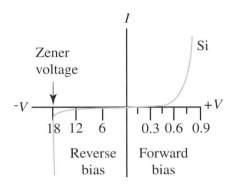

Figure 31-15. Characteristic curve for a zener diode.

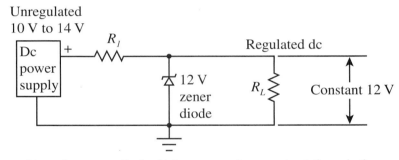

Figure 31-16. Use of a zener diode. Voltage remains constant though the current varies.

Varactor Diode

The variable capacitor (varactor) diode is made with a small amount of doping in the depletion zone. The doping increases farther from the junction. The depletion zone is a better insulator than conductor, and the outer parts of the silicon chip are better conductors. The depletion zone acts as an insulator between two conductive plates (a capacitor). The depletion zone becomes smaller or larger as the bias voltage is changed. This is the same as changing the distance between the plates of a capacitor. Thus, controlling the voltage across a varactor diode makes a variable capacitor in a circuit. The capacitance of the varactor varies greatly compared to an ordinary PN junction diode, **Figure 31-17.** Schematic symbols used for a varactor diode are shown in **Figure 31-18.**

Schottky Diode

Unlike the PN junction diode, the Schottky diode is made using an N-type semiconductor region and a piece of metal such as gold or silver. See **Figure 31-19.** When forward biased, the diode has so many free electrons available that an extremely fast response to the forward bias occurs. This gives the Schottky diode its fast-switching characteristic for use in such high-frequency applications as microwave and digital logic circuits. The schematic symbol for a Schottky diode is shown in **Figure 31-20.**

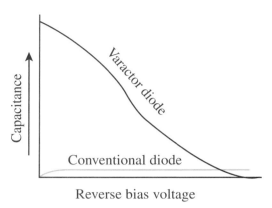

Figure 31-17. The capacitance of a varactor diode changes much more than the capacitance of a conventional diode.

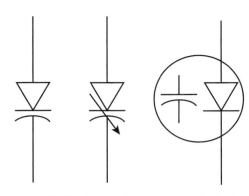

Figure 31-18. Various schematic symbols for a varactor diode.

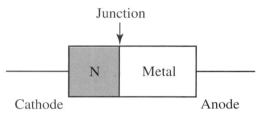

Figure 31-19. The Schottky diode does not have a PN junction but a junction of N-type material and a metal.

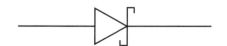

Figure 31-20. Schematic symbol for a Schottky diode.

Tunnel Diode

The tunnel diode is formed by doping the germanium or gallium arsenide device one hundred to several thousand times more than a conventional PN junction diode. The tunnel diode is used in such high-frequency applications as microwave oscillators and converters. **Figure 31-21** shows the schematic symbols for a tunnel diode.

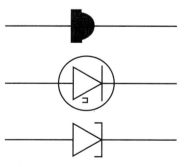

Figure 31-21. Schematic symbols for a tunnel diode.

Negative resistance: The property exhibited when an increase in voltage produces a decrease, rather than an increase, in current.

Valley voltage: In a tunnel diode, the voltage that corresponds to the valley current.

The tunnel diode has a special characteristic known as ***negative resistance.*** As the forward voltage increases, the current through the diode begins to decrease at some point. This defies Ohm's law because as the voltage increases, the current decreases (an increase in resistance). As the forward voltage increases, the forward current decreases until it reaches a minimum value. This forward voltage is the ***valley voltage.*** From the valley voltage point, the current exponentially increases with the forward bias voltage, as does a conventional PN junction diode. See **Figure 31-22.**

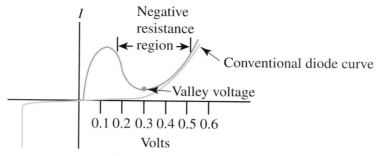

Figure 31-22. Comparison of the characteristics of a tunnel diode to those of a conventional diode.

PIN Diode

The PIN diode has an undoped region between the P and N regions, **Figure 31-23.** When reverse biased, the PIN diode acts as a constant-value capacitor. With a forward bias, the diode acts as a variable resistance. Furthermore, the internal resistance of the PIN diode varies linearly with the forward bias. The PIN diode is used in high-frequency applications.

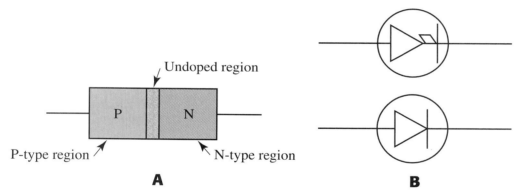

Figure 31-23. Pin diode. A—Construction of the PIN diode. B—Schematic symbols for the PIN diode.

IMPATT Diode

The IMPATT (impact avalanche and transit time) diode is also a negative-resistance diode operated in its reverse breakdown region. The IMPATT diode is a high-power, high-frequency (10 GHz to 100 GHz) diode used in the microwave industry. It usually has the same schematic symbol as a PN junction diode.

Testing Diodes

In a few instances, a diode will be open. The most common defect is a shorted diode, which will cause blown fuses or overheated components. A simple test for a defective diode uses an ohmmeter. An analog meter should be set on the R×100 range, but most digital meters have a diode test position, **Figure 31-24.**

When the negative meter lead is connected to the cathode and the positive lead to the anode, the battery inside the ohmmeter is used to forward bias the junction and a low ohmmeter reading is given. When the meter leads are changed so the positive is on the cathode and the negative is on the anode, the PN junction is reverse biased and the ohmmeter gives an infinity or high resistance indication.

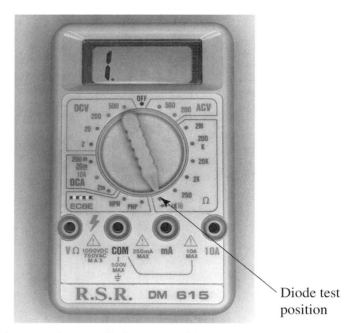

Diode test position

Figure 31-24. Digital meters have a diode test position.

To test a diode, the following procedure can be used:

1. Set the meter to the proper position for diode testing.
2. Connect the positive lead of the ohmmeter to the cathode and the negative lead to the anode. The meter should indicate high resistance or infinity. Zero resistance indicates a shorted diode.
3. Connect the negative lead of the ohmmeter to the cathode and the positive lead to the anode. The meter should indicate low resistance. A zero reading indicates a shorted diode. A high or infinity meter indicates an open diode.

 Caution: Some analog meter polarities are reversed. To check the meter polarity, use a known good diode and connect the meter leads to forward bias the diode. If the negative lead is not on the cathode, the meter reversal must be considered when making a diode test.

Another method of testing a diode is using a curve tracer, a specially made oscilloscope that displays the characteristic curve of the diode under test. See **Figure 31-25.** The circuit in **Figure 31-26** can be used to set up an oscilloscope for displaying diode curves. Adjustment of the X and Y variable gain should be done using a known good diode.

Selecting a Replacement Diode

A defective diode should be replaced with one of the same part number. However, an exact replacement may not always be possible and a substitute must be found. Although many replacement or substitution guides are available, consider the following factors and questions when selecting a replacement diode:

- *Type.* Is the diode junction, zener, high power, low power, etc.?
- *Forward current.* How much current must the diode be able to carry in the forward direction?

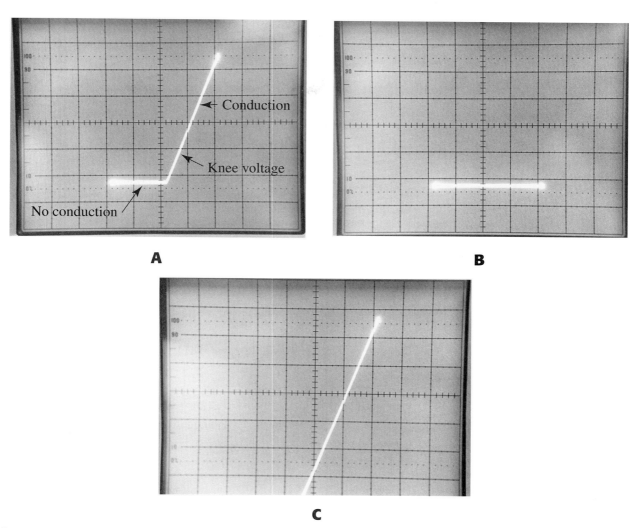

Figure 31-25. Curver tracer displays. A—Good diode. B—Open diode. C—Shorted diode.

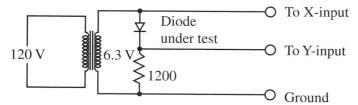

Figure 31-26. Oscilloscope circuit for using an oscilloscope as a curve tracer (in X-Y operation).

- *PRV.* How much voltage must the diode be able to endure in the reverse direction?
- *Power dissipation.* How much power must the diode be able to dissipate?
- *Surge current.* How much current can the diode handle when the equipment power switch is turned on?

Whenever possible, a defective diode should be replaced with one of the same part number. Be certain the replacement diode is connected to the circuit with the cathode in the correct position.

Summary

- Semiconductor devices are made of silicon or germanium.
- Pure semiconductor materials are not good conductors.
- Doping is the process of placing other materials in the crystal structure.
- When a PN junction is forward biased, a current will flow in the circuit. When a PN junction is reverse biased, current will not flow in the circuit.
- Terminals of diodes are called cathode and anode.
- The forward voltage drop of a silicon diode is 0.65 V and 0.3 V for a germanium diode.
- Do not exceed the PRV or PIV rating of a diode or the diode will be destroyed.
- Diodes are used for demodulation or rectification. Rectification is the process of changing alternating current into direct current.
- Diodes can be tested using an ohmmeter and should give a low and high resistance reading as the meter leads are reversed.
- Special purpose diodes are zener, varactor, tunnel, PIN, and IMPATT. The zener diode is used for voltage regulation. The tunnel diode has a negative resistance characteristic.
- When replacing a diode, consider type, forward current, PRV, power dissipation, and surge current ratings.

Important Terms

Do you know the meanings of these terms used in the chapter?

anode	leakage current
avalanche point	N-type
bias voltage	negative resistance
breakdown voltage	P-type
cathode	PN junction
covalent bond	peak reverse voltage (PRV)
depletion zone	rectification
diode	reverse biased
doping	solid-state
forward biased	valley voltage
hole	

Questions and Problems

Please do not write in this text. Write your answers on a separate sheet of paper.

1. How is the crystal structure of semiconductor material changed to enable it to be a good conductor?
2. Determine if the diodes in the circuits in **Figure 31-27** are forward or reverse biased.
3. Identify the cathode and anode of the diodes in **Figure 31-28.**
4. What is the current of each of the circuits in **Figure 31-29?**
5. What are two applications of solid-state diodes?
6. What is the purpose of a zener diode?
7. Draw the schematic symbols for the PN junction, zener, varactor, Schottky, tunnel, and PIN diodes.
8. When tested with an ohmmeter, a good diode will give what readings?
9. Diodes can be tested using an ohmmeter or a(n) _____.
10. What are the factors to consider when selecting a replacement diode?

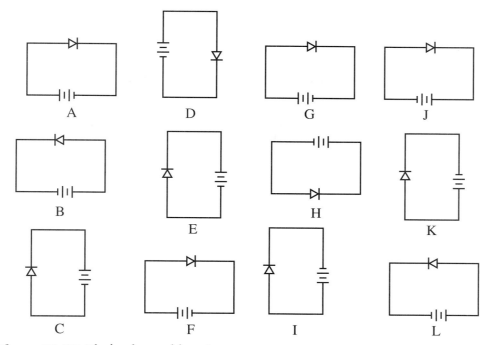

Figure 31-27. Diodes for problem 2.

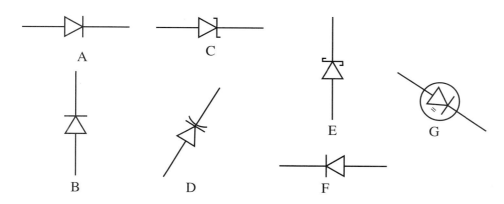

Figure 31-28. Diodes for problem 3.

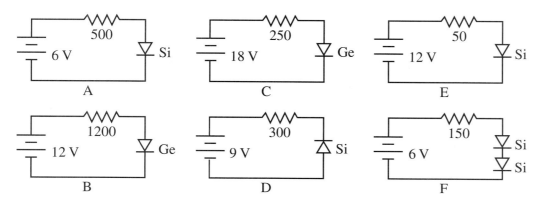

Figure 31-29. Diode circuits for problem 4.

Chapter 32 Graphic Overview

Dc Power Supply — troubleshoot by → Step-by-step Process — and

Step-by-step Process:
- Determine and Analyze Symptoms
- Localize to Specific Area
- Isolate to a Circuit
- Locate Component

Dc Power Supply — consists of:

Voltage Divider — is a → Resistor Network — produces the → Required Voltages

Regulator — maintains → Dc Voltages — at a → Desired Value

Filter — smooths → Pulsing Dc — to → Steady Dc

Rectifier — changes → Ac — to → Dc

Rectifier — may be:
- Half-wave
- Full-wave
- Bridge
- Voltage Doubler

Transformer — changes → High Ac Voltages — to → Higher or Lower → Ac Voltages

DC POWER SUPPLIES 32

Objectives

After studying this chapter, you will be able to:
- ○ Draw a block diagram of a typical dc power supply.
- ○ Draw schematic diagrams of half-wave, full-wave center-tapped, voltage doubler, and bridge power supplies.
- ○ Explain the function of the power supply filter system.
- ○ Explain the function of the regulator.
- ○ Calculate the series resistor for a zener regulator.
- ○ Describe the procedure for efficient troubleshooting.
- ○ Draw a troubleshooting block diagram.
- ○ State the four possible types of system faults.

Introduction

In this chapter, you will begin to put together components to make circuits for a particular electronic function. Recall that rectification is the process of changing ac to dc. The purpose of any dc power supply is to produce the dc voltages necessary to operate equipment circuits. Power companies supply the common 120 V alternating current. A dc power supply circuit will be found in most electronic equipment. This chapter describes how to use the PN junction diode to change ac to dc using various types of dc power supply circuits.

Dc Power Supply Fundamentals

Understanding the basic principles of a typical dc power supply is necessary before studying specific power supply circuits. A typical dc power supply consists of a transformer, rectifier, filter system, regulator, and voltage divider network, **Figure 32-1.** In addition, the dc power supply usually contains an ON/OFF power switch and safety devices, such as fuses or a circuit breaker. You should know the function of each block and be able to draw the block diagram from memory.

These are the functions of the various parts of a dc power supply:
- The transformer changes high ac voltage to a lower or higher value.
- The rectifier changes alternating current from the transformer into direct current.
- The filter system smoothes pulsing dc into a steady dc value.
- The *regulator* maintains the voltage at a desired value.
- The voltage divider is a resistor network that divides the output voltage of the regulator into the required voltages for other circuits.

Regulator: A device or circuit designed to maintain a voltage or current at a predetermined value.

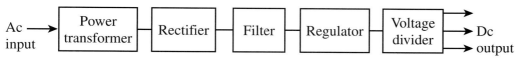

Figure 32-1. Block diagram of a typical dc power supply. Some power supplies may not contain a transformer, regulator, or voltage divider.

Finally, the power supply is connected to various transistor amplifier circuits and other circuits requiring power, **Figure 32-2.** All these circuits are connected in parallel across the power supply. They make up the load in which the power supply must furnish the necessary voltages and currents to operate correctly. The sum of all the load currents is the total load current the power supply must furnish.

Various types of power supplies are in use. They include half-wave, full-wave center-tapped, full-wave bridge, and voltage doubler.

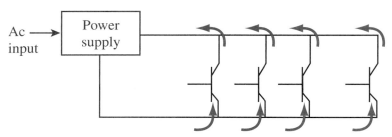

Figure 32-2. Device loads are connected in parallel across a power supply.

Half-wave Rectification

Power supplies use the ability of the PN junctions to allow current flow in only one direction for rectifying or changing alternating current to direct current. The diode is connected in series with a load resistor (R_L) representing all the circuits connected across the power supply. See **Figure 32-3.**

When the current in the transformer is flowing counterclockwise, the voltage polarity of the transformer current is positive on the anode. The diode is forward biased and current flows through the R_L and diode. The current is plotted on the graph in the positive direction during time period t_1.

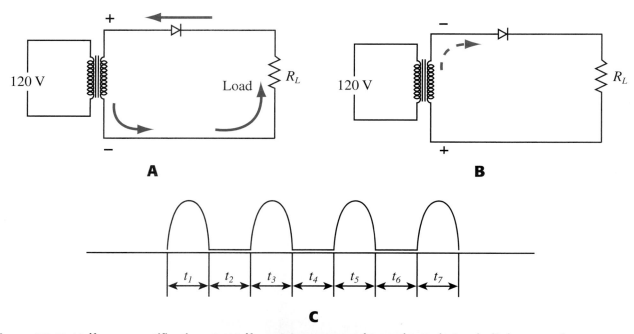

Figure 32-3. Half-wave rectification. A—Half-wave power supply conducts during half the ac cycle. B—During the second half of the ac cycle, the diode is reverse biased and does not conduct. C—The graph shows a diode conducting during the odd time periods and reverse biased during the even time periods.

When the direction of the current reverses in the transformer (shown by dashed lines in Figure 32-3B), the voltage polarity is negative on the anode. This causes the diode to be reverse biased. Since current does not flow in the reverse direction through the diode, no current flows through the load R_L. This is plotted on the graph in Figure 32-3C during time period t_2.

At this point, the alternating current of the transformer reverses again, flowing in the same direction as shown in Figure 32-3A and causing current during t_3. The cycle of current-no-current continues from t_4 through t_6. Although pulses of dc current are produced, the diode causes the current to flow in one direction; thus, the circuit changes the ac at the transformer to dc across the load. Since only half the ac cycle is being changed to dc, the circuit is called a ***half-wave*** power supply. The average output voltage of a half-wave power supply is equal to the peak voltage of the transformer secondary divided by pi (π). The equation is:

$$V_{AVE} = \frac{V_{PEAK}}{\pi}$$

Half-wave: In a power supply, one in which half the ac cycle is rectified into direct current.

The diodes of any power supply operate in the same way, allowing current to flow in just one direction. The only differences between the power supplies are the diode connection, resulting output waveform, and output voltage.

Full-wave Rectification

A full-wave center-tapped power supply is different from a half-wave power supply in that it contains two diodes and a center-tapped transformer. Since the center tap is the reference for the circuit, it is often connected to the chassis, or common ground. Half the full secondary voltage is used for rectification.

When the current is flowing counterclockwise, D_1 is conducting, **Figure 32-4.** This is plotted on the graph in Figure 32-4C during t_1 and other odd-numbered time periods. The current changes direction (dashed lines) in the transformer, Figure 32-4C,

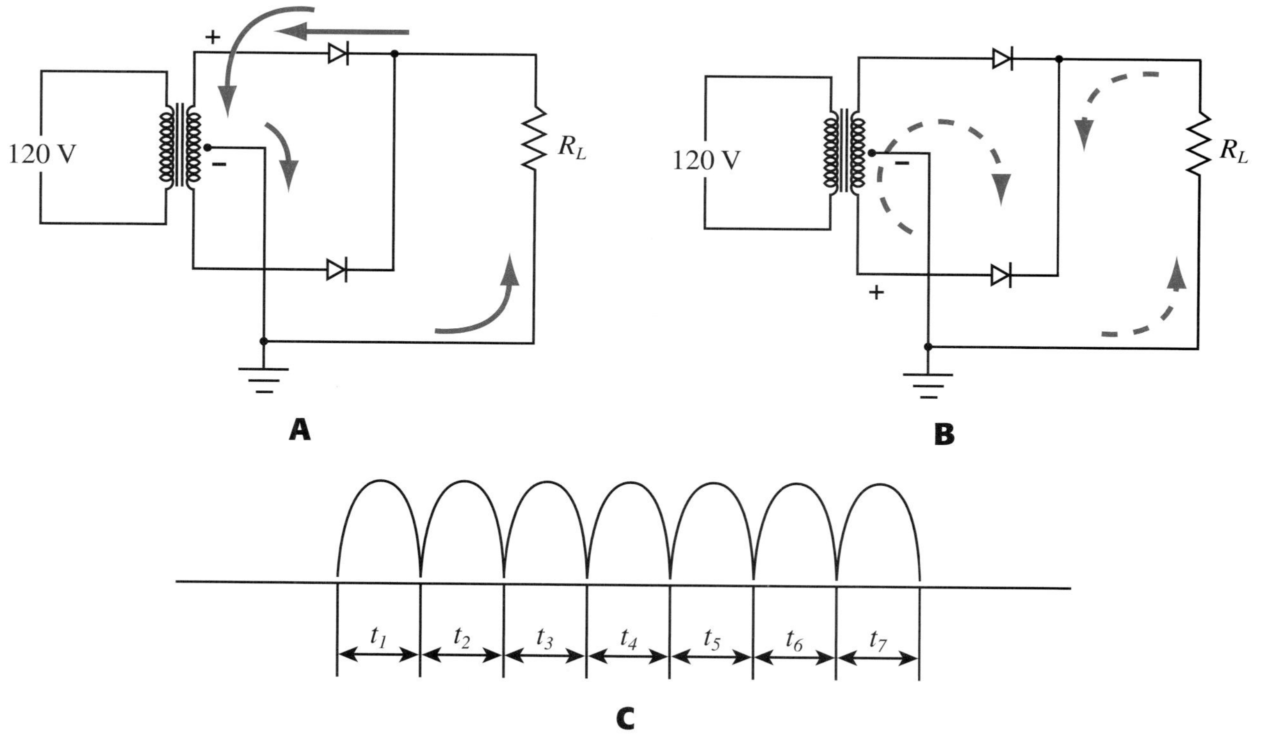

Figure 32-4. Full-wave rectification. A—Conduction of a full-wave center-tapped power supply during half of the ac cycle. B—Conduction during the second half of the ac cycle. C—Graph of the output of a full-wave power supply.

and diode D_2 conducts during t_2 and other even-numbered time periods. Since the whole ac cycle is being changed to dc, the circuit is called a ***full-wave*** power supply. The average output voltage of a full-wave power supply is equal to two times the peak voltage of the transformer secondary divided by pi (π). The equation is:

$$V_{AVE} = 2\, \frac{V_{PEAK}}{\pi}$$

The main advantage of the full-wave power supply over the half-wave is twice as many output pulses. This increases the output voltage and requires less filtering to smooth the current into a steady dc value. The main disadvantage of the full-wave is the extra cost of the transformer due to the center tap.

Bridge Power Supply

The bridge power supply consists of a four-diode circuit, but it eliminates the need for the center tap on the transformer and uses the full secondary voltage. During the first half of the ac cycle, diodes D_1 and D_3 conduct, **Figure 32-5.** Diodes D_2 and D_4 conduct during the second half of the ac cycle, Figure 32-5B. The output waveform is full-wave rectification, and the average output voltage is calculated the same as the full-wave center-tapped power supply.

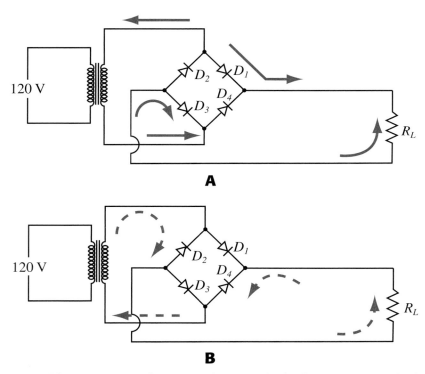

Figure 32-5. Bridge power supply. A—Conduction of a bridge power supply during half the ac cycle. B—Conduction of a bridge power supply during the second half of the ac cycle.

The bridge power supply may have four individual diodes, or all four diodes may be molded into a single device, **Figure 32-6.** Because of its large size, the rectifier usually determines the maximum forward current rating of the bridge. High-current rectifiers require a heat sink to carry away excess heat generated in high-current power supplies, **Figure 32-7.**

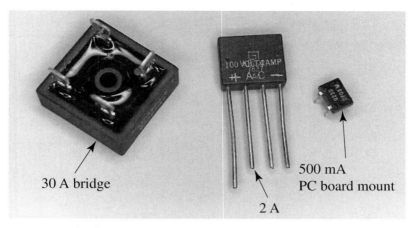

Figure 32-6. Various bridge rectifiers.

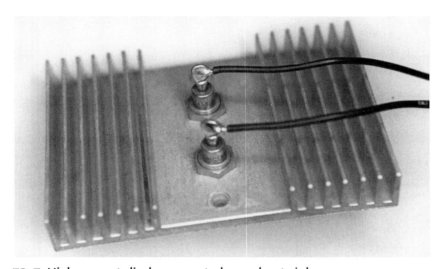

Figure 32-7. High-current diodes mounted on a heat sink.

Half-wave Voltage Doubler

The half-wave *voltage doubler* originally was used because it doubled the line voltage while eliminating the power transformer. It was in common use during the vacuum tube era. The half-wave voltage doubler has been rediscovered by some circuit designers using a power transformer to give voltages suited for solid-state devices.

The half-wave voltage doubler contains two diodes: a capacitor, which has a high capacitance (usually electrolytic), and a transformer. The transformer voltage can be any standard value. Starting with the negative half of the input voltage cycle, the polarity makes the anode of diode D_1 positive. Current flows through the diode, charging the capacitor to the peak input voltage. See **Figure 32-8.** Note the capacitor charge polarity.

During the positive half of the input voltage cycle, the ac current flows through load resistor (R_L) and diode D_2. The charge placed on the capacitor during the first half of the cycle discharges through the transformer. This places the capacitor charge voltage in series with the transformer voltage and doubles the available dc voltage across the load.

The doubler has a major advantage. It uses a small, lightweight transformer yet is able to give twice the dc output voltage. This eliminates the need for a large transformer.

Voltage doubler: A dc power supply that doubles the line voltage without the use of a transformer.

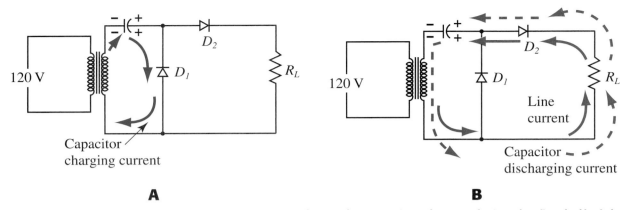

Figure 32-8. Half-wave voltage doubler power supply. A—The capacitor charges during the first half of the ac cycle. B—During the second half of the ac cycle, the source causes current through the load and diode D_2. The capacitor discharges in series with the source, doubling the available voltage.

Full-wave Voltage Doubler

The full-wave voltage doubler provides full-wave rectification with the addition of another capacitor. During the positive half of the transformer input voltage cycle, diode D_1 conducts as shown in **Figure 32-9,** charging capacitor C_2. At the same time, it provides current through the load. Capacitor C_1 was charged on the other half of the cycle and discharges in series with the input voltage in the same manner as the half-wave power supply.

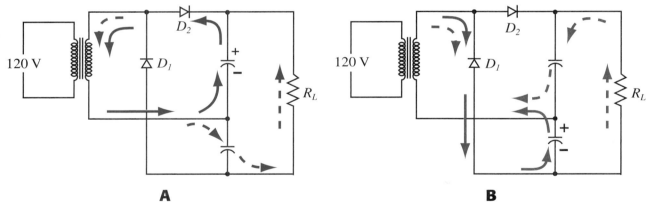

Figure 32-9. Full-wave voltage doubler power supply. A—Full-wave voltage doubler D_1 conducting. B—Full-wave voltage doubler D_2 conducting.

During the negative half of the input voltage cycle, diode D_2 conducts and charges capacitor C_1, again providing current to the load. Refer to Figure 32-9B. Capacitor C_2 discharges in series with the transformer input voltage.

Although the full-wave voltage doubler requires an additional large capacitor, the rectified output waveform is full-wave, giving a smoother dc output voltage and requiring less filtering.

Filter System

The filter system of a power supply consists of an inductor, capacitors, and resistors. These components smooth and steady the dc pulses. A filter capacitor is connected across the output of the half-wave rectifier circuit. The capacitor is charged at the same time current is flowing through the load from the rectifier circuit. See **Figure 32-10.**

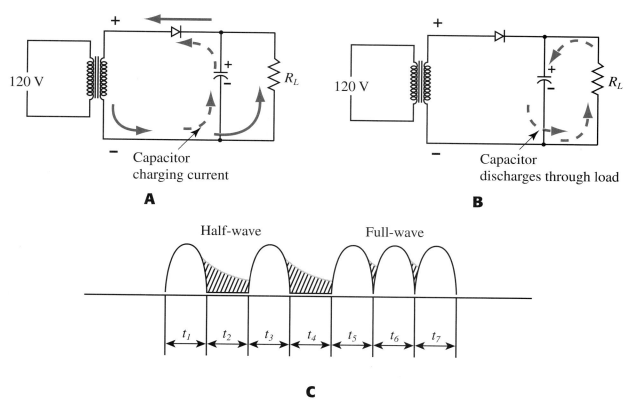

Figure 32-10. Filter system. A—The filter capacitor charges during half the ac cycle. B—During the other half of the ac cycle, the capacitor discharges through the load. During the period the diodes do not conduct, the capacitor provides current by discharging through the load. C—A smaller capacitance is required for filtering the full-wave supply.

When the rectifier circuit is not conducting, the capacitor discharges through the load, Figure 32-10B. This provides current for the load during the negative half of the transformer cycle.

The dc level is maintained by the filter capacitor, Figure 32-10C. In the right-hand portion of the graph, the same is true of the full-wave; however, the capacitor needs to fill in a smaller portion between pulses.

Regulator

Often the ac power line voltage will change by a small amount, typically 110 V to 125 V. The regulator maintains the power supply output voltage at a specific value. That is, it makes the output voltage stable even though the line voltage may change. A regulator circuit can be made from either a zener diode, a transistor circuit (Chapter 34), or an integrated circuit (Chapter 35). For this discussion, a simple zener diode circuit will be used.

The power supply in **Figure 32-11** is a standard bridge supply with a single capacitor C_1 as the filter capacitor. The zener diode is connected across or in parallel with the load R_L. Resistor R_1 drops the excess power supply voltage and provides a resistance that limits the current through the zener diode. Without R_1, the zener diode current would quickly increase to a point at which the diode would burn up. As shown, the sum of the zener diode V_Z voltage and the voltage drop of resistor V_{R_1} must be equal to the source voltage across the filter capacitor V_{C_1}. The equation is:

$$V_{C_1} = V_Z + V_{R_1}$$

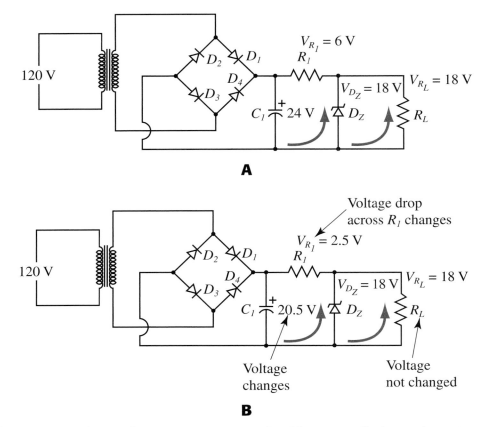

Figure 32-11. The regulator. A—A power supply with a zener diode regulator. B—The voltage across R_1 changes so the voltage across the load remains constant.

During the operation of the circuit, if the power supply output voltage changes slightly, the voltage drop across resistor R_1 also changes. Refer to Figure 32-11B. The zener diode voltage is always constant. This maintains the sum of V_{R_1} and V_Z equal to the input voltage. The load current remains constant and the zener diode current changes to make the shift in the voltage drop across resistor R_1.

How the value of resistor R_1 is determined is a question seldom considered. The resistance is found by the application of Ohm's law. The procedure is as follows:

1. Determine the maximum load current.
2. Calculate the peak voltage across the filter capacitor.
3. Calculate the minimum voltage across the zener diode.
4. Calculate the maximum voltage across the resistor R_1.
5. Calculate the maximum current for the zener diode.
6. Calculate the resistance of R_1.
7. Calculate the power rating necessary for R_1.

▼ *Example 32-1:*

What is the resistance required for R_1 in **Figure 32-12?**

The maximum load current should be determined by connecting the device to an experimenter's power supply and measuring the load current. In this example, it is presumed to be measured at 90 mA.

The peak voltage across the filter capacitor is:

$$V_{C_1} = 12 \text{ V} \times 1.414$$
$$= 17 \text{ V}$$

The minimum voltage across the zener diode is:

$$V_Z(\text{min.}) = 12\text{ V} - (12 \times 5\%)$$
$$= 12\text{ V} - 0.6\text{ V}$$
$$= 11.4\text{ V}$$

The maximum voltage across the resistor R_1 is:

$$V_{R_1}(\text{max.}) = V_{C_1} - V_Z(\text{min.})$$
$$= 17\text{ V} - 11.4\text{ V}$$
$$= 5.6\text{ V}$$

The maximum current for the zener diode is (power rating is presumed 1 W):

$$I_Z(\text{max.}) = \frac{P_Z}{V_Z}$$
$$= \frac{1\text{ W}}{9.1\text{ V}}$$
$$= 110\text{ mA}$$

The resistance of R_1 is:

$$R_1 = \frac{V_{R_1}}{I_Z + I_L}$$
$$= \frac{5.6\text{ V}}{110\text{ mA} + 90\text{ mA}}$$
$$= 28\ \Omega$$

The power dissipation necessary for R_1 is:

$$P_{R_1} = I_T \times V_{R_1}$$
$$= 200\text{ mA} \times 5.6\text{ V}$$
$$= 1.12\text{ W}$$

For the previous example, you would select a resistor with a standard resistance of 30 Ω and power of 2 W. Such a selection would ensure the resistor would not be damaged by excessive heat dissipation.

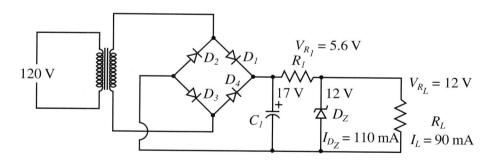

Figure 32-12. Zener diode circuit for Example 32-1.

Voltage Divider Network

In your studies of dc network analysis, resistors were connected to form a voltage divider. The voltage divider is used to make the necessary voltages for different circuits within the equipment. Some equipment may not have a voltage divider network. **Figure 32-13** shows an example of a resistor voltage divider network connected to a regulator. Notice the additional zener diode D_{Z_2} connected at a point within the voltage divider. This provides additional regulation for that part of the circuit.

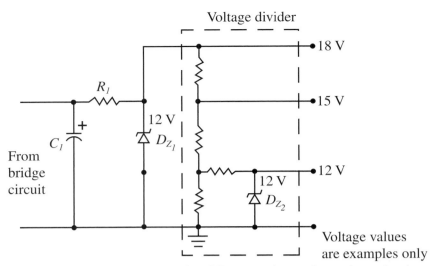

Figure 32-13. A voltage divider provides different voltages for various circuits.

Troubleshooting Power Supplies

When a power supply fails, the equipment it operates stops working or fails in some way. Any time a circuit does not operate correctly, troubleshooting becomes necessary. Before beginning to troubleshoot, you must understand what the circuit or system is supposed to do, and what it does or does not do.

Electronic Systems

A *system* is an assembly of parts to fulfill a specific purpose. A telephone, television receiver, industrial motor control, or any such apparatus is a system. In this chapter the systems to be studied are power supplies.

A power supply system failure will most likely result when the system:
- Does nothing (complete failure).
- Fails to perform one or more functions.
- Does everything right plus something else.
- Fails intermittently.

System: An assembly of parts or circuits to do a specific task.

Troubleshooting Methods

The method of troubleshooting used by a technician determines how rapidly and easily the problem is found. A technician who relies on past experience but lacks sufficient training sees every problem as major. This technician usually resorts to randomly replacing one part after another in an attempt to solve the problem. The approach is like shooting a gun until the target is hit or the gun runs out of bullets. The method is a waste of time, energy, and efficiency.

Before beginning, some technicians instantly reach for a schematic diagram of the system to be tested. In all probability, they are totally familiar with the symptoms and know the exact section in which the problem is likely to be found. Without both experience and knowledge, troubleshooting is very difficult.

A trained technician uses the block diagram, or flowchart method. It provides a logical, organized process for locating and correcting any type of problem. Of all the diagrams used in electronics, the block diagram is the most useful. It shows construction of the device, path of the signals, and interdependence of the sections.

Block diagrams may have several circuits or a complete device in each block of the diagram. A common block diagram can be applied to any type of device, **Figure 32-14.** It consists of input, output, control, and power supply. The power block is often omitted because it is assumed to be in every type of system.

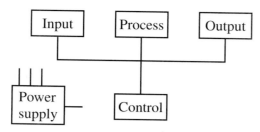

Figure 32-14. A common block diagram consists of input, output, control, and power supply. The power block is often omitted because it is assumed to be in every type of system.

Through proper tests, the problem is isolated to a particular block, then to the assembly or component causing the problem within the block. In some cases, the problem is isolated to a subsystem, then to a section within that subsystem, **Figure 32-15.** Block diagrams can be as complete and detailed as the system demands.

A troubleshooting block diagram, or flowchart, illustrates the thought process a technician uses to solve a problem. See **Figure 32-16.** The flowchart provides a clear, step-by-step methodology that eliminates guesswork. As a result, more work can be completed in less time.

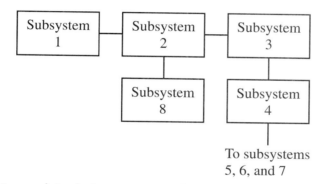

Figure 32-15. A complete device may use subsystems to isolate a problem.

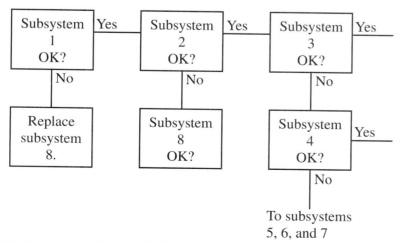

Figure 32-16. Troubleshooting block diagram for subsystems.

Troubleshooting Procedures

Many repair jobs take longer than necessary because the technician fails to correctly observe the symptoms shown by the system. Instead of hurriedly trying one repair, then another, start by carefully observing all the symptoms during your first inspection. Any troubleshooting process involves these general steps:

- Following an organized step-by-step process.
- Observing the symptoms.
- Analyzing the possible causes.
- Limiting the possibilities through tests and measurements.

Efficient Troubleshooting

Whatever the problem, repair of an electronic device should follow the procedure illustrated in **Figure 32-17:**

1. Determine and analyze the symptoms.
2. Determine the suspect section.
3. Obtain proper test equipment.
4. Localize the trouble to a specific area or block of the circuit.
5. Isolate the problem to a component.
6. Replace the defective component and check operation.

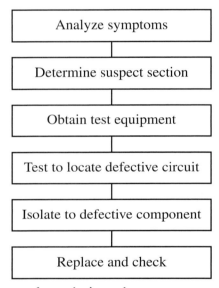

Figure 32-17. Block diagram of a typical repair.

Figure 32-18 shows a flowchart isolating a problem from section to component.

Before any test equipment is used, much can be learned from a simple inspection of the equipment under repair:

- Make a visual inspection to determine whether pilot lamps are on or off. Check that the power is turned on. Look for blistered or blackened parts due to overheating.
- Listen for unusual sounds and account for normal sounds.
- Notice odors and heat. It takes experience to learn the distinctive odors transformers and other components produce. Notice normal and abnormal heat radiating from parts.
- If the equipment operates intermittently, watch for part-time operating parts, temperature-sensitive parts, and poor connections.

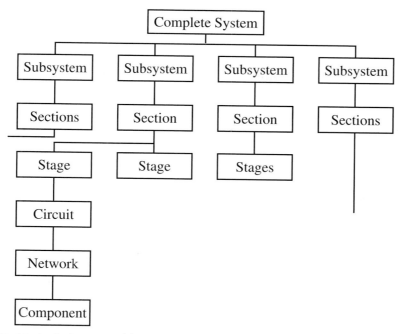

Figure 32-18. Isolating a problem to a single component.

When troubleshooting a power supply, you can make use of the common block diagram with the power supply block deleted. The input is the 120 V ac supply, and the output is the regulator or voltage divider, **Figure 32-19.** The process block is expanded to include three processes: transformer, rectifier, and filter. The control block includes the ON/OFF switch and fuse. Some people may include the regulator in the control section.

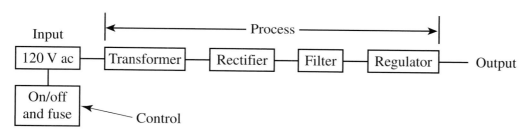

Figure 32-19. Applying the common block diagram to a power supply. The process block is expanded to include the transformer, rectifier, and filter.

Visualize any type of equipment under repair as a common block diagram. This will give you an overall understanding of how any equipment operates.

If a piece of equipment does not operate, the power supply section should be verified first. Use some simple VOM tests to determine the block(s) in which the problem is located.

Figure 32-20 is a troubleshooting flowchart for a power supply. The most common power supply problem is no power with blown fuses. At the top of the flowchart, the first diamond asks you to determine whether the dc voltage is low or zero. If zero, follow the diagram to the diamond that asks if the fuses are blown. If the answer is yes, the items in the rectangular block should be checked. If the answer is no, continue downward on the diagram.

Study the complete flowchart and note the troubleshooting process. The yes-no approach is a simple but organized method for locating a problem on any equipment.

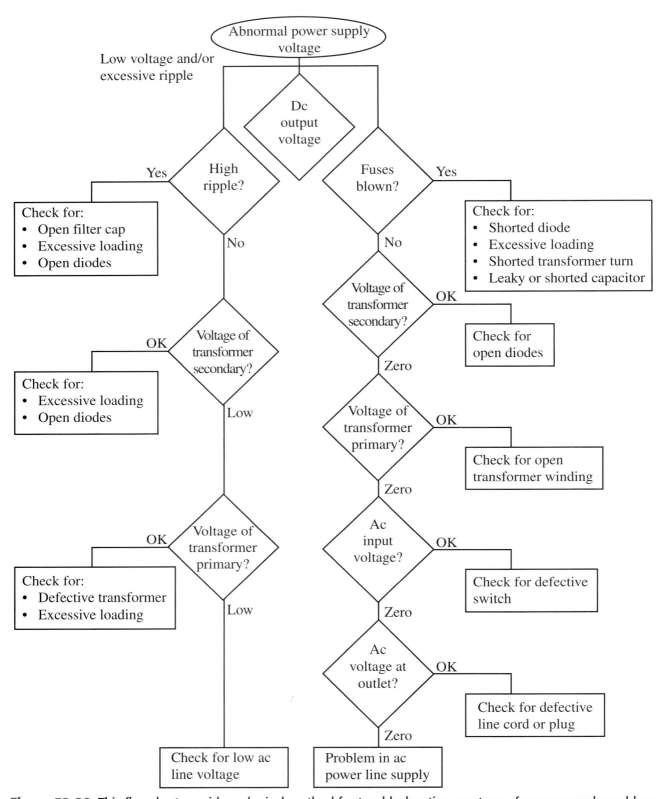

Figure 32-20. This flowchart provides a logical method for troubleshooting any type of power supply problem.

In electronic lab studies, students use single-stage amplifiers and other simple circuits, but when confronted with a complete system, they have difficulty troubleshooting. Although familiar circuits are buried within the system, the troubleshooting principles covered in this chapter can be applied to any system, device, or circuit, no matter how simple or complex.

Summary

- Power supply types are half-wave, full-wave center-tapped, full-wave bridge, and voltage doubler.
- A dc power supply consists of a transformer, rectifier, filter system, regulator, and voltage divider network.
- The filter capacitor turns dc pulses into a constant dc value.
- A regulator maintains the voltage at a specific value.
- A system is an assembly of parts performing a specific action.
- The best method of troubleshooting is to use a block diagram that provides a logical, step-by-step method for locating a problem.
- For any troubleshooting, follow an organized, step-by-step process; observe all the symptoms; analyze the possible causes; and limit the possibilities using tests and measurements.

Important Terms

Do you know the meanings of these terms used in the chapter?

full-wave   system
half-wave   voltage doubler
regulator

Questions and Problems

Please do not write in this text. Write your answers on a separate sheet of paper.

1. Draw the block diagram of a typical dc power supply.
2. Draw the schematic diagram of half-wave, full-wave center tapped, bridge, half-wave voltage doubler, and full-wave voltage doubler power supplies.
3. What is the purpose of the filter system?
4. Draw the output waveforms of a half-wave and full-wave rectifier.
5. Calculate the resistance of R_1 in **Figure 32-21.**
6. Define a *system.*
7. List several types of systems that can be found around your home.
8. Draw a common block diagram.
9. List four general steps in the troubleshooting process.
10. List four possible system faults.
11. If a half-wave power supply has a secondary voltage of 24 V_{RMS}, what is the average output voltage?
12. If the power supply in problem 11 were a full-wave power supply, what would be the average output voltage?

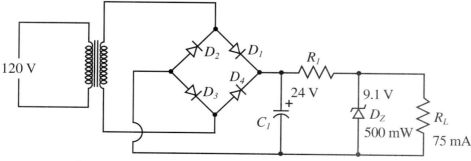

Figure 32-21. Circuit for problem 5.

Chapter 33 Graphic Overview

TRANSISTOR FUNDAMENTALS

Objectives

After studying this chapter, you will be able to:
- ○ Explain the relationship between the three currents involved in a bipolar junction transistor (BJT).
- ○ Explain the three operating states of a BJT.
- ○ Calculate the current gain and voltage gain of an amplifier.
- ○ Explain the operation of a BJT and metal oxide semiconductor field effect transistor (MOSFET).
- ○ Draw the schematic symbols for the BJT, junction field effect (JFET), and MOSFET.
- ○ Explain how to use an ohmmeter to test a transistor.

Introduction

In Chapter 31, you learned a diode is formed by the junction of P-type and N-type semiconductor materials. This chapter discusses the bipolar transistor. Although new electronic equipment designs use integrated circuits and surface mounted devices, transistors will continue to be used because semiconductor devices have a long useful life. Many designs use special transistors to do specific jobs for which integrated circuits are not easily adapted. JFET, MOSFET, and VMOS transistors are also discussed in this chapter. Transistors are used in switching circuits and amplifiers.

Bipolar Junction Transistor (BJT)

With the invention of the transistor in 1948, the electronics field changed faster than anytime since the invention of the vacuum tube. From satellites, to the military, to computer systems, to consumer electronics, we are still seeing the effects of the transistor on the world around us.

Junction Transistor Construction

A bipolar transistor is constructed like a diode. Unlike the diode, the transistor has three doped regions, **Figure 33-1.** Bipolar transistors can be made with either germanium or silicon. Silicon is more commonly used because it has better temperature characteristics.

Two bipolar transistor combinations are possible. In one combination, a thin, lightly doped P-type region is produced between two N-type regions. This is known as an *NPN transistor.* The regions are base, emitter, and collector. The arrow of the schematic symbol points toward the N-type region and the negative polarity of the power supply.

The second bipolar transistor combination is constructed with the opposite material types of the NPN. The *base* region is the N-type between two P-type regions. It is called a *PNP transistor.* In the schematic symbol, the arrow is pointing toward the N-type material.

NPN transistor: A transistor with a P-type base and N-type collector and emitter.

Base: The controlling element of a bipolar transistor.

PNP transistor: A transistor with an N-type base and a P-type emitter and collector.

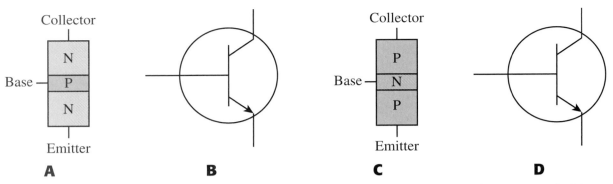

Figure 33-1. Bipolar junction transistor. A—Construction of a NPN transistor. B—Schematic symbol for a NPN transistor. Sometimes the circle is omitted. C—Construction of a PNP transistor. D—Schematic symbol for a PNP bipolar transistor.

Transistor Current Flow

An NPN transistor will be used in this discussion of transistor current flow. The currents for a PNP would be the same, except the direction of the currents would be reversed. When the symbols were designed, engineers used the conventional current theory; therefore, the arrow of the symbol points toward the direction of conventional current flow. As stated earlier, this text uses electron current flow.

When the base is made positive with respect to the emitter, the emitter-base PN junction becomes forward biased and electrons flow into the base material. See **Figure 33-2.** Since the base material is very thin, most of the electrons are attracted to the **collector.** If the collector is connected to a positive supply voltage, the electrons will flow from the collector to the positive terminal of the power supply. In the NPN transistor, the **emitter** gives off, or "emits," electrons while the collector "collects" them.

Collector: The main transistor electrode through which the primary current flows.

Emitter: The electrode of a transistor that carries the total current.

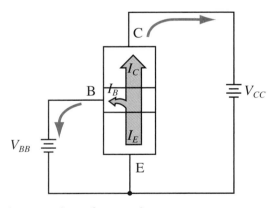

Figure 33-2. Basic operation of a transistor.

Beta: The current gain of a transistor amplifier connected in a common-emitter configuration.

Three currents are involved in a bipolar transistor: base current (I_B), collector current (I_C), and emitter current (I_E). The emitter current is the sum of the base current and the collector current. The emitter carries the total current. The equation is:

$$I_E = I_B + I_C$$

In transistor operation, the base current controls the collector current. An increase in base current will cause an increase in collector current. Likewise, if the base current is reduced to zero, the collector current is zero. The base current is controlled by the amount of bias voltage on the base. A small change in the base current causes a large change in the collector current. This results in a dc current gain, or **beta.** It is represented by the Greek letter beta (β). For example, if the base current is 50 μA and the collector current is 2 mA, the current gain of the transistor is 40.

Thus:

$$\beta_{dc} = \frac{I_C}{I_B}$$
$$= \frac{2 \times 10^{-3}}{50 \times 10^{-6}}$$
$$= 40$$

The current gain characterizes the transistor as an amplifier. For low-power transistors, the β is typically 20 to 300. However, the β can vary greatly between transistor types and sometimes is found using a transistor data reference manual.

Transistor Operating States

The three states of operation in a transistor are cutoff, saturation, and linear. With no bias voltage present on the base of the transistor, the transistor is off, or in the ***cutoff state.*** When the bias voltage is increased until no further increase in the collector current takes place, the transistor is on, or in the ***saturation state.***

Cutoff state: The point at which the collector current stops flowing.

Cutoff State

The base is connected to the supply voltage (V_{CC}) through variable resistor R_B, **Figure 33-3.** The collector is connected to V_{CC} by collector resistor R_C. When the resistance of R_B is adjusted high enough to cause the transistor to be cut off, the collector current is zero. The transistor essentially presents an open to the circuit.

With the transistor at cutoff, no current flows through collector resistor R_C, and no voltage drop appears across resistor R_C. The transistor and R_C are in series across the power supply, so the sum of the voltage across R_C and the voltage across the transistor must equal the supply voltage (V_{CC}). In the cutoff state, the collector voltage V_C is equal to the supply voltage V_{CC}.

Thus:

$$V_C = V_{CC} - V_{R_C}$$
$$= 12\text{ V} - 0\text{ V}$$
$$= 12\text{ V}$$

Saturation state: The point of maximum collector current or drain current of a transistor.

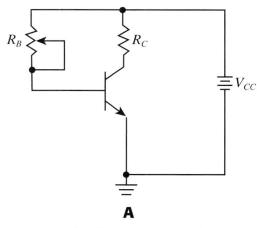
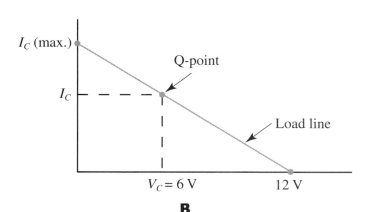

Figure 33-3. Cutoff state. A—Potentiometer R_B controls the amount of bias available. B—The load line is drawn from saturation to cutoff.

Saturation State

When base resistor R_B is adjusted until maximum collector current flows through collector resistor R_C, the transistor is in the saturation state. Essentially, the transistor presents a short to that part of the circuit. With maximum collector current flowing, all the supply voltage is dropped across R_C and the collector voltage is zero. A few millivolts will be present because of the internal resistance of the silicon material.

Thus:

$$V_C = V_{CC} - V_{R_C}$$
$$= 12 \text{ V} - 12 \text{ V}$$
$$= 0 \text{ V}$$

These points are marked on the graph in Figure 33-3B where I_C is maximum and V_C is zero. A line drawn from the point where I_C and V_C are maximum is called a *load line.*

Load line: A line drawn across the collector characteristic curves of a transistor from saturation to cutoff.

In saturation, a transistor has an excess base current, and the base has an excess of current carriers. When the transistor tries to come out of saturation, a small delay occurs called *saturation delay time* (or *storage time*). For example, the 2N3904 transistor has a storage time of 900 ns. After the base drive signal is removed, it takes 900 ns for the transistor to come out of saturation. Replacing a transistor with one having a slower storage time can cause problems in a high-speed switching circuit.

Saturation delay time: The time required for a transistor to come out of a saturated state. Also called **storage time**.

Linear State

Linear state: In a transistor amplifier, where the collector voltage is half the supply voltage.

If R_B is adjusted until the collector voltage V_C is approximately half the supply V_{CC} voltage, the transistor is in the *linear state.* A transistor uses the linear state when the transistor circuit is used as an audio amplifier. In the linear state, the collector current will be approximately half the maximum collector current. As shown in Figure 33-3B, dashed lines drawn through these points will come together at the center of the load line. This point on the load line is called the *quiescent,* or "Q" point. The *Q-point* is the inactive, or no-signal, dc operating point of the transistor. How the Q-point position on the load line affects an amplifier is discussed in Chapter 34.

Quiescent: The condition of a circuit at rest when no input signal is being applied. Also called the **Q-point**.

Transistor Circuits and Measurements

The circuit of **Figure 33-4** is the same as that in Figure 33-3, except an ac source and capacitor are added to the base circuit. The ac source is the input signal to be amplified. When building or troubleshooting a transistor circuit, three circuits are considered: base, collector, and emitter.

Common emitter: A transistor amplifier circuit in which the emitter is connected to a common ground.

The ac source (input), power supply, and emitter are connected with the common ground. The output, taken from the collector, is also referenced to ground. With the emitter at common ground, the transistor circuit is called a *common-emitter* amplifier. Dc voltage measurements V_B and V_C are also indicated. Other types of transistor amplifier circuits are covered in Chapter 34.

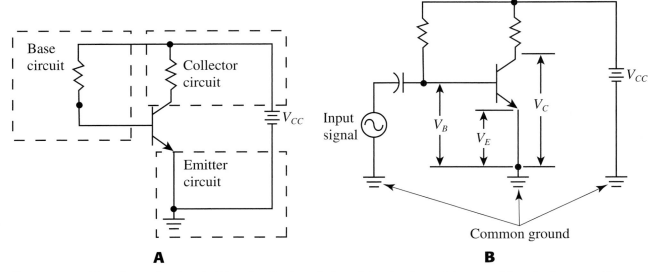

A   **B**

Figure 33-4. Transistor circuits. A—The circuit of each transistor lead must be considered when working with the circuit. B—A transistor circuit using common grounding.

Applying an Ac Signal to the Base

In the circuit in **Figure 33-5,** the resistance of R_B is adjusted so the E-B junction is forward biased with the base voltage V_B at about 0.7 V, but not in saturation. This causes a base current presumed to be 50 µA and makes the Q-point on the base current graph as shown in Figure 33-5B.

Notice the base voltage is across the emitter-base PN junction. Thus, the emitter-base voltage V_{BE} is simply the PN junction voltage drop. In this case, the transistor is silicon and the V_{BE} voltage is 0.7 V. If the transistor were germanium, the V_{BE} voltage would be 0.3 V.

In Figure 33-5C, the ac signal is set at 0.05 V and applied to the base through capacitor C_1 (capacitors pass ac and block dc). As the graph shows, this causes the base current I_B to increase and decrease from 30 µA to 70 µA, a 40 µA change in the base current. Since I_B changes with the input signal, so does the collector current I_C. However, since I_C is β times larger than I_B, the ac signal output at the collector is also β times larger.

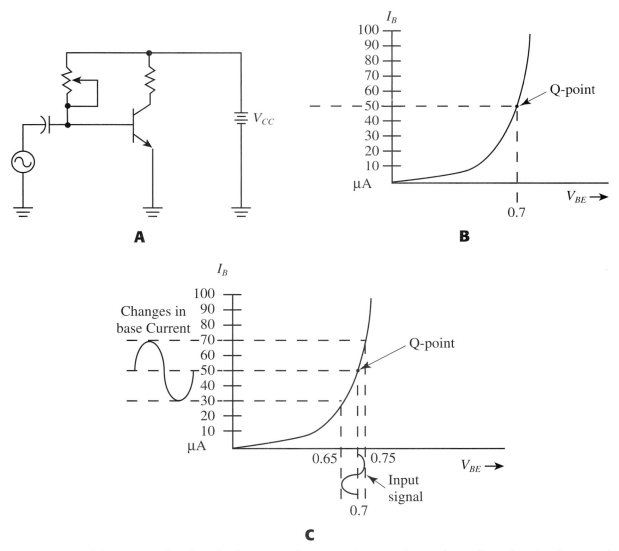

Figure 33-5. Applying an ac signal to the base. A—The potentiometer is used to adjust the circuit operating point. B—To eliminate distortion, the Q-point is adjusted to the center of the curve. C—An input signal causes the base voltage and base current to change.

The changing collector current flows through R_C, developing a changing voltage drop across R_C and the transistor. The voltage across the transistor is the output voltage, which is also β times the input voltage. Thus, the input voltage has been amplified by the transistor circuit.

Voltage Gain

Voltage gain: The ratio of the output voltage to the input voltage.

If the output signal voltage of an amplifier is divided by the input signal voltage, the result is the ***voltage gain*** of the amplifier. For example, in the block diagram in **Figure 33-6,** the input voltage to the base is 1.5 V and the output voltage (collector) is 12 V.

The voltage gain is:

$$A_V = \frac{V_{OUT}}{V_{IN}}$$
$$= \frac{12\ V}{1.5\ V}$$
$$= 8$$

No unit is given to the voltage gain. It is simply an amplification factor. In this case, it is a gain of 8.

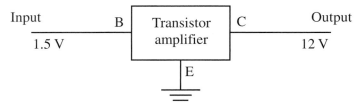

Figure 33-6. Due to voltage gain, the output signal of an amplifier is higher than the input signal voltage.

Transistor Packages

Transistors are available for various applications, such as general purpose, audio, radio frequency (RF), power, and switching. Numbering systems vary widely depending on the manufacturer. Some common numbers are 2N2222, 2N3904, MPS30, 2SB1688 or 2SC2920, and TIP31C.

Packaging varies with the type of circuit and power dissipation of the transistor. The electronics industry has adopted a set of standard transistor outline (TO) numbers, **Figure 33-7.**

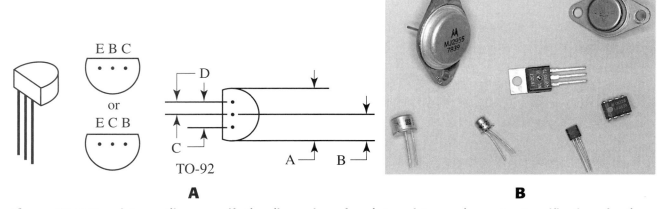

Figure 33-7. Transistor outlines specify the dimension of each transistor package. A— Specifications for the commonly used TO-92 package. B—Other common transistor packages.

Transistor Data Sheet

The data sheet for the 2N3904 transistor is given in Appendix G. The maximum ratings are the absolute limits for collector-emitter voltage, collector current, total device power dissipation, and other quantities. These are the main specifications to consider when a replacement transistor is selected.

Heat Sinks

Some transistors, especially power transistors, have high collector currents and must be mounted on heat sinks to keep them from overheating. A *heat sink* absorbs the heat from the transistor and dissipates it into the air, **Figure 33-8.** Silicone grease is used between the transistor and heat sink to aid in transferring excess heat.

Heat sink: A mass of metal used for carrying heat away from a component.

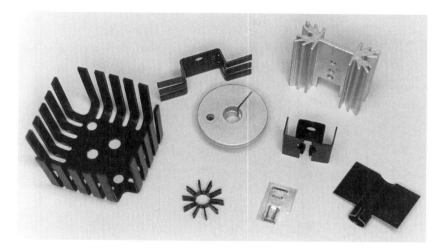

Figure 33-8. Heat sinks take heat away from transistors and increase their power rating.

Field Effect Transistors (FET)

The bipolar junction transistor uses an input (base) current to control the output (collector) current. Field effect transistors (FET) use an input (gate) voltage to control the output (drain) current. The gate is electrically insulated from the other parts of the transistor. Field effect transistors are of two types: junction field effect (JFET) and metal oxide semiconductor field effect (MOSFET).

Junction Field Effect Transistor (JFET)

The low input impedance disadvantage of the BJT is not suitable in some applications. To solve this problem, the semiconductor industry developed the field effect transistor, which has a high input impedance without special circuitry. Unlike the BJT, the leads of the JFET are named. They are *source* (emitter), *gate* (base), and *drain* (collector), **Figure 33-9.**

JFET Construction

A JFET starts out as a piece of P-type or N-type silicon. The source and drain leads are then connected to the ends of the piece of silicon, **Figure 33-10.** No PN junction exists at either the source or drain ends of the silicon. P-type material is diffused into both sides of the N-type piece of silicon to form PN junctions. These two P-type regions are joined with a connection to make the gate of the transistor.

Source: A lead or electrode of a field-effect or MOSFET transistor that corresponds to the emitter of a bipolar transistor.

Gate: The controlling element of a JFET or a MOSFET transistor.

Drain: 1. The current taken from a voltage source. 2. The electrode of a field effect transistor that corresponds to the collector of a bipolar transistor.

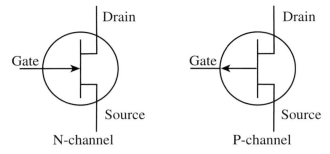

Figure 33-9. Schematic symbols for junction field effect transistors.

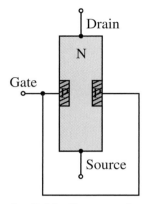

Figure 33-10. Construction of a field-effect transistor.

JFET Operation

Figure 33-11 shows a JFET connected to a dc power source. For now, the gate is not connected to any power source. The source is connected to the negative end, and the drain is connected to the positive end of the power source V_{DD} through the load resistance R_L. The source-to-drain area is a piece of N-doped silicon, rich in electrons. The drain current I_D flows from the negative terminal to the source, through the piece of silicon, and out the drain to the positive source terminal. Notice the drain current is limited by the load resistance and resistance of the piece of silicon.

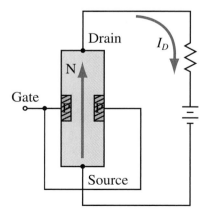

Figure 33-11. Operation of a field-effect transistor. No gate voltage is connected in this figure.

As the drain current I_D flows through the JFET, a depletion region is set up around the gate PN junctions, **Figure 33-12.** Since a depletion region is void of current carriers, drain current cannot flow in the depletion regions. Due to the action of the depletion regions, the drain current is confined to the space between the depletion regions called the *channel.*

As drain-source voltage is increased, the drain current increases. The increase in drain current is not linear because of the gate depletion region action on the drain current. When the drain current increases, the depletion region becomes larger, extending further into the channel, Figure 33-12B. As the drain current increases, it causes a decrease in the channel width, which opposes any further increase in the drain current.

Increasing drain-source voltage further causes the depletion regions to nearly come together, Figure 33-12C. When this condition exists, an increase in drain-source voltage causes no further increase in the drain current. This is the saturation state. The point where the drain current reaches its saturated value with zero gate-to-source voltage is the drain saturation current level and specified as I_{DDS}. If the drain-source voltage is increased too far, a breakdown of the transistor occurs and causes a very high current conduction value.

Channel: The conductive path between the source and drain of a field-effect transistor.

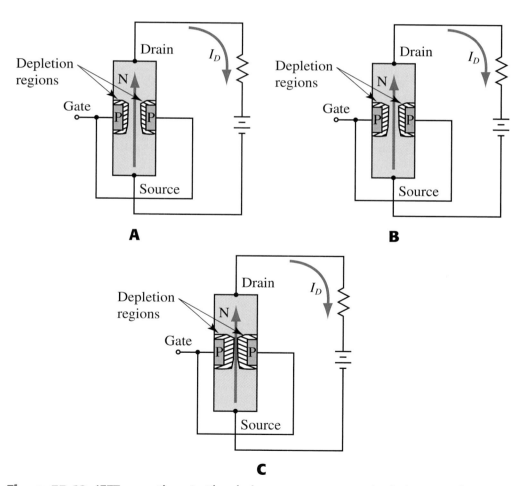

Figure 33-12. JFET operation. A—The drain current causes a depletion zone between the source and gate. B—As the drain current increases, the depletion zone increases. C—The drain current increases the depletion zone until saturation occurs.

In **Figure 33-13,** the gate is connected to the negative terminal of the gate power supply V_{GG} with the positive terminal on the source. This reverse biases the gate-source PN junction, and the load resistance is of a value that allows a moderate drain current. As the gate-to-source voltage is increased, making the gate more negative, the depletion regions are extended further into the channel, Figure 33-13B. This results in reduction of the drain current. The gate voltage can eventually cause the depletion regions to come together and pinch off the channel so no drain current remains. This condition is called *pinch-off voltage.* Thus, the drain current can be controlled by the gate voltage.

In the previous discussion, the drain current went through N-type material, so the transistor is called an N-channel JFET. Its possible to construct a P-channel JFET using P-type material and N-type gate regions.

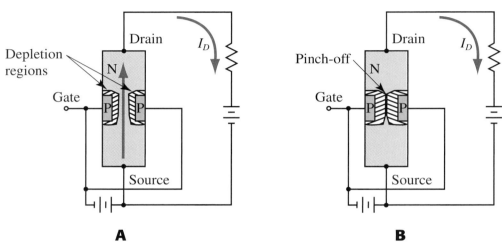

Figure 33-13. JFET gate. A—The gate is connected to a power source. B—The gate voltage increases until pinch-off occurs.

Metal Oxide Semiconductor Field Effect (MOSFET)

The JFET previously discussed uses PN junctions at the gate. Unlike the JFET, the MOSFET transistor gate is insulated from the rest of the device by a metal oxide and is called a *metal oxide semiconductor field effect transistor,* or MOSFET. The MOSFET transistor can be either *depletion-mode* or *enhancement-mode* type.

Depletion-mode MOSFET

Figure 33-14 shows the construction of an N-channel depletion-mode MOSFET. The N-channel is formed on a P-type substrate to which a substrate lead is connected. A very thin layer of silicon dioxide is placed on top of the N-channel. A metal plated area, to which the gate lead is connected, is formed on the silicon dioxide.

A P-channel MOSFET is constructed in the same manner with the channel made of P-type material and the substrate of N-type material. In the depletion-mode, the MOSFET operation is similar to the JFET, **Figure 33-15.** With no gate voltage, the drain current flows through the channel from source-to-drain as shown by the $V_{GS} = 0$ curve. As the gate voltage goes negative, the drain current decreases.

Unlike the JFET, the depletion-mode MOSFET can have a forward biased gate-to-source junction because the silicon dioxide insulates the gate from the rest of the transistor. This prevents any dc currents from flowing in or out of the gate. A positive bias voltage on the gate causes an increase in the drain current.

Pinch-off voltage: The gate voltage of a field effect transistor that blocks the current for all source-drain voltages below the junction breakdown value.

Metal oxide semiconductor field effect transistor: A type of field effect transistor in which the gate is insulated from the rest of the device by a metal oxide.

Depletion-mode: In MOSFET transistors, the operation in which current flows when the gate-source voltage is zero and is increased or decreased by changing the gate-source voltage.

Enhancement-mode: In MOSFET transistors, the operation in which no current flows when the gate-source voltage is zero and the current is increased by increasing the gate-source voltage.

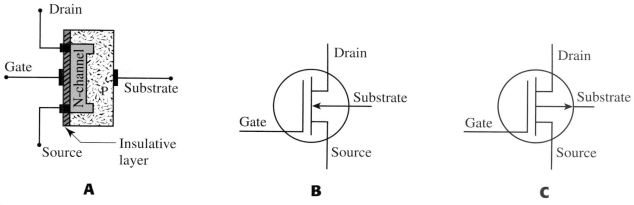

Figure 33-14. MOSFET. A—Construction of a D-MOSFET. B—Schematic symbol for an N-channel, D-MOSFET. C—Schematic symbol for a P-channel, D-MOSFET.

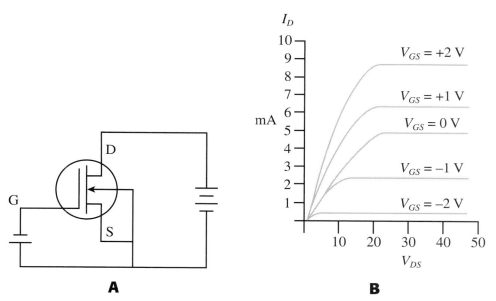

Figure 33-15. Depletion-mode MOSFET. A—Correct biasing of a MOSFET. B—Typical MOSFET characteristic curves. A positive bias voltage on the gate causes an increase in the drain current.

Enhancement-mode MOSFET

An N-channel enhancement-mode MOSFET in **Figure 33-16** does not have a channel. With no gate voltage, the drain current is zero and the transistor is in the off state. When the gate is made positive, electrons are attracted from the substrate. This causes an N-channel to be created between the source and drain and the transistor to be turned on. Any additional increase in the gate voltage causes an increase in both the size of the channel and the drain current. The enhancement-mode MOSFET transistor is a normally "off" device. A P-channel enhancement-mode MOSFET is constructed the same way as the N-channel, using P-type material and the substrate of N-type material.

Due to its normally off characteristic, the enhancement-mode MOSFET can be used as a switch. Turned on by a gate voltage and off by the absence of gate voltage, the MOSFET is used for digital circuits.

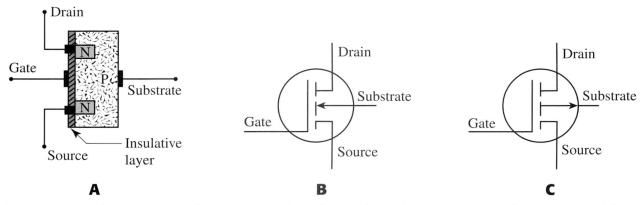

Figure 33-16. Enhancement mode. A—Construction of an N-channel, E-MOSFET. B—Schematic symbol for an N-channel, E-MOSFET. The dashed line indicates no channel exists between the source and drain. C—Schematic symbol for a P-channel, E-MOSFET.

MOSFETs and ESD

Because the silicon dioxide layer is so thin, it can be destroyed by excessive gate voltage. It is vital the maximum gate-to-source voltage be observed. Electrostatic discharge (ESD) from fingers or an instrument probe can easily puncture the silicon dioxide layer. MOSFETs should be shipped in sealed ESD-approved packaging to prevent charges from building up on the leads of the transistor. For shipment, a shorting clip may be used to short the transistor leads. If possible, do not remove the clip until after the transistor is soldered into the circuit. More information on ESD can be found in Chapter 12.

VMOS Transistors

The VMOS transistor is an enhancement-mode MOSFET modified to handle currents and voltages larger than a conventional MOSFET. Before the invention of the VMOS transistor, MOSFETs could not be used in circuits requiring large power ratings, such as BJT circuits.

In the VMOSFET transistor, a V-groove is cut into the structure and forms the gate, **Figure 33-17.** During operation, the gate is made positive, causing electrons to be attracted into the P-region surface under the gate. Above a certain voltage, the P-type silicon surface inverts its polarity, forming an N-channel and creating a low resistance from source-to-drain.

A major advantage of the VMOS transistor is lack of thermal runaway. The VMOS transistor has a negative temperature coefficient. As the temperature of the transistor increases, the drain current decreases, reducing the power dissipation. Unlike the BJT, the VMOS transistor cannot go into thermal runaway. Also, to increase the load power, the VMOS transistor can be connected in parallel without current hogging. If one of the transistors begins to conduct more current, the negative temperature coefficient reduces it and maintains approximately equal currents through the parallel transistors.

The VMOS transistor provides a high input impedance, nanosecond switching time, lack of storage time, linear characteristics, low on-state voltage, and temperature stability. VMOS transistors are durable and require only simple circuitry. They are well-suited for RF power amplifier circuits and interconnections between types of integrated circuits and high power loads.

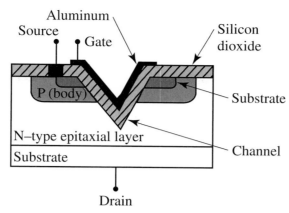

Figure 33-17. Construction of VMOS transistor. The epitaxial is a grown or deposited crystal layer with the same crystal orientation as the original substrate. The substrate is the supporting material on which the device is made.

Troubleshooting

When troubleshooting a transistor circuit, you will find the problem is often the transistor itself. Transistors usually develop shorts between the emitter-base junction or from the emitter to the collector. Such shorts cause major changes in the E, B, and C (S, D, and G) voltages around a transistor. In addition, opens within the transistor, high leakage currents, or changes in other transistor characteristics may occur.

Once a transistor is suspected of causing a circuit problem, it can be tested using a transistor tester, curve tracer, or ohmmeter. A transistor tester or curve tracer may not be readily available; therefore, only the ohmmeter test is discussed here.

Ohmmeter Test for BJTs

An easy test of the quality of a bipolar junction transistor can be made using an ohmmeter. The procedure is as follows:

1. Identify the leads of the transistor.
2. Set an analog meter to the R×100 range or the diode test position on a digital meter.
3. Connect one of the test probes to the base of the transistor, **Figure 33-18.**
4. Connect the other lead to the collector. You should get a low or high indication, depending on the type of transistor (NPN or PNP).
5. Connect the lead in step 4 to the emitter. You should get the same low or high reading.
6. Exchange the test probe on the base and repeat steps 4 and 5. You should get the opposite readings from before. For example, if the readings were low, you should get high readings now.
7. Connect one of the probes to the emitter and the other probe to the collector, **Figure 33-19.** You should get a high reading unless the transistor is a JFET type.

 Caution: When an analog meter is on the R×1 range, some transistors can be damaged. Also, due to ESD, use extra care when testing MOSFETs.

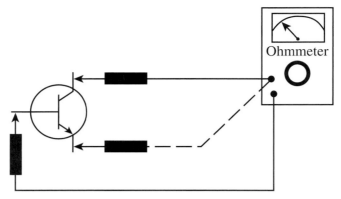

Figure 33-18. Step 3 in transistor ohmmeter test.

Figure 33-19. Step 7 in transistor ohmmeter test.

Ohmmeter Test for JFETs

An ohmmeter can be used to determine whether a short or open exists within a JFET transistor. The same high-low E-B and C-B ohmmeter readings will be found for gate-to-source and gate-to-drain ohmmeter readings. The main difference is the source-to-drain resistance, which is between 100 Ω and 1000 Ω for a JFET.

Ohmmeter Test for MOSFETs

An ohmmeter can be used to determine whether a short or open exists within a MOSFET transistor. Unlike the other types of transistors, the MOSFET should have an infinity resistance gate-to-source and gate-to-drain in both directions because of the silicon dioxide insulation at the gate of the transistor.

Other ohmmeter readings around the MOSFET will depend on whether it is a depletion-mode or enhancement-mode MOSFET. These readings are taken between the source and drain, substrate and source, and substrate and drain as follows:

- Depletion:
  Source-to-drain = 200 to 1000
  Substrate-to-source and drain = 100 to 500 in one direction, infinity in the other
- Enhancement:
  Source-to-drain = infinity
  Substrate-to-source and drain = 500 in one direction, infinity in the other

Summary

- A bipolar transistor can be made of germanium or silicon.
- The leads of a bipolar transistor are the base, collector, and emitter.
- Bipolar transistors can be NPN or PNP types.
- The electron current flows from the emitter to the collector. The emitter current is equal to the sum of the base current and collector current.
- The base current controls the collector current.
- The transistor operating states are cutoff, saturation, and linear. In the cutoff state, no collector current flows. In the saturation state, maximum collector flows. In the linear state, the transistor collector current is approximately half the maximum value.
- An ac signal applied to the base of a transistor causes corresponding changes in the collector current, which is beta times greater.
- The voltage gain of a transistor stage equals the ratio of the output signal to the input signal voltages.
- The most important ratings are the maximum ratings given in a data sheet.
- Heat sinks carry heat away from the transistor and increase the power dissipation of the transistor.
- Field effect transistor leads are source, gate, and drain. The transistors come in both P-channel and N-channel. MOSFET transistors may be either depletion or enhancement mode.
- MOSFET transistors can be damaged by ESD.
- An ohmmeter can be used to test transistors.

Important Terms

Do you know the meanings of these terms used in the chapter?

base
beta
channel
collector
common emitter
cutoff state
depletion-mode
drain
emitter
enhancement-mode
gate
heat sink
linear state

load line
metal oxide semiconductor field effect
   transistor (MOSFET)
NPN transistor
pinch-off voltage
PNP transistor
quiescent (Q-point)
saturation state
saturation delay time
source
storage time
voltage gain

Questions and Problems

Please do not write in this text. Write your answers on a separate sheet of paper.

1. What are two uses of transistors?
2. Identify the leads (emitter, base, collector) of the transistors in **Figure 33-20.**
3. If a transistor has a base current of 100 µA and a collector current of 1.5 mA, what is the emitter current?
4. If a transistor circuit has an I_B of 40 µA and an I_C of 3 mA, what is the beta?
5. What are the three states of operation of a transistor?
6. Using **Figure 33-21,** draw the load line for the circuit and locate the Q-point for linear operation.
7. On the circuit diagram of **Figure 33-22,** indicate the points of measurement for each: V_C, V_B, V_{RL}, and V_E.
8. What is the voltage gain of **Figure 33-23?**
9. List several important specifications given on a transistor data sheet.
10. What are the leads for the JFET?
11. Draw the schematic symbol for the JFET and MOSFET transistors.
12. Identify the leads of the transistors in **Figure 33-24.**
13. What transistor must be protected from ESD?
14. Why should the R×1 range on an analog meter *not* be used for an ohmmeter test of transistors?
15. What one measurement gives an indication of the operation of a transistor circuit?

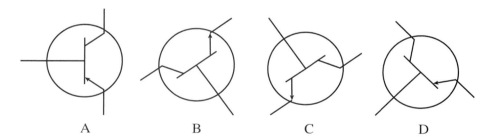

A   B   C   D

Figure 33-20. Transistor leads for problem 2.

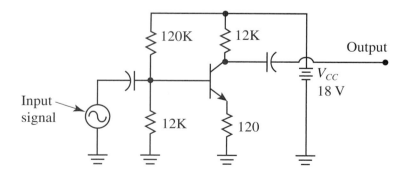

Figure 33-21. Circuit for problem 6.

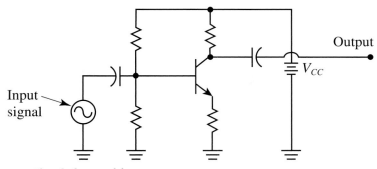

Figure 33-22. Circuit for problem 7.

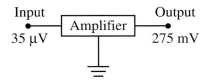

Figure 33-23. Circuit for problem 8.

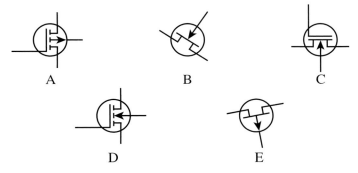

Figure 33-24. Symbols for problem 12.

Transistor Amplifiers

tests may be:
- Dynamic — determines — How Well — a — Circuit — is — Performing
- Static — made in — State of Rest — or — No Signal Condition

may be — Power Amplifier — types are:
- Single-ended
- Push-pull
- Darlington
- Complementary Symmetry

have — Interstage Coupling — can be:
- Transformer
- RC
- Direct

are classified:
- Class A
- Class B
- Class C
- Class AB
- Class D
- Class S

are a — Stage — which is — All Components — making up the — Amplifier Circuit

TRANSISTOR AMPLIFIERS

Objectives

After studying this chapter, you will be able to:
- ○ Draw a block diagram of a typical amplifier.
- ○ Draw schematic diagrams for common-emitter, common-base, and common-collector amplifiers.
- ○ Identify methods of biasing.
- ○ Explain the function of an emitter bypass capacitor.
- ○ Identify methods of amplifier coupling.
- ○ Identify types of power amplifiers.
- ○ Explain the difference between static tests and dynamic tests.
- ○ Explain how to troubleshoot a defective stage.

Introduction

You were introduced to the common-emitter amplifier in Chapter 33. This chapter expands on that circuit and also covers common-base, common-collector, and power amplifiers, along with characteristics of the various amplifiers. The coupling, or inter-connection, of two or more amplifiers is also discussed.

Amplifier Fundamentals

An *amplifier* is a circuit, or group of circuits, that causes the output to be greater than the input. If an amplifier circuit has an input signal of 1 V, and the output is 10 V, the input has been amplified by a factor of 10. The amplifier has a voltage gain of 10. See **Figure 34-1.**

Figure 34-2 shows an amplifier system with two stages and a power supply. A *stage* is an amplifying device (a transistor or integrated circuit) plus the individual components that make the circuit operate. The dc power supply provides dc for the transistor amplifier circuits to operate.

Amplifier: A device or electronic circuit that increases the magnitude of the current or voltage.

Stage: A transistor or other device and the components necessary to make it operate.

Figure 34-1. Block diagram of an amplifier.

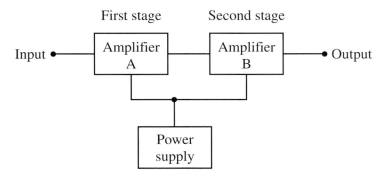

Figure 34-2. Block diagram of a two-stage amplifier. From a signal viewpoint, any amplifier is a series circuit.

Transistor Circuits

Transistor amplifier circuits can be classified in many ways. The design of the circuit largely depends on the intended job of the circuit, operating frequency, and point of biasing with reference to cutoff.

Common-emitter Amplifier

The common-emitter amplifier is the most commonly used transistor amplifier circuit. Operating in the linear state, the input signal causes small changes in the base current. In turn, the base current causes the collector current to change. However, the change in collector current is much greater in value than the base current.

The input to the amplifier is applied to the base, and the output is taken from the collector (with reference to ground), **Figure 34-3.** Although the circuit is shown with separate power supplies for the base and collector, the transistor lead connected to common ground is the emitter. For this reason, the circuit is called a ***common-emitter*** amplifier. Due to the E-B junction, the common-emitter amplifier has a low input impedance and a fairly high output impedance. The output signal is 180° out-of-phase as compared to the input signal.

Common emitter: A transistor amplifier circuit in which the emitter is connected to a common ground.

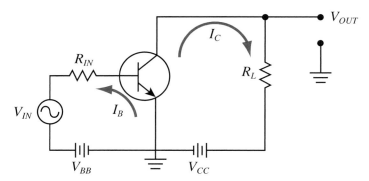

Figure 34-3. In a common-emitter amplifier circuit, the emitter is connected to the common ground.

Common-base Amplifier

The input to the common-base amplifier is applied to the emitter, and the output is taken from the collector (with reference to ground), **Figure 34-4.** The base is connected to the common ground; therefore, the circuit is called a ***common-base*** amplifier.

Any input signal causes variations in the emitter current I_E. The emitter current is equal to the sum of the collector current and the base current ($I_E = I_B + I_C$). Since the collector current is always less than the emitter current, the variations in the collector current are also slightly less. Therefore, the common-base amplifier has a current gain of slightly less than 1. The voltage and power gain of a common-base amplifier can be as high as 200.

Common base: A transistor circuit in which the base is connected to a common ground.

The common-base amplifier provides a low input impedance and a high output impedance. The low-high impedance characteristic of the amplifier makes the common-base amplifier suitable for matching a low impedance source to a high impedance load. In addition, a common-base amplifier may be used where a signal is picked up by an antenna because it provides a high voltage gain to the weak signal. The low impedance prevents any noise signals from being amplified along with the desired signal. A common-base amplifier is often used as the first amplifier in the tuning circuit of communications equipment, amplifying the weak signal coming in from the antenna.

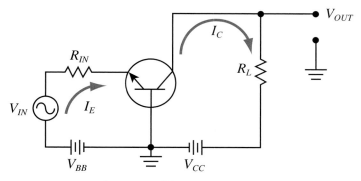

Figure 34-4. In a common-base amplifier circuit, the base is connected to the common ground.

Common-collector Amplifier

The input to the common-collector amplifier is applied to the base, and the output is taken across a load resistance in the emitter circuit (with reference to ground). See **Figure 34-5.** Capacitor C_2 is required to prevent dc current from flowing through the load to the emitter of the transistor. The collector is not connected directly to common ground but is connected to ac ground through capacitor C_1. Thus, the circuit is called a *common-collector* amplifier.

Common collector: A transistor circuit in which the collector is connected to a common ac ground by connecting a capacitor from the collector to ground.

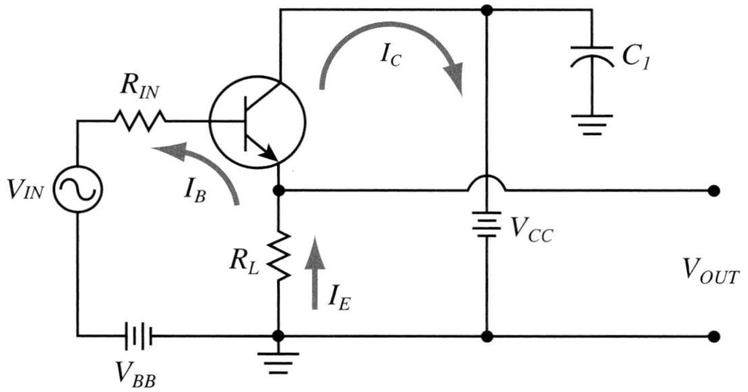

Figure 34-5. Because of the necessary dc voltage, the collector is connected to ac ground through a capacitor in the common-collector amplifier circuit.

The input signal causes changes in the emitter current, which causes a changing voltage drop across emitter resistor R_E. This voltage drop is the output signal. The common-collector amplifier always has a voltage gain of slightly less than 1, but it can provide a current gain of 100 or more. It has a high input impedance and a low output impedance. The common-collector circuit is often used to match a high-impedance device to a low-impedance device. Because the output signal is in phase with the input signal, the common-collector circuit is called an *emitter follower;* the output follows the input.

Whatever the required characteristic, the common-emitter, common-base, and common-collector circuits can be designed to meet the circuit needs. The characteristics of the three circuits are summarized in **Figure 34-6.** The values shown are for typical small signal circuits.

Emitter follower: A transistor circuit in which the output is taken from across a resistance in series with the emitter or source of the transistor.

Circuit	Characteristics		
Common emitter	Input Z	Low	1K
	Output Z	High	50K
	Current gain	High	100
	Voltage gain	High	100
	Power gain	Highest	10,000
	Phase	180°	
Common base	Input Z	Lowest	40
	Output Z	Highest	1 Meg
	Current gain	Unity	0.98
	Voltage gain	High	200
	Power gain	Medium	200
	Phase	0°	
Common collector	Input Z	Highest	300K
	Output Z	Lowest	50
	Current gain	High	50
	Voltage gain	Unity	0.98
	Power gain	Lowest	100
	Phase	0°	

Figure 34-6. Characteristics of the basic transistor amplifier circuits.

Methods of Biasing

The separate bias supplies of the previous circuits are costly and impractical. Using a single power supply for both the collector and base bias voltages is more practical. Common bias methods are the fixed bias and voltage divider method.

Fixed Bias Method

Figure 34-7 is the same common-emitter circuit shown in Chapter 33. The variable resistor is replaced with a fixed resistor connected from the base to the power supply. This circuit provides a simple way of biasing the base from a single power supply. The value of the base current is almost equal to the power supply voltage, minus a 0.7 V drop across the E-B junction (silicon junction voltage drop), divided by the resistance of R_1. The equation is:

$$I_B = \frac{V_{CC} - 0.7}{R_1}$$

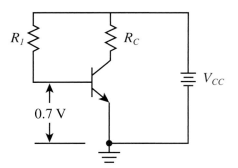

Figure 34-7. Biasing of the common-emitter amplifier.

Since the base bias voltage is determined by the fixed values of the power supply and the base resistor R_1, the circuit is a *fixed bias.*

A disadvantage of the fixed bias circuit is the changes in E-B junction impedance with the input signal. This results in changes in the base current and Q-point. The fixed bias circuit can also be unstable, and temperature changes can cause a shift in the bias.

Voltage Divider Method

In **Figure 34-8,** two resistors are connected across a source, forming a voltage divider. The voltage across R_2 is calculated at 1.2 V. Any circuit connected between ground and the junction of R_1 and R_2 would have a voltage of 1.2 V. If the base of a transistor is connected to this point, the 1.2 V becomes the base bias voltage and sets the bias for the circuit. The resistance of R_2 determines the base bias voltage.

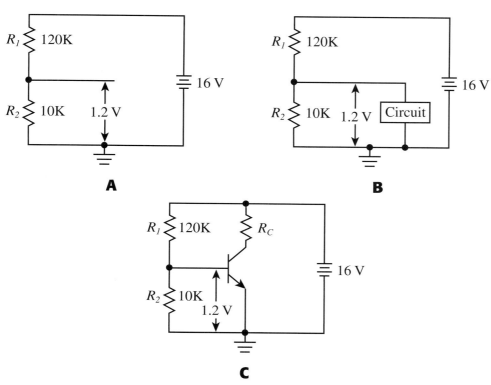

Figure 34-8. Voltage divider method. A—Two resistors form a voltage divider. B—The voltage divider becomes the voltage source for any circuit connected to it. C—The voltage divider sets the base bias voltage for the transistor.

The *voltage divider* method makes a transistor circuit more stable. If the base current increases due to a rise in temperature, the voltage across R_1 will increase. Since the sum of the two voltage divider voltages must always be equal to the power supply voltage, the voltage drop across R_2 decreases. The voltage across R_2 sets the bias, so the reduced bias voltage lowers the emitter current and the base current returns to its original value. If the temperature decreases, the bias voltage increases and returns the base current to its original value.

Value of R_2

Resistor R_2 is in parallel with the E-B junction, which has a relativity small resistance. If a high resistance is used for R_2, the circuit will operate as though R_2 were not there. No stability is provided for the circuit. If the resistance of R_2 is small, the resistance will shunt much of the input signal to ground, making the circuit a poor amplifier with a very low input impedance.

Temperature Stabilization

As previously discussed, if the operating temperature of a transistor increases, its internal resistance decreases. This causes a shift in the base current, making the circuit unstable. Although the voltage divider bias method improves stability, the addition of a small resistance in the emitter circuit improves circuit temperature stability even more.

Emitter Resistor R_E

All the transistor current goes through the emitter resistor R_E and develops a corresponding voltage drop, **Figure 34-9.** Measured with reference to ground, this is the emitter voltage V_E.

If the emitter current increases, the voltage drop across R_E increases, making the emitter more positive with reference to ground. Remember, the E-B junction bias is the voltage between the base and emitter. If the emitter becomes more positive, it has the effect of making the base more negative. In effect, this reduces the base operating bias voltage and reduces the collector current. The effect is called *negative feedback* or *degenerative feedback.* Although the addition of R_E reduces the amplifier gain, the advantage of increased temperature stability outweighs the loss in gain.

Negative feedback: Part of the output signal is fed back to the input 180° out of phase, causing reduced gain. Also called **degenerative feedback.**

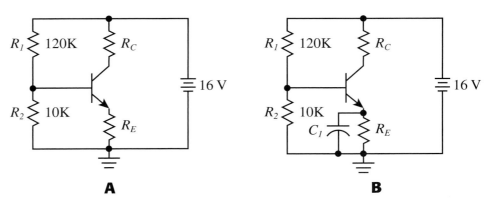

Figure 34-9. Temperature stabilization. A—A resistor in the emitter circuit stabilizes the amplifier for changes in temperature. B—Degeneration produced by the emitter resistor is reduced by adding a bypass capacitor across the resistor.

Emitter Bypass Capacitor

When an ac signal is applied to the base, the base current changes with the signal. This, in turn, changes the collector current. Since the emitter current is the sum of the collector and base currents, the emitter current changes and causes the voltage drop across R_E to change, producing the degenerative feedback effect. This is undesirable with the ac signal because it reduces the gain of the amplifier. Every time the signal tries to increase the collector current, the degenerative feedback action reduces it. If the signal reduces the collector current, the feedback increases it. Therefore, the action of the degenerative feedback effect reduces the gain, leaving the signal with less amplification than it should have.

To reduce the effects of degenerative feedback while maintaining stability, a capacitor is placed in parallel with emitter resistor R_E. Refer to Figure 34-9B. As the voltage across R_E tries to change, the capacitor charges and discharges (a filter action) through the resistor and maintains a constant dc voltage across R_E. This capacitor is usually called a *bypass capacitor* because it bypasses the ac signal component around the resistor. For low-frequency signals, the reactance of C_1 should be low enough to bypass the signal. The value of C_1 is chosen so its reactance is 1/10 the value of R_E, at the lowest frequency to be amplified. Therefore, resistor R_E provides only dc feedback for circuit stabilization.

Bypass capacitor: A capacitor that provides a low-impedance ac path around a circuit or component.

Voltage Feedback

Another method of stabilizing the temperature of a transistor amplifier is shown in **Figure 34-10.** The base R_1R_2 bias circuit is connected directly to the collector instead of the power supply V_{CC}. As the temperature of the transistor increases, the collector current increases. With increased collector current, the voltage across the collector load resistor R_C increases. As the resistor voltage drop increases, the collector voltage decreases.

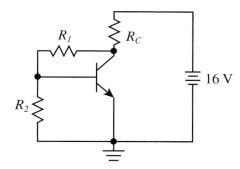

Figure 34-10. R_1 connected to the collector provides a voltage feedback.

With the bias circuit connected to the collector, the reduced collector voltage causes a decrease in the base bias voltage and base current. This brings the base current back to its original value. Since the base voltage and current is controlled by the collector voltage, this circuit provides voltage feedback and counteracts any increase or decrease in temperature change.

Amplifier Operation Classification

The class of operation of an amplifier is determined by the length of time the collector current flows during the signal input cycle. The length of time is determined by the Q-point of the amplifier and classifies the amplifier as Class A, B, or C.

Class A Amplifier

For Class A operation, the transistor must be biased so the collector current flows during the full input signal cycle. The Q-point is set at the linear (straight) portion of the I_B-I_C characteristics curve, **Figure 34-11.** Class A amplifiers are often called *linear amplifiers* and are used in audio systems.

Linear amplifier: An amplifier in which the output is always an amplified replica of the input signal.

Changing the Q-point

Recall from Chapter 33 the saturation and cutoff points are the two extremes on the load line graph. See **Figure 34-12.** If the resistance of R_B is adjusted so the Q-point is *lower* on the load line, the input signal drives the amplifier currents beyond the lower limit of the circuit (cutoff). This causes clipping distortion of the output signal, Figure 34-12A.

The same clipping distortion of the output is obtained if the Q-point is *higher* on the load line, Figure 34-12B. The amplifier operates close to saturation and causes clipping of the signal in the other direction.

The same type of clipping distortion will occur if the operation of the amplifier is taken into the nonlinear portion of the I_B-I_C curve. Refer to Figure 34-12C.

Another cause of signal clipping can occur because of excessive input signal. In Figure 34-12D, the Q-point is centered on the load line, but the input signal is adjusted too high.

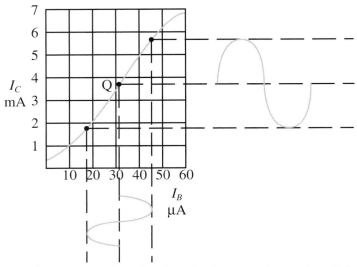

Figure 34-11. For Class A operation, the Q-point is set at the center of the linear portion of the curve.

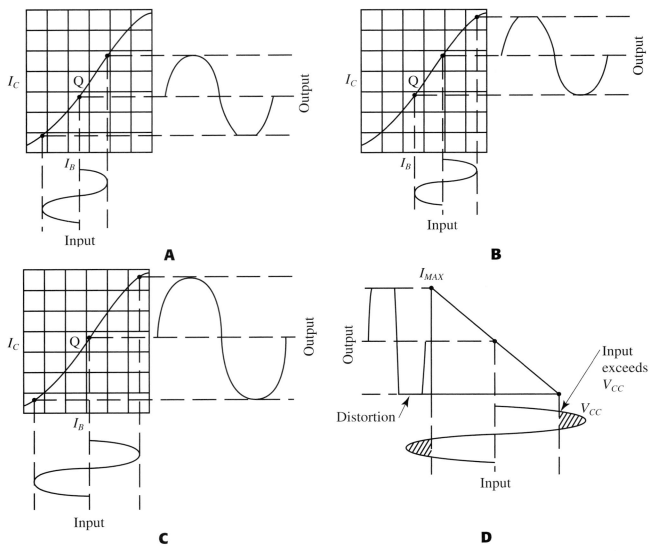

Figure 34-12. Changing the Q-point. A—Clipping caused by lowering the Q-point. B—Clipping caused by a higher Q-point. C—Nonlinear distortion of the output signal. D—Excessive input signal can cause an amplifier to go beyond its limits, clipping the output signal.

The amplifier output is forced to go beyond the limits of the circuit (saturation and cutoff). The amplifier is said to be "overdriven" and both positive and negative clipping occurs.

Class B Amplifier

The Q-point of a Class B amplifier is placed almost at cutoff in **Figure 34-13.** The transistor is biased so the collector current flows for approximately half, or 180°, of the input cycle. Since the collector and base currents cannot become less than zero, both the input and output waveforms are clipped, causing an output pulse rather than a sine wave. Class B operation is used in a two-transistor power amplifier circuit called a push-pull amplifier (discussed under power amplifiers). It is found in many audio amplifiers.

Class C Amplifier

In the Class C amplifier, the Q-point is placed beyond cutoff, **Figure 34-14.** The transistor is biased so collector current flows for less than half the input cycle. Again, the collector and base currents cannot become less than zero, so both the input and output waveforms are clipped, and the output is a pulse rather than a sine wave. Class C operation is used in many RF-tuned circuits for transmitters and other high-frequency equipment.

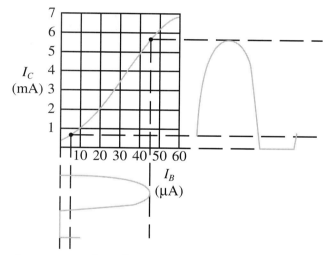

Figure 34-13. In class B operation, the Q-point is near cutoff.

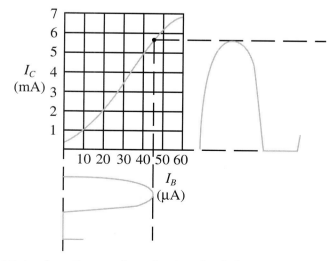

Figure 34-14. In class C operation, the Q-point is beyond cutoff.

Interstage Coupling

In most cases, one transistor amplifier stage cannot provide enough amplification. Additional stages must be connected for the required gain. *Coupling* refers to how the two stages are connected to transfer the signal from one stage to the other. The output of the first stage is coupled to the input of the second stage. The output of the second stage is coupled to the input of the third stage. The output of the third stage is coupled to the fourth. When a signal passes from one stage to another in a series sequence, the stages are *cascaded.* The most common methods of interstage coupling are transformer, resistance-capacitance (RC), and direct.

Transformer Coupling

Figure 34-15 shows an amplifier that uses transformer coupling. The signal is applied to the first stage using the primary of transformer T_1, whose secondary is connected to the base of transistor Q_1. The output of Q_1 is coupled to the second stage Q_2 by transformer T_2. The output of Q_2 is coupled to the output of the amplifier by transformer T_3.

Using a single power source, the transistor collectors obtain dc through the transformer primaries, while resistor combinations R_1R_2 and R_4R_5 provide voltage divider biasing. Resistors R_3 and R_6 are emitter stabilizing resistors with bypass capacitors C_2, C_3. Capacitors C_1, C_4, and C_5 are bypass capacitors, which bypass the voltage divider resistors. This allows the transformer secondary to operate to ground rather than through the voltage divider resistor.

These transformers are usually the step-down type with a turns ratio of 7:1 or 10:1 and matching 50K Ω to 1K Ω. Disadvantages of transformer coupling include cost, size, weight, and frequency response limitations.

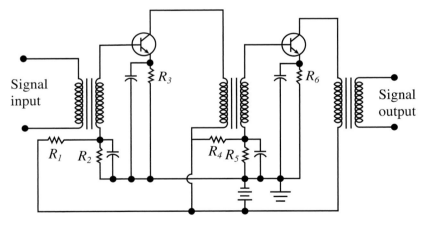

Figure 34-15. Transformer coupling of two transistor stages.

RC Coupling

Figure 34-16 shows a two-stage amplifier using RC coupling. Although resistor R_6 is part of the bias voltage divider, R_6 and capacitor C_3 make up the coupling network. The signal (changing collector current) at Q_1 is coupled to transistor Q_2 by the charging and discharging of C_3.

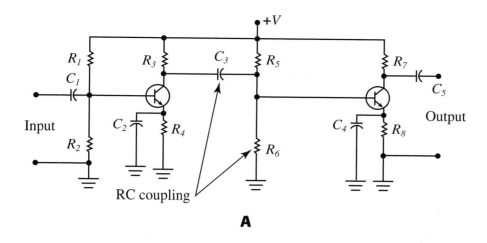

A

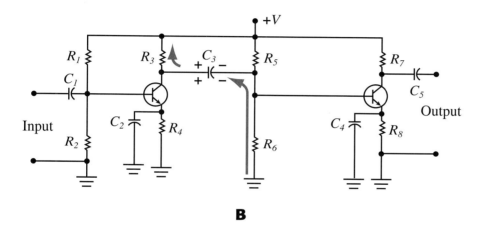

B

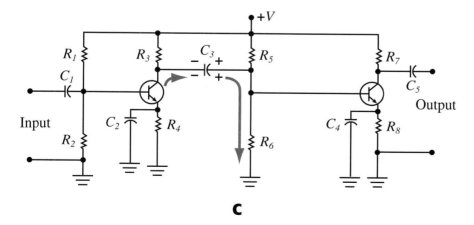

C

Figure 34-16. RC coupling. A—Two amplifier stages connected by RC coupling. B—Charging of the coupling capacitor. C—Discharging of the coupling capacitor.

To examine the RC circuit operation, presume a sine wave is applied to the base of Q_1. With the negative alternation first, the negative signal alternation decreases the forward bias of Q_1. This causes the base current of Q_1 to decrease and the collector current to decrease as well. A decrease in collector current causes the voltage drop across R_3 to decrease. This increases the collector voltage of Q_1 and makes the capacitor C_3 plate more positive, charging capacitor C_3 through resistor R_6. Do not confuse the positive-negative signs of the capacitor with the positive and negative voltages of the transistors.

During the positive half of the input signal, the base becomes more positive, causing an increase in the Q_1 base current. The Q_1 collector current and voltage drop across resistor R_3 increase while the Q_1 collector voltage decreases. The reduced Q_1 collector voltage causes C_2 to discharge through resistor R_6.

From the signal point of view, C_2 and R_6 are a series RC network. RC coupling has the disadvantage of being frequency selective. C_2 and R_6 control the low-frequency response and can cause some signal distortion. The advantages of RC coupling include small size, low cost, and less weight.

Direct Coupling

At low-signal frequencies, the frequency response disadvantage of RC coupling can be eliminated by direct coupling. The collector of Q_1 is connected directly to the base of Q_2, **Figure 34-17.** Although the coupling is a simple connection compared to other methods, component tolerance values are more critical. Without the capacitor to block dc between the amplifier stages, the circuit design must consider the base and collector voltages and currents for both stages. Direct coupling has one major disadvantage. If one of the transistors has a short circuit, all the transistors directly connected must be replaced in most cases.

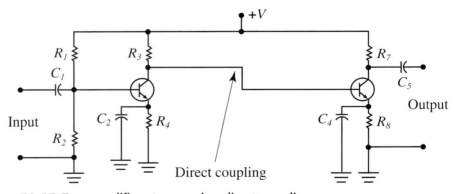

Figure 34-17. Two amplifier stages using direct coupling.

Power Amplifiers

The previously discussed amplifiers are called voltage amplifiers because their output voltage is larger than the input. After several stages of voltage amplification, the signal variations use the entire load line. Any additional voltage increase will cause a clipping distortion of the signal. Since the load line cannot be exceeded, additional amplification must be made by power amplification.

Power amplifiers are used to deliver the required power to a speaker or antenna. A power transistor must have a much higher collector current than a voltage amplifier. The high collector current flowing through the internal resistance of a power transistor causes heat to be generated within the transistor. Power amplifier stages use larger transistors and a heat sink to dissipate heat generated by the higher currents.

Single-ended Amplifier

A one-stage power amplifier is sufficient to power a speaker in a tape player, radio, or television receiver. One such amplifier is shown in **Figure 34-18,** where the amplifier consists of a single power transistor transformer coupled to the speaker. The transistor current must never be high enough to damage the voice coil of the speaker. This limits the power delivered to the speaker. In Figure 34-18B, the transformer has been eliminated by simply using a capacitor to couple the speaker to the collector. In Figure 34-18C, the speaker has been placed in series with the emitter circuit.

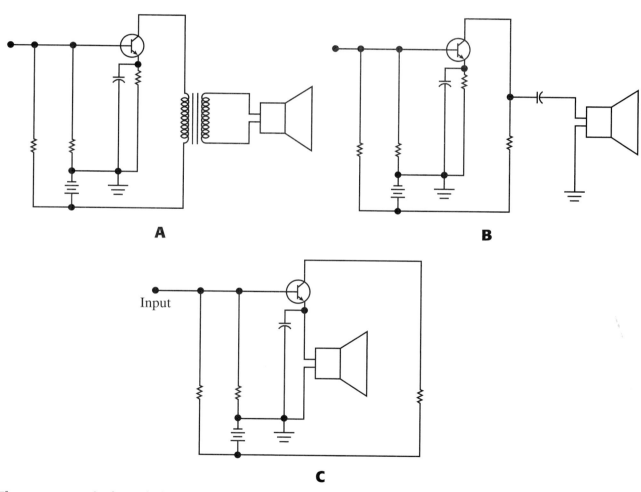

Figure 34-18. Single-ended amplifier. A—Single-ended power amplifier using transformer coupling for the speaker. B—An amplifier that uses capacitor coupling for the speaker. C—In this circuit, the speaker is connected in series with the emitter of the transistor. Excessive emitter current can damage the speaker voice coil.

Darlington Amplifier

Darlington is a widely used type of power amplifier, **Figure 34-19.** With the emitter of Q_1 connected to the base of Q_2, the Q_1 emitter current is the base current of Q_2. Transistor Q_1 should have an emitter current of 0.1 to 0.01 times the emitter current of transistor Q_2. Since I_{E_1} is equal to I_{B_2}, the total current gain of the Darlington pair is the product of the beta of Q_1 and Q_2. The equation is:

$$\beta = \beta_1 \times \beta_2$$

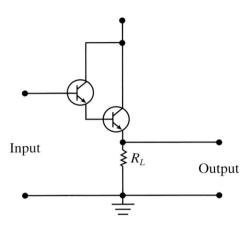

Figure 34-19. Darlington amplifier connection.

Example 34-1:
If the beta of Q_1 and Q_2 is 50, what is the current gain?

$$\begin{aligned}
\beta &= \beta_1 \times \beta_2 \\
&= 50 \times 50 \\
&= 2500
\end{aligned}$$

▲

Often Darlington transistors will be identical or come in pairs inside a single package. Sometimes a third transistor is used to make a Darlington triplet for increased power output. The Darlington has a higher input impedance than a single transistor amplifier. As a result, the Darlington has less loading effect on the driving circuits.

Push-pull amplifier:
An amplifier with two identical signal branches connected to operate in phase opposition.

Push-pull Amplifier

The ***push-pull amplifier*** uses two or more transistors in the output stage. Advantages of the push-pull amplifier include increased output power, greater efficiency, reduction of even harmonic distortion, and lower noise level.

Crossover distortion:
Distortion that occurs in a push-pull amplifier at the points where the signal crosses over the zero reference.

The input signals of Q_1 and Q_2 are 180° out of phase with each other due to the center tap of the transformer, **Figure 34-20.** During the first half of the input cycle, Q_1 conducts. During the second half, Q_2 conducts. A problem with the push-pull amplifier is shown in Figure 34-20C. With the transistors biased close to cutoff, the output becomes distorted as the transistors switch from one to the other. This ***crossover distortion*** occurs because amplification decreases and the curve becomes nonlinear near cutoff.

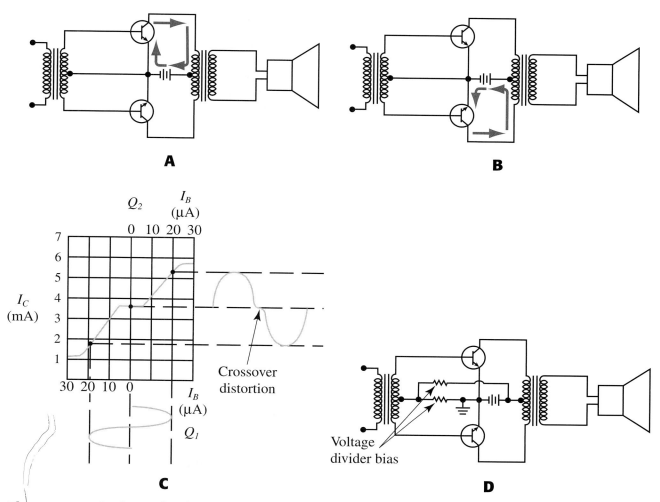

Figure 34-20. A—First cycle of a push-pull power amplifier. B—Second cycle of a push-pull power amplifier. C—When the push-pull amplifier changes from one transistor to another, a crossover distortion can occur. D—Bias resistors eliminate crossover distortion.

Crossover distortion can be reduced by designing a small amount of forward bias into the transistor amplifiers. Resistors R_1 and R_2 form a voltage divider to create the necessary bias.

Complementary Symmetry Amplifier

The typical complementary symmetry amplifier contains two opposite types of transistors, one NPN and one PNP, connected in series (stacked). See **Figure 34-21.** This design eliminates the need for transformers.

The input signal is connected to both transistors by coupling capacitor C_1. Since the transistor types are opposite, Q_1 conducts on the positive half of the signal. This charges the output coupling capacitor C_2. On the negative half of the cycle, Q_2 conducts and charges C_2 in the other direction. The increased I_E in Q_1 makes the capacitor charge higher in that direction. The increased I_E in Q_2 makes the capacitor discharge. The opposite half-cycle charge-discharge of capacitor C_2 produces the load current in the speaker. From a transistor point of view, the circuit is still a push-pull amplifier.

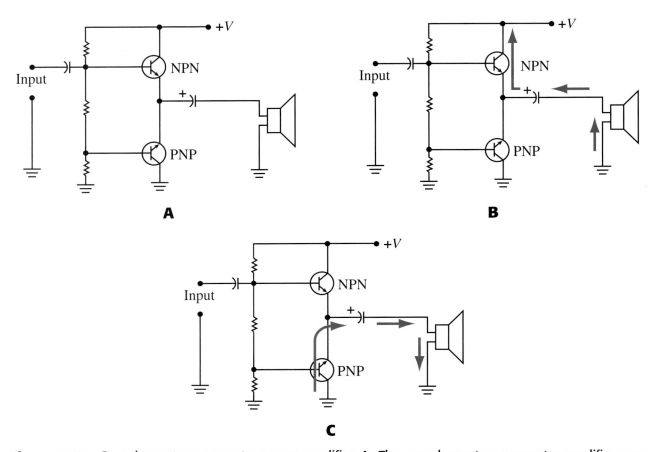

Figure 34-21. Complementary symmetry power amplifier. A—The complementary symmetry amplifier uses one NPN and one PNP transistor. B—Complementary symmetry when Q_1 is conducting. C—Complementary symmetry when Q_2 is conducting.

Troubleshooting Amplifiers

Static test: Testing performed without power or signal being processed.

Tests performed for troubleshooting amplifier circuits may be static or dynamic. *Static tests* are made when a circuit is in a state of rest. For example, an amplifier with no signal applied to the input is in a state of rest. Static tests include voltage tests to determine the condition of the operating voltages when the power is turned on. With the power off, continuity and resistance tests indicate the quality of fuses or absence of short circuits.

Dynamic test: Testing performed while circuits or equipment are operating.

Dynamic tests determine how well a circuit or system is performing. Dynamic testing provides information about the presence, absence, or character of the signal at various points in the circuits under test.

The signal can be thought of as the common denominator of all electronic systems. It is the one factor all systems have in common. Regardless of the complexity of the equipment, every electronic device must respond to, pass, amplify, or in some way properly process the signal. Therefore, any circuit defect will have an effect on the signal, causing it to operate improperly. In nearly every case, troubleshooting is most efficient when some form of dynamic testing is used to quickly locate the problem in a subsystem, section, or circuit. Dynamic testing consists of either signal tracing or signal injection.

Signal Tracing

The process of *signal tracing* locates the problem to a stage within the amplifying device. In **Figure 34-22,** if an oscilloscope measurement indicates an input signal at point 1, but no output at point 3, something between the two points is causing the problem. It becomes a yes-no process. (Is there signal or not?) This question must be answered each time a measurement is taken. A measurement taken at point 2 will determine if output comes from the first stage. If there is a signal at point 2, the problem exists with the second stage. If no signal is present at point 2, the first stage is at fault.

The process of signal tracing is fundamental to troubleshooting. Although an oscilloscope was used in the situation just described, the signal tracing process is the same with other types of test equipment.

Signal tracing: The process of tracing a signal through a circuit or piece of equipment.

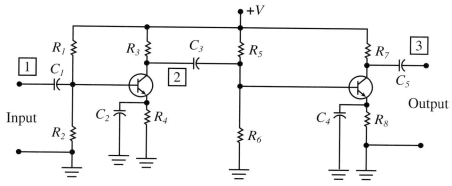

Figure 34-22. Signal tracing is determining if a signal is present at different points in a circuit.

Signal Injection

Signal injection is testing that determines how many stages or sections the signal will pass through and still have normal output. For the signal injection tests in **Figure 34-23,** a signal generator is placed at point 2 and the result is noted at output point 3. If the signal appears normal, the signal generator is placed at point 1, and the output is rechecked for normal signal. No matter how large or small the system, the defective section, stage, or subsystem can be located with just a few test points.

Signal injection: The process of putting a signal into a circuit or piece of equipment to see if it can pass through normally.

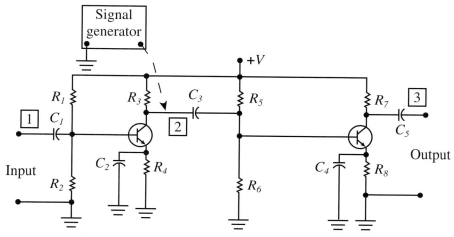

Figure 34-23. Using signal injection, amplifier performance can be determined by how well the amplifier passes a known signal.

Summary

- An amplifier is an electronic circuit that causes the output to be larger than the input.
- A stage is an electronic circuit, such as an amplifier and all the components necessary for the circuit to operate.
- Transistor amplifier circuits may be common-emitter, common-base, or common-collector, depending on which transistor lead is connected to common ground.
- The two methods of biasing transistor amplifiers are fixed and voltage divider.
- Transistor amplifiers must be stabilized for temperature using a small resistor in the emitter circuit or voltage feedback from the collector.
- The emitter bypass capacitor prevents degenerative feedback of the signal.
- Amplifiers are classified as Class A, B or C. The Class A amplifier is sometimes called a linear amplifier.
- The interconnection of amplifiers is called coupling. Coupling may be done by transformer, RC, or direct.
- Power amplifier types include single-ended, Darlington, push-pull, and complementary symmetry.
- Testing may be static or dynamic. Static tests are made with or without power and no signal. Dynamic tests are made under normal operating conditions, with a signal processed by the circuits.
- Signal tracing or signal injection should be used to locate a defective stage.

Important Terms

Do you know the meanings of these terms used in the chapter?

amplifier	emitter follower
bypass capacitor	fixed bias
cascaded	linear amplifier
common base	negative feedback
common collector	push-pull amplifier
common emitter	signal injection
coupling	signal tracing
crossover distortion	stage
degenerative feedback	static test
dynamic test	voltage divider bias

Questions and Problems

Please do not write in this text. Write your answers on a separate sheet of paper.

1. Draw a two-stage amplifier using a voltage amplifier and power amplifier.
2. Draw schematic diagrams of common-emitter, common-base, and common-collector amplifiers.
3. What type of amplifier is an emitter follower?
4. What is the base current in **Figure 34-24?**
5. In **Figure 34-25,** what is the result if the resistance of R_2 is extremely low?
6. What is the effect of an open emitter bypass capacitor on a common-emitter amplifier?
7. Identify two methods of temperature stabilization in a common-emitter amplifier.
8. Name three types of amplifier couplings.
9. Name four types of common power amplifier circuits.

10. Draw a schematic diagram of a complementary symmetry power amplifier.
11. Explain the difference between static tests and dynamic tests.
12. What type of troubleshooting process should be used to quickly locate a defective stage?

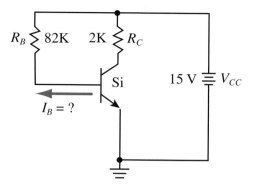

Figure 34-24. Circuit for problem 4.

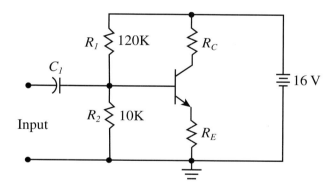

Figure 34-25. Circuit for problem 5.

Chapter 35 Graphic Overview

MISCELLANEOUS DEVICES

35

Objectives

After studying this chapter, you will be able to:

○ Draw schematic symbols for the SCR, DIAC, TRIAC, unijunction transistor, LED, photodiode, phototransistor, and opto-isolator.

○ Explain the process of testing an SCR and TRIAC.

○ Draw the schematic symbol for a unijunction transistor.

○ Calculate the resistance value for a current-limiting resistor in an LED circuit.

Introduction

This chapter discusses more devices that use PN junctions, including thyristors, silicon controlled rectifiers, TRIACs, and unijunction transistors. Light-emitting diodes, photodiodes, phototransistors, and opto-isolators are also covered. Many useful applications are possible using other circuits with these components.

Thyristors

Thyristors are a special family of semiconductor devices that serve as switches in controlling circuits. They are replacing many electromechanical devices, such as relays, because thyristors are smaller, consume less power, and require less maintenance. Thyristors are used in lamp dimmers, flashers, motor speed controls, automotive ignition systems, communications, and industrial systems. The three most widely used thyristors are the silicon controlled rectifier, bidirectional diode thyristor, and three-terminal thyristor.

Silicon Controlled Rectifier (SCR)

The *silicon controlled rectifier* (SCR) is a four-layer, three-terminal PNPN semiconductor device, **Figure 35-1.** An SCR acts like an ordinary PN junction diode. Current flows only in the forward direction, but it must be triggered into operation by the gate terminal. Once triggered, the SCR acts as a latched (locked) ON switch, and the gate no longer controls the current. To return the SCR to the off state, the current flow through the SCR must be stopped.

While in operation, the center PN junctions act as they would under a zener voltage. The outer PN junctions are already forward biased by the power source, **Figure 35-2.** No gate voltage is connected to the SCR. If the zener voltage across the SCR is increased, the center junction becomes saturated with current carriers to the avalanche point. At this point, maximum current flows through the SCR and the lamp is on. The current is limited by the resistance of the lamp. The current continues to flow until it decreases to a level unable to support the avalanche. Then, the SCR turns off. The minimum current required to support avalanche is called the *holding current.*

The gate is connected to a source through a N.O. momentary switch, Figure 35-2B. If the switch is pressed, allowing a momentary voltage on the gate, a small gate current sufficient to trigger the avalanche process is produced. The lamp

Silicon controlled rectifier. A four-layer PNPN semiconductor device that, in its normal state, blocks current flow. Once turned on, current continues to flow after the control signal is removed.

Holding current: The minimum current required to keep an SCR in conduction or a relay pulled in an active state.

turns on and remains on, even after the gate voltage is removed. The lamp stays on until the current is stopped by opening the circuit or reducing the current below the holding current level.

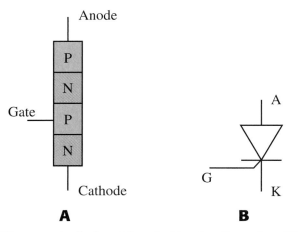

Figure 35-1. Silicon controlled rectifier. A—Construction of an SCR. B—Schematic symbol for an SCR.

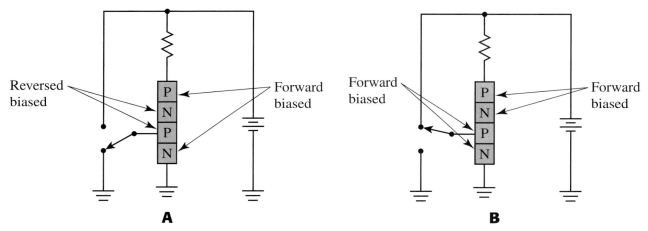

Figure 35-2. Construction and operation of an SCR. A—Biasing of the SCR junctions. B—When the gate is momentarily forward biased, the SCR conducts.

Testing an SCR

An SCR can be tested using an analog ohmmeter on the R×1 range or the diode test position on a digital meter. The following procedure is used:

1. Use a data book to identify the leads as cathode, anode, and gate. If a data book is not available, or the SCR is not listed, use your best guess to assign the terminal identification.
2. Connect the ohmmeter negative lead to the cathode and the positive lead to the anode. Alligator or other clip leads are useful with these connections. The ohmmeter should indicate infinity. A low resistance would indicate cathode-anode leakage. Zero resistance would indicate a short.
3. Connect a jumper from the anode to the gate, **Figure 35-3.** The ohmmeter reading should fall to a low or zero reading.

4. Remove the jumper from the gate. The ohmmeter should stay on the low or zero reading. If the ohmmeter goes high when the jumper is disconnected, the power source of the meter is not high enough to produce the necessary holding current.

5. If the lead identification in step 1 was incorrect, the SCR will not test correctly. Reassign the cathode, anode, and gate lead identifications and begin the test over. If the test does not work again, reassign the leads and retest.

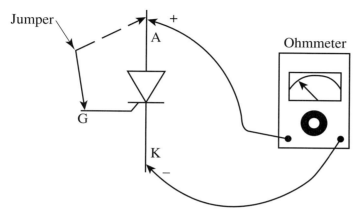

Figure 35-3. Testing an SCR. The jumper momentarily touches the anode to turn on the SCR.

Bidirectional Diode Thyristor (DIAC)

The bidirectional diode thyristor, or **DIAC,** is a three-layer, two-junction semiconductor device, **Figure 35-4.** Similar to a bipolar transistor, it has only two leads and resembles two zener diodes connected back to back.

When a positive voltage is applied to the DIAC, junction A will be forward biased while junction B will be reverse biased. See **Figure 35-5.** As the applied voltage increases, breakover voltage is reached. This allows junction B to develop avalanche breakdown and current to flow through the DIAC.

When a negative voltage is applied to the DIAC, junction A is reverse biased and junction B is forward biased. This time junction A will develop the avalanche breakdown when the breakover voltage is exceeded.

DIAC: A two-lead alternating current semiconductor switch.

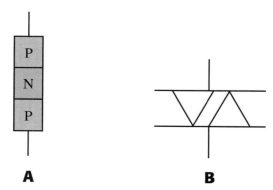

Figure 35-4. Bidirectional diode thyristor. A—Construction of a DIAC. B—Schematic symbol for a DIAC.

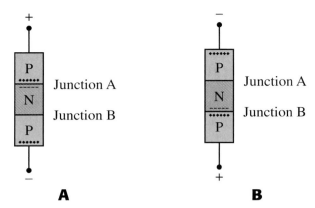

Figure 35-5. DIAC operation. A—Forward bias on junction A. B—Exceeding the breakover voltage, junction B is forward biased and the DIAC conducts.

Due to the negative resistance of the DIAC, the voltage drop across the DIAC will be less than the breakover voltage point. See **Figure 35-6.** For reliable circuit design, the forward current of a DIAC should be half the maximum operating power dissipation current. DIACs can be used as alternating voltage limiters but are primarily used to produce a sharp trigger pulse for other types of thyristors, such as the TRIAC.

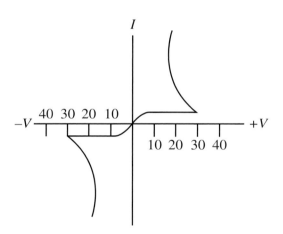

Figure 35-6. Conduction characteristics curve for a typical DIAC.

Three-terminal Thyristor (TRIAC)

TRIAC: Three-terminal, solid-state device that acts like a bidirectional switch.

A ***TRIAC*** is a multilayer, three-terminal thyristor that conducts in both directions, **Figure 35-7.** It can be considered a pair of SCRs connected in parallel, but in opposite directions, and controlled by one gate common to both SCRs. The terminals are named main terminal 1 (MT_1), main terminal 2 (MT_2), and gate. (Since it conducts in both directions, cathode and anode are not appropriate names).

The TRIAC is not really two SCRs because the gate makes contact with both a P-material and N-material region. This allows the TRIAC to be triggered with current of either polarity. Since the TRIAC conducts in both directions, it is better suited for ac than is the SCR. The procedure for testing a TRIAC is the same as for testing an SCR, except the TRIAC must be tested in both directions.

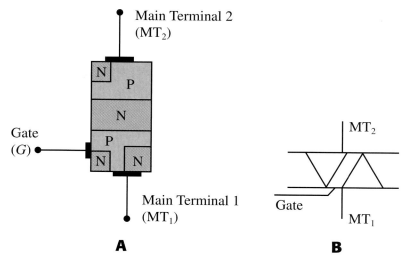

Figure 35-7. Three-terminal thyristor. A—Construction of a TRIAC. B—Schematic symbol for a TRIAC.

Unijunction Transistor

The unijunction ("uni" means one) transistor (UJT) was developed before FET technology and used as a voltage-controlled switch. Although not completely forgotten by today's designers, the UJT is considered obsolete by many. FET technology provides faster switching capability and higher stability. The UJT is not a member of the thyristor family as sometimes thought.

The UJT is constructed using a bar of N-type silicon to which a small amount of P-type of material is added to make a small P-region. See **Figure 35-8.** Unlike other transistors, the UJT has two base leads. A base terminal is connected to each end of the N-type bar material and labeled base 1 (B_1) and base 2 (B_2). The P-type region is closer to B_2 than B_1. The emitter lead is connected to the P-type region.

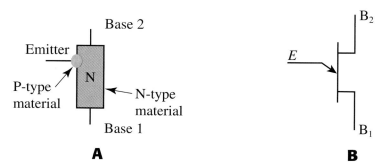

Figure 35-8. Unijunction transistor. A—Construction of a UJT. B—Schematic symbol for a UJT.

B_2 is normally connected to a positive voltage, and B_1 is connected to common ground, **Figure 35-9.** The N-type bar is lightly doped and has a resistance of 5K Ω to 10K Ω. Made of silicon, the emitter-base junction has a voltage drop of 0.65 V. The internal structure of the UJT is represented by a continuous resistance with the emitter-base junction (represented by a diode) connected at the center.

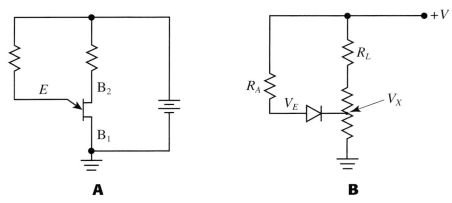

Figure 35-9. UJT connection. A—Connecting the UJT to a source voltage. B—Internal structure of the UJT.

The load resistor R_L limits the maximum current through the UJT, while R_A limits the emitter current. The emitter junction is reverse-biased and will not conduct until the voltage (V_E) at the emitter terminal is equal to or greater than V_X, plus the voltage drop of the emitter junction $(V_E = V_X + V_D)$. When this occurs, the PN junction conducts heavily in avalanche and causes the resistance of the emitter-B_1 half to decrease. This causes an increase in current through the UJT.

Contrary to Ohm's law, the emitter-to-ground voltage in **Figure 35-10** has decreased, yet the current has increased. This is a negative resistance phenomenon as discussed in your study of diodes. The UJT will return to the off condition when the forward bias on the PN junction is removed and the emitter current falls below the minimum holding value.

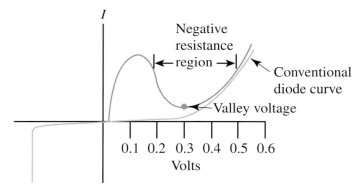

Figure 35-10. Like certain diodes, the unijunction transistor has negative resistance characteristics.

Light-emitting Diode (LED)

Many light-emitting diodes (LED) are constructed with gallium arsenide phosphide (GaArP) and emit a red light at approximately 660 nanometers. The LED contains an LED semiconductor chip in a plastic container, **Figure 35-11.** Like diodes, the leads of the LED are named cathode and anode. To locate the cathode, the plastic case has a flat edge on one side of the case.

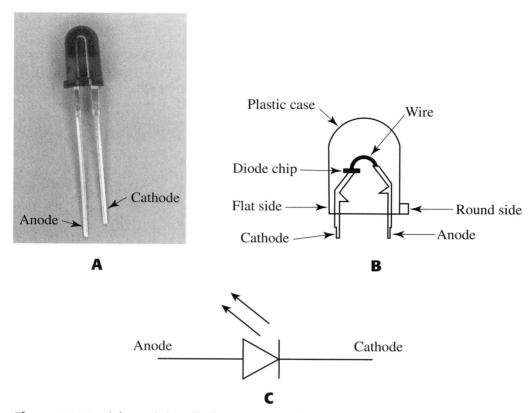

Figure 35-11. Light-emitting diode. A—LED packages can be of any color. B—Construction of an LED. C—Schematic symbol for an LED.

The LED chip is constructed using a GaAs substrate, a GaAsP, N-type layer to which an insulative layer is added. See **Figure 35-12.** A window is etched into the insulator and a P-type layer is made by diffusion through the window. During operation, photons of light are emitted from the window. With the addition of an aluminum contact, the anode connection is made. The cathode is connected to the gold contact.

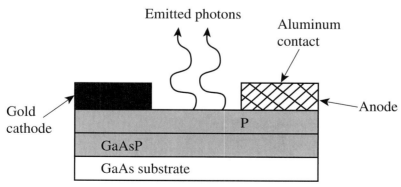

Figure 35-12. Construction details of the LED chip. Visible light is produced when using gallium arsenide phosphide, while gallium arsenide produces infrared (invisible) light.

The plastic case serves not only as a container but as a lens and magnifier to conduct light away from the LED. See **Figure 35-13.** The shape of the plastic lens controls the amount of diffusion. The position of the LED chip controls the shape of the radiation pattern.

The forward bias of an LED must increase to approximately 1.2 V before the forward current begins to flow, **Figure 35-14.** The forward voltage drop of a typical LED is 1.6 V (some are 2.2 V). Consult a reference manual for the maximum forward current of a specific LED. The forward current of most common LEDs is approximately 20 mA but can vary widely.

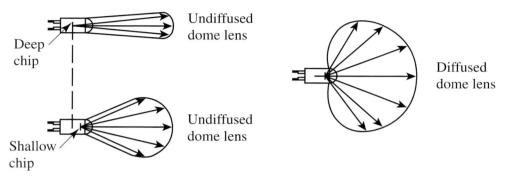

Deep chip

Shallow chip

Undiffused dome lens

Undiffused dome lens

Diffused dome lens

Figure 35-13. Emission characteristics of undiffused, diffused LED lens and chip position.

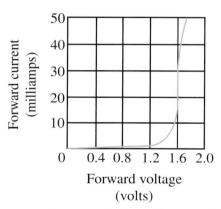

Forward current (milliamps)

Forward voltage (volts)

Figure 35-14. Forward current does not begin to flow until the forward bias of the LED increases to approximately 1.2 V.

 Caution: Some LEDs can be damaged by electrostatic discharge. ESD precautions should be observed when working with LEDs.

LEDs are often used as a "power on" indicating lamp. Operating as a power indicator, the current must be limited to a safe value for the LED. A current-limiting resistor must be calculated for the LED circuit, **Figure 35-15.** The LED circuit is to be connected to a power source of 18 V, and the current is to be limited to 20 mA. If the LED has a forward voltage drop of 1.6 V, then 16.4 V must be dropped across the limiting resistor.

The resistance of R_1 is calculated as:

$$R_1 = \frac{16.4 \text{ V}}{20 \text{ mA}}$$

$$= 820 \ \Omega$$

Thus, an 820 Ω resistor would be used in the circuit to limit the current at 20 mA.

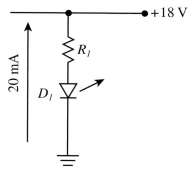

Figure 35-15. An LED requires a current-limiting resistor to prevent burnout.

Testing LEDs

Obviously, a defective LED will not light. To verify proper operation of an LED, simply bridge the defective LED with a good one. Also, the LED may be removed from the circuit, and a test circuit made with a power supply and limiting resistor.

Photodiodes

A photodiode is constructed similarly to a normal PN junction diode, except the junction is doped to respond to light. See **Figure 35-16.** A small glass window allows light to strike the silicon anode or PN junction. Normally, the diode is slightly reverse biased in the circuit. The action of the light causes the junction to become forward biased and current flows through the diode. Photodiodes are used for data transmission on fiber-optic cables.

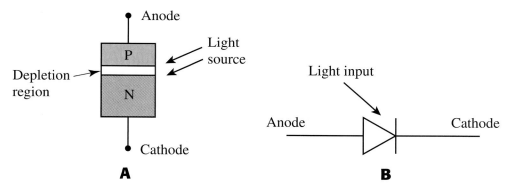

Figure 35-16. Photodiode. A—Construction of a photodiode. B—Schematic symbol for a photodiode.

Phototransistors

A phototransistor operates much like a photodiode, except it has current amplification. A window allows the light to strike one of the junctions, causing the current to flow through the transistor as though a signal had been applied to the base of the transistor. See **Figure 35-17.** The base lead can be left open (floating), or it can be biased to a specific level and provide more control over the circuit operation. Phototransistors are used in automatic lighting systems for turning lights off and on. They are also used in opto-isolator devices to isolate one circuit from another.

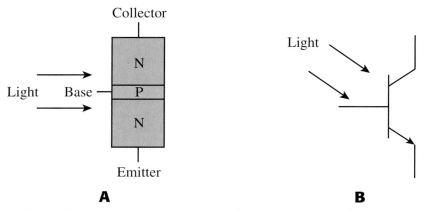

Figure 35-17. Phototransistor. A—Construction of a phototransistor. B—Schematic symbol for a phototransistor.

Opto-isolator

An opto-isolator (or opto-coupler) combines the photodiode or phototransistor with an LED enclosed in a single package. See **Figure 35-18.** Opto-isolators are used to electrically isolate one circuit from another. For most opto-isolators, the resistance of limiting resistor R_I should allow 1.5 V across the LED and a minimum current of 10 mA. A current lower than 10 mA will make the operation of the transistor circuit unreliable.

Other opto-controlled devices are photoFETs, SCRs, and TRIACs. Opto-isolator devices are used in industrial controls and fiberoptic communications.

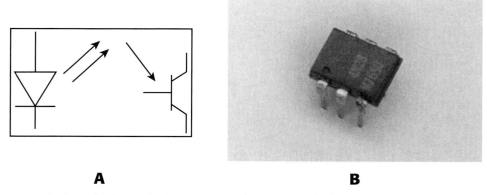

Figure 35-18. Opto-isolator. A—An opto-coupler, or opto-isolator, contains a photodiode and phototransistor. B—A common package for the opto-coupler is a six-pin DIP.

Summary

- The thyristor family includes the SCR, DIAC, and TRIAC. The minimum current flowing in an SCR or TRIAC is called the holding current.
- Thyristors can be tested using an ohmmeter.
- An SCR allows current in only one direction. A DIAC or TRIAC allows current in both directions.

- The DIAC is a bidirectional thyristor used to produce a sharp trigger pulse for TRIACs.
- The unijunction transistor has a negative resistance.
- The LED package serves as a lens or magnifier. The package also diffuses the light transmitted by the LED. A limiting resistor must be in series with the LED.
- Light controls the current flowing through a photodiode or phototransistor. The photodiode and phototransistor are combined to form an opto-isolator.

Important Terms

Do you know the meanings of these terms used in the chapter?

DIAC

holding current

silicon controlled rectifier

TRIAC

Questions and Problems

Please do not write in this text. Write your answers on a separate sheet of paper.

1. Draw the schematic symbols for SCR, DIAC, TRIAC, and UJT.
2. What is a holding current?
3. What is the main difference between the lead identification of a bipolar transistor and unijunction transistor?
4. Calculate the resistance of the limiting resistor in **Figure 35-19.**
5. What would be the resistance in problem 4 if the supply voltage were 48 V?
6. Draw the schematic symbols for LED, photodiode, phototransistor, and opto-isolator.
7. Calculate the limiting resistor R_1 for **Figure 35-20.**

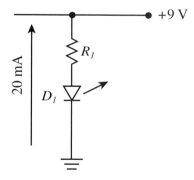

Figure 35-19. LED circuit for problem 4.

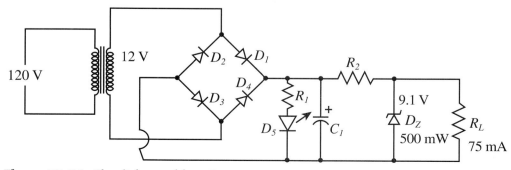

Figure 35-20. Circuit for problem 7.

Chapter 36 Graphic Overview

Integrated Circuits

- used for → **Dedicated Applications**
- **Operational Amplifiers**
  - circuit types are → **Comparator**
    - tests for:
      - Amplitude
      - Current
      - Frequency
      - Phase
      - Voltage
      - Waveform Type
      - Numerical Value
      - Error-sensing
    - Inverting
    - Noninverting
    - Voltage Follower
    - Summing
    - Difference
    - BIFET
      - a → **Bipolar Circuit**
        - with → **JFET input**
  - have → **Inverting** and **Noninverting Inputs**
- **Power Amplifiers**
- have
  - **Package Types**
    - are:
      - DIP
      - SIP
      - Flat-pack
      - TO-220
      - LAN
  - **Package Material**
    - made of:
      - Plastic
      - Metal
      - Ceramic

LINEAR INTEGRATED CIRCUITS

36

Objectives

After studying this chapter, you will be able to:

- ○ Identify the types of integrated circuit packages.
- ○ Calculate the gain of an amplifier stage.
- ○ Explain the differences between an open-loop and closed-loop operational amplifier circuit.
- ○ Identify inverting, noninverting, voltage follower, summing, and difference op-amp circuits.
- ○ Explain the input and output signal voltages of a comparator circuit.
- ○ Describe the differences between bipolar, BIFET, and MOSFET operational amplifiers.
- ○ Explain how to locate a defective integrated circuit.

Introduction

During the 1950s, the electronic industry began using packaged circuits containing resistors and capacitors. The integrated circuit (IC) was introduced in 1958. An integrated circuit combines transistors, resistors, capacitors, and other components into one package. The development of the integrated circuit allowed the electronics industry to expand at fast rate, and it continues to do so. The rapid expansion of the electronics industry is a result of miniaturization and numerous applications of integrated circuits.

Integrated Circuit Fabrication

A detailed description of construction of an integrated circuit is beyond the scope of this textbook, but the process can be summarized. Silicon doped with a P-type impurity is melted and formed into a round bar. The bar is sliced into 10-mil-thick wafers that form the substrate, **Figure 36-1.**

The surface is oxidized and coated with a photoresist for photolithography processing. The wafers are exposed to a photomask and developed. The unhardened photoresist is washed away. N-type doping is injected into the unexposed areas, forming PN junctions. This process is repeated until the required PN junctions are made in the substrate.

Many IC chips are formed on one wafer, refer to Figure 36-1B. They are placed on the mounting, and wires are welded to the required points on the chip. Finally, a case is made around the IC chip assembly.

Integrated Circuit Packages

Plastic, metal, and ceramic are three materials used to package integrated circuits. Many ICs are packaged with plastic because it is inexpensive. Under high temperature and pressure, the plastic is molded around the silicon chip and leads. The plastic case is not hermetically sealed and has no protection against moisture or other airborne particles. It is not suitable for high-temperature power dissipation because plastic does not carry heat away from the chip.

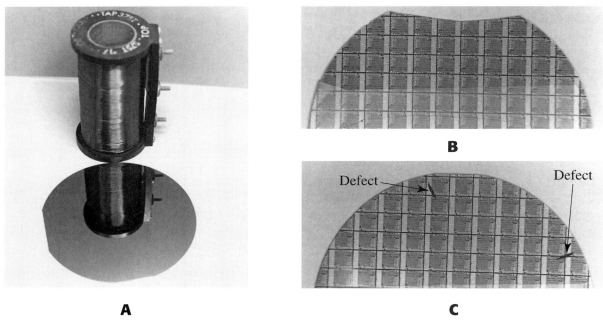

B

A   **C**

Figure 36-1. Integrated circuit fabrication. A—Integrated circuits start from a silicon wafer. The coil is shown to illustrate the disc surface is polished to a mirror finish. B—Integrated chips are inspected and cut from the wafer. C—The wafer is inspected under a microscope and defects are marked.

A metal can package costs more to manufacture but provides a better seal against the environment. It can be welded (soldered) to ensure it is sealed. The metal enclosure shields the silicon chip.

A ceramic package provides the best protection against the environment. It has the highest temperature dissipation and can be used at extreme temperatures.

Package Types

Integrated circuits are placed in various types of packages. The dual in-line package (DIP) is one of the most popular. It has two rows of 14, 16, or 64 pins. A small version, called the mini-DIP, has eight pins. Pin numbering starts at the identification key mark and progresses counterclockwise, **Figure 36-2.**

Another popular IC package is the single in-line (SIP) package. It contains one row of pins numbered from the identification key mark, **Figure 36-3.** A metal can package shields the IC chip and is used in high-frequency applications, **Figure 36-4.**

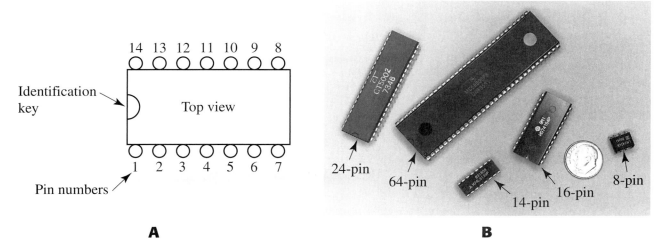

A   **B**

Figure 36-2. Dual-in-line package. A—The pins are counted in a counterclockwise direction (top view). B—Various DIP packages. (Dime shown for comparison.)

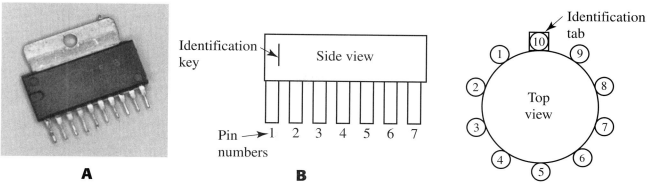

Figure 36-3. Single-in-line package. A—The SIP has one row of pins. B—Method of counting SIP pins (top view).

Figure 36-4. Metal can package.

The flat-pack package is used extensively in surface mounted device technology, **Figure 36-5.** Large integrated circuits, such as the PLCC (plastic leaded chip carrier), may have 28 to 48 pins. The PGA (pin grid array) may have 68 to 200 pins. These packages may contain thousands of transistors and complete subassemblies.

Other transistor packages, such as the TO-3, TO-5, and TO-220 are used for integrated circuits, **Figure 36-6.** They are often used for voltage regulators in power supplies.

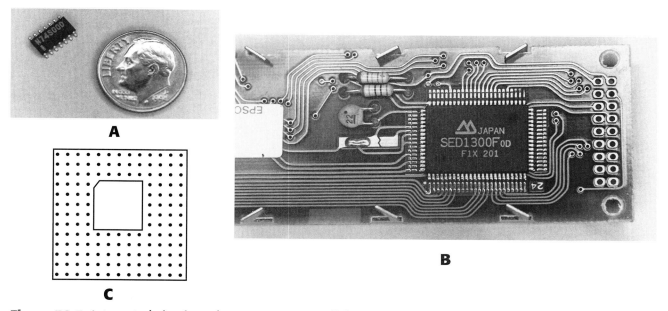

Figure 36-5. Integrated circuit package types. A—Small flat-pack packages are a product of miniaturization for the space program. B—Large surface mount ICs greatly reduce the number of components on a circuit board. C—PGA (pin grid array) ICs may contain several thousand circuits.

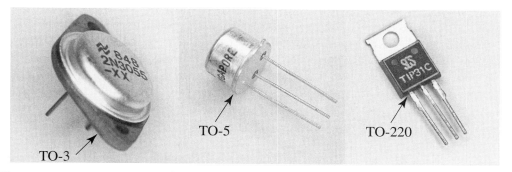

Figure 36-6. TO-3, TO-5, and TO-220 transistor packages for integrated circuits.

Operational Amplifiers

Integrated circuits are widely used in operational amplifiers (op-amps). In the 1940s, operational amplifiers were developed for use in analog computers. These op-amps were high-performance dc amplifiers that used vacuum tubes. They were also large and wasted electrical power. In 1967, Fairchild Semiconductor introduced the first integrated circuit op-amp. Today's IC op-amps are far superior to the early vacuum tube circuits. They are small, inexpensive, and require only a few microwatts of power.

Operational Amplifier Operation

The operational amplifier is a multiple-use, high-performance linear amplifier, often packaged in a mini-DIP. The op-amp has two inputs, inverting (–) and noninverting (+), and one output, **Figure 36-7.** The polarity of a signal applied to the inverting input is reversed at the output. A signal applied to the noninverting input has the same polarity at the output. Other DIP packaging may have four (quad) op-amps in one 14-pin integrated circuit.

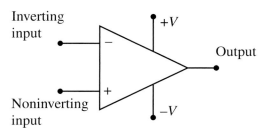

Figure 36-7. Schematic symbol for an operational amplifier.

Additional Op-amp Components

You cannot change the inside of an IC; however, understanding its internal operation will help you know the additional components required for operation. A typical op-amp consists of a differential amplifier, voltage amplifier, and output amplifier, **Figure 36-8.**

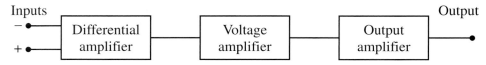

Figure 36-8. Block diagram of an operational amplifier.

All op-amps have a differential amplifier for the input stage. A basic differential amplifier has two input (V_1 and V_2) and two output ($\pm V_{OUT}$) terminals, **Figure 36-9.** The amplifier output is between the two output terminals. Emitter resistor R_E is common to both emitters and ensures the voltage across the emitter-base junction of both transistors is equal. The collector resistors R_C and base resistors R_B provide biasing. The output of the amplifier is the difference between the two inputs. The output voltage taken between the two collectors is equal to the difference between the two input voltages times the amplifier voltage gain. The equation is:

$$V_{OUT} = A_V(V_1 - V_2)$$

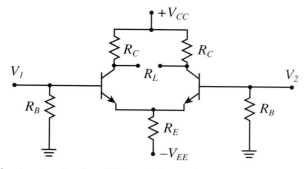

Figure 36-9. The basic circuit of a differential amplifier has two input and two output terminals.

The voltage amplifier is a special high-gain amplifier that provides a buffer between the differential and output amplifiers. The output amplifier of the op-amp is usually a combination of Darlington pairs and a complementary symmetry amplifier.

Operational Amplifier Power Supply

Most op-amp circuits operate using a dual or bipolar power supply with a range of ±22 V, **Figure 36-10.** Some circuit designs can be operated from a single polarity power supply less than 3 V. For experiments, two 9 V transistor batteries can be used if no other power supply is available. However, the maximum power supply voltage for a specific op-amp must be checked. The power supply leads should be kept as short as possible. To prevent the op-amp circuit from becoming an oscillator, power supplies are usually bypassed with capacitors of 0.1μf to 10 μf, **Figure 36-11.** Many problems with op-amp operation are due to poor grounding or lack of bypassing. When working with a lab experimenter, place a 100 μf capacitor across both the positive and negative power supplies and leave them connected for all future op-amp experiments. When you are working with op-amps, *never* reverse the power supply polarity. Doing so, even momentarily, will destroy the op-amp. Furthermore, *never* apply an input signal when the power supply is off.

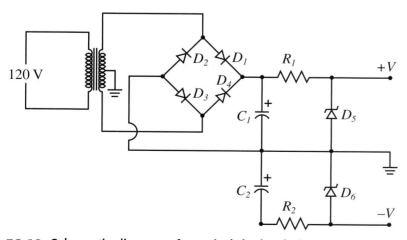

Figure 36-10. Schematic diagram of a typical dual-polarity power supply for op-amps.

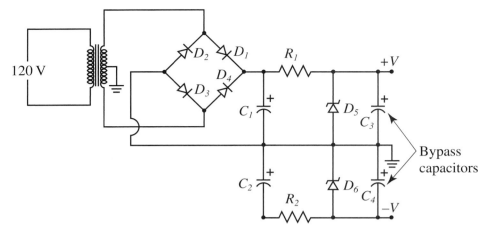

Figure 36-11. Bypass capacitors across a power supply eliminate many stability problems with op-amps.

Characteristics of Operational Amplifiers

The electrical characteristics of an op-amp are specified for a given supply voltage and temperature. Other characteristics can have additional conditions.

Op-Amp Impedances

Input and output impedances are important characteristics of op-amps. Op-amps have an input impedance (Z_I) ranging from 5K to 20M, depending on the type of op-amp and any external resistances. The output impedance (Z_O) varies from a few ohms to several hundred ohms.

Op-Amp Gain

The gain of an op-amp is independent of the supply voltage; however, the maximum output voltage cannot exceed the total supply. For example, if the circuit power supply is ±12 V, the maximum output voltage cannot be greater than 24 V.

If the input voltage is increased beyond the limits of the power supply, a clipping distortion of the output will result. See **Figure 36-12.**

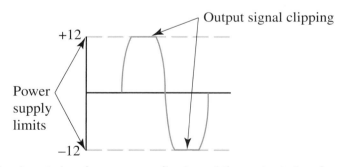

Figure 36-12. Excessive input signal can cause clipping of the output signal.

Closed-loop Circuit

The gain, or amount of amplification, in an op-amp is determined by a feedback resistor (R_F) that feeds some of the amplified signal from the output back to the input. See **Figure 36-13.** A lower feedback resistance provides more feedback and causes a lower amplifier gain. This op-amp circuit arrangement has output-to-input feedback and is called a ***closed-loop***. The gain can be approximated using the equation:

$$\text{Gain} = -\frac{R_F}{R_{IN}}$$

Closed-loop: A circuit in which the output is continuously fed back to the input.

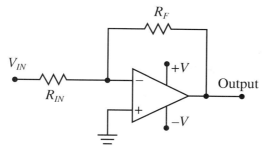

Figure 36-13. A feedback resistor makes a closed-loop circuit.

▼ *Example 36-1:*

If feedback resistor R_F has a value of 100K and the input resistor is 1K, what is the approximate amplifier gain?

$$Gain = -\frac{R_F}{R_{IN}}$$
$$= -\frac{100K}{1K}$$
$$= -100$$

For a voltage gain of 100, the output will be approximately 100 times greater than the input. The output has a negative sign because the amplifier output signal is 180° out of phase with the input signal. ▲

Open-loop Circuit

Although the method is seldom used, the feedback resistor can be eliminated, resulting in no loop for feedback. See **Figure 36-14.** This arrangement is called an ***open-loop*** circuit. The open-loop voltage gain (A_{VOL}) depends on the characteristics of the individual op-amp and operating frequency. A data sheet should be consulted for the characteristics of a specific op-amp. The open-loop circuit is not normally used. The gain is so high the circuit becomes unstable and unpredictable. The only common op-amp circuit that can use the open-loop method is a comparator, which is discussed later in this chapter.

Open-loop: In an operational amplifier, the condition in which no feedback resistance is used in the circuit.

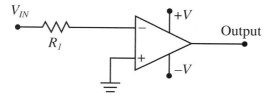

Figure 36-14. Operation of an open-loop circuit can be unpredictable.

Common-mode Rejection Ratio (CMRR)

If identical signals are applied to both inputs, a perfectly balanced differential amplifier produces 0 V at the output. ***Common-mode rejection ratio*** (CMRR) is a measure of the ability of the op-amp to reject signals simultaneously present at both inputs. Expressed in decibels, CMRR is the ratio of the common-mode input voltage to the generated output voltage. The equation is:

Common-mode rejection ratio: A measure of the ability of the op-amp to reject signals simultaneously present at both inputs.

$$CMRR = \frac{V_{CM}}{V_{OUT}}$$

Where:

V_{CM} is the differential voltage

Slew Rate

Slew rate is a measure of how well an op-amp output voltage follows the input voltage. It is measured in volts per microsecond (V/μS). The slew rate is caused by internal or external capacitances. A certain amount of time is required to charge and discharge a capacitance. Internal coupling and frequency compensation capacitors cause a delay in processing a signal through the integrated circuit. The slew rate (SR) is calculated by the ratio of the voltage change to the time change, **Figure 36-15.** The equation is:

$$SR = \frac{\Delta V}{\Delta t}$$

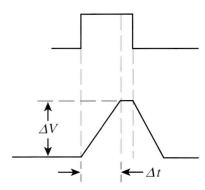

Figure 36-15. Two values are needed to calculate the slew rate: voltage change to time change.

▼ **Example 36-2:**

What is the slew rate in **Figure 36-16?**

$$SR = \frac{\Delta V}{\Delta t}$$

$$= \frac{14 \text{ V}}{7 \text{ μS}}$$

$$= \frac{2 \text{ V}}{\text{μS}}$$

▲

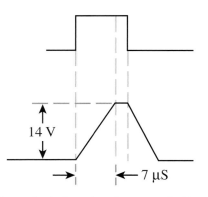

Figure 36-16. Waveform for calculating the slew rate in Example 36-2.

▼ *Example 36-3:*

What is the slew rate in **Figure 36-17?** Notice the sine wave input. The output waveform must be a triangular edge before the slew rate can be calculated. Using points A and B for the measurement, the slew rate is calculated as:

$$SR = \frac{\Delta V}{\Delta t}$$

$$= \frac{12\ V}{30\ \mu S}$$

$$= \frac{0.4\ V}{\mu S}$$ ▲

Manufacturers normally use a square or rectangular wave to measure the slew rate. Maintaining consistent measurement standards is difficult when using a sine wave from more than one manufacturer.

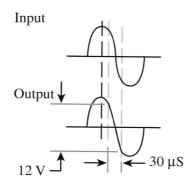

Figure 36-17. Waveform for calculating the slew rate in Example 36-3.

Dc Output Offset Voltage

When the inputs of an op-amp are grounded, the output should be zero, but a small output voltage will exist in practical op-amps. This is called the *dc output offset voltage.* To compensate for the output offset voltage, manufacturers have made pin connections for adjusting the output to zero. See **Figure 36-18.** The offset potentiometer is adjusted for zero output voltage.

Dc output offset voltage: In an op-amp circuit, the dc output voltage measured between the output and ground when both inputs are grounded.

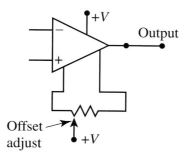

Figure 36-18. Many op-amp circuits require a dc offset adjustment.

Types of Operational Amplifier Circuits

The following paragraphs describe various op-amp circuits and special characteristics of each. Included are the inverting amplifier, noninverting amplifier, voltage follower, summing amplifier, and difference amplifier.

Inverting Amplifier

The output signal of the inverting amplifier is 180° out of phase with the input signal, **Figure 36-19.** As discussed earlier, the gain of the amplifier is controlled by the R_F/R_I ratio. The equation is:

$$A_V = -\frac{R_F}{R_I}$$

The negative sign indicates the output signal is an inverse of the input signal. From a practical standpoint, the input impedance of the inverting amplifier is equal to the individual input resistor.

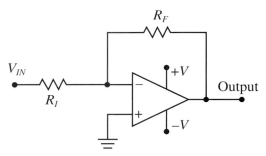

Figure 36-19. An op-amp inverting amplifier has 180° phase difference between the signal input and output.

Noninverting Amplifier

The op-amp in **Figure 36-20** is connected as a noninverting amplifier with the input signal applied to the noninverting input (+). Since the amplifier does not invert the input, the output signal is in phase. The phase difference between the input and the output is zero. The gain of the amplifier is found by the equation:

$$A_V = 1 + \frac{R_F}{R_I}$$

The input impedance of the noninverting amplifier does not depend on external components as in the inverting amplifier. Therefore, the input impedance of the noninverting amplifier is very high.

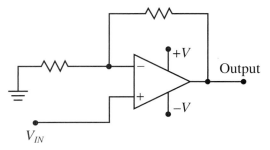

Figure 36-20. A noninverting amplifier has no input/output phase difference.

Voltage Follower

The voltage follower is simply a unity-gain noninverting amplifier, **Figure 36-21.** No input or feedback resistor is used, but the output is connected directly to the inverting input of the amplifier. Like the noninverting amplifier, the voltage follower has a high input impedance. A voltage follower op-amp circuit is used as a buffer between two circuits.

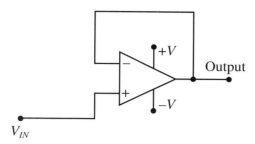

Figure 36-21. A voltage follower amplifier can provide isolation between two circuits without changing the signal.

Summing Amplifier

In a summing amplifier, voltages V_1, V_2, and V_3 are added, **Figure 36-22.** The output is calculated by the equation:

$$V_O = -R_F \left(\frac{V_1}{R_1} + \frac{V_2}{R_2} + \frac{V_3}{R_3} \right)$$

If the feedback resistor R_F and the input resistances are equal (precision), the output voltage is calculated by the equation:

$$V_O = -(V_1 + V_2 + V_3)$$

The input voltages may be dc, sine wave, or square wave. Increasing the feedback resistor R_F will provide an increase in the gain. The input impedance is simply the value of the individual input resistors.

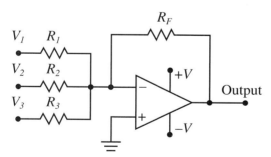

Figure 36-22. In a summing amplifier circuit, the output is the sum of the input voltages.

Difference Amplifier

The difference amplifier output is the difference between the two input voltages, **Figure 36-23.** The input voltages are applied simultaneously to the inverting and noninverting inputs of the op-amp. With unequal resistors and R_F controlling the gain, the output voltage is calculated by the equation:

$$V_O = -\left(\frac{R_F}{R_1} V_1 \right) + \left(1 + \frac{R_F}{R_1} \right) \left(\frac{R_3}{R_2 + R_3} \right) V_2$$

For equal (precision) resistors, the output is calculated as:

$$V_O = V_2 - V_1$$

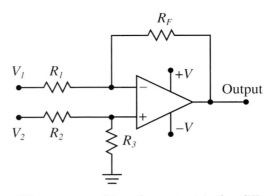

Figure 36-23. In a difference amplifier, the output is the difference between the two input voltages.

A major advantage of the difference amplifier is its ability to sense a small differential voltage within a larger signal. The difference amplifier can measure *and* amplify the small signal.

Comparators

Comparator: A circuit that compares two or more signals and indicates whether they are matched in some way.

As mentioned earlier, a ***comparator*** is a circuit that compares two or more signals and indicates whether they are matched in some way. Comparators may test the signals for amplitude, current, frequency, phase, voltage, waveform type, numerical value, or error sensing. Since a comparator evaluates the input signals, any feedback loop would make an input invalid for comparison to any other signal. Therefore, a comparator is operated in the open-loop mode.

The inputs of a comparator can accept any analog waveform voltage. The output voltage is digital and binary. This means the output can have only one of two possible voltage levels corresponding to the two power supply voltages (+ and –). These volt-

Rail voltages: The power supply input voltages of a comparator or operational amplifier.

ages are called the ***rail voltages***. The term "rail voltage" was originally used only for comparators; however, it has found its way into op-amp operation as well.

With two different voltages applied to the inputs, the inverting input changes the sign of its voltage, and the sign of the noninverting voltage stays the same. See **Figure 36-24.** The circuit compares the two voltages and uses the sign of the larger value. If the sign of the larger value is positive, the comparator output will switch to the positive rail. If the sign of the larger value is negative, the comparator will switch to the negative rail. Rail voltages may be either positive or negative, or positive and ground (0 V).

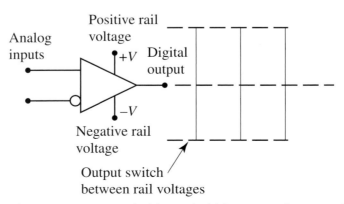

Figure 36-24. The comparator symbol has a bubble on one input to distinguish it from the op-amp symbol.

Comparators are used extensively in the electronics field in such applications as:
- Zero-crossing detectors
- Window detectors
- Ladder comparators
- Schmitt triggers
- Timing circuits
- Pulse-width modulators

A simple test for a defective comparator is to connect a jumper across the reference resistor and force the comparator to its opposite rail voltage. If both output states are obtained, the IC and power supply voltage are correct. The reference voltage or the input signal is at fault. If the comparator will not switch states, check the power supply, reference voltage, and input signal. If these are correct, the IC should be replaced.

MOSFET and BIFET Op-Amps

Special hybrid op-amps are available for applications in which the input impedance must be higher than an ordinary op-amp circuit can provide. A FET is used for the input stage, and bipolar transistors are used for the rest of the circuitry.

A MOSFET input is used when the operating frequency is higher than an ordinary op-amp can process. Also, the MOSFET op-amp has an input bias and offset current much smaller than bipolar op-amps. Unfortunately, the MOSFET op-amp is susceptible to electrostatic discharge. The output offset voltage and internal noise tends to be much higher than with bipolar op-amps.

The BIFET op-amp is a bipolar circuit with a JFET input stage. Its input characteristics are similar to the MOSFET. The offset and bias currents are a little smaller, and the input impedance is slightly lower. Although the BIFET does experience noise generation, its input stage is as durable as a bipolar stage. See **Figure 36-25.**

Numerous types of circuits, such as differentiator, integrator, active filter, and wave generating, can be constructed using op-amps. Op-amps are also widely used to connect the real (analog) world to digital systems.

	Bipolar	BIFET	BIMOS
	741C	LF–351	3130
Input impedance	1 Meg	10^6 M	10^6 M
Bandwidth	1 MHz	4 MHz	15 MHz
CMRR	90 dB	100 dB	90 dB
Power dissipation	500 mW	500 mW	600 mW

Figure 36-25. Comparison of some common operational amplifiers.

Integrated Circuit Power Amplifiers

Many types of IC audio power amplifiers are available. The power amplifiers require only a volume control, speaker, power supply, and some resistors and capacitors connected externally to complete the circuit. These external components provide speaker coupling, stage stability to prevent oscillation, power supply bypassing, high-frequency bypassing, and feedback to control the gain and additional stability. The power amplifiers described next have power output levels of 20 W or less.

LM386 Audio Power Amplifier

The LM386 power amplifier is designed for use in low-voltage consumer products. To keep the number of external parts low, the gain is set internally at 20. The gain can be adjusted up to 200 by connecting an external resistor and capacitor to pins one and eight. Packaged in an eight-pin mini-DIP, the LM386 appears to be an op-amp but is a power amplifier with a choice of inputs. The inputs are referenced to ground, and the output is automatically biased to half the supply voltage. Besides applications in consumer audio products, the LM386 is used as a line driver, small servo driver, or power converter. See **Figures 36-26** and **36-27**.

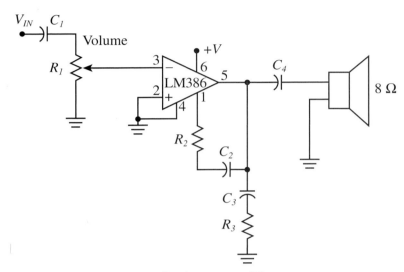

Figure 36-26. LM386 audio power amplifier.

Device Number	Power Supply Voltage
LM386N	15
LM386–1	6
LM386–3	9
LM386–4	22

Figure 36-27. Various device numbers and supply voltages for the LM386 audio power amplifier.

LM383 Audio Power Amplifier

The LM383 is an 8 W audio power amplifier that comes in a five-pin TO-220 package, **Figure 36-28**. It is intended for automotive audio applications. Its high-current (3.5 A) capability enables the device to drive low-impedance loads (speakers) with low distortion. It has short-circuit protection from current and temperature. Version LM383A can withstand 40 V transients on its supply voltage.

LM1875 Audio Power Amplifier

The LM1875 power amplifier has low distortion and high-quality performance for consumer audio applications. It delivers 20 W into a 4 Ω or 8 Ω load using dual polarity power supplies. The amplifier is internally compensated and stable for gains of 10 or greater. Outstanding features include a fast slew rate and high-output voltage swing. It is available in either the TO-220 or a stereo SIP package designed to be bolted directly to a heat sink. See **Figure 36-29**.

Working with Integrated Circuits

When working with ICs, a 30 W or less soldering iron should be used to avoid excessive heat. If the IC is sensitive to electrostatic discharge, the work station must be ESD-protected.

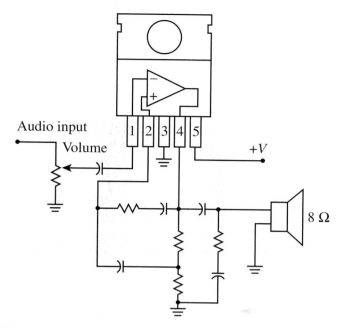

Figure 36-28. LM383 audio power amplifier uses the TO-220 transistor package.

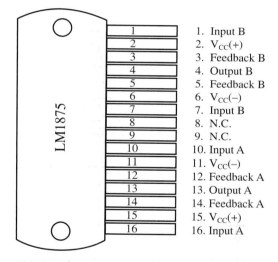

1. Input B
2. $V_{CC}(+)$
3. Feedback B
4. Output B
5. Feedback B
6. $V_{CC}(-)$
7. Input B
8. N.C.
9. N.C.
10. Input A
11. $V_{CC}(-)$
12. Feedback A
13. Output A
14. Feedback A
15. $V_{CC}(+)$
16. Input A

Figure 36-29. The LM1875 is a stereo audio power amplifier packaged in a single in-line package.

When replacing an integrated circuit, install the IC with the keyway index in the correct position. Make sure pins do not bend or go outside the socket. A badly bent IC pin can break off when straightened or later, causing a technician to waste time searching for the cause of the circuit failure. If one pin is excessively bent, carefully straighten it with a small needle-nose pliers. To straighten the pins of an IC, place the IC on its side against a flat surface and bend the pins into proper alignment. See **Figure 36-30.**

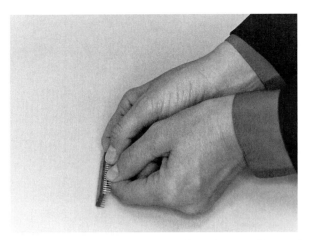

Figure 36-30. To straighten the pins of an IC, place it on a flat surface and carefully press the pins into proper alignment.

Testing Integrated Circuits

Integrated circuit problems usually consist of:
- No power to the IC.
- No input or output signal.
- Internally open pins.
- Pins internally shorted to ground or power supply.
- Miscellaneous shorts between pins or a solder bridge.
- Output pins stuck at a high or low voltage level.

The process of testing an integrated circuit includes:
- Verifying correct power is applied to the IC.
- Checking for proper input and output signal.
- Checking the correct voltages are on the various other pins.

Integrated circuits are found in all areas of electronic equipment, including audio, radio, television, and digital circuits. Troubleshooting ICs is a signal-tracing process.

▶| Summary

- Plastic, metal, and ceramic are used to package ICs.
- Integrated circuits can have DIP, SIP, metal can (TO-5), flat-pack, or standard transistor (TO-220 and TO-3) packaging.
- The operational amplifier is a major application of the integrated circuit. Op-amps have an inverting (−) and noninverting (+) input. Many are designed for dual (±) power supply connections and must be capacitor-bypassed.
- Most op-amp circuits have a feedback resistor R_F (called a closed-loop circuit) that controls the gain of the amplifier.
- The dc offset voltage is a voltage at the output when both inputs are grounded.
- The slew rate of an op-amp measures the ability of an output signal to follow the input signal changes.
- Op-amp circuits may include inverting, noninverting, voltage follower, summing, and difference amplifiers.
- A comparator is a circuit that compares two or more signals for matching.
- The comparator output (rail) voltage is a high or low voltage.
- BIFET and MOSFET op-amps have FET transistors in the input stages, and the balance of the transistors are bipolar, making the input impedance much higher.

- Integrated circuits are used for audio power amplifiers.
- When working with integrated circuits, make sure the pins are straight when installing the IC in a socket. Observe ESD precautions.
- Integrated circuit problems usually are no power to the IC, no input or output signal, open pins, shorted pins, or output pins stuck high or low.

Important Terms

Do you know the meanings of these terms used in the chapter?

closed-loop
common-mode rejection ratio
comparator
dc output offset voltage

open-loop
rail voltages
slew rate

Questions and Problems

Please do not write in this text. Write your answers on a separate sheet of paper.

1. What are three types of packaging materials for integrated circuits?
2. What type of integrated circuit package material is preferred for extreme temperature applications?
3. List six types of integrated circuit packages.
4. What type of integrated circuit package is used for surface mounted devices?
5. Draw a block diagram of the three internal parts of an operational amplifier.
6. What causes an operational amplifier to become an oscillator?
7. What is the op-amp gain in **Figure 36-31?**
8. What is the slew rate in **Figure 36-32?**
9. Draw diagrams of the following op-amp circuits:
   a. inverting
   b. noninverting
   c. voltage follower
   d. summing
   e. difference
10. The input of a comparator is _____, and the output is _____ type voltage.
11. What is the difference between the input impedance of a BIFET op-amp and bipolar op-amp?
12. What is the best way to straighten the pins of an IC?
13. What precautions should be observed when working with integrated circuits?
14. Outline the process for testing an integrated circuit.
15. What process is used to locate a defective integrated circuit?

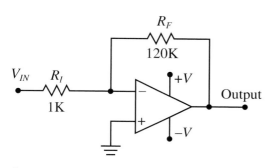

Figure 36-31. Circuit for problem number 7.

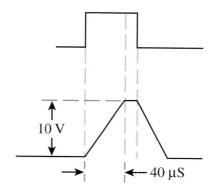

Figure 36-32. Waveform for problem 8.

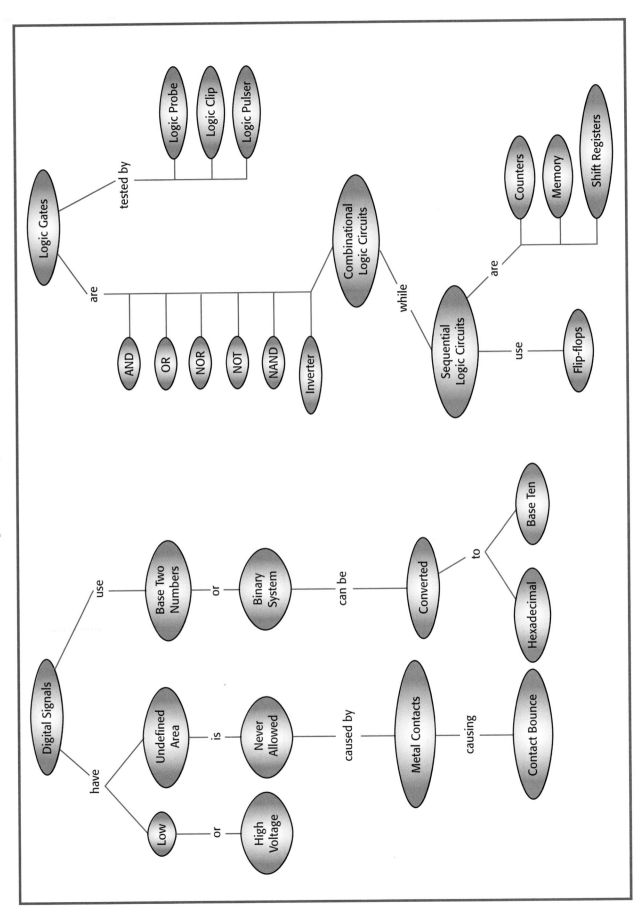

Objectives

After studying this chapter, you will be able to:
- ◯ Explain the differences between analog and digital signals.
- ◯ Count in binary from one to 15.
- ◯ Convert decimal numbers to binary and binary numbers to decimal.
- ◯ Convert decimal numbers to hexadecimal and hexadecimal numbers to decimal.
- ◯ Convert binary numbers to hexadecimal and hexadecimal numbers to binary.
- ◯ Explain the relationship between low and high logic levels and voltage.
- ◯ Draw the schematic symbols and truth tables for AND, OR, NOT (inverter), NAND, NOR, XOR, and XNOR logic gates.
- ◯ Identify equipment for testing digital circuitry.

Introduction

For many years, the application of digital electronics was limited to computers and calculators. Today, digital electronics are found in telecommunications systems, audio systems, radio control, industrial electronics, radar, cash registers, and video games. The advancement of integrated circuits has made digital application easy. Anyone involved in the electricity and electronic field must have an understanding of basic digital electronics.

What Is Digital?

You are familiar with circuits such as the one in **Figure 37-1.** As the arm of the potentiometer moves upward, the voltage from point X to ground increases. As the arm moves down, the voltage decreases. The output voltage changes or varies continuously and is called an **analog signal.**

If the switch in the circuit is closed, the voltage at point X jumps from 0 V to 5 V, **Figure 37-2.** If the switch is opened, the voltage jumps from 5 V to 0 V. Unlike the potentiometer, there are no in-between voltages. Signals are either low or high and are called **digital signals.**

Analog signal: A signal that continuously changes amplitude.

Digital signal: A signal that has distinct steps: on or off, voltage or no voltage.

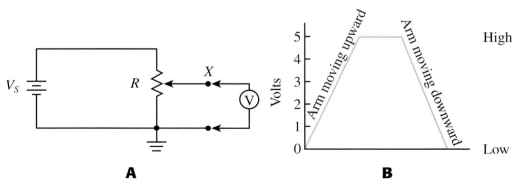

Figure 37-1. Analog signals. A—The potentiometer operates as a voltage divider. B—Analog output of a potentiometer. As the arm of the potentiometer moves upward, the voltage increases. As the arm moves down, the voltage decreases.

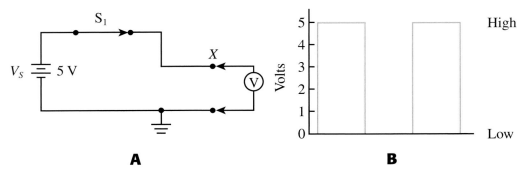

A

B

Figure 37-2. Digital signals. A—When the switch is open, the output voltage is zero. When the switch is closed, the output voltage is 5 V. B—Voltage is either low or high with none in between.

Logic circuit: A circuit that operates using logic functions.

Circuits use transistors or integrated circuits as switches for generating digital signals. *Logic circuits* use or process digital signals. Early logic circuits used resistors, diodes, and transistors. They were referred to as RTL (resistor-transistor logic) or DTL (diode-transistor logic). Advances in logic circuits brought about transistor-transistor logic (TTL) and complementary metal-oxide semiconductor (CMOS) circuitry. Although TTL has disappeared from manufacturing, its ruggedness (compared to CMOS) has kept it in student training labs. CMOS, TTL, and other logic devices have different input and output characteristics. A high value (5V) for a TTL device input is not a high value for a CMOS device.

Characteristics of a Digital Signal

The input characteristics of a TTL device are shown in **Figure 37-3.** TTL devices operate on a standard 5 V for the power supply. To be low, the input voltage must be 0.8 V or less. To be high, the input voltage must be greater than 2 V. The area between 0.8 V and 2 V is called the ***undefined area*** and should not be allowed. If a voltage of an undefined value should occur, the TTL device will not know whether the voltage is high or low.

Undefined area: The voltage levels between low and high levels of a binary 0 or 1.

For the output to be low, the voltage must be 0.4 V or less. For the output to be high, the voltage must be greater than 2.4 V. The area between 0.4 V and 2.4 V is the undefined area. The high output level depends on the load resistance placed on the output. The greater the load current, the lower the output voltage level.

Digital CMOS devices, such as the 4000 and 74C00 series, have different input and output characteristics compared to TTL devices. They have a wide range of power supply values, ranging from less than 3 V to 15 V. As a general rule, a low will be any

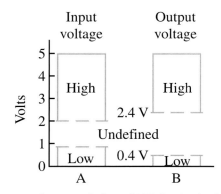

Figure 37-3. Input-output characteristics of TTL logic devices. In an undefined area, the circuit does not know if the digital signal is high or low.

voltage within 0% to 30% of V_{DD} and a high between 70% to 100% of V_{DD}. When using a 10 V supply voltage, the input voltage must be 3 V or less to be low. To be high, the input voltage must be greater than 7 V, **Figure 37-4.** For the output to be low, the voltage must be 0.05 V or less. For the output to be high, the voltage must be greater than 9.95 V.

Using a 10 V supply voltage in the 74HC00 series, the input voltage must be 1 V or less to be low. To be high, the input voltage must be greater than 3.5 V. See **Figure 37-5.** For the output to be low, the voltage must be 0.1 V or less. For the output to be high, the voltage must be greater than 4.9 V. Refer to a data manual for the input and output characteristics of the specific device you are using.

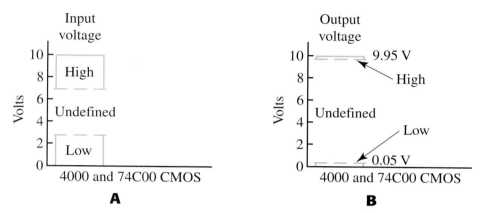

Figure 37-4. Digital CMOS device, 4000 and 74C00 series. A—Input characteristics. B—Output characteristics.

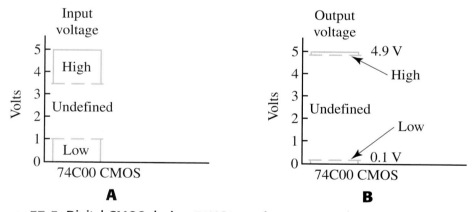

Figure 37-5. Digital CMOS device, 74HC00 series. A—Input characteristics. B—Output characteristics.

Number Systems for Digital Electronics

If you said you saw seven cars filling up at the gas station, most people would understand the meaning of the word seven. In digital electronics, electric circuits do not understand the number seven or any other similar number. Digital circuits only understand the meaning of a high or low voltage. Therefore, another number system is necessary in digital electronics.

Decimal Number System

Any number system is a code that uses symbols to represent a quantity of items. The decimal system uses 10 symbols: 0, 1, 2, 3, 4, 5, 6, 7, 8, and 9. See **Figure 37-6.** To use the system, start with the first column and count from 0 to 9. To count any further, the first column goes back to 0 and the second column is increased from 0 to 1, Figure 37-6A.

The first column is again increased to 9, then returned to 0, Figure 37-6B. The second column is increased by 1 to 2. This process continues until both columns are at 9, or the count is 99, Figure 37-6C. Adding one more to column 1 makes both columns go to 0. The third column is increased from 0 to 1, and the count is 100, Figure 37-6D.

The decimal number system is a base 10 system. The ***base*** of a number system tells you two things. First, it indicates how many symbols are in the number system. In the decimal system there are 10 symbols. Second, the base determines the ***place value.***

Base: The total number of symbols used in a numbering system.

Place value: The value of each column in a number system found by multiplying the previous place value by the base of the number system.

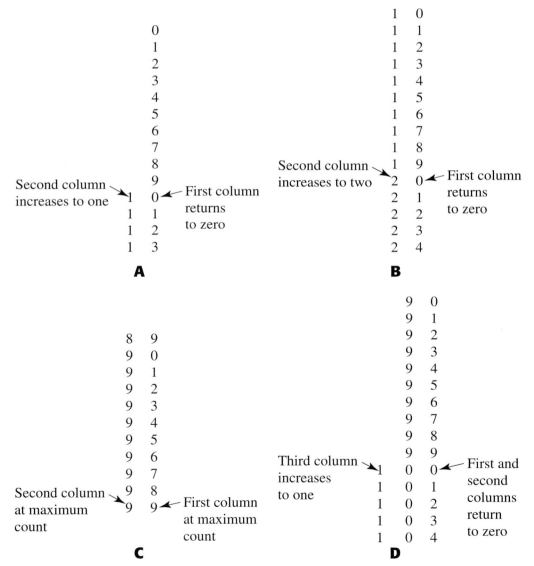

Figure 37-6. Counting sequence in the 10-symbol decimal system. A—The first column returns to zero, and the second column goes to one. B—The first column increases to nine, then returns to zero. The second column increases to two. C—Counting to maximum count in the first two columns. D—The first two columns return to zero, and the third column is increased by one.

Each place value is found by multiplying the previous place value by the base of the number system. See **Figure 37-7.** Starting at the decimal point, the first column represents the ones; the second column represents tens (1 times 10); the third column represents hundreds (10 times 10); the fourth column represents thousands (100 times 10), and so on.

The total value is found by multiplying each number symbol by the place value and adding the place values. For example, in Figure 37-7, the total value is found by figuring 5 times 100, plus 4 times 10, plus 7 times 1 for a total of 547. Determining the values of the place value columns is the same process regardless of the base of the number system.

The alphabet can also be used to explain counting, **Figure 37-8.** The count begins in the first column with A and increases to Z, the maximum count for that column. The first column returns to A, and the second column increases to A. The two columns increase in count as AA, AB, AC, AD . . . AX, AY, AZ. When the first column reaches Z again, the second column increases by 1 to B, and the first column starts over again at A. This continues until both columns are at ZZ. The next count would be AAA.

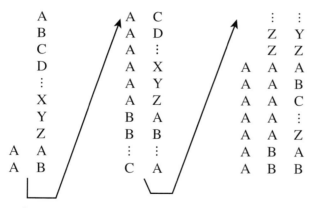

Figure 37-7. Finding the total value of the place values in a base 10 number system.

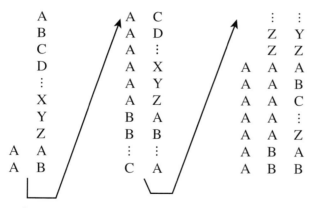

Figure 37-8. A counting system using the alphabet.

Binary Number System

The *binary* system is a base 2 (bi) number system that uses two symbols, 0 and 1. By multiplying the base times the ones place value, the other place values become 2, 4, 8, 16, 32, 64, 128, 256, 512, and so on. The base (2) is multiplied by each previous place value. See **Figure 37-9.**

Binary: A numbering system using a base of 2.

512	256	128	64	32	16	8	4	2	1

Figure 37-9. Binary (base 2) place values.

A number written in binary has only zeros and ones. The number 1101 would read, "One, one, zero, one," (not "one thousand, one hundred one"). To find its decimal value, repeat the same process used to find the value 547 in the decimal number system (refer to Figure 37-7). Take each number symbol times its corresponding place value and add them together. For the binary number 1101, the decimal number is 8 + 4 + 0 + 1 = 13, **Figure 37-10.** For the binary number 111, the decimal number is 4 + 2 + 1 = 7.

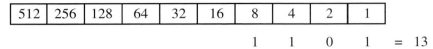

512	256	128	64	32	16	8	4	2	1

$$1 \quad 1 \quad 0 \quad 1 \quad = 13$$

Figure 37-10. Decimal number 13 written as binary number 1101.

Converting Decimal to Binary

Two methods can be used to convert a decimal number to binary: dividing by the base of the number system and using place value subtraction.

Dividing by the base. To convert a decimal number to binary, divide the decimal number and its quotients by the base of the number system (2) and write down the remainders. Divide until the last quotient is zero.

▼ *Example 37-1:*

Convert the decimal number 11 to binary.

    11 ÷ 2 = 5 with a remainder of 1
    5 ÷ 2 = 2 with a remainder of 1
    2 ÷ 2 = 1 with a remainder of 0
    1 ÷ 2 = 0 with a remainder of 1
           = 1 0 1 1

The binary equivalent of decimal 11 is 1011. The decimal point is to the right of the last digit unless otherwise indicated.   ▲

▼ *Example 37-2:*

Convert the decimal number 8 to binary.

    8 ÷ 2 = 4 with a remainder of 0
    4 ÷ 2 = 2 with a remainder of 0
    2 ÷ 2 = 1 with a remainder of 0
    1 ÷ 2 = 0 with a remainder of 1
          = 1 0 0 0

The binary equivalent of decimal 8 is 1000.   ▲

▼ *Example 37-3:*

Convert the decimal number 27 into binary.

    27 ÷ 2 = 13 with a remainder of 1
    13 ÷ 2 = 6 with a remainder of 1
    6 ÷ 2 = 3 with a remainder of 0
    3 ÷ 2 = 1 with a remainder of 1
    1 ÷ 2 = 0 with a remainder of 1
           = 1 1 0 1 1

The binary equivalent of decimal 27 is 11011.   ▲

Place value subtraction. To convert a decimal number to binary using the place value subtraction method, simply write down a series of base 2 place values. Subtract the highest place value possible and write 1 in that place value. Continue this process, subtracting the highest place value from the balance until 0 is reached. When finished, any place value that does not have 1 is filled in with 0. This subtraction method will not work for other number bases because the number of symbols used in other bases is greater than 1.

▼ *Example 37-4:*
Convert the decimal number 11 to binary.

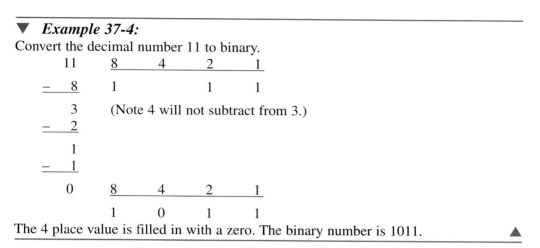

The 4 place value is filled in with a zero. The binary number is 1011.   ▲

Hexadecimal Number System

The *hexadecimal* number system is a base 16 system using the symbols 0 through 9 plus A, B, C, D, E, and F. See **Figure 37-11.** The letter A stands for 10, B stands for 11, and so on. The hexadecimal system is used extensively in programming microprocessors and is easily converted to binary numbers.

To convert decimal numbers to hexadecimal numbers, divide the decimal number by base 16 and write down the remainder.

Hexadecimal: A number system with a base of 16. The sumbols 0 through 9 and A through F are used.

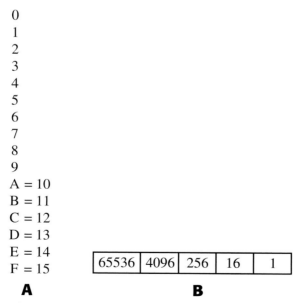

Figure 37-11. Hexadecimals. A—The hexadecimal system uses numbers and letters. B—Hexadecimal place values.

▼ *Example 37-5:*
Convert decimal number 23 to a hexadecimal number.

    23 ÷ 16 = 1 with a remainder of 7
     1 ÷ 16 = 0 with a remainder of 1
            = 1 7

The hexadecimal equivalent of decimal number 23 is 17.   ▲

▼ *Example 37-6:*
Convert decimal number 26 to a hexadecimal number.

    26 ÷ 16 = 1 with a remainder of 10 (10 = A)
     1 ÷ 16 = 0 with a remainder of 1
            = 1 A

The hexadecimal equivalent of decimal number 26 is 1A.   ▲

▼ *Example 37-7:*
Convert decimal number 2895 to a hexadecimal number.

    2895 ÷ 16 = 180 with a remainder of 15 (15 = F)
     180 ÷ 16 = 11 with a remainder of 4
      11 ÷ 16 = 0 with a remainder of 11 (11 = B)
              = B 4 F

The hexadecimal equivalent of decimal number 2895 is B4F.   ▲

To convert hexadecimal numbers to decimal numbers, multiply each hexadecimal digit by the place value and add.

▼ *Example 37-8:*
Convert the hexadecimal number 3FC to a decimal number. Place values are 256 (3), 16 (F), and 1 (C).

    256 × 3 = 768
     16 × F = 240 (F = 15)
      1 × C = 12 (C = 12)
    ──────────────
    Sum   = 1020   ▲

▼ *Example 37-9:*
Convert the hexadecimal number 1D5 to a decimal number.

    256 × 1 = 256
     16 × D = 208 (D = 13)
      1 × 5 = 5
    ──────────────
    Sum   = 469   ▲

Hexadecimal to binary, or binary to hexadecimal conversion is very simple. For hexadecimal to binary, each hexadecimal digit is converted directly to a four-bit binary number and lined up beside the other binary digits.

▼ *Example 37-10:*
Convert hexadecimal number A6 to binary.

    A = 1010
    6 = 0110 (The zeros must be included.)
10100110 is the binary number for A6.   ▲

To convert binary to hexadecimal, starting at the right side, separate or break the binary number into groups of four digits, and write the hexadecimal number for each group.

▼ *Example 37-11:*

Convert binary number 1011011010100010 to a hexadecimal number.

   1011 0110 1010 0010   (Break the number into groups of four.)

     B   6   A   2   (Write the hexadecimal for each group.)

The hexadecimal for 1011011010100010 is B6A2.   ▲

Binary Numbers and Voltage

The binary system lends itself to digital circuitry because the 0 and 1 symbols can be represented by a low or high voltage. See **Figure 37-12.** When the voltage is 0 V, 0 is represented. When the voltage is 5 V, 1 is represented.

A mechanical switch is used to make the voltage go high and low. When the metal contacts of any mechanical switch close, the contacts "bounce" up and down briefly and cause the voltage to bounce momentarily. To prevent switch *contact bounce,* a circuit is used with the switch to "debounce" the switch.

Contact bounce: The uncontrolled making and breaking of contacts when relays or switches are closed.

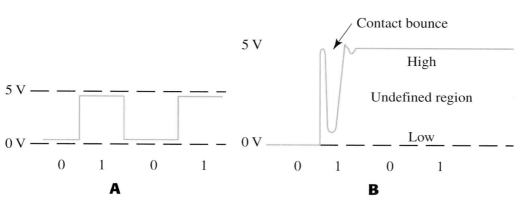

Figure 37-12. High and low voltages can represent binary numbers. A—0 V represents binary 0 and 5V represents binary 1. B—The mechanical bouncing of switch contacts can cause a series of digital pulses.

Logic Gates

In Chapter 30 you used the logic functions AND, OR, NOR, NOT, NAND, and memory. In this chapter, those logic functions are expanded with truth tables and integrated circuit equivalents. The logic functions within the integrated circuits process the 1s and 0s, making digital circuits extremely useful. A wide variety of logic gates are available in standard integrated circuit packages, making them easily applicable to many types of electronic applications.

AND Gate

The AND gate must have *both* switches closed before the lamp will light. As with the light circuit, the logic gate must have input to both A and B before getting output at C. See **Figure 37-13.** The internal operation of the gates is not important because nothing can be changed. However, it is important to understand the inputs and resulting output of a logic circuit. Some integrated circuits have pins with certain controls on the inputs or output. These need to be understood also.

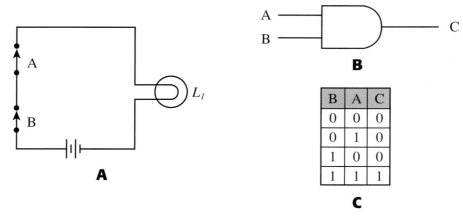

Figure 37-13. AND gate requires both switches to be closed. A—Switches A and B must be closed to complete the circuit. B—Schematic symbol for the AND gate. C—Truth table for the AND gate.

Truth table: A table that shows all possible inputs and resulting output of a digital or other logic-type circuit.

Control pins: The pins on a relay or integrated circuit that control the device.

A *truth table* is a chart that shows all possible inputs and the resulting output, Figure 37-13C. Technicians and engineers use truth tables when working with logic circuits. If the integrated circuit is a special type of digital device, the truth table will also contain the operation of any control pins. For example, some ICs have pins whose output goes high or low or is forced to do nothing, regardless of the input conditions, when a high or low voltage is placed on them. These pins are called the *control pins* of the integrated circuit.

The two left columns (A and B) of the truth table indicate all possible input conditions of the gate. The right column (C) shows the resulting output for each input combination. Output exists only when inputs A and B are 1 or have a high voltage.

Another way to express a logic function is through Boolean algebra. For the AND gate, the Boolean expression is:

$$A \cdot B = C$$

In Boolean algebra, A • B stands for A *and* B.

OR Gate

The OR gate can have one *or* all switches closed before the lamp will light, **Figure 37-14.** As with the light circuit, the logic gate will have output if A *or* B switches are closed. Output will also occur if both switches are closed.

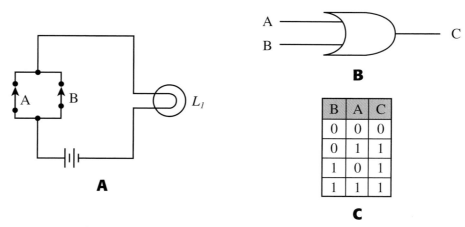

Figure 37-14. OR gate operates when either or both switches are closed. A—Switch A or B must be closed for the lamp to light. B— Schematic symbol for the OR gate. C—Truth table for the OR gate.

The Boolean expression for an OR gate is:

$$A + B = C$$

In Boolean algebra, A + B stands for A *or* B (not A plus B).

Inverter

Unlike other digital gates, the inverter has only one input. It is also called a NOT gate and sometimes not considered a gate at all. The inverter symbol, the bubble, is never alone but is always connected with another symbol, usually the triangle, **Figure 37-15.** The inverter does what its name implies; the output is the inverse of the input. If the input is 1, the output is 0. If the input is 0, the output is 1. Whatever the input *is,* the output is *not.*

The Boolean expression for the inverter is:

$$A = \overline{A}$$

In Boolean algebra, $\overline{A}$ is *not* A; A equals $\overline{A}$.

Another symbol that resembles the inverter is the noninverting **buffer/driver.** The noninverting buffer has no logical purpose but is used to isolate (buffer) one circuit from another, or to supply a greater drive current at its output. Buffer/drivers are available in both inverting and noninverting forms and are used for driving LEDs, lamps, relays, and other devices.

Buffer/driver: A circuit that isolates one electronic circuit from another and provides the necessary power to drive another circuit.

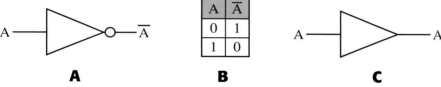

Figure 37-15. Inverter or NOT gate produces an output opposite the input. A—Schematic symbol for the inverter. B—Truth table for the inverter. C—Symbol for a noninverting buffer, not to be confused with the inverter.

NAND Gate

Any digital circuit can be produced using the AND, OR, and NOT logic gates. Other gate circuits are produced by combining the three basic gates, making even greater circuit combinations possible. The NAND gate is a combination of the NOT and AND gates, or NOT-AND, **Figure 37-16.** The Boolean expression for the NAND gate is:

$$A \cdot B = \overline{A \cdot B}$$

NOR Gate

The NOR gate is made by combining the inverter with the OR gate, **Figure 37-17.** The Boolean expression for the NOR gate is:

$$A + B = \overline{A + B}$$

Exclusive OR Gate

The exclusive OR (XOR) gate has a limited output compared to the ordinary OR gate, **Figure 37-18.** It has output when only one of the inputs is high. The XOR gate excludes output from the 00 and 11 inputs. The Boolean expression for the XOR is:

$$A + B = A \oplus B$$

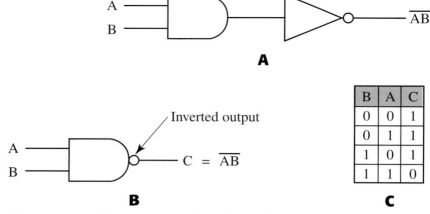

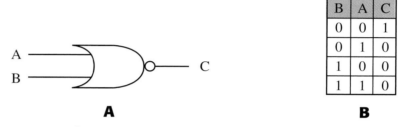

Figure 37-16. The NAND gate is an inverted AND. A—Separate NOT and AND gates combined to produce the NAND logic function. B—Schematic symbol for the NAND gate. Notice the inversion symbol on the right side. C—Truth table for the NAND gate.

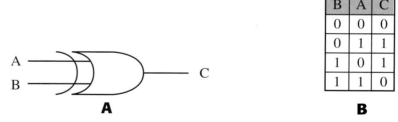

Figure 37-17. The NOR gate is an inverted OR. A—Schematic symbol for the NOR gate. B—Truth table for the NOR gate. Notice how the output is an inversion of the OR.

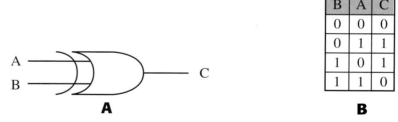

Figure 37-18. XOR gate. A—Schematic symbol for the XOR gate. B—Truth table for the XOR gate.

Exclusive NOR Gate

The output of an exclusive NOR (XNOR) gate is simply an inverted XOR output, **Figure 37-19.** The Boolean expression for the exclusive OR is:

$$A + B = \overline{A \oplus B}$$

Combinational logic: A logic circuit that contains both series and parallel logic circuits.

Sequential logic: A logic device or circuit that has more than one input and output.

Combinational-Sequential Circuits

Logic circuits are classified in two groups. Logic circuits using AND, OR, and NOT gates are called **combinational logic** circuits. The other group consists of **sequential logic** circuits. Sequential logic circuits are constructed using flip-flops (FF). Flip-flops may involve timing and memory devices, and other circuits, such as counters and shift registers.

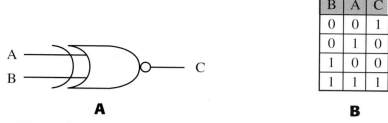

B	A	C
0	0	1
0	1	0
1	0	0
1	1	1

A   **B**

Figure 37-19. The XNOR gate is an inverted XOR. A—Schematic symbol for the XNOR gate. B—Truth table for the XNOR gate.

Combinational and sequential logic circuits are combined to produce many types of digital equipment. Consumer products, sophisticated test equipment, computer systems, and military and scientific equipment are some applications.

Digital Test Equipment

In addition to the VOM and oscilloscope, the electronic industry has developed specialized test equipment for digital circuitry. These test instruments include the logic probe, logic clip, and logic pulser.

Logic Probe

One of the most useful digital test instruments is the logic probe, **Figure 37-20.** The power and ground connections are connected to the circuits under test, and the probe is used as a signal tracer. Probe lights indicate whether the test point is at a high or low voltage level. A light may be provided to indicate a pulsing level. The probe may have a switch to select either TTL-type or CMOS-type integrated circuits and a switch to catch a pulse (memory) so a very short pulse can be picked up.

Figure 37-20. A logic probe is one of the vital test instruments used when troubleshooting digital circuits.

Logic Clip

A logic clip allows you to see the action of all the IC pins at the same time, **Figure 37-21.** Because it is difficult for the human eye to follow so many indicator lights at once, the logic clip is best used when the IC inputs and output are changing very slowly or are stepped by a logic pulser.

Logic Pulser

A logic pulser can generate a single pulse or a set of pulses, **Figure 37-22.** The pulser is connected to an input, and the output is monitored to see if the circuit is able to pass the pulse properly.

Figure 37-21. A logic clip shows the state of all the IC pins at the same time.

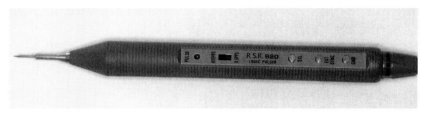

Figure 37-22. A logic pulser injects a pulse signal.

◢▷ Summary

- A signal may be classified as analog or digital.
- Circuits that process digital signals are called logic circuits.
- A digital signal must be either high or low. An undefined area should never occur. The digital families have different characteristics for specifying high or low for a digital signal.
- Numbers for digital systems depend on the base of the system. The base controls the number of symbols and place value in the number system.
- The binary number system is a base 2 system.
- To convert a decimal number to binary, use either the "divide by 2" or "place value" subtraction methods.
- The hexadecimal number system is a base 16 system.
- Binary numbers 1 and 0 are represented by a high or low voltage.
- Combinational logic gates include AND, OR, NOR, NAND, XOR, and XNOR.
- Sequential logic circuits are made from flip-flop circuits.
- Digital test equipment includes a logic probe, logic clip, and logic pulser.

◢▷ Important Terms

Do you know the meanings of these terms used in the chapter?

analog signal	digital signal
base	hexadecimal
binary	logic circuit
buffer/driver	place value
combinational logic	sequential logic
contact bounce	truth table
control pins	undefined area

⇥ Questions and Problems

Please do not write in this text. Write your answers on a separate sheet of paper.

1. Draw graphs of an analog signal and digital signal.
2. What are the typical low and high logic levels for TTL and CMOS gates?
3. Why can TTL and CMOS devices *not* be directly connected together?
4. Convert the following binary numbers to decimal numbers.
   a. 101
   b. 1101
   c. 100110
   d. 10001
   e. 101011
   f. 11111111
5. Convert the following decimal numbers to binary numbers.
   a. 5
   b. 24
   c. 50
   d. 128
   e. 189
   f. 1001
6. Convert the following decimal numbers to hexadecimal numbers.
   a. 175
   b. 188
   c. 206
   d. 26
   e. 90
   f. 378
7. Convert the following hexadecimal numbers to decimal numbers.
   a. BB
   b. 4BC
   c. C7
   d. 1A4
   e. 32
   f. DAC
8. Convert the following hexadecimal numbers to binary numbers.
   a. AE6
   b. F7
   c. E6A
   d. 5DC
   e. D4
   f. EE
   g. B7C
   h. 2A6
9. Convert the following binary numbers to hexadecimal numbers.
   a. 10111101
   b. 1101
   c. 11011101
   d. 00111010
   e. 11000101
   f. 10001111
10. What are the place values for an octal (8) number system?
11. Draw the schematic symbols for AND, OR, NOT (inverter), NAND, NOR, XOR, and XNOR logic gates.
12. Draw truth tables for each gate in problem 11.
13. Logic circuits are classified into two groups: _____ and _____.
14. What are the three types of digital troubleshooting equipment described?
15. In what manner is a logic probe used?

Appendix A
SCHEMATIC SYMBOLS

Electrical and electronic diagrams use schematic symbols to represent various components. Although there has been an attempt to standardize schematic symbols, some manufacturers still insist on using their own. The following symbols are approved by the Institute of Electrical and Electronic Engineers (IEEE).

General Symbols

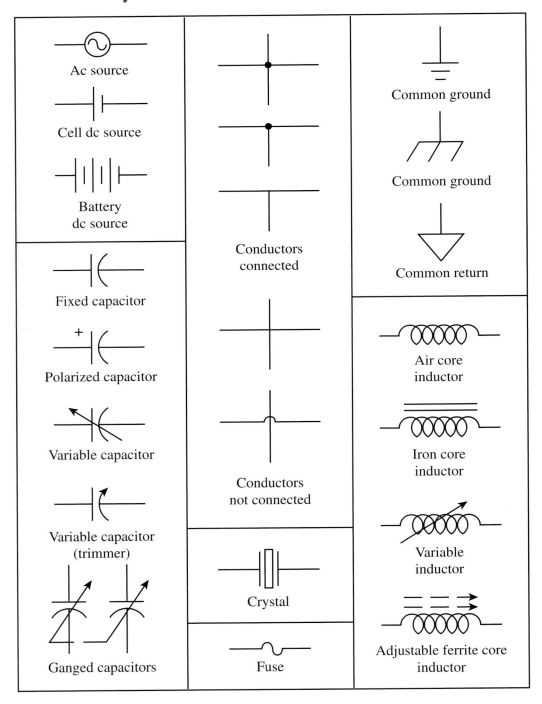

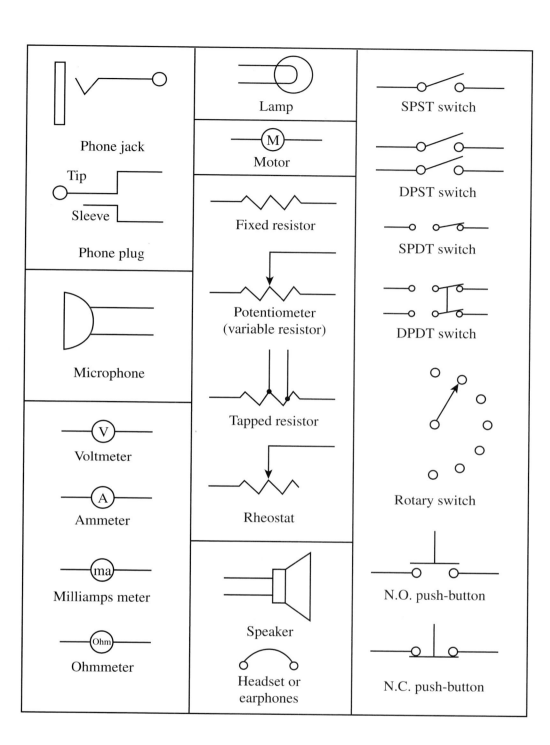

Phone jack

Tip

Sleeve

Phone plug

Microphone

Voltmeter

Ammeter

Milliamps meter

Ohmmeter

Lamp

Motor

Fixed resistor

Potentiometer
(variable resistor)

Tapped resistor

Rheostat

Speaker

Headset or
earphones

SPST switch

DPST switch

SPDT switch

DPDT switch

Rotary switch

N.O. push-button

N.C. push-button

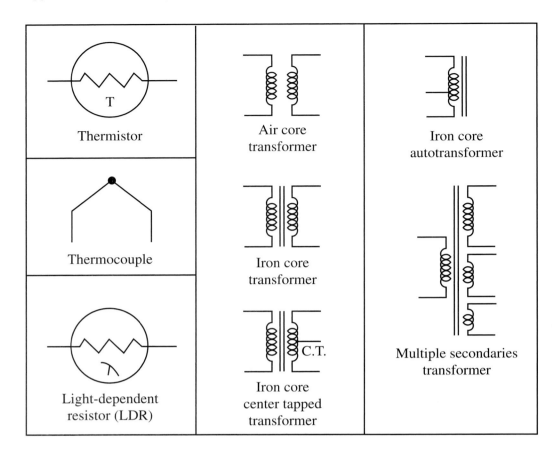

Semiconductor Symbols

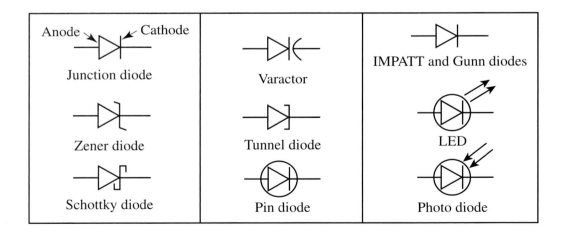

Semiconductor Devices

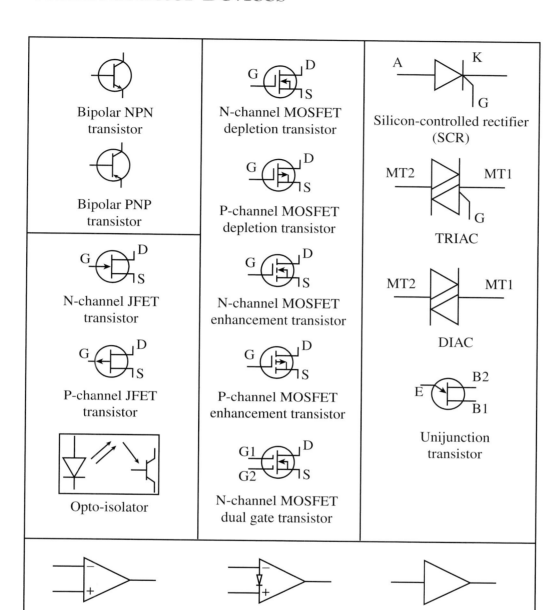

Bipolar NPN
transistor

Bipolar PNP
transistor

N-channel JFET
transistor

P-channel JFET
transistor

Opto-isolator

N-channel MOSFET
depletion transistor

P-channel MOSFET
depletion transistor

N-channel MOSFET
enhancement transistor

P-channel MOSFET
enhancement transistor

N-channel MOSFET
dual gate transistor

Silicon-controlled rectifier
(SCR)

TRIAC

DIAC

Unijunction
transistor

Operational amplifier

Norton op-amp

Single-ended amplifier

Logic Symbols

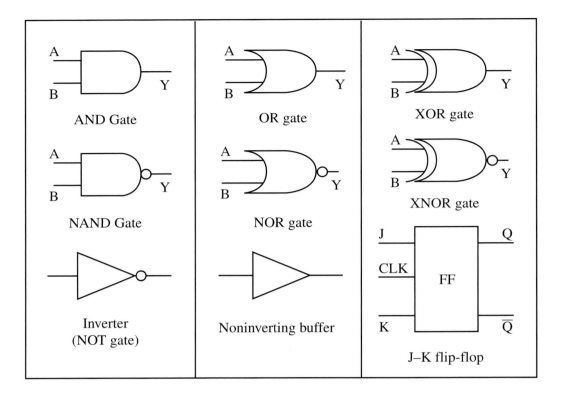

Appendix B
THE GREEK ALPHABET

A	α	alpha
B	β	beta
Γ	γ	gamma
Δ	δ	delta
E	ε	epsilon
Z	ζ	zeta
H	η	eta
Θ	θ	theta
I	ι	iota
K	κ	kappa
Λ	λ	lambda
M	μ	mu
N	ν	nu
Ξ	ξ	xi
O	o	omicron
Π	π	pi
P	ρ	rho
Σ	σ	sigma
T	τ	tau
Υ	υ	upsilon
Φ	ϕ	phi
X	χ	chi
Ψ	ψ	psi
Ω	ω	omega

Appendix C
LETTER SYMBOLS AND STANDARD ABBREVIATIONS

A	Unit of current, ampere
A_I	Current gain
A_P	Power gain
A_V	Voltage amplification gain
α	Temperature coefficient of resistance
α, h_{FB}	Common-base current gain
B	Magnetic flux density
BW	Bandwidth
β, h_{FE}	Common-emitter current gain
°C	Unit of temperature, degrees Celsius
C	Capacitance
C	Unit of charge of electrons, coulombs
$\cos \theta$	Power factor, cosine of the phase angle
dB	Unit of gain or loss in decibels
Δ	Change in
E, V	Voltage
e, v	Instantaneous voltage
E_{AVG}, V_{AVG}	Average voltage
E_{cemf}	Cemf voltage
E_{EFF}, E_{RMS}	Effective or root-mean-square voltage
E_{MAX}, E_{PEAK}	Maximum or peak voltage
E_P	Primary voltage
$E_{P\text{-}P}$	Peak-to-peak voltage
E_S	Source voltage
E_S	Secondary voltage
eV	Electron volt (1.602×10^{-19})
ε	Permitivity
ε_r	Relative permitivity
ε_o	Permitivity of free space (8.85×10^{-12} F/m)
°F	Unit of temperature, degrees Fahrenheit
F	Unit of capacitance, farad
f	Frequency
f_r	Resonant frequency
G	Conductance
G	Unit of magnetic flux density, gauss
Gb	Unit of magnetomotive force, gilbert
H	Unit of inductance, henry
H	Magnetic field intensity
Hz	Unit of frequency, hertz
I	Current
i	Instantaneous current
I_B	Transistor dc base current

I_C	Transistor dc collector current
I_{CBO}	Transistor collector cutoff current
I_E	Transistor dc emitter current
I_{EFF}, I_{RMS}	Effective, or root-means-square, current
I_{MAX}, I_{PEAK}	Maximum or peak current
I_p	Primary current
I_s	Secondary current
L	Inductance
λ	Ac wavelength
L_M	Mutual inductance
M_X	Unit of magnetic flux, maxwell
μ	Permeability
μ_r	Relative permeability
μ_o	Permeability of free space ($4\pi \times 10^{-7}$)
N_p	Number of turns in the primary
N_s	Number of turns in the secondary
Ω	Unit of resistance, reactance, or impedance, ohm
ω	Angular velocity, or radian frequency
P	Power, true power
p	Instantaneous power
P_D	Power dissipation
$P_{D_{MAX}}$	Maximum power dissipation rating
PIV, PRV	Peak inverse (reverse) voltage
PRR	Pulse repetition rate
PRT	Pulse repetition time
Φ	Magnetic flux
Q	Electrical charge, quality factor, or figure of merit, reactive power
R	Resistance
R_{EQ}	Equivalent resistance
R_i	Input resistance or internal resistance
R_L	Load resistance
$\mathscr{R}$	Reluctance
ρ	Electrical resistivity
S	Unit of conductance, siemen
σ	Conductivity
T	Temperature
T	Time period of AC waveform, motor torque
T	SI unit of magnetic flux density, tesla
t	Time constant, time
t_d	Pulse duration time
t_f	Fall time
t_r	Rise time
θ	Phase angle
V	Unit of voltage, volt
VA	Unit of apparent power, volt-ampere
VAR	Unit of reactive power, volt-ampere reactive
V_B	Base-to-ground transistor dc voltage
V_C	Collector-to-ground transistor dc voltage
V_{CC}	Transistor collector supply voltage
V_{CE}	Transistor collector-to-emitter dc voltage
V_{DD}	Transistor drain supply voltage
V_{EFF}, V_{RMS}	Effective or root-means-square ac voltage

V_f	Feedback voltage
V_{MAX}, V_{PEAK}	Maximum or peak voltage
W	Unit of power, watt
Wb	SI unit of magnetic flux, weber
X	Net reactance
X_C	Capacitive reactance
X_L	Inductive reactance
Z	Impedance

Appendix D
RELAY CONTACT ARRANGEMENTS

Three words used to describe the contact configurations of a relay are pole, throw, and break. *Pole* describes the number of isolated circuits that can pass through a relay at one time. *Throw* indicates how many different circuits each pole controls. *Break* is the number of separate contacts a relay uses to open or close each individual circuit. Because of the many possible contact combinations, the National Association of Relay Manufacturers (NARM) developed a set of code numbers and a standard set of contact arrangements as shown here.

Contact Code and NARM Designation			
1–SPST–NO	9–DPST–NO–DB	17–4PDT	25–7PST–NC
2–SPST–NC	10–DPST–NC–DB	18–5PST–NO	26–7PDT
3–SPST–NO–DM	11–DPDT	19–5PST–NC	27–8PST–NO
4–SPST–NC–DB	12–3PST–NO	20–5PDT	28–8PST–NC
5–SPDT	13–3PST–NC	21–6PST–NO	29–8PDT
6–SPDT–DB	14–3PDT	22–6PST–NC	
7–DPST–NO	15–4PST–NO	23–6PDT	
8–DPST–NC	16–4PST–NC	24–7PST–NO	

NARM Basic Contact Arrangements

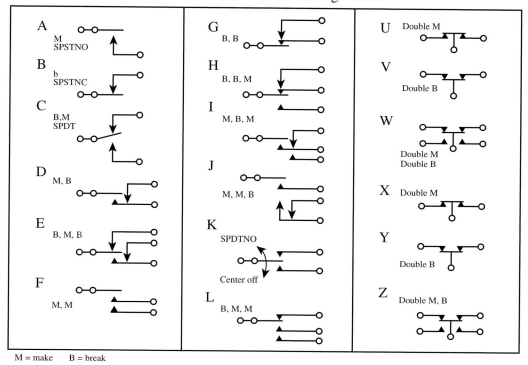

M = make   B = break

Appendix E
DECIBEL EQUATIONS

The number of decibels corresponding to a given power ratio is 10 times the common logarithm of the ratio. The equation is:

$$dB = 10 \log \frac{P_2}{P_1}$$

The number of decibels corresponding to a given voltage or current ratio is 20 times the common logarithm of the ratio. When the impedances across which the signals are being measured are equal, the equations are:

$$dB = 20 \log \frac{E_2}{E_1}$$

and

$$dB = 20 \log \frac{I_2}{I_1}$$

If the impedances are unequal, the equations become:

$$dB = 20 \log \frac{E_2 \sqrt{Z_1}}{E_1 \sqrt{Z_2}}$$

and

$$dB = 20 \log \frac{E_2 \sqrt{Z_2}}{E_1 \sqrt{Z_1}}$$

The following decibel table lists some of the more common power and voltage ratios with their decibel values. If a more complete table is required, consult an engineering or electronic reference.

dB	Gain Power	Gain Voltage	Loss Power	Loss Voltage
0	1	1	1	1
0.5	1.12	1.06	0.89	0.94
1	1.26	1.12	0.79	0.89
2	1.59	1.26	0.63	0.79
4	2.51	1.59	0.40	0.63
8	6.31	2.51	0.16	0.40
10	10	3.16	0.10	0.32
15	31.6	5.62	0.03	0.18
20	100	10	0.01	0.1
25	220	18	0.005	0.06
30	1000	31.6	0.001	0.03
35	2200	56	0.0005	0.02
40	10,000	100	0.0001	0.01
50	100,000	316	0.00001	0.003

Appendix F
MINIATURE LAMP DATA

The table lists some common miniature lamps and their characteristics. For a more complete listing, consult an engineering or lamp manufacturer's reference manual.

Lamp Number	Volts	Amps	Bead Color	Base
PR2	2.4	0.50	Blue	Flange
PR3	3.6	0.50	Green	Flange
PR4	2.3	0.27	Yellow	Flange
PR6	2.5	0.30	Brown	Flange
PR7	3.8	0.30		Flange
PR12	5.95	0.50	White	Flange
PR13	4.75	0.50		Flange
PR18	7.2	0.55		Flange
PR20	8.6	0.50		Flange
12	6.3	0.15		2-pin
41	2.5	0.50	White	Screw
43	2.5	0.50	White	Bayonet
44	6.3	0.25	Blue	Bayonet
45	3.2	0.35	Green	Bayonet
46	6.3	0.25	Blue	Screw
47	6.3	0.15	Brown	Bayonet
48	2.0	0.06	Pink	Screw
49	2.0	0.06	Pink	Bayonet
50	6.3	0.20	White	Screw
51	6.3	0.20	White	Bayonet
53	14.4	0.12		Bayonet
55	6.3	0.40	White	Bayonet
57	14.0	0.24	White	Bayonet
67	13.5	0.59		Bayonet

MOTOROLA
Semiconductors
BOX 20912 ● PHOENIX, ARIZONA 85036

Designers Data Sheet

"SURMETIC"▲ RECTIFIERS

. . . subminiature size, axial lead mounted rectifiers for general-purpose low-power applications.

Designers Data for "Worst Case" Conditions

The Designers▲ Data Sheets permit the design of most circuits entirely from the information presented. Limit curves — representing boundaries on device characteristics — are given to facilitate "worst case" design.

1N4001 thru 1N4007

LEAD MOUNTED SILICON RECTIFIERS

**50-1000 VOLTS
DIFFUSED JUNCTION**

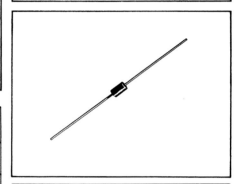

*MAXIMUM RATINGS

Rating	Symbol	1N4001	1N4002	1N4003	1N4004	1N4005	1N4006	1N4007	Unit
Peak Repetitive Reverse Voltage Working Peak Reverse Voltage DC Blocking Voltage	V_{RRM} V_{RWM} V_R	50	100	200	400	600	800	1000	Volts
Non-Repetitive Peak Reverse Voltage (halfwave, single phase, 60 Hz)	V_{RSM}	60	120	240	480	720	1000	1200	Volts
RMS Reverse Voltage	$V_{R(RMS)}$	35	70	140	280	420	560	700	Volts
Average Rectified Forward Current (single phase, resistive load, 60 Hz, see Figure 8, T_A = 75°C)	I_O				1.0				Amp
Non-Repetitive Peak Surge Current (surge applied at rated load conditions, see Figure 2)	I_{FSM}				30 (for 1 cycle)				Amp
Operating and Storage Junction Temperature Range	T_J, T_{stg}				−65 to +175				°C

*ELECTRICAL CHARACTERISTICS

Characteristic and Conditions	Symbol	Typ	Max	Unit
Maximum Instantaneous Forward Voltage Drop (i_F = 1.0 Amp, T_J = 25°C) Figure 1	v_F	0.93	1.1	Volts
Maximum Full-Cycle Average Forward Voltage Drop (I_O = 1.0 Amp, T_L = 75°C, 1 inch leads)	$V_{F(AV)}$	—	0.8	Volts
Maximum Reverse Current (rated dc voltage) T_J = 25°C T_J = 100°C	I_R	0.05 1.0	10 50	μA
Maximum Full-Cycle Average Reverse Current (I_O = 1.0 Amp, T_L = 75°C, 1 inch leads	$I_{R(AV)}$	—	30	μA

* Indicates JEDEC Registered Data.

MECHANICAL CHARACTERISTICS

CASE: Void free, Transfer Molded
MAXIMUM LEAD TEMPERATURE FOR SOLDERING PURPOSES: 350°C, 3/8'' from case for 10 seconds at 5 lbs. tension
FINISH: All external surfaces are corrosion-resistant, leads are readily solderable
POLARITY: Cathode indicated by color band
WEIGHT: 0.40 Grams (approximately)

▲Trademark of Motorola Inc.

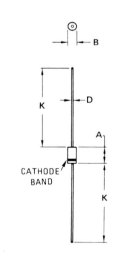

DIM	MILLIMETERS		INCHES	
	MIN	MAX	MIN	MAX
A	5.97	6.60	0.235	0.260
B	2.79	3.05	0.110	0.120
D	0.76	0.86	0.030	0.034
K	27.94	—	1.100	—

CASE 59-04
Does Not Conform to DO-41 Outline.

© MOTOROLA INC., 1975   DS 6015 R3

1N4001 THRU 1N4007

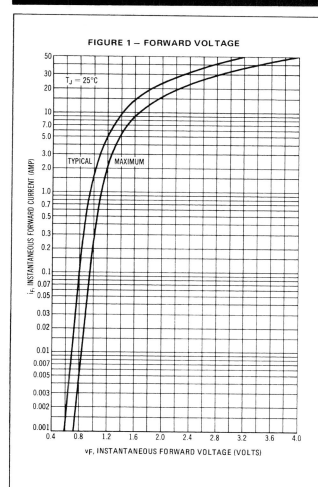

FIGURE 1 — FORWARD VOLTAGE

v_F, INSTANTANEOUS FORWARD VOLTAGE (VOLTS)

i_F, INSTANTANEOUS FORWARD CURRENT (AMP)

$T_J = 25°C$

TYPICAL   MAXIMUM

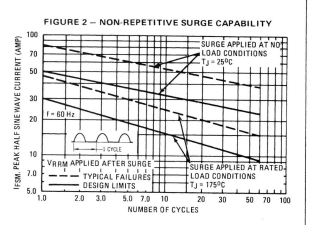

FIGURE 2 — NON-REPETITIVE SURGE CAPABILITY

I_{FSM}, PEAK HALF SINE WAVE CURRENT (AMP)

NUMBER OF CYCLES

SURGE APPLIED AT NO LOAD CONDITIONS $T_J = 25°C$

f = 60 Hz

1 CYCLE

V_{RRM} APPLIED AFTER SURGE
- - - TYPICAL FAILURES
―――― DESIGN LIMITS

SURGE APPLIED AT RATED LOAD CONDITIONS $T_J = 175°C$

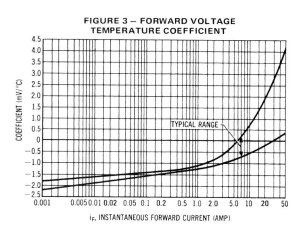

FIGURE 3 — FORWARD VOLTAGE TEMPERATURE COEFFICIENT

COEFFICIENT (mV/°C)

i_F, INSTANTANEOUS FORWARD CURRENT (AMP)

TYPICAL RANGE

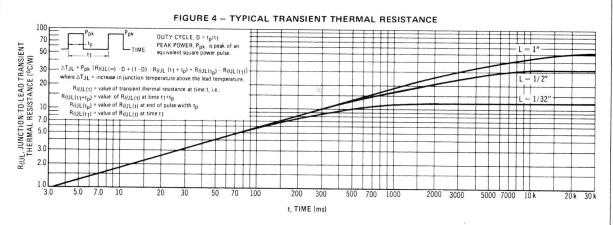

FIGURE 4 — TYPICAL TRANSIENT THERMAL RESISTANCE

$R_{\theta JL}$, JUNCTION-TO-LEAD TRANSIENT THERMAL RESISTANCE (°C/W)

t, TIME (ms)

P_{pk} P_{pk}
t_p
t_1
TIME

DUTY CYCLE, D = t_p/t_1
PEAK POWER, P_{pk}, is peak of an equivalent square power pulse.

$\Delta T_{JL} = P_{pk} [R_{\theta JL}(\infty) \cdot D + (1-D) \cdot R_{\theta JL}(t_1 + t_p) + R_{\theta JL}(t_p) - R_{\theta JL}(t_1)]$
where ΔT_{JL} = increase in junction temperature above the lead temperature.

$R_{\theta JL}(t)$ = value of transient thermal resistance at time t, i.e.:
$R_{\theta JL}(t_1 + t_p)$ = value of $R_{\theta JL}(t)$ at time $t_1 + t_p$
$R_{\theta JL}(t_p)$ = value of $R_{\theta JL}(t)$ at end of pulse width t_p
$R_{\theta JL}(t_1)$ = value of $R_{\theta JL}(t)$ at time t_1

L = 1"
L = 1/2"
L = 1/32"

The temperature of the lead should be measured using a thermocouple placed on the lead as close as possible to the tie point. The thermal mass connected to the tie point is normally large enough so that it will not significantly respond to heat surges generated in the diode as a result of pulsed operation once steady-state conditions are achieved. Using the measured value of T_L, the junction temperature may be determined by:

$$T_J = T_L + \Delta T_{JL}.$$

 MOTOROLA *Semiconductor Products Inc.*

May 1997

National Semiconductor

LM555/LM555C
Timer

General Description

The LM555 is a highly stable device for generating accurate time delays or oscillation. Additional terminals are provided for triggering or resetting if desired. In the time delay mode of operation, the time is precisely controlled by one external resistor and capacitor. For astable operation as an oscillator, the free running frequency and duty cycle are accurately controlled with two external resistors and one capacitor. The circuit may be triggered and reset on falling waveforms, and the output circuit can source or sink up to 200 mA or drive TTL circuits.

Features

- Direct replacement for SE555/NE555
- Timing from microseconds through hours
- Operates in both astable and monostable modes
- Adjustable duty cycle
- Output can source or sink 200 mA
- Output and supply TTL compatible
- Temperature stability better than 0.005% per °C
- Normally on and normally off output
- Available in 8 pin MSOP package

Applications

- Precision timing
- Pulse generation
- Sequential timing
- Time delay generation
- Pulse width modulation
- Pulse position modulation
- Linear ramp generator

Schematic Diagram

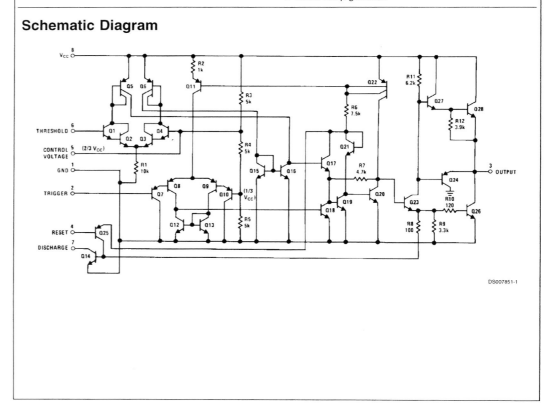

DS007851-1

www.national.com

Absolute Maximum Ratings (Note 2)

If Military/Aerospace specified devices are required, please contact the National Semiconductor Sales Office/ Distributors for availability and specifications.

Supply Voltage	+18V
Power Dissipation (Note 3)	
LM555H, LM555CH	760 mW
LM555, LM555CN	1180 mW
LM555CMM	613 mW
Operating Temperature Ranges	
LM555C	0°C to +70°C
LM555	−55°C to + 125°C

Storage Temperature Range	−65°C to +150°C
Soldering Information	
Dual-In-Line Package	
Soldering (10 Seconds)	260°C
Small Outline Packages	
(SOIC and MSOP)	
Vapor Phase (60 Seconds)	215°C
Infrared (15 Seconds)	220°C

See AN-450 "Surface Mounting Methods and Their Effect on Product Reliability" for other methods of soldering surface mount devices.

Electrical Characteristics (Notes 1, 2)

($T_A = 25°C$, $V_{CC} = +5V$ to $+15V$, unless othewise specified)

Parameter	Conditions	LM555 Min	LM555 Typ	LM555 Max	LM555C Min	LM555C Typ	LM555C Max	Units
Supply Voltage		4.5		18	4.5		16	V
Supply Current	$V_{CC} = 5V$, $R_L = \infty$		3	5		3	6	mA
	$V_{CC} = 15V$, $R_L = \infty$		10	12		10	15	mA
	(Low State) (Note 4)							
Timing Error, Monostable								
Initial Accuracy			0.5			1		%
Drift with Temperature	$R_A = 1k$ to $100 \, k\Omega$,		30			50		ppm/°C
	$C = 0.1 \, \mu F$, (Note 5)							
Accuracy over Temperature			1.5			1.5		%
Drift with Supply			0.05			0.1		%/V
Timing Error, Astable								
Initial Accuracy			1.5			2.25		%
Drift with Temperature	R_A, $R_B = 1k$ to $100 \, k\Omega$,		90			150		ppm/°C
	$C = 0.1 \, \mu F$, (Note 5)							
Accuracy over Temperature			2.5			3.0		%
Drift with Supply			0.15			0.30		%/V
Threshold Voltage			0.667			0.667		x V_{CC}
Trigger Voltage	$V_{CC} = 15V$	4.8	5	5.2		5		V
	$V_{CC} = 5V$	1.45	1.67	1.9		1.67		V
Trigger Current			0.01	0.5		0.5	0.9	μA
Reset Voltage		0.4	0.5	1	0.4	0.5	1	V
Reset Current			0.1	0.4		0.1	0.4	mA
Threshold Current	(Note 6)		0.1	0.25		0.1	0.25	μA
Control Voltage Level	$V_{CC} = 15V$	9.6	10	10.4	9	10	11	V
	$V_{CC} = 5V$	2.9	3.33	3.8	2.6	3.33	4	V
Pin 7 Leakage Output High			1	100		1	100	nA
Pin 7 Sat (Note 7)								
Output Low	$V_{CC} = 15V$, $I_7 = 15$ mA		150			180		mV
Output Low	$V_{CC} = 4.5V$, $I_7 = 4.5$ mA		70	100		80	200	mV

Electrical Characteristics (Notes 1, 2) (Continued)

(T_A = 25˚C, V_{CC} = +5V to +15V, unless otehwise specified)

Parameter	Conditions	Limits						Units
		LM555			LM555C			
		Min	Typ	Max	Min	Typ	Max	
Output Voltage Drop (Low)	V_{CC} = 15V							
	I_{SINK} = 10 mA		0.1	0.15		0.1	0.25	V
	I_{SINK} = 50 mA		0.4	0.5		0.4	0.75	V
	I_{SINK} = 100 mA		2	2.2		2	2.5	V
	I_{SINK} = 200 mA		2.5			2.5		V
	V_{CC} = 5V							
	I_{SINK} = 8 mA		0.1	0.25				V
	I_{SINK} = 5 mA					0.25	0.35	V
Output Voltage Drop (High)	I_{SOURCE} = 200 mA, V_{CC} = 15V		12.5			12.5		V
	I_{SOURCE} = 100 mA, V_{CC} = 15V	13	13.3		12.75	13.3		V
	V_{CC} = 5V	3	3.3		2.75	3.3		V
Rise Time of Output			100			100		ns
Fall Time of Output			100			100		ns

Note 1: All voltages are measured with respect to the ground pin, unless otherwise specified.

Note 2: Absolute Maximum Ratings indicate limits beyond which damage to the device may occur. Operating Ratings indicate conditions for which the device is functional, but do not guarantee specific performance limits. Electrical Characteristics state DC and AC electrical specifications under particular test conditions which guarantee specific performance limits. This assumes that the device is within the Operating Ratings. Specifications are not guaranteed for parameters where no limit is given, however, the typical value is a good indication of device performance.

Note 3: For operating at elevated temperatures the device must be derated above 25˚C based on a +150˚C maximum junction temperature and a thermal resistance of 164˚C/W (T0-5), 106˚C/W (DIP), 170˚C/W (S0-8), and 204˚C/W (MSOP) junction to ambient.

Note 4: Supply current when output high typically 1 mA less at V_{CC} = 5V.

Note 5: Tested at V_{CC} = 5V and V_{CC} = 15V.

Note 6: This will determine the maximum value of R_A + R_B for 15V operation. The maximum total (R_A + R_B) is 20 MΩ.

Note 7: No protection against excessive pin 7 current is necessary providing the package dissipation rating will not be exceeded.

Note 8: Refer to RETS555X drawing of military LM555H and LM555J versions for specifications.

Connection Diagrams

Metal Can Package

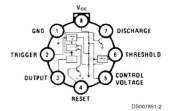

DS007851-2

Top View
Order Number LM555H or LM555CH
See NS Package Number H08C

Dual-In-Line, Small Outline
and Molded Mini Small Outline Packages

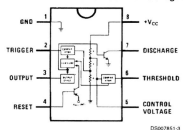

DS007851-3

Top View
Order Number LM555J, LM555CJ,
LM555CM, LM555CMM or LM555CN
See NS Package Number J08A, M08A, MUA08A or
N08E

Typical Performance Characteristics

Minimuim Pulse Width Required for Triggering

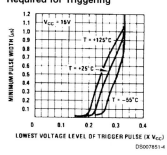

DS007851-4

Supply Current vs Supply Voltage

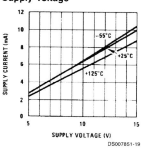

DS007851-19

High Output Voltage vs Output Source Current

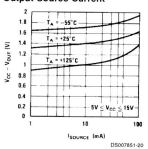

DS007851-20

Low Output Voltage vs Output Sink Current

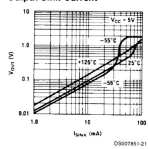

DS007851-21

Low Output Voltage vs Output Sink Current

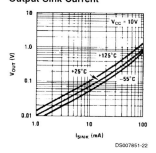

DS007851-22

Low Output Voltage vs Output Sink Current

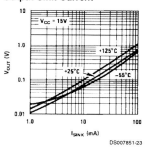

DS007851-23

Output Propagation Delay vs Voltage Level of Trigger Pulse

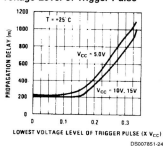

DS007851-24

Output Propagation Delay vs Voltage Level of Trigger Pulse

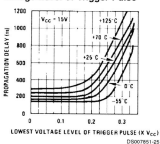

DS007851-25

Discharge Transistor (Pin 7) Voltage vs Sink Current

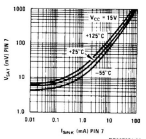

DS007851-26

Typical Performance Characteristics (Continued)

Discharge Transistor (Pin 7)
Voltage vs Sink Current

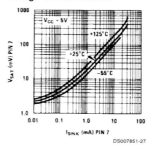

DS007851-27

Applications Information

MONOSTABLE OPERATION

In this mode of operation, the timer functions as a one-shot (*Figure 1*). The external capacitor is initially held discharged by a transistor inside the timer. Upon application of a negative trigger pulse of less than $1/3 \, V_{CC}$ to pin 2, the flip-flop is set which both releases the short circuit across the capacitor and drives the output high.

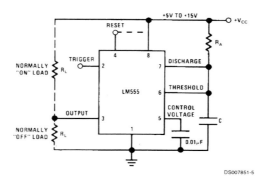

DS007851-5

FIGURE 1. Monostable

The voltage across the capacitor then increases exponentially for a period of $t = 1.1 \, R_A \, C$, at the end of which time the voltage equals $2/3 \, V_{CC}$. The comparator then resets the flip-flop which in turn discharges the capacitor and drives the output to its low state. *Figure 2* shows the waveforms generated in this mode of operation. Since the charge and the threshold level of the comparator are both directly proportional to supply voltage, the timing internal is independent of supply.

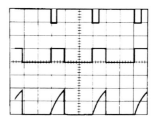

DS007851-6

V_{CC} = 5V	Top Trace: Input 5V/Div.
TIME = 0.1 ms/DIV.	Middle Trace: Output 5V/Div.
R_A = 9.1 kΩ	Bottom Trace: Capacitor Voltage 2V/Div.
C = 0.01 µF	

FIGURE 2. Monostable Waveforms

During the timing cycle when the output is high, the further application of a trigger pulse will not effect the circuit so long as the trigger input is returned high at least 10 µs before the end of the timing interval. However the circuit can be reset during this time by the application of a negative pulse to the reset terminal (pin 4). The output will then remain in the low state until a trigger pulse is again applied.

When the reset function is not in use, it is recommended that it be connected to V_{CC} to avoid any possibility of false triggering.

Figure 3 is a nomograph for easy determination of R, C values for various time delays.

NOTE: In monostable operation, the trigger should be driven high before the end of timing cycle.

Applications Information (Continued)

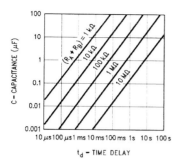

FIGURE 3. Time Delay

DS007851-7

ASTABLE OPERATION

If the circuit is connected as shown in *Figure 4* (pins 2 and 6 connected) it will trigger itself and free run as a multivibrator. The external capacitor charges through $R_A + R_B$ and discharges through R_B. Thus the duty cycle may be precisely set by the ratio of these two resistors.

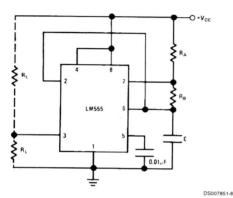

FIGURE 4. Astable

DS007851-8

In this mode of operation, the capacitor charges and discharges between 1/3 V_{CC} and 2/3 V_{CC}. As in the triggered mode, the charge and discharge times, and therefore the frequency are independent of the supply voltage.

Figure 5 shows the waveforms generated in this mode of operation.

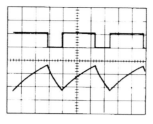

DS007851-9

V_{CC} = 5V   Top Trace: Output 5V/Div.
TIME = 20 µs/DIV.   Bottom Trace: Capacitor Voltage 1V/Div.
R_A = 3.9 kΩ
R_B = 3 kΩ
C = 0.01 µF

FIGURE 5. Astable Waveforms

The charge time (output high) is given by:
$$t_1 = 0.693 (R_A + R_B) C$$
And the discharge time (output low) by:
$$t_2 = 0.693 (R_B) C$$
Thus the total period is:
$$T = t_1 + t_2 = 0.693 (R_A + 2R_B) C$$
The frequency of oscillation is:

$$f = \frac{1}{T} = \frac{1.44}{(R_A + 2 R_B) C}$$

Figure 6 may be used for quick determination of these RC values.

The duty cycle is:

$$D = \frac{R_B}{R_A + 2R_B}$$

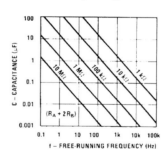

FIGURE 6. Free Running Frequency

DS007851-10

FREQUENCY DIVIDER

The monostable circuit of *Figure 1* can be used as a frequency divider by adjusting the length of the timing cycle. *Figure 7* shows the waveforms generated in a divide by three circuit.

Applications Information (Continued)

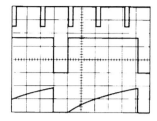

DS007851-11

V_{CC} = 5V   Top Trace: Input 4V/Div.
TIME = 20 µs/DIV.   Middle Trace: Output 2V/Div.
R_A = 9.1 kΩ   Bottom Trace: Capacitor 2V/Div.
C = 0.01 µF

FIGURE 7. Frequency Divider

PULSE WIDTH MODULATOR

When the timer is connected in the monostable mode and triggered with a continuous pulse train, the output pulse width can be modulated by a signal applied to pin 5. *Figure 8* shows the circuit, and in *Figure 9* are some waveform examples.

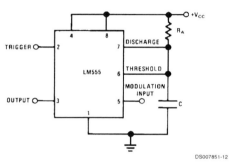

DS007851-12

FIGURE 8. Pulse Width Modulator

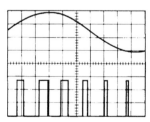

DS007851-13

V_{CC} = 5V   Top Trace: Modulation 1V/Div.
TIME = 0.2 ms/DIV.   Bottom Trace: Output Voltage 2V/Div.
R_A = 9.1 kΩ
C = 0.01 µF

FIGURE 9. Pulse Width Modulator

PULSE POSITION MODULATOR

This application uses the timer connected for astable operation, as in *Figure 10*, with a modulating signal again applied to the control voltage terminal. The pulse position varies with

the modulating signal, since the threshold voltage and hence the time delay is varied. *Figure 11* shows the waveforms generated for a triangle wave modulation signal.

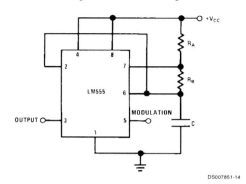

DS007851-14

FIGURE 10. Pulse Position Modulator

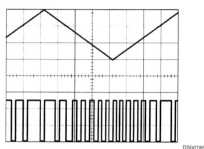

DS007851-15

V_{CC} = 5V   Top Trace: Modulation Input 1V/Div.
TIME = 0.1 ms/DIV.   Bottom Trace: Output 2V/Div.
R_A = 3.9 kΩ
R_B = 3 kΩ
C = 0.01 µF

FIGURE 11. Pulse Position Modulator

LINEAR RAMP

When the pullup resistor, R_A, in the monostable circuit is replaced by a constant current source, a linear ramp is generated. *Figure 12* shows a circuit configuration that will perform this function.

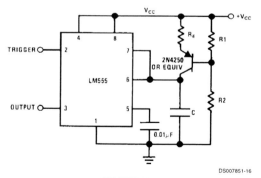

DS007851-16

FIGURE 12.

Applications Information (Continued)

Figure 13 shows waveforms generated by the linear ramp. The time interval is given by:

$$T = \frac{2/3 \, V_{CC} \, R_E \, (R_1 + R_2) \, C}{R_1 \, V_{CC} - V_{BE} \, (R_1 + R_2)}$$

$$V_{BE} \cong 0.6V$$

$$V_{BE} \cong 0.6V$$

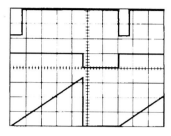

DS007851-17

V_{CC} = 5V   Top Trace: Input 3V/Div.
TIME = 20 µs/DIV.   Middle Trace: Output 5V/Div.
R_1 = 47 kΩ   Bottom Trace: Capacitor Voltage 1V/Div.
R_2 = 100 kΩ
R_E = 2.7 kΩ
C = 0.01 µF

FIGURE 13. Linear Ramp

50% DUTY CYCLE OSCILLATOR

For a 50% duty cycle, the resistors R_A and R_B may be connected as in *Figure 14*. The time period for the output high is the same as previous, t_1 = 0.693 R_A C. For the output low it is t_2 =

$$\left[(R_A \, R_B)/(R_A + R_B) \right] C \, \ell n \left[\frac{R_B - 2R_A}{2R_B - R_A} \right]$$

Thus the frequency of oscillation is

$$f = \frac{1}{t_1 + t_2}$$

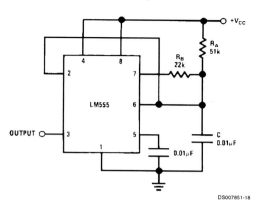

DS007851-18

FIGURE 14. 50% Duty Cycle Oscillator

Note that this circuit will not oscillate if R_B is greater than 1/2 R_A because the junction of R_A and R_B cannot bring pin 2 down to 1/3 V_{CC} and trigger the lower comparator.

ADDITIONAL INFORMATION

Adequate power supply bypassing is necessary to protect associated circuitry. Minimum recommended is 0.1 µF in parallel with 1 µF electrolytic.

Lower comparator storage time can be as long as 10 µs when pin 2 is driven fully to ground for triggering. This limits the monostable pulse width to 10 µs minimum.

Delay time reset to output is 0.47 µs typical. Minimum reset pulse width must be 0.3 µs, typical.

Pin 7 current switches within 30 ns of the output (pin 3) voltage.

May 1998

N *National Semiconductor*

LM741
Operational Amplifier

General Description

The LM741 series are general purpose operational amplifiers which feature improved performance over industry standards like the LM709. They are direct, plug-in replacements for the 709C, LM201, MC1439 and 748 in most applications.

The amplifiers offer many features which make their application nearly foolproof: overload protection on the input and output, no latch-up when the common mode range is exceeded, as well as freedom from oscillations.

The LM741C/LM741E are identical to the LM741/LM741A except that the LM741C/LM741E have their performance guaranteed over a 0˚C to +70˚C temperature range, instead of −55˚C to +125˚C.

Schematic Diagram

DS009341-1

Offset Nulling Circuit

DS009341-7

Absolute Maximum Ratings (Note 1)

If Military/Aerospace specified devices are required, please contact the National Semiconductor Sales Office/Distributors for availability and specifications.

(Note 6)

	LM741A	LM741E	LM741	LM741C
Supply Voltage	±22V	±22V	±22V	±18V
Power Dissipation (Note 2)	500 mW	500 mW	500 mW	500 mW
Differential Input Voltage	±30V	±30V	±30V	±30V
Input Voltage (Note 3)	±15V	±15V	±15V	±15V
Output Short Circuit Duration	Continuous	Continuous	Continuous	Continuous
Operating Temperature Range	−55°C to +125°C	0°C to +70°C	−55°C to +125°C	0°C to +70°C
Storage Temperature Range	−65°C to +150°C	−65°C to +150°C	−65°C to +150°C	−65°C to +150°C
Junction Temperature	150°C	100°C	150°C	100°C
Soldering Information				
N-Package (10 seconds)	260°C	260°C	260°C	260°C
J- or H-Package (10 seconds)	300°C	300°C	300°C	300°C
M-Package				
Vapor Phase (60 seconds)	215°C	215°C	215°C	215°C
Infrared (15 seconds)	215°C	215°C	215°C	215°C

See AN-450 "Surface Mounting Methods and Their Effect on Product Reliability" for other methods of soldering surface mount devices.

ESD Tolerance (Note 7)	400V	400V	400V	400V

Electrical Characteristics (Note 4)

Parameter	Conditions	LM741A/LM741E			LM741			LM741C			Units
		Min	Typ	Max	Min	Typ	Max	Min	Typ	Max	
Input Offset Voltage	$T_A = 25°C$										
	$R_S \leq 10\ k\Omega$					1.0	5.0		2.0	6.0	mV
	$R_S \leq 50\Omega$		0.8	3.0							mV
	$T_{AMIN} \leq T_A \leq T_{AMAX}$										
	$R_S \leq 50\Omega$			4.0							mV
	$R_S \leq 10\ k\Omega$						6.0			7.5	mV
Average Input Offset Voltage Drift				15							µV/°C
Input Offset Voltage Adjustment Range	$T_A = 25°C, V_S = ±20V$	±10				±15			±15		mV
Input Offset Current	$T_A = 25°C$		3.0	30		20	200		20	200	nA
	$T_{AMIN} \leq T_A \leq T_{AMAX}$			70		85	500			300	nA
Average Input Offset Current Drift				0.5							nA/°C
Input Bias Current	$T_A = 25°C$		30	80		80	500		80	500	nA
	$T_{AMIN} \leq T_A \leq T_{AMAX}$			0.210			1.5			0.8	µA
Input Resistance	$T_A = 25°C, V_S = ±20V$	1.0	6.0		0.3	2.0		0.3	2.0		MΩ
	$T_{AMIN} \leq T_A \leq T_{AMAX},$ $V_S = ±20V$	0.5									MΩ
Input Voltage Range	$T_A = 25°C$							±12	±13		V
	$T_{AMIN} \leq T_A \leq T_{AMAX}$				±12	±13					V

Electrical Characteristics (Note 4) (Continued)

Parameter	Conditions	LM741A/LM741E			LM741			LM741C			Units
		Min	Typ	Max	Min	Typ	Max	Min	Typ	Max	
Large Signal Voltage Gain	$T_A = 25°C$, $R_L \geq 2$ kΩ										
	$V_S = \pm20V$, $V_O = \pm15V$	50									V/mV
	$V_S = \pm15V$, $V_O = \pm10V$				50	200		20	200		V/mV
	$T_{AMIN} \leq T_A \leq T_{AMAX}$,										
	$R_L \geq 2$ kΩ,										
	$V_S = \pm20V$, $V_O = \pm15V$	32									V/mV
	$V_S = \pm15V$, $V_O = \pm10V$				25			15			V/mV
	$V_S = \pm5V$, $V_O = \pm2V$	10									V/mV
Output Voltage Swing	$V_S = \pm20V$										
	$R_L \geq 10$ kΩ	±16									V
	$R_L \geq 2$ kΩ	±15									V
	$V_S = \pm15V$										
	$R_L \geq 10$ kΩ				±12	±14		±12	±14		V
	$R_L \geq 2$ kΩ				±10	±13		±10	±13		V
Output Short Circuit Current	$T_A = 25°C$	10	25	35		25			25		mA
	$T_{AMIN} \leq T_A \leq T_{AMAX}$	10		40							mA
Common-Mode Rejection Ratio	$T_{AMIN} \leq T_A \leq T_{AMAX}$										
	$R_S \leq 10$ kΩ, $V_{CM} = \pm12V$				70	90		70	90		dB
	$R_S \leq 50\Omega$, $V_{CM} = \pm12V$	80	95								dB
Supply Voltage Rejection Ratio	$T_{AMIN} \leq T_A \leq T_{AMAX}$,										
	$V_S = \pm20V$ to $V_S = \pm5V$										
	$R_S \leq 50\Omega$	86	96								dB
	$R_S \leq 10$ kΩ				77	96		77	96		dB
Transient Response	$T_A = 25°C$, Unity Gain										
Rise Time			0.25	0.8		0.3			0.3		µs
Overshoot			6.0	20		5			5		%
Bandwidth (Note 5)	$T_A = 25°C$	0.437	1.5								MHz
Slew Rate	$T_A = 25°C$, Unity Gain	0.3	0.7			0.5			0.5		V/µs
Supply Current	$T_A = 25°C$					1.7	2.8		1.7	2.8	mA
Power Consumption	$T_A = 25°C$										
	$V_S = \pm20V$		80	150							mW
	$V_S = \pm15V$					50	85		50	85	mW
LM741A	$V_S = \pm20V$										
	$T_A = T_{AMIN}$			165							mW
	$T_A = T_{AMAX}$			135							mW
LM741E	$V_S = \pm20V$										
	$T_A = T_{AMIN}$			150							mW
	$T_A = T_{AMAX}$			150							mW
LM741	$V_S = \pm15V$										
	$T_A = T_{AMIN}$					60	100				mW
	$T_A = T_{AMAX}$					45	75				mW

Note 1: "Absolute Maximum Ratings" indicate limits beyond which damage to the device may occur. Operating Ratings indicate conditions for which the device is functional, but do not guarantee specific performance limits.

Electrical Characteristics (Note 4) (Continued)

Note 2: For operation at elevated temperatures, these devices must be derated based on thermal resistance, and T_j max. (listed under "Absolute Maximum Ratings"). $T_j = T_A + (\theta_{jA} P_D)$.

Thermal Resistance	Cerdip (J)	DIP (N)	HO8 (H)	SO-8 (M)
θ_{jA} (Junction to Ambient)	100°C/W	100°C/W	170°C/W	195°C/W
θ_{jC} (Junction to Case)	N/A	N/A	25°C/W	N/A

Note 3: For supply voltages less than ±15V, the absolute maximum input voltage is equal to the supply voltage.

Note 4: Unless otherwise specified, these specifications apply for $V_S = \pm15V$, $-55°C \le T_A \le +125°C$ (LM741/LM741A). For the LM741C/LM741E, these specifications are limited to $0°C \le T_A \le +70°C$.

Note 5: Calculated value from: BW (MHz) = 0.35/Rise Time(μs).

Note 6: For military specifications see RETS741X for LM741 and RETS741AX for LM741A.

Note 7: Human body model, 1.5 kΩ in series with 100 pF.

Connection Diagram

Metal Can Package

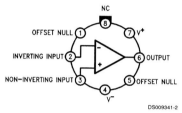

DS009341-2

Note 8: LM741H is available per JM38510/10101

**Order Number LM741H, LM741H/883 (Note 8),
LM741AH/883 or LM741CH
See NS Package Number H08C**

Ceramic Dual-In-Line Package

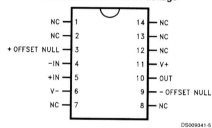

DS009341-5

Note 9: also available per JM38510/10101
Note 10: also available per JM38510/10102

**Order Number LM741J-14/883 (Note 9),
LM741AJ-14/883 (Note 10)
See NS Package Number J14A**

Dual-In-Line or S.O. Package

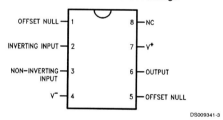

DS009341-3

**Order Number LM741J, LM741J/883,
LM741CM, LM741CN or LM741EN
See NS Package Number J08A, M08A or N08E**

Ceramic Flatpak

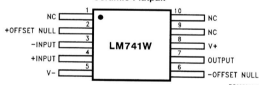

DS009341-6

**Order Number LM741W/883
See NS Package Number W10A**

MOTOROLA SEMICONDUCTORS

P.O. BOX 20912 • PHOENIX, ARIZONA 85036

2N3903
2N3904

NPN SILICON ANNULAR TRANSISTORS

. . . designed for general purpose switching and amplifier applications and for complementary circuitry with types 2N3905 and 2N3906.

- High Voltage Ratings — $V_{(BR)CEO}$ = 40 Volts (Min)
- Current Gain Specified from 100 μA to 100 mA
- Complete Switching and Amplifier Specifications
- Low Capacitance — C_{ob} = 4.0 pF (Max)

NPN SILICON SWITCHING & AMPLIFIER TRANSISTORS

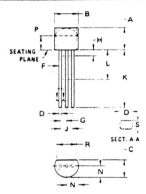

MAXIMUM RATINGS

Rating	Symbol	Value	Unit
*Collector-Emitter Voltage	V_{CEO}	40	Vdc
*Collector-Base Voltage	V_{CBO}	60	Vdc
*Emitter-Base Voltage	V_{EBO}	6.0	Vdc
*Collector Current — Continuous	I_C	200	mAdc
**Total Device Dissipation @ T_A = 25°C Derate above 25°C	P_D	625 5.0	mW mW/°C
Total Power Dissipation @ T_A = 60°C	P_D	450	mW
**Total Device Dissipation @ T_C = 25°C Derate above 25°C	P_D	1.5 12	Watts mW/°C
**Operating and Storage Junction Temperature Range	T_J, T_{stg}	−55 to 150	°C

THERMAL CHARACTERISTICS

Characteristic	Symbol	Max	Unit
Thermal Resistance, Junction to Case	$R_{\theta JC}$	83.3	°C/W
Thermal Resistance, Junction to Ambient	$R_{\theta JA}$	200	°C/W

*Indicates JEDEC Registered Data.
**Motorola guarantees this data in addition to the JEDEC Registered Data.

EQUIVALENT SWITCHING TIME TEST CIRCUITS

FIGURE 1 — TURN-ON TIME

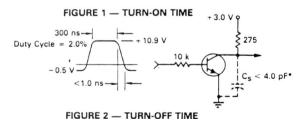

FIGURE 2 — TURN-OFF TIME

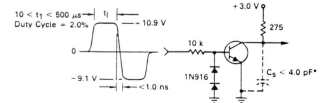

*Total shunt capacitance of test jig and connectors

NOTES.
1. CONTOUR OF PACKAGE BEYOND ZONE "P" IS UNCONTROLLED.
2. DIM "F" APPLIES BETWEEN "H" AND "L". DIM "D" & "S" APPLIES BETWEEN "L" & 12.70 mm (0.5") FROM SEATING PLANE. LEAD DIM IS UNCONTROLLED IN "H" & BEYOND 12.70 mm (0.5") FROM SEATING PLANE.

DIM	MILLIMETERS		INCHES	
	MIN	MAX	MIN	MAX
A	4.32	5.33	0.170	0.210
B	4.44	5.21	0.175	0.205
C	3.18	4.19	0.125	0.165
D	0.41	0.56	0.016	0.022
F	0.41	0.48	0.016	0.019
G	1.14	1.40	0.045	0.055
H	–	2.54	–	0.100
J	2.41	2.67	0.095	0.105
K	12.70	–	0.500	–
L	6.35	–	0.250	–
N	2.03	2.67	0.080	0.105
P	2.92	–	0.115	–
R	3.43	–	0.135	–
S	0.36	0.41	0.014	0.016

All JEDEC dimensions and notes apply.

CASE 29-02
(TO-226AA)

✻ELECTRICAL CHARACTERISTICS (T_A = 25°C unless otherwise noted.)

Characteristic		Symbol	Min	Max	Unit
OFF CHARACTERISTICS					
Collector-Emitter Breakdown Voltage[1] (I_C = 1.0 mAdc, I_B = 0)		$V_{(BR)CEO}$	40	—	Vdc
Collector-Base Breakdown Voltage (I_C = 10 µAdc, I_E = 0)		$V_{(BR)CBO}$	60	—	Vdc
Emitter-Base Breakdown Voltage (I_E = 10 µAdc, I_C = 0)		$V_{(BR)EBO}$	6.0	—	Vdc
Collector Cutoff Current (V_{CE} = 30 Vdc, $V_{EB(off)}$ = 3.0 Vdc)		I_{CEX}	—	50	nAdc
Base Cutoff Current (V_{CE} = 30 Vdc, $V_{EB(off)}$ = 3.0 Vdc)		I_{BL}	—	50	nAdc
ON CHARACTERISTICS[1]					
DC Current Gain		h_{FE}			—
(I_C = 0.1 mAdc, V_{CE} = 1.0 Vdc)	2N3903		20	—	
	2N3904		40	—	
(I_C = 1.0 mAdc, V_{CE} = 1.0 Vdc)	2N3903		35	—	
	2N3904		70	—	
(I_C = 10 mAdc, V_{CE} = 1.0 Vdc)	2N3903		50	150	
	2N3904		100	300	
(I_C = 50 mAdc, V_{CE} = 1.0 Vdc)	2N3903		30	—	
	2N3904		60	—	
(I_C = 100 mAdc, V_{CE} = 1.0 Vdc)	2N3903		15	—	
	2N3904		30	—	
Collector-Emitter Saturation Voltage		$V_{CE(sat)}$			Vdc
(I_C = 10 mAdc, I_B = 1.0 mAdc)			—	0.2	
(I_C = 50 mAdc, I_B = 5.0 mAdc)			—	0.3	
Base-Emitter Saturation Voltage		$V_{BE(sat)}$			Vdc
(I_C = 10 mAdc, I_B = 1.0 mAdc)			0.65	0.85	
(I_C = 50 mAdc, I_B = 5.0 mAdc)			—	1.0	
SMALL-SIGNAL CHARACTERISTICS					
Current-Gain — Bandwidth Product		f_T			MHz
(I_C = 10 mAdc, V_{CE} = 20 Vdc, f = 100 MHz)	2N3903		150	—	
	2N3904		200	—	
Output Capacitance (V_{CB} = 5.0 Vdc, I_E = 0, f = 100 kHz)		C_{obo}	—	4.0	pF
Input Capacitance (V_{BE} = 0.5 Vdc, I_C = 0, f = 100 kHz)		C_{ibo}	—	8.0	pF
Input Impedance		h_{ie}			kΩ
(I_C = 1.0 mAdc, V_{CE} = 10 Vdc, f = 1.0 kHz)	2N3903		0.5	8.0	
	2N3904		1.0	10	
Voltage Feedback Ratio		h_{re}			X 10^{-4}
(I_C = 1.0 mAdc, V_{CE} = 10 Vdc, f = 1.0 kHz)	2N3903		0.1	5.0	
	2N3904		0.5	8.0	
Small-Signal Current Gain		h_{fe}			—
(I_C = 1.0 mAdc, V_{CE} = 10 Vdc, f = 1.0 kHz)	2N3903		50	200	
	2N3904		100	400	
Output Admittance (I_C = 1.0 mAdc, V_{CE} = 10 Vdc, f = 1.0 kHz)		h_{oe}	1.0	40	µmhos
Noise Figure		NF			dB
(I_C = 100 µAdc, V_{CE} = 5.0 Vdc, R_S = 1.0 kΩ, f = 10 Hz to 15.7 kHz)	2N3903		—	6.0	
	2N3904		—	5.0	

SWITCHING CHARACTERISTICS

			Symbol	Min	Max	Unit
Delay Time	(V_{CC} = 3.0 Vdc, $V_{BE(off)}$ = 0.5 Vdc,		t_d	—	35	ns
Rise Time	I_C = 10 mAdc, I_{B1} = 1.0 mAdc)		t_r	—	50	ns
Storage Time		2N3903	t_s	—	800	ns
	(V_{CC} = 3.0 Vdc, I_C = 10 mAdc,	2N3904		—	900	
Fall Time	I_{B1} = I_{B2} = 1.0 mAdc)		t_f	—	90	ns

(1) Pulse Test: Pulse Width ≤ 300 µs, Duty Cycle ≤ 2.0%.

 MOTOROLA *Semiconductor Products Inc.*

GLOSSARY OF IMPORTANT TERMS

A

Absolute number: A number having a definite value, such as 0, 6, 17, 89, etc.

Absolute permeability: The permeability of free space. It is equal to 1.26×10^{-6}. It is represented by μ_o.

Acid flux: A special flux used in soldering sheet metal or plumbing. Acid flux is never used in electrical soldering.

Active device: An electronic component whose output responds to its input. Active devices include transistors and integrated circuits.

Adjacent side: The side of a triangle connected to the angle.

Admittance: The reciprocal of impedance. Admittance is used in the calculation of complex parallel RCL circuits. It is represented by Y.

Alloy: A mixture of two or more metals.

Alternating current: A current of electrons that moves first in one direction, then the other. It is abbreviated ac.

Ampere: The unit of measurement for current.

Amplifier: A device or electronic circuit that increases the magnitude of the current or voltage.

Analog meter: A meter that uses a magnetically operated pointer with a scale to indicate an electrical quantity.

Analog signal: A signal that continuously changes amplitude.

AND (gate): A logic gate that produces an output only when all of its inputs are at a high level.

Angstrom: A value equal to 10^{-10}.

Angular velocity: The rate at which an angle changes. It is expressed in radians per second.

Anode: The positive terminal of a device.

Antilogarithm: The inverse of logarithm.

Apparent power: Electrical power equal to the voltage times the current. It is expressed in either volt-amperes or watts.

Armature: The movable part of a generator, alternator, or relay.

Atom: The smallest particle of an element that retains the properties of the element.

Atomic number: The number of protons in the nucleus of an atom.

Attenuation: The decrease in amplitude or intensity of a signal.

Attenuator: A resistive network used to decrease the amplitude of a signal.

Audio taper: A control that has a nonlinear decrease in resistance and is used in audio applications.

Autotransformer: A tapped-point, single-winding transformer that operates as both a primary and secondary.

Avalanche point: *See* breakdown voltage.

Average value: The value obtained by dividing the sum of a number of quantities by the number of quantities.

B

Balanced atom: An atom with the same number of electrons as protons.

Balanced bridge: A circuit whose components are adjusted so the output voltage is zero across the bridge.

Bandwidth: The given band of frequencies effectively passed by a circuit or device and measured at the half-power points of maximum amplitude.

Base: 1. A number raised to a power. 2. The controlling element of a bipolar transistor. 3. The total number of symbols used in a numbering system.

Base current: The current that flows in the base lead of a bipolar transistor.

Battery: A dc voltage source consisting of two or more cells. It converts chemical energy into electrical energy.

Beta: The current gain of a transistor amplifier connected in a common emitter configuration. It is represented by the Greek letter β (beta).

B-H curve: A curve plotted on a graph to show successive states during the magnetization of a ferromagnetic material.

Bias voltage: The voltage across a semiconductor P-N junction.

Binary: A numbering system using a base of 2.

Birdcaging: The separation of a stranded conductor.

Breadboarding: Using a flat plastic board to mount components.

Break-before-make: Refers to contacts that open before another set of contacts close.

Breakdown voltage: 1. The voltage at which a dielectric fails and a current flows through the dielectric. 2. The voltage across a semiconductor that causes maximum current in the reverse direction through the semiconductor material. Also called avalanche point.

Breaks: The number of separate contacts a relay uses to open or close individual circuits.

Bridge: 1. An instrument using a bridge circuit to measure accurate quantities. 2. An unwanted solder conductive path between two points.

British thermal units: The quantity of heat needed to raise one pound of water one degree Fahrenheit.

Brush: A piece of conductive material that rides on the commutator of an electric motor.

Buffer/driver: A circuit that isolates one electronic circuit from another and provides the necessary power to drive another circuit.

Bypass capacitor: A capacitor that provides a low-impedance ac path around a circuit or component.

C

Cable: A stranded conductor or group of conductors insulated from each other.

Capacitance: The ability to store an electric charge in the form of electrons. Capacitance is measured in farads and represented by C.

Capacitive reactance: The opposition or resistance to alternating current by a capacitor or component that has a capacitive quality. Capacitive reactance is measured in ohms and represented by X_C.

Cartesian coordinate system: A rectangular coordinate system using the X-axis for the horizontal direction and the Y-axis for the vertical direction.

Cascaded: Connecting two or more circuits or amplifying stages so the output of one provides the input of the next.

Cathode: The negative terminal of a semiconductor diode.

Cathode ray tube: A vacuum tube in which an electron beam strikes the surface of the screen and produces a particular pattern of light. It is abbreviated CRT.

Cell: 1. A chemical type of voltage source. 2. A single unit that produces a direct current voltage by converting chemical energy into electrical energy. Cells are either primary or secondary.

Channel: The conductive path between the source and drain of a field-effect transistor.

Circuit breaker: A safety device that automatically opens a circuit if it is overloaded.

Circular-mil: A universal term used to define the cross-sectional area of a conductor. It is abbreviated cmil.

Circumference: The total distance around a circle.

Closed circuit: A circuit that has a complete current path.

Closed-loop: A circuit in which the output is continuously fed back to the input.

Coaxial cable: A concentric transmission line in which one conductor completely surrounds another conductor.

Coefficient: 1. The ratio of change of a quantity under such specified conditions as temperature, volume, or length. 2. The number that precedes an unknown in an algebraic equation. In $6x + 4y$, 6 and 4 are the coefficients.

Coefficient of coupling: The amount of coupling between two components or circuits. It is represented by k.

Cold soldered joint: An undesirable soldered connection formed by insufficient heat.

Collector: The main transistor electrode through which the primary current flows.

Combinational logic: A logic circuit that contains both series and parallel logic circuits.

Common base: A transistor circuit in which the base is connected to a common ground.

Common collector: A transistor circuit in which the collector is connected to a common ac ground by connecting a capacitor from the collector to ground. Also called an *emitter follower* circuit.

Common emitter: A transistor amplifier circuit in which the emitter is connected to a common ground.

Common-mode rejection ratio (CMRR): A measure of the ability of the op-amp to reject signals simultaneously present at both inputs. It is the ratio of the common-mode input voltage to the generated output voltage. It is expressed in decibels.

Commutator: Part of an armature to which the coils of a motor are connected. It consists of copper pieces insulated from each other. To provide an electrical connection between the armature and the power source, the motor brushes ride on the outer portion of the commutator.

Comparator: A circuit that compares two or more signals and indicates whether they are matched in some way.

Complex number: A two-part number containing a real number and imaginary number (also known as a *j*-number). In ac circuit analysis, the real number is the resistance and the imaginary number is the reactance.

Complex plane: A rectangular method of graphing using an imaginary axis and real axis.

Composite: A waveform with many different voltages and frequencies.

Compound: A substance made of two or more chemical elements.

Conductance: The ability of a circuit or conductor to carry an electrical current. It is measured in siemens (formerly mho) and represented by G.

Conductor: Any material that contains free electrons and allows electrical current to flow.

Cone: The moving part of a speaker that causes the surrounding air to vibrate and creates audible sound.

Connector: A coupling device that provides an electrical and mechanical connection or disconnection between one or more wires or between a cable and a piece of equipment.

Constant-k: Refers to a filter in which the values of inductance and capacitance are designed so the product of X_L and X_C is constant at all frequencies.

Contact bounce: The uncontrolled making and breaking of contacts when relays or switches are closed.

Continuity: An uninterrupted path for electrical current.

Continuity test: A test performed to ensure a complete or continuous current path exists through a device, conductor, or circuit.

Control circuit: A circuit that controls various motor functions such as speed.

Control pins: The pins on a relay or integrated circuit that control the device.

Controlled circuit: A circuit controlled by either automatic or manual switching.

Coulomb: The unit of measurement for a quantity of electrons. It is abbreviated C.

Counter electromotive force (cemf): The voltage generated by a magnetic field around a coil. The cemf is always in opposite polarity to the source ($180°$ out of phase).

Coupling: Connecting two or more circuits so power or signal is transferred from one to another.

Coupling transformer: A transformer used to connect one circuit to another.

Covalent bond: A type of linkage between atoms in which each atom contributes one electron to be shared with the others.

Crossover distortion: Distortion that occurs in a push-pull amplifier at the points where the signal crosses over the zero reference. Crossover distortion happens between the time one transistor stops conduction and the other begins.

Crossover network: A filter for a multispeaker system that separates the output signal of an amplifier into two or more frequency bands.

Cutoff frequency: The frequency at which the gain of an amplifier or circuit falls below 0.707 times the maximum gain (3 dB).

Cutoff state: The point at which the collector current stops flowing. It is caused by the absence of an emitter-base bias voltage.

D

Damped wave: A wave in which each cycle constantly decreases in amplitude.

Dc offset: Refers to a waveform above (+) or below (−) the baseline by a given value of dc voltage.

Dc output offset voltage: In an op-amp circuit, the dc output voltage measured between the output and ground when both inputs are grounded.

Dc pulses: Waveform pulses that do not cross the X-axis.

Decibel: The standard unit of measurement for gain or loss in voltage or power levels. It is abbreviated dB.

Deflection plates: Two pairs of parallel electrodes set at right angles to each other. An electrostatic field applied to the plates causes an electron beam to move in a cathode ray tube.

Degenerative feedback: A negative feedback voltage in which part of the output is fed into the input, or a circuit action that causes a reduced gain.

Delta: 1. Letter of the Greek alphabet used for indicating "a change in." 2. Type of transformer connection used in three-phase ac power.

Denominator: The bottom number in a fraction.

Depletion-mode: In MOSFET transistors, the operation in which current flows when the gate-source voltage is zero and is increased or decreased by a changing the gate-source voltage.

Depletion zone: The area extending on both sides of the junction in which the current carriers have been removed by a voltage. The region is depleted of current carriers.

Desoldering: The process of removing solder from a soldered connection.

DIAC: A two-lead alternating current semiconductor switch.

Diamagnetic: A material that is less magnetic than air or in which the intensity of magnetism is negative. Bismuth is the leading material in this category.

Diameter: The distance across the middle of a circle.

Dielectric: The insulation between the plates of a capacitor.

Dielectric constant: In a capacitor, the ratio of capacitance with a given dielectric to one with a dielectric of air.

Dielectric strength: The maximum voltage a dielectric can withstand without puncturing.

Difference in potential: The voltage difference between two points. *See* electromotive force.

Differentiator: A circuit in which the output voltage is proportional to the rate of change of the input voltage.

Digital meter: A meter that provides a digital readout of a measurement.

Digital signal: A signal that has distinct steps: on or off, voltage or no voltage.

Diode: A semiconductor device that conducts current in only one direction.

Direct current: Current that flows in one direction. It is abbreviated dc.

Directly proportional: The value of two numbers increases or decreases at the same rate.

Dissipated energy: Electrical energy lost in the form of heat.

Domain: Areas within a material made up of atoms with the same magnetic polarity.

Domain theory: Suggests each electron in an atom is spinning on its own axis as well as orbiting the nucleus of the atom.

Doping: Adding P-type or N-type materials to a semiconductor material.

Double-pole: In switch or relay contacts, indicates two circuits are controlled at the same time.

Double-throw: In switch or relay contacts, indicates switching will occur in two directions.

Drag soldering A soldering technique used in multipin devices.

Drain: 1. The current taken from a voltage source. 2. The electrode of a field effect transistor that corresponds to the collector of a bipolar transistor.

Dross: The film found on the top surface of molten solder in a solder pot caused by the surface of the solder combining with oxygen.

Dry cell: A voltage-generating cell in which the electrolyte is absorbed into a paper product or made into a paste.

Duty cycle: 1. In a waveform, the ratio of the time a pulse is high to the complete time period of the waveform. 2. In equipment, the ratio of the amount of time a piece of equipment is idle to the amount of time it operates.

Dynamic test: Testing performed while circuits or equipment are operating.

E

Eddy currents: Currents induced in the core of a transformer. Also called Foucault currents.

Effective value: The amount of alternating current that will produce the same amount of work as that of direct current. Also called the *RMS value*.

Efficiency: The ratio of the useful output to the total input of a device or circuit.

Electrical current: The flow of electrons through a conductor.

Electromagnet: A temporary magnet made from a coil of wire with or without a core of iron. A magnetic field exists only when current flows in the conductor.

Electromagnetism: Magnetism produced by an electrical current.

Electromotive force: The force that causes electrons to move through a conductor. It is measured in volts and abbreviated emf. Also called the *difference in potential*.

Electron: A small negative particle that revolves around the nucleus of an atom.

Electron beam: A stream of electrons that strike the inner surface of a cathode ray tube.

Electron gun: A mechanical assembly at the back of a cathode ray tube that produces an electron beam.

Electrostatic discharge: The movement of charges from one point to another. It is abbreviated ESD.

Electrostatic field: The electric force field around or between two points, such as between the plates of a capacitor or between a person and a component lying on a bench top.

Element: A material that cannot be broken down into a simpler form without destroying its identity.

Emitter: The electrode of a transistor that carries the total current.

Emitter follower: A transistor circuit in which the output is taken from across a resistance in series with the emitter or source of the transistor. Also called a *common collector* or a common drain (in a FET) circuit.

Energized: Electrically connected to a voltage source.

Engineering notation: A notation using powers of ten in which the exponent is always a multiple of three: $\times^0$, $\times^3$, $\times^6$, $\times^9$, $\times^{12}$, $\times^{-3}$, $\times^{-6}$, $\times^{-9}$, etc.

Enhancement-mode: In MOSFET transistors, the operation in which no current flows when the gate-source voltage is zero and the current is increased by increasing the gate-source voltage.

Equation: Two expressions separated by an equal sign.

Eutectic alloy: A mixture of metals with the lowest possible melting point.

Excitation current: The current that flows in the primary of a transformer when the secondary is open.

Exponent: The number of times a number is to be multiplied by itself. For example, 5^3 (3 is the exponent) equals $5 \times 5 \times 5$.

F

Fall time: Time in which a pulse drops from 90% to 10% of maximum amplitude. It is abbreviated T_F.

Farad: The unit of measurement for capacitance. It is represented by F.

Ferrite: A powdered, compressed magnetic material made mostly of iron oxide and combined with one or more other metals and ceramic. It has a high resistance, making eddy currents low at high frequencies. It is used in the cores of such inductive devices as transformers and chokes.

Ferromagnetic: Refers to materials that are easily magnetized. "Ferro" means the material is based on iron.

Field (winding): The winding in a motor that produces the main magnetic field.

Figure of merit: *See* Q-factor.

Fillet: The solder that solidifies between two or more items being soldered. A good fillet is concave.

Fixed bias: A bias voltage that is constant in value.

Fluorescent screen: The screen surface of a cathode ray tube. It is coated with a material that causes light to be emitted when bombarded with electrons.

Flux: A liquid or solid substance applied to metals before soldering to remove oxides and protect the hot metal from contamination by atmospheric oxygen.

Flux density: The number of magnetic lines of flux per square meter.

Forward biased: A voltage in such polarity across a PN junction that current flows through the junction.

Free electron: An electron that is free to move from one atom to another.

Frequency: The number of times per second (or other specified duration) that a particular action takes place. In alternating current, it is represented by f. The unit of measurement is Hertz.

Frequency spectrum: The entire range of frequencies of electromagnetic radiation.

Full-wave: In a dc power supply, one in which both halves of the ac input cycle are changed to direct current.

Function switch: The main switch on a meter that indicates what is to be measured, such as voltage or resistance.

G

Gate: The controlling element of a JFET or a MOSFET transistor.

Graticule: The graduated scale on the screen of an oscilloscope.

H

Half-power point: The frequency limits at which the maximum voltage or current drops 70.7% (3 dB).

Half-wave: In a power supply, one in which half the ac cycle is rectified into direct current.

Hard ground: A connection directly to ground.

Hard soldering: Soldering in which metals are heated to over 800°F (427°C).

Heat sink: A mass of metal used for carrying heat away from a component.

Henry: The unit of measure for inductance. It is abbreviated H.

Hexadecimal: A number system with a base of 16. Zero through 9 and A through F are used.

Holding current: The minimum current required to keep an SCR in conduction or a relay pulled in an active state.

Hole: A vacancy caused by the absence of an electron in the electronic valence structure of a semiconductor.

Horsepower: A measure of power used in motors. One horsepower is equal to 746 watts.

Hypotenuse: The longest side of a right triangle or the diagonal (corner-to-corner) dimension of a rectangle.

Hysteresis: The amount of magnetism that lags behind the magnetizing force due to molecular friction.

Hysteresis loop: A graphic curve showing the relationship between a magnetizing force and the resulting magnetic flux.

Hysteresis loss: The power loss caused by molecular friction.

I

Imaginary number: A number pertaining to the square root of a negative number and used to represent a reactance.

Impedance: The total resistance toward alternating current. When inductance or capacitance is included with resistance, impedance is the vector sum of the two quantities. Impedance is represented by Z and measured in ohms.

Improper fraction: A fraction in which the numerator is equal to or greater than the denominator.

In phase: A condition that occurs when the waveforms of two or more traces cross the X-axis at the same time and in the same direction.

Inductance: That property of a coil that tends to oppose any change in the current. It is represented by L and measured in henrys.

Induction: 1. Producing an electromotive force and current in a conductor using variations in the magnetic field affecting the conductor. 2. Producing an electric charge or magnetic field in a substance by the proximity of an electrified source, a magnet, or magnetic field.

Inductive kick: The high voltage produced by the collapsing field in a coil when the current through the coil is cut off.

Inductive reactance: The opposition to the flow of alternating or changing current by an inductor. It is represented by X_L and measured in ohms.

Inductor: A conductor wound in a spiral or coil to increase its inductance.

Infinity: A hypothetical measurement so large or small no number can be assigned to it.

Input resistance: The resistance or impedance offered to a signal applied to an input circuit as "seen" by the signal source.

Instantaneous value: The magnitude of a changing value at any given moment.

Insulator: A material with few or no free electrons.

Integrator: A device or circuit in which the output is proportional to the integral of the input signal.

Inversely proportional: The value of one number increases as the value of another number decreases or vice versa.

K

Kirchhoff's current law: In a parallel circuit, the algebraic sum of all the branch currents, including the source, is equal to zero.

Kirchhoff's voltage law: In a series circuit, the algebraic sum of all the voltages around the circuit, including the source, is equal to zero.

L

Law of magnetic poles: Like poles repel each other and unlike poles attract each other.

Leading edge: 1. The edge of a waveform first produced by the oscilloscope. 2. The first edge of the waveform from the left when reading a waveform from left to right.

Leakage current: 1. An undesirable small current that flows between two points. 2. In a capacitor, a small value stray current that flows through or across the surface of an insulator or the dielectric of a capacitor.

Left-hand rule of coils: If you hold a coil in your left hand so your fingers point to the direction of current flow, your thumb will point toward the north pole.

Left-hand rule of conductors: If you hold a conductor with your left hand so your thumb points in the direction of electron flow, your fingers will point in the direction of the magnetic field.

Linear: Pertaining to a straight line. Something that changes equally.

Linear amplifier: An amplifier in which the output is always an amplified replica of the input signal.

Linear state: In a transistor amplifier, where the collector voltage is half the supply voltage.

Lissajous method: The measurement of frequency or phase difference using Lissajous figures (patterns) produced on the screen of a cathode-ray tube.

Literal number: A number with no specified value, symbolized by a letter of the alphabet or some other special character.

Litz wire: A conductor made of fine separately insulated strands, woven together to make one conductor. Litz wire reduces the skin effect at high frequencies.

Load: The device or circuit that consumes power.

Load line: A line drawn across the collector characteristic curves of a transistor from saturation to cutoff.

Locked-rotor: A condition in which the rotor is not moving.

Lodestone: A natural magnet consisting mostly of a magnetic iron oxide called magnetite.

Logarithm: The exponent to which a base number is raised to obtain the number.

Logic: A science dealing with the basic principles and applications of truth tables, switching, and gating.

Logic circuit: A circuit that operates using logic functions.

Logic function: A method of controlling how a circuit operates, such as AND, OR, and NOT gates, or a combination of these.

M

Magnetic field: The area around a natural, permanent, or electromagnet where magnetic forces can be detected.

Magnetic flux: The total magnetic lines of force in a magnetic field.

Magnetic leakage: Passage of magnetic flux outside the path along which it can do useful work, causing a loss of energy.

Magnetic saturation: In an iron core, the point where any further increase of magnetizing force produces little or no increase in the magnetic lines of force in the core.

Magnetic shielding: A metallic enclosure to protect devices or systems against magnetic fields.

Magnetomotive force: The force that produces flux in a magnetic circuit. It is abbreviated mmf.

Make-before-break: In switch or relay contacts, one set of contacts makes or closes its circuit before the other set breaks or opens. Also called *shorting* contacts.

Matter: Anything that occupies space and has mass.

Maximum power transfer: The condition that exists when the resistance of the load equals the internal resistance of the source.

Maxwell: The centimeter-gram-second (cgs) unit of magnetic flux where 1 weber equals 10^8 maxwells. It is abbreviated Mx.

M-derived (filter): A modified form of the constant-k filter, based upon the ratio of the filter cutoff frequency to the frequency of infinite attenuation.

Memory: A motor control circuit that remembers the last instruction given.

Midrange frequencies: Audio frequencies that span approximately 1 kHz to 4 kHz.

Millman's theorem: A proven theory that provides a shortcut for finding the common voltage across any number of parallel branches with different sources.

Molecule: A chemically bonded group of two or more atoms.

Momentary switch: A switch that automatically returns to its original position after being activated.

Metal oxide semiconductor field effect transistor: A type of field effect transistor in which the gate is insulated from the rest of the device by a metal oxide.

Motor controller: A circuit that controls the on/off and speed of a motor.

Motor starter: A device or circuit that enables a motor to start.

Multimeter: A meter that measures volts, ohms, and amps.

Mutual inductance: The inductance between two or more conductors or inductors. It is represented by L_M.

N

NAND: The logic gate function NOT AND where the output is an inverted AND.

Negative temperature coefficient: An increase in temperature causes a decrease in resistance or vice versa.

Negative feedback: Part of the output signal is fed back to the input 180° out of phase, causing reduced gain. Circuit conditions can occur that produce a degenerative effect. Also called *degenerative feedback.*

Negative going edge: The edge of a waveform that decreases toward the negative direction.

Negative resistance: The property exhibited when an increase in voltage produces a decrease, rather than an increase, in current.

Net reactance: The difference between the inductive reactance X_L and the capacitive reactance X_C ($X = X_L - X_C$). It is represented by X and measured in ohms.

Network: An electrical combination of two or more components.

Neutron: One of three elementary particles. The neutron has a neutral charge.

Node: A point where two or more components are connected.

Nonlinear: Not the same throughout a given distance, time, resistance, or temperature.

Nonshorting: In switch or relay contacts, contacts that break (open) completely before another set of contacts make (close).

NOR: The logic gate function NOT OR. The output is an inverted OR.

Normally closed: A type of switch or relay contact that opens when activated. It is abbreviated N.C.

Normally open: A type of switch or relay contacts that closes when activated. It is abbreviated N.O.

Norton's theorem: Looking back from the terminals referenced, the current in any impedance connected to two terminals of a network is as though it were connected to a constant-current source.

NOT: The logic function of inverse.

NPN transistor: A transistor with a P-type base and N-type collector and emitter.

N-type (material): A material in which electrons have been added so the electrons serve as the majority current carrier.

Nucleus: The center part of an atom containing protons and neutrons.

Numerator: The top number in a fraction.

O

Ohm: The unit of measurement for resistance, inductive reactance, capacitive reactance, and impedance. It is represented by the Greek letter Ω (omega).

Ohm's adjust: The control switch on an analog meter used to adjust the pointer to infinity or another designated mark.

Ohm's law: The mathematical relationship of voltage, current, and resistance.

Ohm's law for magnetic circuits: Magnetic flux is equal to magnetomotive force divided by reluctance.

Ohms per volt: The unit of measurement for sensitivity of an analog meter.

One cycle: Measurement taken from one point to a corresponding point on a waveform.

Open circuit: A circuit that does not have a complete path for current to flow.

Open-loop: 1. In an operational amplifier, the condition in which no feedback resistance is used in the circuit. 2. A control system that does not provide a self-correcting action for errors in the operation of the system.

Opposite side: The side of a right triangle across from or opposite the angle being referenced.

OR: The logic gate function OR.

Origin: The intersection of the X and Y axes in a rectangular coordinate system.

Out of phase: When two or more traces do not cross the X-axis of the oscilloscope graticule at the same time or in the same direction.

Overshoot: When the waveform trace goes beyond the maximum amplitude point in either the positive or negative direction.

P

Parallel circuit: A circuit that contains more than one path for current to flow.

Paramagnetic: Describes a material having a permeability slightly greater than a vacuum.

Passive device: An idle component that may control or change, but does not create or amplify, energy.

Peak reverse voltage (PRV): The peak ac voltage a PN junction can withstand in the reverse direction. Also called peak inverse voltage (PIV).

Peak-to-peak value: The value of alternating current from the positive peak to the negative peak.

Peak value: The maximum instantaneous value of voltage, current, or power.

Permanent magnet: A piece of hardened steel or other magnetic material that indefinitely retains its magnetism.

Permeability: Measure of how well a material conducts magnetic lines of force. It is represented by μ (mu).

Positive going edge: The edge of a waveform that goes toward the positive direction.

Preshoot: A waveform distortion where the trace momentarily goes in the opposite direction before making the main portion of the trace.

Phosphor material: A material placed upon the inner surface of a cathode ray tube. The electron gun at the back of the CRT produces electrons at high velocity that strike the phosphor material and produce light.

Pi network: A group of three components connected in the form of the Greek letter pi (π).

Piezoelectric effect: 1. Characteristic of certain natural and synthetic crystals to produce a voltage when subjected to mechanical stress, such as compression, expansion, or twisting. 2. The mechanical movement of certain natural and synthetic crystals under the influence of an electric field.

Piezoelectric speaker: A speaker in which the mechanical movements are produced by piezoelectric action. Also called a crystal speaker.

Pinch-off voltage: The gate voltage of a field effect transistor that blocks the current for all source-drain voltages below the junction breakdown value. Pinch-off occurs when the depletion zone completely fills the area of the device.

Place value: The value of each column in a number system found by multiplying the previous place value by the base of the number system.

PN junction: The region of transition between P-type and N-type material in a single semiconductor crystal.

PNP transistor: A transistor with an N-type base and a P-type emitter and collector.

Polar notation: In trigonometry, it is written as the radius and angle (r, $\angle$).

Polarization: The increased resistance of an electrolytic cell as the potential of an electrode changes during electrolysis.

Pole: 1. One end of a magnet. 2. One electrode of a battery. 3. An output terminal on a switch or relay.

Positive temperature coefficient: An increase in temperature causes an increase in resistance or vice versa.

Potentiometer: A variable resistor.

Power: How fast energy is used or converted to another form of energy. It is represented by P and measured in watts (W) or joules (j).

Power factor: The cosine of the phase angle used when calculating true power.

Primary: The input winding of a transformer.

Primary cell: A nonrechargable electric cell that produces electric current through a chemical reaction.

Proper fraction: A fraction in which the numerator is smaller than the denominator.

Proportion: A relationship of four quantities where the product of the first and fourth quantities equals the product of the second and third quantities, or the first ratio is equal to the second ratio.

Proton: The positively charged particle of an atom.

P-type (material): A material in which doping causes an excess of holes.

Pulse width: Time duration of the pulse measured at 50% of peak amplitude of a waveform.

Push-pull amplifier: An amplifier with two identical signal branches connected to operate in phase opposition.

Q

Q-factor: 1. In an inductor or capacitor, the ratio of reactance to effective series resistance at a given frequency. 2. A measure of the sharpness of resonance or frequency selectivity. Also called figure of merit or Q.

Quadrants: The divisions of a circle. Four quadrants of 90° make up a complete circle.

Quiescent: The condition of a circuit at rest when no input signal is being applied. Also called Q-point.

R

Radian: A unit of circular measure. The measured angle intercepts an arc on the circumference equal to the radius of the circle. One radian equals approximately 57.3°.

Radius: The distance from the center of a circle to a point on the circumference of the circle. It is equal to half the diameter.

Rail voltages: The power supply input voltages of a comparator or operational amplifier.

Ramp: A gradual, linear, positive or negative transition in a waveform, which can be a positive ramp or negative ramp.

Range switch: The switch on a meter or other instrument used to select the maximum range of measurement.

Rarefaction: To make a signal more difficult to define or distinguish.

RC circuit: A series, parallel, or series-parallel circuit made up of a resistance and capacitance.

RC time constant: The product of resistance times the capacitance of a resistor-capacitor circuit ($t = RC$). The time required for the current or voltage to decrease or increase 63.2% of maximum value.

Reactance: The resistance an inductor or a capacitor exhibits toward alternating current.

Reciprocal: The reciprocal of a number is found by dividing one by that number ($1/X$).

Rectangular notation: A rectangular coordinate system using two axes specified by X, Y.

Rectification: The process of changing alternating current to direct current.

Reflowed: To remelt a soldered connection.

Regulator: A device or circuit designed to maintain a voltage or current at a predetermined value.

Relative permeability: The permeability of a material compared to air. It is represented by μ_r.

Relay: 1. An electromechanical switch in which contacts are open or closed by an electromagnet. 2. A switching device that uses solid-state semiconductor components.

Reluctance: The resistance a material has to magnetic lines of force. It is represented by $\mathfrak{R}$.

Residual magnetism: Magnetism that remains in a device after the magnetizing force is removed.

Resistance: The property of conductors or materials that opposes the flow of current. It is measured in ohms and represented by R.

Resistivity: A measure of the resistance of a material to electric current. It is represented by the Greek letter ρ (rho).

Resistor: Electrical component used to oppose the flow of electrical current.

Resonance: In a tuned circuit, the condition when X_L and X_C are equal in value.

Resonant circuit: A circuit containing both inductance and capacitance and tuned to resonance.

Resonant frequency: The frequency in which inductive reactance is equal to capacitive reactance.

Response curve: A graph that plots the output in relation to the frequency of a circuit or device.

Retentivity: Property of a material that measures its ability to stay magnetized after the magnetizing force is removed.

Retrace period: Time period during which the electron beam returns from the right side of the cathode ray tube screen to the left side to begin the next trace.

Reverse biased: A voltage applied to a semiconductor junction in such a way that no current (except leakage current) flows.

Revolving-armature alternator: Generates an output voltage by turning the armature conductors in a magnetic field.

Revolving-field alternator: Generates an output voltage by rotating the magnetic field while the output is taken from the stationary conductors.

Rheostat: A variable resistor with two terminals, one fixed and one movable.

Ribbon cable: A flat, ribbon-shaped set of conductors.

Right triangle: A triangle that has a 90° angle.

Ringing: A waveform defect that resembles a type of damped wave on a portion of a waveform.

Rise time: Time required for the leading edge of a pulse to rise from 10% to 90% of its maximum value. It is represented by T_R.

RMS value: The value of alternating current that will do the same work as that same value in direct current. It is equal to 0.707 times the peak value. RMS stands for root-mean-square.

Rosin flux: A common flux used for electrical soldering. It is inert at room temperature but becomes a mild acid when hot.

Rotor: The moving part of a motor.

S

Saturation delay time: The time required for a transistor to come out of a saturated state.

Saturation state: The point at which no further increase of a quantity will make a change on another. The point of maximum collector current or drain current of a transistor. *See also* magnetic saturation.

Sawtooth: A waveform shape that resembles the teeth of a saw blade.

Scanning: The process of moving an electron beam across a cathode ray tube.

Schematic: An electrical/electronic diagram that uses symbols to show how various components are electrically connected.

Scientific notation: Writing numbers using powers of ten and maintaining one digit to the left of the decimal point.

Secondary: The output winding of a transformer.

Secondary cell: A rechargeable electric cell.

Selectivity: Measure of how well a circuit or piece of equipment can distinguish one frequency from another.

Self-induction: The property that causes a counter-electromotive force to be produced in a conductor or coil when the magnetic field collapses or expands with a change in the amplitude of the current.

Semiconductor: A solid or liquid conductor with a resistivity between that of metals and insulators. Common semiconductor materials include selenium, silicon, and germanium.

Sensitivity: A measure of how much current it takes to move the meter pointer to the full-scale position.

Sequential logic: A logic device or circuit that has more than one input and output.

Series circuit: A circuit with only one current path.

Series-parallel circuit: A circuit made up of both series and parallel circuit combinations.

Shaded pole motor: A single-phase induction motor with one or more auxiliary short-circuited stator windings magnetically displaced from the main winding.

Short circuit: An unwanted path of current flow.

Shorting: In switch or relay contacts, the new contacts make continuity before the old contacts break continuity.

Shunt damping resistor: A resistor in parallel with a parallel RCL circuit.

SI units: Standard International units of measurement.

Signal injection: The process of putting a signal into a circuit or piece of equipment to see if it can pass through normally.

Signal tracing: The process of tracing a signal through a circuit or piece of equipment.

Signed number: A number preceded by a positive or negative sign.

Silicon controlled rectifier. A four-layer PNPN semiconductor device that, in its normal state, blocks current flow. Once turned on, current continues to flow after the control signal is removed. It is abbreviated SCR.

Simultaneous equation: Two or more equations that contain the same variables.

Single-phasing: The tendency of a rotor to continue to rotate when one winding is open and the other remains excited.

Single-pole: A switch or relay arrangement in which only one circuit can be controlled.

Single-throw: A switch or relay term used to indicate the number of different circuits a pole can control.

Skin effect: The tendency of high radio frequency currents to flow near the surface of a conductor.

Slew rate: Maximum rate of change of the output voltage of a closed-loop amplifier under large-signal conditions. Large signal causes saturation of the amplifier stage.

Soft ground: A connection to ground through an impedance high enough to limit the current to a safe value.

Soft soldering: Soldering done below 800°F (427°C).

Solder: A metal mixture used to connect metal parts or components.

Soldering: The process of joining two metals.

Solenoid: A device consisting of an electromagnet and a movable core used for moving valves, electric switches, and mechanical levers.

Solid-state: Any component, device, or study based on semiconductor theory.

Source: A lead or electrode of a field-effect or MOSFET transistor that corresponds to the emitter of a bipolar transistor.

Source (voltage): A device that supplies voltage to a circuit or piece of equipment.

Speaker: An energy converter that changes electrical energy to mechanical energy. Sound is caused by mechanical energy moving the air.

Specific gravity: The weight of a substance compared to the weight of the same volume of water at the same temperature.

Spider: A thin flexible piece of membrane connecting the cone of a speaker to its frame.

Split-phase motor: A single-phase induction motor with two field windings. Starting torque is developed by phase displacement between the field windings. One winding is usually switched open when the motor has reached a certain speed.

Square wave: A square or rectangular periodic wave that alternately assumes two fixed values for equal lengths of time.

Squirrel-cage motor: A motor that uses a rotor resembling a cage.

Stage: A transistor or other device and the components necessary to make it operate.

Static test: Testing performed without power or signal being processed.

Stator: The stationary part of a motor.

Step-down transformer: A transformer that has fewer turns in the secondary than in the primary.

Step-up transformer: A transformer that has more turns in the secondary than in the primary.

Storage time: 1. The time needed to turn off a transistor that has been driven into saturation. 2. After the input current or voltage has been removed, the time during which the output current or voltage of a pulse is falling from maximum to zero.

Stranded conductor: Consists of several small conductors twisted into a single conductor.

Stray capacitance: Capacitance introduced into a circuit by the leads and wires connecting the circuit components.

Stray inductance: Inductance introduced into a circuit by the leads and wires connecting the circuit components.

Stripping: The process of removing the insulation from a conductor.

Sulfation: The accumulation of lead sulfate upon the plates of a lead-acid storage battery.

Superconductor: A conductor where resistance decreases as its temperature is reduced to near absolute zero.

Superposition theorem: When a number of voltages in a linear network are simultaneously applied to the network, the current that flows is the sum of the component currents that would flow if the same voltages had acted individually.

Susceptance: The reciprocal of reactance. It is represented by B.

Switch: A control device that either prevents or allows the flow of current in a circuit.

System: An assembly of parts or circuits to do a specific task.

T

Tank circuit: A parallel LC or RCL circuit.

Tapered: In a potentiometer, the resistance of the resistive element does not vary by the same amount throughout its range.

Tee network: A combination of components connected to form a "T" shape.

Temperature coefficient: A factor that indicates whether electrical resistance will increase or decrease with temperature. It is represented by α.

Thermistor: A resistive device in which resistance changes with temperature.

Theta: A letter of the Greek alphabet used to indicate a phase angle. It is represented by θ.

Thevenin's theorem: States any network can be reduced to a single series resistance and a single voltage source.

Three-phase: Related to a power circuit or device powered by a three-terminal source $120°$ apart in phase.

Three-pole: A switch or relay contact arrangement that can control three separate circuits at the same time.

Throws: The number of different contact positions per pole available on the relay.

Tilt: 1. A distortion of a square wave in which the horizontal flat portion of the wave becomes tilted and forms a ramp. 2. A measure of the amount of slope to the full amplitude.

Time constant: The time required for an exponential quantity to change by an amount equal to 63.2% of the total maximum value.

Time delay: 1. The amount of time required for a signal to travel between two points. 2. The elapsed time required for something to take place.

Time period: The time required for one cycle.

Tinning: Coating conductors with molten solder.

Tolerance: The allowable percentage of deviation over or under the specified value.

Toroid: A very efficient type of coil wound upon a "doughnut-shaped" core.

Torque: A measure of the circular twisting motion of a shaft.

Trace period: The period of time that a waveform is made on the screen of a cathode ray tube.

Trailing edge: The edge toward the right to be made on a waveform.

Triangular wave: A waveform that has rise and fall distinctly shaped like an equal triangle.

Transducer: A device that converts one form of energy to another.

Transient response: Refers to how a circuit responds in a short period of time. For example, transient response elements of waveforms are rise time, fall time, overshoot, and ringing.

TRIAC: Three-terminal, solid-state device that acts like a bi-directional solid-state switch.

Triboelectric effect: Rubbing two materials together to generate static electricity.

Triggering: The start of the oscilloscope trace period at a predetermined point on the signal cycle.

Trigonometric function: The resulting number when one side of a triangle is divided into another side.

True power: The actual power consumed by a circuit or device which considers the voltage-current phase angle in the calculation.

Truth table: A table that shows all possible inputs and resulting output of a digital or other logic-type circuit.

Turns ratio: The ratio of the primary to secondary turns in a transformer.

Tweeter: A small speaker designed to reproduce high frequencies.

U

Unbalanced bridge: A voltage or current exists between the two branches of a bridge circuit.

Undefined area: The voltage levels between low and high levels of a binary 0 or 1.

Undershoot: A waveform distortion in which the trace momentarily goes in the opposite direction before making the actual waveform trace.

Unity: Any number raised to the zero power is equal to 1.

V

Valence shell: The outermost electron shell or level of an atom.

Valley voltage: In a tunnel diode, the voltage that corresponds to the valley current.

Vector: An arrowed line indicating a certain magnitude and direction.

Vector diagram: A diagram using vectors to indicate the vector relationship of the quantities.

Vector sum: In alternating current circuit applications, the addition of quantities using vectors. Mathematically, the square root of the sum of the squares.

Voice coil: A coil of small diameter wire that produces magnetic fields corresponding to the electrical currents representing the sound. The magnetic field produced by the voice coil reacts with the permanent magnetic field of the speaker in which it is installed.

Volt: The unit of measurement for voltage, EMF, or difference in potential. It is abbreviated V.

Voltage: Electrical pressure that causes electrons to move (current).

Voltage divider bias: The bias on the base of a transistor developed by a voltage divider resistor network.

Voltage divider equation: A special equation used for series circuits to find a voltage across any single resistance.

Voltage doubler: A dc power supply that doubles the line voltage without the use of a transformer.

Voltage gain: The ratio of the output voltage to the input voltage. It is expressed as a number without any unit of measurement.

Volume unit: The unit of measurement for the power level of voice waves. Zero volume units equal +4 dBm.

W

Watt: The unit of measurement for power. It is abbreviated W.

Wavelength: The length in distance of one cycle. It is found by dividing the velocity of radio waves by the frequency.

Weber: The unit of measurement of magnetic flux. One weber equals 10^8 maxwells. It is abbreviated Wb.

Wet cell: A voltage source cell that contains a liquid electrolyte.

Wetting: The attraction between the solder and a base metal.

Wheatstone bridge: A type of bridge circuit used to measure unknown resistance.

Wicking: 1. The process of solder being drawn underneath the wire insulation. 2. The process of removing solder from a connection.

Wire wrap: Making an electrical connection by spinning (wrapping) the wire around a terminal and making the connection without soldering.

Woofer: A speaker used to reproduce the lower frequencies.

Wye: Type of transformer connection used in three-phase ac power.

X

X-axis: The horizontal (left-to-right) axis of a graph.

Y

Y-axis: The vertical (up-and-down) axis of a graph.

Z

Zero adjust: An adjustment to an electronic instrument so it is calibrated to a zero value.

INDEX

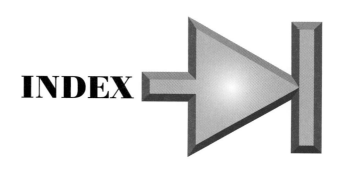